Chlorinated Organic Compounds in the Environment

Regulatory and Monitoring Assessment

Sub Ramamoorthy
Sita Ramamoorthy

Lewis Publishers
Boca Raton   New York

Acquiring Editor: Ken McCombs
Project Editor: Andrea Demby
Marketing Manager: Arline Massey
Cover design: Denise Craig
PrePress: Greg Cuciak
Manufacturing: Carol Royal

Library of Congress Cataloging-in-Publication Data

Ramamoorthy, S.
    Chlorinated organic compounds in the environment: regulatory and monitoring assessment/by Sub Ramamoorthy and Sita Ramamoorthy.
        p. cm.
    Includes bibliographical references and index.
    ISBN 1-56670-041-8 (alk. paper)
    1. Organochlorine compounds—Environmental aspects—Handbooks, manuals, etc. I. Ramamoorthy, Sita. II. Title.
TD195.C5R36 1997
363.738′4—dc21
                                                                                                                                                                                                 97-6799
   CIP

This book contains information obtained from authentic and highly regarded sources. Reprinted material is quoted with permission, and sources are indicated. A wide variety of references are listed. Reasonable efforts have been made to publish reliable data and information, but the author and the publisher cannot assume responsibility for the validity of all materials or for the consequences of their use.

Neither this book nor any part may be reproduced or transmitted in any form or by any means, electronic or mechanical, including photocopying, microfilming, and recording, or by any information storage or retrieval system, without prior permission in writing from the publisher.

The consent of CRC Press LLC does not extend to copying for general distribution, for promotion, for creating new works, or for resale. Specific permission must be obtained in writing from CRC Press LLC for such copying.

Direct all inquiries to CRC Press LLC, 2000 Corporate Blvd., N.W., Boca Raton, Florida 33431.

© 1997 by CRC Press LLC
Lewis Publishers is an imprint of CRC Press LLC

No claim to original U.S. Government works
International Standard Book Number 1-56670-041-8
Library of Congress Card Number 97-6799
Printed in the United States of America 1 2 3 4 5 6 7 8 9 0
Printed on acid-free paper

Preface

This book is intended to provide up-to-date information on chlorinated organic compounds in the environment that can be used for monitoring, impact assessment, and regulatory decision-making processes by environmental managers, regulatory personnel, scientists, and students. The information presented in this book will help in predicting the potential for environmental contamination as well as the critical medium of exposure to the health of the ecosystem and humans. The toxicity profile provided on each chemical will assist the readers to evaluate short-term and long-term effects on the ecosystem as well as on human health. The information on environmental residues and the worldwide regulations will allow comparisons to be made between a given jurisdiction and other areas on the extent of chlorinated organic compound contamination and the associated regulatory approaches. The complexities and diversities of environmental problems require broader understanding and expand our knowledge into different disciplines such as chemistry, life sciences, and engineering. The effective implementation of recommendations requires an empathy for social, political, economic, and regulatory factors.

There is first an introduction in the book on chlorinated organic compounds in the environment, followed by fate processes and their environmental migration based on their physical and chemical properties and processes. Next, the book deals with chlorinated aliphatic hydrocarbons, chlorinated aromatic hydrocarbons — monocyclic and polycyclic, followed by chlorinated biocides, chlorinated phenols, and chlorinated dioxins and furans in the environment. The last two chapters deal with prioritization for regulatory and monitoring assessment and regulatory decision-making processes. Chapters three to eight cover the North American and International regulations and advisories in the management of chlorinated organic compounds in every matrix of exposure to the ecosystem and humans. Ample references are provided at the end of each chapter. The book closes with a glossary of terms and a subject index.

Authors

Sub Ramamoorthy received his Ph.D. in 1969 in physical chemistry from the University of Madras, India. He was awarded his D.Sc. degree in 1984 for research work on "Heavy Metals in the Environment-Interdisciplinary Research Studies." In 1970, Dr. Ramamoorthy joined the Department of Environment-Inland Waters Directorate in Ottawa, Canada as a National Research Council Fellow. He was an adjunct professor at the University of Ottawa while employed at the Division of Biological Sciences, Ecological Kinetics Group, National Research Council of Canada, Ottawa from 1976 to 1979. In 1979, Dr. Ramamoorthy joined the Alberta Department of Environment as the Head of the Limnology Section, Animal Sciences Division at the Alberta Environmental Centre in Vegreville, Alberta. He became a Senior Manager at the Environmental Regulatory Services, Alberta Environmental Protection, Edmonton, Alberta, which he joined in 1986. Dr. Ramamoorthy is a co-author of five books on Environmental Management and has authored more than 70 publications in primary journals in addition to over 30 reports in the field of environmental chemistry and toxicology. Dr. Ramamoorthy was the co-chair of the 15th International Symposium on Chlorinated Dioxins and Related Compounds that was held in Edmonton, Alberta in 1995. He is an active member of the International Society of Regulatory Toxicology and Pharmacology, the International Society of Environmental Toxicology and Chemistry, and the International Humic Substances Society. He is listed in the *American Men and Women of Sciences*, 13th ed., and in the CNTC *Directory of Toxicological Expertise in Canada.* Dr. Ramamoorthy has served on federal/provincial environmental committees, QA/QC committees, a panel of referees for the International Association on Water Quality, England (IAWQ), *Journal of Water Science and Technology,* (U.K.), and *Water Research* (U.K.). Dr. Ramamoorthy is currently with Chemex Labs, Inc., Edmonton, Alberta, now known as Maxxam Analytics, Inc. – Partnership with Novamann International.

Sita Ramamoorthy obtained her M.A. in economics in 1970 from the University of Madras, India. In the 1970s, she worked in the data processing field in Ottawa, Ontario, Canada. In 1980, she joined the Alberta Department of Environment as a research assistant in the Toxicology Branch, Animal Sciences Division, at the Alberta Environmental Centre in Vegreville, Alberta. In that capacity, Sita Ramamoorthy carried out extensive literature searches on mammalian toxicity, particularly inhalation studies. Later she moved to the Aquatic Biology Branch and conducted critical analyses of literature data and performed aquatic fish bioassays. In 1987, she joined the Department of Occupational Health and Safety in Edmonton, Alberta, and is currently with Workplace Health, Safety and Strategic Services Division, Alberta Department of Labour. She has published bibliographies on chemical toxicants in the aquatic environment.

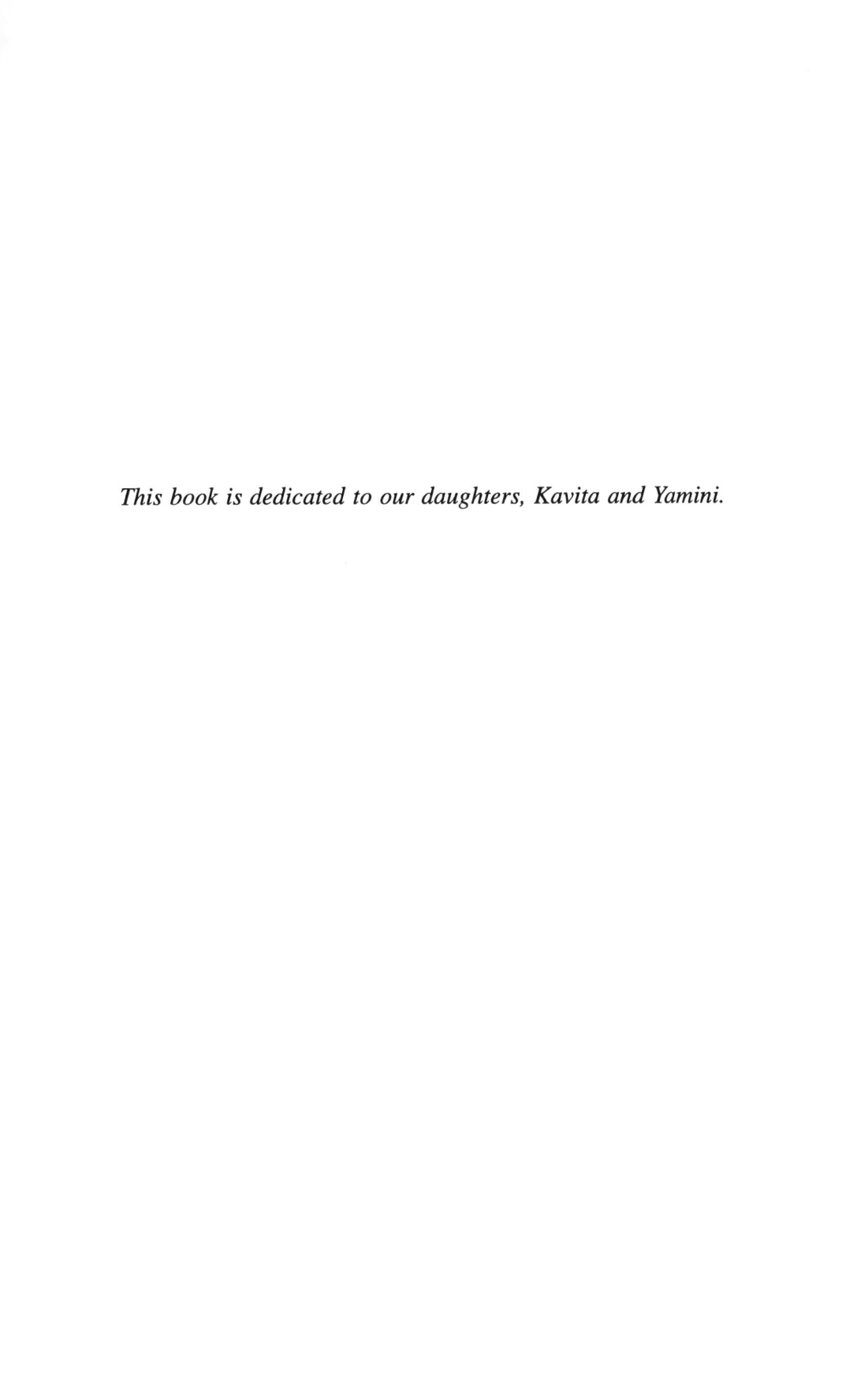

This book is dedicated to our daughters, Kavita and Yamini.

Table of Contents

1. Introduction 1

2. Fate Processes and Environmental Migration of Chlorinated Organic
   Compounds 3
   Physico-chemical Properties 3
       Solubility in Water 3
       Estimates of Water Solubility of Chloroorganic Compound 4
       Vapor Pressure 6
       Octanol-Water Partition Coefficient 7
   Physico-chemical Processes 8
       Volatilization 8
   Volatilization from Soil 13
   Sorption 14
   Bioconcentration 16
   Transformation Processes 19
   References 24

3. Chlorinated Aliphatic Hydrocarbons 27
   Properties 27
   Production and Use Pattern 31
   Sources in the Environment (Natural and Anthropogenic) 33
   Environmental Residues and Exposure Routes (Air, Water, Soil, and
   Other Media) 40
       Chloromethane 40
       Dichloromethane (Methylene Dichloride) 42
       Chloroform 43
       Carbon Tetrachloride 44
       Bromoform and Chlorodibromomethane 44
       Chloroethane 45
       1,1-Dichloroethene 45
       1,1-Dichloroethane 45
       1,2-Dichloroethane 46
       1,1,1-Trichloroethane 46
       1,2-Dichloroethene 47
       1,1,2-Trichloroethane 47
       Trichloroethylene 48
       1,1,2,2-Tetrachloroethane 50
       Tetrachloroethylene 50
       Hexachloroethane (HCE) 52
       Vinyl Chloride 53
   Toxicity Profile 54
   Regulations 63
   References 80

4. Chlorinated Aromatic Hydrocarbons — Monocyclic 93
   Properties 93
   Anthropogenic Sources 93
   Production and Use Pattern 94
   Sources in the Environment 95
   Environmental Residues of Chlorobenzenes and Exposure Routes 96
       Air 96
       Water 96
       Soil 104
       Hexachlorobenzene and Other Chlorobenzenes 106
       Dietary Exposure of Hexachlorobenzene 109
   Toxicity Profile of Chlorobenzene 110
   Regulations 115
   References 116

5. Chlorinated Aromatic Hydrocarbons — Polycyclic 125
   Physical and Chemical Properties 126
   Production and Use Pattern 127
       Production 127
       Use Pattern 131
   Sources in the Environment 132
       Natural 132
       Anthropogenic 132
   Environmental Residues and Exposure Routes 134
       Air 134
       Water 135
           Precipitation Samples 136
       Soils and Sediments 136
       Aquatic Plants and Invertebrates 139
       Fish 142
       PCB Levels in Food Items 146
       Microbial Transformations 148
   Toxicity Profile 150
       Invertebrates 150
       Birds 153
       Mammals 154
       Nonlethal Effects 156
       Key Toxicity Studies 156
           Inhalation Route 156
           Oral Route 157
           Dermal Route 158
           Populations at High Risk 158
   Regulations 162
   References 163

6. Chlorinated Biocides 177
    Production and Use Pattern 177
        DDT, DDE, and DDD 179
        Hexachlorocyclohexane 179
        Chlordane 179
        Aldrin and Dieldrin 179
        Heptachlor and Heptachlor Epoxide 183
        Toxaphene 183
        Endrin 183
        Endosulfan 183
        Mirex and Chlordecone 183
        Chloropyrifos 184
        Dichlorvos 184
        Chlorfenvinphos 184
    Environmental Residues and Exposure Routes 184
        DDT, DDE, and DDD 184
        γ-HCH 186
        Chlordane 191
        Aldrin and Dieldrin 197
        Heptachlor 199
        Toxaphene 200
        Endrin 202
        Endosulfan 203
        Mirex 204
        Methoxychlor 206
        Chloropyrifos 208
        Dichlorvos 209
        Chlorfenvinphos 210
    Toxicity Profile 212
    Regulations 213
    References 223

7. Chlorinated Phenols 235
    Properties 235
    Production and Use Pattern 235
    Sources in the Environment 239
        Natural Sources 239
        Anthropogenic Sources 240
        Primary Routes 240
    Environmental Residues and Exposure Routes 241
        Air 244
        Water 245
        Other Media 247
    Pentachlorophenol 248

>   Levels Monitored or Estimated in the Environment 248
>   Air 248
>   Water 248
>   Other Environmental Media 249
>   Fish 249
>   Biotransformations 252
>   Microbial Metabolism 252
>   Metabolism of Pentachlorophenol 253
>   Photolysis 254
>   Toxicity Profile 255
>   Acute Toxicity 255
>   Plants 257
>   Embryo Lethality 257
>   Chronic Effects 259
>   Mammalian Toxicity 261
>   Regulations 262
>   References 263
>
> 8. Chlorinated Dioxins and Furans 275
>   Physical and Chemical Properties 275
>   Production and Use Pattern 280
>   Sources in the Environment 280
>   Natural Sources 280
>   Anthropogenic Sources 281
>   Thermal Processes 284
>   Metal Refining Processes 286
>   Other Industrial Processes 286
>   Environmental Residues and Exposure Routes 288
>   Air 288
>   Water 289
>   Soils and Sediments 293
>   Fish 297
>   Toxicity Profile 297
>   Invertebrates 297
>   Fish 297
>   Health Effects of 2,3,7,8-T4CDD 304
>   Health Effects of PCDFs 306
>   Relative Toxicity of Chlorinated Dioxins and Furans 306
>   Regulations 307
>   References 309
>
> 9. Prioritization for Regulatory and Monitoring Assessment 321
>   Hazard Identification 321
>   Basic Aspects 322
>   Specific Aspects 322
>   Aqueous Solubility 322

  Biodegradation 324
  Carbon Dioxide Evolution Test Methods.................................. 324
  Algal Bioassay 324
 Toxicity Testing 324
  Aquatic Toxicity Test... 324
  Avian Toxicity Test.. 325
  Mammalian Toxicity Test 326
  Subchronic Mammalian Toxicity Test 326
  Chronic Mammalian Toxicity Test.. 326
  Mammalian Developmental Toxicity 327
  Carcinogenicity Hazard Identification....................................... 327
  Mutagenicity Hazard Identification... 327
 Predictive Capability of Fate Processes 328
  Volatilization 328
  Sorption.. 330
  Water Solubility 331
  Bioconcentration 331
  Photolysis 332
 Exposure Assessment 335
  Influence of Fate Processes on Environmental Pathways........ 337
 Ranking Protocols... 337
 References... 340

10. The Regulatory Decision-Making Process....................................... 343
 Risk Assessment 343
 Strength and Boundary of Information 345
  Laboratory Data 345
  Data on Chemical 345
 Risk Communication 345
  Risk Comparison 347
 Risk Management and Public Participation................................... 348
  Participants.. 348
  Participatory Approaches.. 348
 Acceptable Risk... 349
  Acceptable Level 350
 References... 352

 Glossary 353

 Index 357

CHAPTER 1

Introduction

Increased analytical capabilities have changed the priorities in environmental research, monitoring, and regulations. In the early 1970s, we were largely concentrating on gross pollution of air and water. Although chemically related diseases were being noticed, this area of research and monitoring did not get the attention it deserved. As a result of episodes such as Minamata Disease (arising from consuming methylmercury-contaminated fish from Minamata Bay in Japan and receiving industrial discharges); Times Beach, Missouri (contamination of the entire town from chemical residues spread on roads); Love Canal, New York (contamination of homes and schools from buried chemicals); Seveso, Italy (accidental release of chemicals injuring thousands of residents); and Bhopal, India (accidental chemical release killing more than 2,000 residents), the general population has an awareness of possible hazards from exposure to hitherto unknown chemicals. Since then, the focus has shifted to understanding toxic chemicals, possible routes of exposures, and their toxic potential to cause short-term and long-term adverse effects to the ecosystem and to humans.

The numerous chlorinated organic compounds detected in the environment are either due to anthropogenic or natural sources of input or both. These chemicals have some benefits to society, but most of them have been known to persist in the environment exerting adverse effects. Chlorinated organic compounds used or produced in industrial processes are identified and grouped by chemical class. Each group has been assessed for physical and chemical properties, production and use pattern, bioaccumulation, environmental residues and their exposure routes, and toxicity profile. Each chapter includes regulations for every possible route of exposure from North America and other countries around the world.

Chlorinated organic compounds represent an important class of compounds and many of them have been produced in large quantities and sometimes are indispensable from technological and societal dependency. They may be released during their production and use in a variety of ways ranging from emissions, effluents, and disposal by consumers. Spills and accidental releases contaminate the environment unintentionally or as unwanted by-products of chemical production and combustion processes (e.g., chlorinated dioxins and furans). Natural processes also contribute to the environmental load of chlorinated organic compounds.

Regulatory agencies have tightened regulations on the use and release of chlorinated organic compounds by society, from production to final disposal. The regulations range from outright bans on the production to restrictions on the use of the chlorinated chemicals of concern. The scientific understanding of the persistence of chlorinated organic compounds, their behaviour in the environment, and their potential to cause adverse effects on the ecosystem and on human health, provide the stimulus for regulatory actions on these compounds.

The objective of this book is to provide the readers with a thorough description of environmental properties, environmental behavior, environmental residues and their exposure routes, and toxicity profiles of the chlorinated organic compounds. Also included are the North American and International regulations for the management of chlorinated organic compounds in the environment, including the workplace, ambient air, water, effluent, aquatic organisms, and consumer food items. Permissible cleanup levels in residential and industrial soil are also included in each chapter.

This book is intended to aid environmental scientists, managers, academics, regulatory personnel, and consultants in generating valid and reliable environmental management strategies and remediation options.

CHAPTER 2

Fate Processes and Environmental Migration of Chlorinated Organic Compounds

The physico-chemical properties of chloroorganic compounds, such as solubility, vapor pressure, and partition coefficient, determine their concentration and residence time in water and the subsequent fate processes, such as physico-chemical and biological processes. The migration of a chloroorganic compound is largely dependent upon the physico-chemical interactions with other components of the aquatic environment. Such components include suspended solids, sediments, and biota.

PHYSICO-CHEMICAL PROPERTIES

Solubility in Water

Among the various properties that influence the fate and transport processes of chlorinated organic compounds, water solubility is a critical one. Water solubility is more important for liquids and solids and of much less environmental importance for gasses. The latter is governed by Henry's law constant. Highly water-soluble chemicals have long residence time in the aquatic media and may undergo the biodegradation (microbial or otherwise), photolytic, and hydrolytic processes more readily. These chemicals have low sorption coefficients for soils and sediments and have relatively low BCF values. Factors such as temperature, salinity, dissolved organic matter, and pH play a significant role in the aqueous solubility of organic chemicals.

No organic chemical is totally insoluble in water. Solubility ranges from extremely low values ($\leq$ 1 ppb) to high values (100,000 ppm) at ambient temperatures. The solubility of a chemical in water may be defined as the maximum amount of the chemical that will dissolve in pure water at a specified temperature. Above this limit of solubility, further addition of the solute (chemical) to solvent (water) will create two distinctly separate phases, namely the saturated aqueous solution and the insoluble solute (solid or liquid).

The different approaches discussed in literature include (1) regression equations using octanol–water partition coefficient (K_{ow}), sorption coefficient, normalized organic carbon (k_{oc}), and bioconcentration factor (BCF); (2) atomic fragments addition method; and (3) theoretical equations using estimated activity coefficients, (Lyman et al. 1982). An overview of available estimation methods and their pathways for estimating water solubility of organic chemicals is given in the same reference.

There are about 18 different regression equations available to correlate water solubility (WS) with n-octanol/water partition efficient (k_{ow}) (Lyman et al. 1982). It can be concluded that most equations cover two-thirds of the chemicals within a factor of 10. Many of the large errors (5 to 14% of the estimates were more than a factor of 100) were associated with the nitrogen-containing compounds and almost all were overestimated.

$$\log 1/WS = 1.214 \log k_{ow} - 0.850 \qquad (2.1a)$$

where WS = mol/L, n = 140, r^2 = 0.914 covering a large variety of chemicals;

$$\log 1/WS = 1.339 \log k_{ow} - 0.978 \qquad (2.1b)$$

where WS = mol/L, n = 156, r^2 = 0.874 covering a large variety of chemicals.

The two outliers in Figure 2.1, 1,3,5,-triazo- 1,3,5,-trinitrocyclohexane and hexachloro-1,3,-butadiene, were not included in the regression to avoid increasing the errors in the estimate. Although correlations cover most of the chemicals, it is not universal in coverage.

They were relatively quite accurate when limited to liquids. Seventy-seven percent of the chemicals were within a factor of 10 and 93% within a factor of 100.

Chloro compounds used in the regression equation (Figure 2.1) are carbon tetrachloride, chloroethane, 2-chloroethylether, chloroform, o-chlorophenol, o-dichlorobenzene, m-dichlorobenzene, p-dichlorobenzene, 3,3'-dichlorobenzidine, 1,2-dichloroethane, hexachloroethane, pentachlorobenzene, pentachloroethane, 1,2,3,5-tetrachlorobenzene, 1,1,2,2-tetrachloroethane, tetrachloroethylene, 1,1,1-trichloroethane, trichloroethylene and vinyl chloride.

Estimates of Water Solubility of a Chloroorganic Compound

1. Obtain the octanol/water partition coefficient, k_{ow}, value for the compound either from the literature (Hansch and Leo 1979; Leo et al. 1971) or by estimation.
2. If the compound is a solid at 25°C, determine the melting point, t_m (°C).
3. Select the most appropriate regression equation(s).
4. Use k_{ow} and t_m (if needed) of the compound under study to calculate its water solubility, S at ≈ 25°C.
5. If two or more regression equations were used, calculate the geometric mean of all the estimated solubility values (converted to the same units).

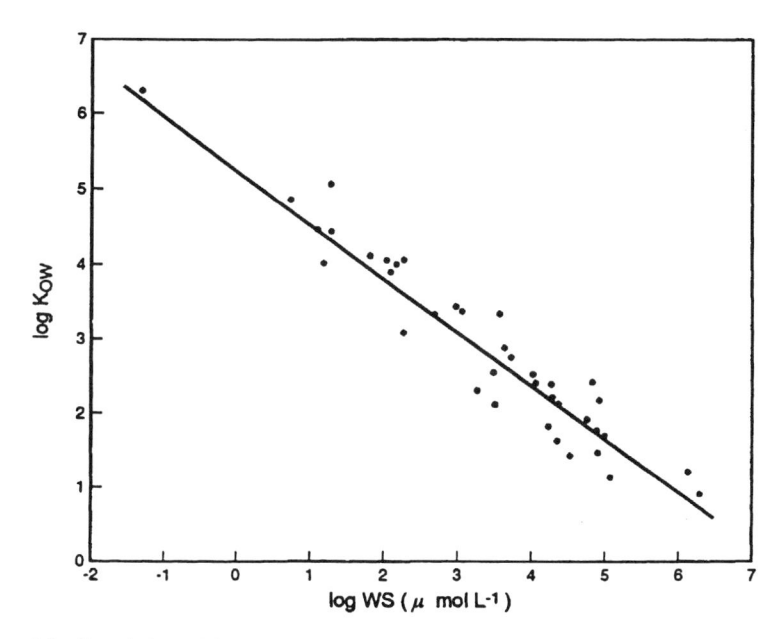

Figure 2.1 Correlation of log k_{ow} with WS for a mixed class of aromatics and chlorinated hydrocarbons, using the regression equation log WS = −1.37 log k_{ow} + 7.26, where WS = µmol/L, n = 41 and r^2 = 0.903. (Reprinted with permission from Banerjee, S. et al. 1980, copyright (1980), American Chemical Society).

Measurements of water solubility of organic chemicals still remain elusive for many compounds. In fact some of the data on aqueous solubility are no more than estimates. The problem is aggravated by the extremely low solubility of many environmentally significant contaminants. For example, the reported solubility of polychlorinated biphenyls (PCBs) varies by a factor of 2 to 4, depending on the procedures used (Haque and Schemedding 1975; Wallhofer et al. 1973). Several techniques are documented in the literature, but the one by Moriguchi (1975) seems to be simple and sound. This technique is based on factoring water solubility of an organic compound into two intrinsic components: free molecular volume and hydrophilic effect of polar groups.

Moriguchi chose Quayle's parachor (molecular volume) over five other additive parameters relating to molecular volume for predictive purposes (Quayle 1953). The molecular volume is calculated by considering the molecules as a sum of its functional groups, and each group is assigned a certain empirical value. The second component, hydrophilicity of polar groups, which accounts for solute–solvent and solvent–solvent interactions, is calculated for various functional groups (Moriguchi 1975). The aqueous solubility of an organic compound can be estimated, using Quayle's parachor and the appropriate hydrophilic reference factor as given below:

$$\log_{10}(1/S) = (1.50)(P_r)(10^2) - (1.51)(E_w) - 1.01 \tag{2.2}$$

where S = aqueous solubility in molal concentration, P_r = Quayle's parachor, and E_w = hydrophilic group factor.

The validity of this equation was tested with 156 compounds of known solubility, and the correlation coefficient was found to be 0.962 (Moriguchi 1975).

Vapor Pressure

Vapor pressure is an important chemical-specific property in predicting the behavior and fate of chloroorganic compounds that are introduced into the environment. Environmental persistence of chlorinated organic compounds in compartments, such as soil and water, and their rate of evaporation is determined by their vapor pressure.

In simple terms, equilibrium vapor pressure can be interpreted as the partitioning of the compound into air from the liquid phase. For some organic compounds, the solids possess a finite vapor pressure that should also be considered in the evaluation of its behavior in the environment. This is particularly important for compounds of low solubility. Vapor pressure at 25°C can be calculated directly where the constants are available at that temperature or by interpolation from other temperatures. The equation of Weast (1974) can be used for several priority pollutants for the calculation of vapor pressure.

$$\log_{10} P = (-0.2185 \, A/K) + B \tag{2.3}$$

where P = vapor pressure in torr, A = molar heat of vaporization, K = temperature in degrees Kelvin, and B = constant.

For a given compound, values of A and B are constant over a moderate range of temperature. Values of A and B for several pollutants have been listed by Weast (1974) and can be used to calculate vapor pressure directly. Where A and B are not available and two or more vapor pressure values are given for temperatures bracketing 25°C, this equation can be used to calculate A and B from the known two sets of ordered pairs (K_1, P_1) and (K_2, P_2). Then the constants A and B are substituted in the equation to calculate P at T = 298 K (= 25°C).

In cases of insufficient data, tables in Dreisbach (1952) can be used from the knowledge of the boiling point at 760 torr and the chemical family to which the compound belongs to develop the "Cox Chart" chemical families using Antoines equation. Charts are available for several organic compounds, such as naphthalenes, halobenzenes with side chains, and phenols. The appropriate table is then referred to for an estimated vapor pressure.

Vapor pressure can be calculated by using the Clausius–Clapeyron equation (2.4). Calculations from this equation provide only a rough estimate of vapor pressure.

$$\ln \frac{P_2}{P_1} = \frac{-\Delta H_v}{R} \frac{T_2 - T_1}{(T_1 T_2)} \tag{2.4}$$

where P = vapor pressure in torr, ΔH_v = molal heat of vaporization, T = temperature in degrees Kelvin, R = gas law constant (1.99 cal/mol K), and subscripts 1 and 2 refer to two different temperatures.

FATE PROCESSES AND ENVIRONMENTAL MIGRATION

The solution of this equation requires the knowledge of boiling point and that of vaporization. This equation provides only a rough estimate of vapor pressure.

Octanol–Water Partition Coefficient

Octanol–water partition coefficient, referred to as K_{ow} is the ratio of a chemical's concentration in the octanol phase to its concentration in the aqueous phase of a two-phase octanol–water system.

$$K_{ow} = \frac{\text{concentration in octanol phase}}{\text{concentration in aqueous phase}}$$

The values of K_{ow} are unitless and usually measured at room temperature (20 or 25°C). Literature shows that temperature has minimal effect on K_{ow} (0.001–0.01 log K_{ow} unit/°C)

K_{ow} is **not** the same as the ratio of solubility of a chemical in pure octanol and pure water because the organic and aqueous phases of the octanol and water systems are not totally immiscible.

The organic compound under study is added to a mixture of octanol and water with the volume ratio adjusted to the expected value of K_{ow}. Very pure octanol and water must be used and the concentration of the organic compound should not exceed 0.01 mol/L. The system is shaken for 15 min — 1 hr in order to reach equilibrium. The mixture is centrifuged and the separated phases are analyzed for the organic compound's concentration in each phase. A rough estimate of K_{ow} may be determined by measuring the chemical's retention time in a high-pressure liquid chromatograph. Numerous studies (Lyman et al. 1982) have shown the linear relationship between K_{ow} and log retention time.

Pharmaceutical and pharmacological groups were the early users of K_{ow} in relating it to structure — activity of drug chemicals. The observed relationship of K_{ow} was used to predict the effect of a new drug whose K_{ow} was measured or estimated.

K_{ow} is a critical parameter in evaluating the fate of organic chemicals since it is well correlated to water solubility, soil and sediment sorption coefficients and bioconcentration factors for aquatic organisms. Measured values of K_{ow} for chloroorganic compounds range from 10^1 to 10^7 (log K_{ow} 1–7) with an uncertainty of ≤ 0.2 log K_{ow} units. Chloroorganic compounds with K_{OW} values less than 1 may be considered to be very hydrophilic with longer residence time in water, low soil/sediment sorption coefficients, and low bioconcentration values. Lyman et al. (1982) have over-viewed the various methods available for estimating K_{ow} and recommend two methods.

(1) Leo's fragment constant method which uses empirically derived atomic or group fragment constants (f) and structural factors (F).

$$\log K_{ow} = \text{sum of fragments (f)} + \text{factors (F)} \tag{2.5}$$

Fragment values for over 100 atoms or atom groups are available in literature. About 200 f values are available due to different f values for a fragment depending on the type of structure (aliphatic or aromatic) to which it is bonded. Fourteen different structural factors (F) values are available for consideration depending on unsaturation, degree of halogenation, branching, molecular flexibility, etc. This method is a fairly powerful one and proven to estimate K_{ow} values for a high percentage of synthetic organic chemicals.

(2) Estimation of K_{ow} from other solvent–water partition coefficients.

Solvent–water partition coefficients (K_{ow}) have been reported for many organic chemicals in solvents such as n-butanol, chloroform, benzene, ethyl ether, cyclohexane, etc. (Leo et al. 1971). K_{ow} can be calculated from a knowledge of K_{ow} for a given organic chemical with the correlation equation for the solvent.

PHYSICO-CHEMICAL PROCESSES

Volatilization

Volatilizational loss of chemicals from water to air is an important mass-transfer pathway for chemicals with low aqueous solubility and low polarity. Many chemicals, despite their low vapor pressure, can volatilize rapidly owing to their high activity coefficients in solution. Volatilizational loss from surfaces is a significant transport process and is a function of physical and chemical properties of the chloroorganic compound under study, as well as on the physical properties such as flow velocity, depth, and turbulence of the water body. The critical factors for volatilization are solubility, molecular weight and vapor pressure of the chloroorganic compound in question, and the nature of the water–air interface. Volatilization of organic chemicals from the soil surface is complicated by other variables. There is no simple laboratory measurement that will reliably extrapolate itself to the field for the soil situation.

Calculation of volatilization rate constant :

$$R_v = -\frac{d[C_w]}{dt} = k_v[C_w] \tag{2.6}$$

where

$$k_v = -\frac{1}{L}\left[\frac{1}{k_\ell} + \frac{RT}{H_c k_g}\right]^{-1} \tag{2.7}$$

and

R_v = volatilization rate of a chemical C (mol L^{-1} hr^{-1})
C_w = aqueous concentration of C (mol L^{-1} (= M))
k_v = volatilization rate constant (hr^{-1})

FATE PROCESSES AND ENVIRONMENTAL MIGRATION

L = depth (cm)
k_ℓ = mass transfer coefficient in the liquid phase (cm hr^{-1})
H_c = Henry's law constant (torr M^{-1})
k_g = transfer coefficient in the gas phase (cm hr^{-1})
R = gas constant (liter-atm-mol^{-1} degree^{-1})
T = absolute temperature (degrees kelvin)

In both phases:

$$k_l = D_l/d_l \text{ and } k_g = D_g/\delta_g$$

where

D = diffusion coefficient
δ = Boundary layer thickness

Calculation of the volatilization loss of an organic chemical

$$\left(k_v^c\right)_{env} = \left(k_v^c/k_v^o\right)_{lab}\left(k_v^o\right)_{env} \qquad (2.8)$$

where

k_v^c = volatilization rate constant for the chemical (hr^{-1}); and
k_v^o = oxygen reaeration constant (hr^{-1}) in the laboratory or environment.

The volatilizational process represents the physical transport of organic chemicals from the waterbodies to the atmosphere. Theoretical concepts of volatilization of chemicals (or chemical compounds) from water to atmosphere have been presented by several researchers (Mackay and Wolkoff 1973; Liss and Slater 1974; Mackay and Leinonen 1975; Chiou and Freed 1977; Smith et al. 1980). A review of these theoretical approaches and their limitations have been discussed (Lyman et al. 1982). The transport across a two-layer system can also be expressed on the assumption that concentrations close to either side of the interface are in equilibrium as expressed by a Henry's law constant (H) (Cohen et al. 1978):

$$F = K_L (C_L - C_s) \qquad (2.9)$$

where

$$C_s = P_v/H \qquad (2.10)$$

$$\frac{1}{K_L} = \frac{1}{k_L} + \frac{RT}{Hk_g} \qquad (2.11)$$

in which K_L is the overall mass transfer (or liquid film) coefficient based on the liquid phase (meters/day); R = gas constant (atm·m³/mol.K); T = temperature (in K) (Rathburn and Tai, 1982); P_v = atmospheric partial pressure (std. atmospheric units) and C_v = concentration in the liquid phase in equilibrium with P_v (or saturation concentration) (Figure 2.2, Gowda and Lock 1985).

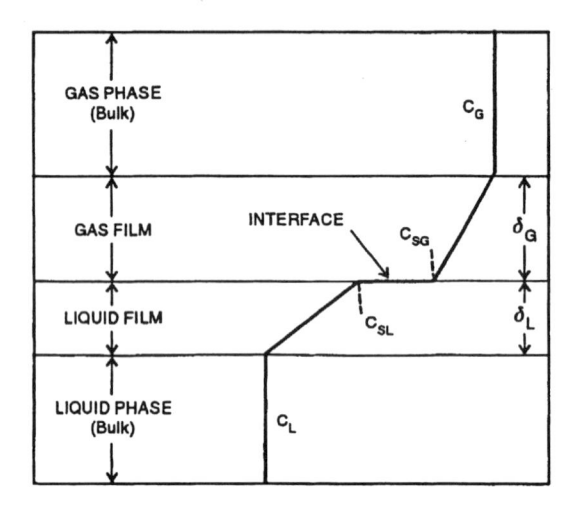

Figure 2.2 Schematic illustration of a two-film gas transfer model. (Reprinted from Gowda and Lock 1985. *J. Environ. Eng.* 111:757. With permission of ASCE.)

Equation 2.9 can be rewritten to express phase resistances to mass transfers:

$$\frac{r_L}{r_T} = \frac{1}{1+(r_G/r_L)} \quad (2.12)$$

$$\frac{r_G}{r_T} = \frac{1}{1+(r_L/r_G)} \quad (2.13)$$

where

$$r_L = \frac{1}{k_L} \qquad r_G = Rt/Hk_G$$

Equations 2.12 and 2.13 can be utilized to determine the relative magnitudes of mass transfer resistances in the liquid and gas phase, respectively. The smallest percent resistance can be predicted for the largest k_L and the smallest k_G. Liquid film resistance can be evaluated as a function of Henry's law constant (H) by utilizing known data on k_L, k_G, and H for organic compounds (Rathburn and Tai, 1982) (Figure 2.2). For average conditions, more than 90% of the resistance is in the liquid film for compounds with H values of about 10^{-3} M atm·m³/g·mol. This profile developed

FATE PROCESSES AND ENVIRONMENTAL MIGRATION

for ethylene and propane (Figure 2.2) can be used to identify organic compounds with similar volatilization characteristics.

Equations have been developed and evaluated to predict values of k_L for streams and rivers in Ontario, Canada (Gowda and Lock 1985). The observed and predicted values of k_L indicated that the predictions from equations 2.14 and 2.15 seem to agree fairly with the observed values (Table 2.1).

Table 2.1 Observed and Computed K_L for Ontario Streams

		Equation 2.14		Equation 2.15	
	Observed K_L	K_{LC} (m/d)	K_{LC} and K_L	K_{LC} (m/d)	K_{LC} and K_L
	2.66	2.10	0.79	2.51	0.95
	2.39	2.22	0.93	2.66	1.11
	3.57	3.34	0.94	4.117	1.17
	3.04	3.01	0.99	3.19	1.05
	1.80	3.80	2.11	4.22	2.35
	4.64	3.14	0.68	3.65	0.79
	8.07	5.57	0.69	6.31	0.78
	2.63	2.57	0.98	2.43	0.92
	2.26	2.60	1.15	2.86	1.27
	0.97	1.87	1.93	2.57	2.64
	3.01	3.32	1.10	3.07	1.02
	3.10	1.97	0.64	2.39	0.77
	3.74	3.15	0.84	2.72	0.73
	5.73	2.51	0.44	3.01	0.52
	4.33	4.28	0.99	3.48	0.80
AV.	3.46	3.03	1.01	3.28	1.12

Note: K_{LC} = computed values; K_L = observed value; m/d = meters per day; AV = average value.

$$K_L = 2.70_g \; 0.35_v - 1.85_{Dm} \; 1.89_B \; 0.4_Z - 0.25_U \; 0.26_{(LS)} \; 0.16 \quad (2.14)$$

$$K_L = 7.16_g \; 0.23_v - 1.51_{Dm} \; 1.59_Z - 0.15_U \; 0.46 \quad (2.15)$$

where

K_L	= liquid film coefficient
B	= channel width
Z	= average velocity
U	= mean velocity
LS	= drop in height
L	= length of reach
S	= bed slope
g	= acceleration due to gravity
D_m	= molecular diffusivity of organic compounds
$v = \mu/\rho$	= kinematic viscosity of water at the instream temperature
ρ	= density of water t° C
μ	= absolute viscosity of water

(From Gowda and Lock 1985. *J. Environ. Eng.* 111:757. With permission of ASCE.)

Volatilization rates from lakes, ponds, or rivers were estimated for the 114 organic chemicals from the EPA priority pollutants list using volatilization rate constant equation (Smith et al. 1978). The estimates with values selected for the parameters in the equation are given in Figures 2.3 and 2.4. The plots suggest that H_c is the major factor in determining the magnitude of volatilization rate constant. It should also be noted that for values of H_c greater than about 40 torr·M^{-1} (roughly corresponding to 25% liquid phase mass transfer resistance), the volatilizational half-life of the chemical will be less than 10 days.

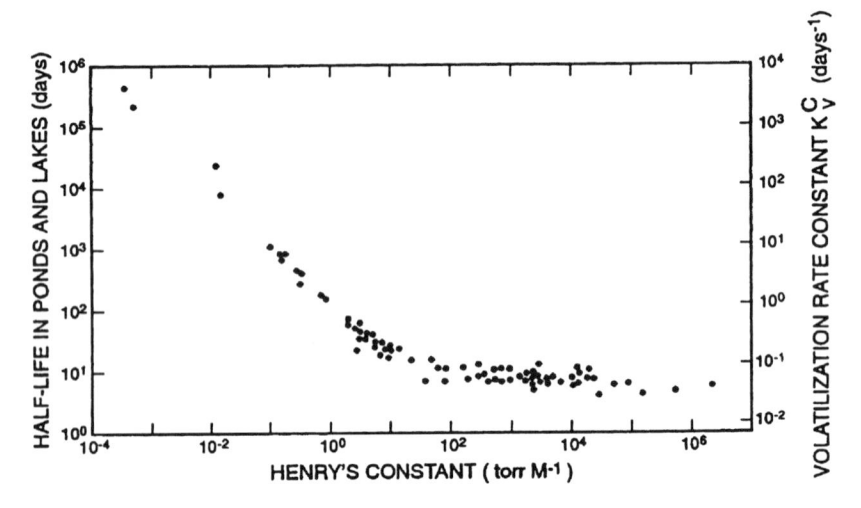

Figure 2.3 Estimated half-lives vs. Henry's Law constant for the 114 priority pollutants in lakes and ponds. Values used in equation: L = 200 cm, k^{ol} = 8.0 cm hr^{-1}, k_g^w = 2100 cm hr^{-1}, n = 1, m = 0.7. (Reprinted from Smith, Bomberger, and Haynes, *Chemosphere*, 10(3):289, 1981. With kind permission from Elsevier Science Ltd., The Boulevard, Langford Lane, Kidlington OX5 1GB, U.K.)

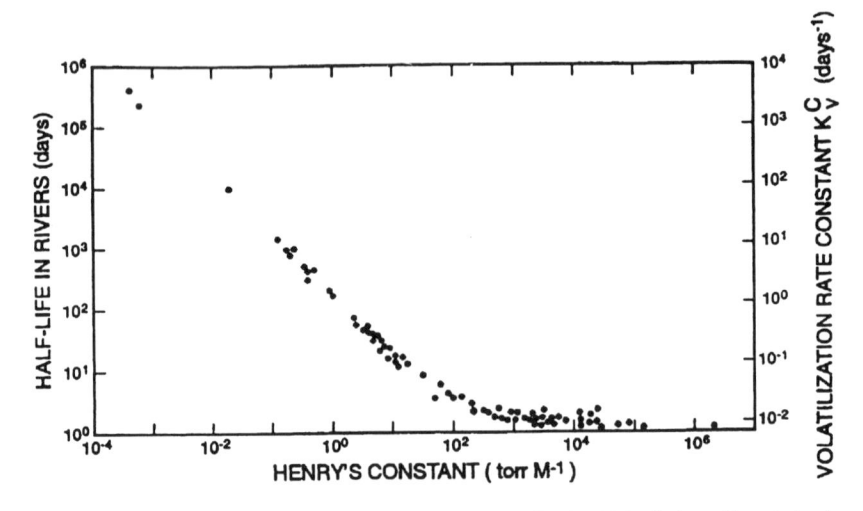

Figure 2.4 Estimated half-lives vs. Henry's Law constant for the 114 priority pollutants in rivers. Equation and values are same as given for Figure 2.3 except k_g^o = 8.0 cm hr^{-1} and n = m = 0.7. (Reprinted from Smith, Bomberger, and Haynes, *Chemosphere*, 10(3):289, 1981. With kind permission from Elsevier Science Ltd., The Boulevard, Langford Lane, Kidlington OX5 1GB, U.K.)

FATE PROCESSES AND ENVIRONMENTAL MIGRATION

Comparison of theoretical loss rates with the measured loss rates are given in Table 2.2 for a few pairs of pesticides (Dobbs et al. 1984). The theoretical values were derived from the literature values for vapor pressures (P) and molecular masses (M) and assuming that loss rate was proportional to $P\sqrt{M}$. The agreement was within a factor of 10 which was anticipated in earlier studies (Dobbs and Cull 1982). If high melting chemicals with vapor pressure difference greater than 10 were excluded, then the agreement was within a factor of 3, indicating the validity of the estimation method.

Table 2.2 Theoretical Loss vs. Measured Loss For Some Pesticides

Chemicals	Theoretical Ratio (Tr)	Measured Ratio (Mr)	Mr/Tr
Parathion ethyl/parathion methyl	0.60	0.64	1.1
Dinoseb/dimethoate	6.2	7.7	1.2
Dibutyl phthalate/hexachlorobenzene	1.2	1.0	0.83
Hexachlorobenzene*/dieldrin*	5.0	3.0	0.60
Trifluralin/dieldrin	40.0	21.0	0.53
Atrazine+/dieldrin	0.091	0.037	0.41
Parathion ethyl/dieldrin	1.7	0.62	0.36
Dinoseb/parathion methyl	5.8	2.0	0.34
Dibutyl phthalate/dieldrin	6.0	0.74	0.12
Atrazine/di(–2 ethylhexyl) phthalate+	0.31	1.1	3.5
Trifluralin/dimethoate	15.0	39.0	2.6
Atrazine*/monuron+	0.62	1.9	3.1
Picloram+/Dieldrin	0.018	0.14	7.8

* Average value from two experiments.
+ Only a small amount of chemical was volatilized during the experiment and the log (amount remaining) — time regression had large 95% confidence limits, therefore the loss rate ratio may be suspect.

Reprinted from Dobbs et al. 1984, *Chemosphere*, 13:691. With kind permission from Elsevier Science Ltd., The Boulevard, Langford Lane, Kidlington OX5 1GB, U.K.

Surfactants can reduce volatilization of an organic chemical from water surface.

Volatilization of Chloroorganic Compounds as a Function of H, Henry's Law Constant

$H, > 10^{-6}$	=	Nonvolatile
$H, 10^{-6}$ to 10^{-5}	=	Volatilizes slowly controlled by air phase resistance
$H, 10^{-5}$ to 10^{-3}	=	Significant volatilization controlled by both air and aqueous phases
$H, < 10^{-3}$	=	Rapid volatilization controlled by aqueous phase resistance

VOLATILIZATION FROM SOIL

Volatilization from soil into air is an important indicator for the persistence of chloroorganic compounds in soil either applied with purpose or in a spill situation. Volatilization of chloroorganic compounds from soil depends upon (1) chemical structure, (2) sorption capacity, (3) soil–water content, (4) air flow across soil surface, (5) humidity, (6) temperature and (7) bulk properties of soil such as porosity, clay content, organic matter content, etc. The above factors determine the movement of the chloroorganic compound to the evaporating soil surface, migration from soil to air, and its vapor density (concentrating capacity) in air.

Various methods available in literature for estimating volatilization of an organic chemical from soil are discussed by Lyman et al. (1982) for their conditions of applicability in calculating the soil half-life of chloroorganic compounds. The kind of data required on the chemical and environmental conditions for the different methods is also discussed (Lyman et al. 1982).

Sorption

The term *sorption* used here covers both adsorption and absorption, which are difficult to distinguish in most situations. Many metals and organic chemicals sorb strongly to sediments, suspended solids, and soils. This determines the fraction that is available for other fate processes.

The understanding and quantification of exchange processes occurring at the sediment–water interface is important in formulating a model for the speciation and transport of heavy metals and organic chemicals. The sediment is a complex mixture with four main components: silica, clays, organic matter, and oxides of iron and manganese. The release of sorbed chemicals into the bulk water is dependent on partition coefficients, which in turn are related to sediment characteristics, the type of chemical, and other environmental parameters. Desorption may be a slow process, posing a long-term problem even after the sources of pollution are eliminated (Ramamoorthy and Rust 1978).

The sorption coefficient (K_{oc}) is the concentration of chemical sorbed by the sediment or soil on an organic carbon basis divided by the concentration of the chemical in the surrounding water column. Expressing sorption on an organic carbon basis instead of a total sediment–soil basis renders a valid comparison of sorption coefficients. Sorption coefficients are relatively constant at low aqueous concentrations of the chemical, but tend to decrease as the concentration of the chemical in the water column is increased, especially for chemicals with high aqueous solubilities. However, most sorption studies are done at concentrations low enough to minimize the variation that could arise from this factor.

Sorption of chemicals to organic matter depends upon the pH and the type of chemical interaction, either ionic or non-ionic. Interaction of non-ionics will not be greatly affected by changes in pH. Whereas, ionic compounds will be repelled by the sorbent surfaces at high pH values due to repulsion of like-charges of the sorbent and sorbate. Chemicals like 2,4-dichlorophenoxy acetic acid (2,4-D), 2,4,5-trichlorophenoxy acetic acid (2,4,5-T), dicamba, chloramben, and picloram behave in this manner. The sorption of these chemicals increase with a decrease in pH leading to the formation of an un-ionized surface and un-ionized form of the chemical. These compounds are also sorbed strongly by hydrated ion and aluminum oxides at low pH values.

Sorption coefficients based on the organic carbon content of the sorbent provide a good basis for relating to other accumulation parameters such as n-octanol–water partition coefficients and bioconcentration factors for biota. Sorption coefficients also provide a measure of the leachability of chemicals which is valuable in environmental impact assessment. Compounds having a K_{oc} value equal to or greater than 1000 are quite strongly bound to organic matter of sediment/soil and are considered immobile. Chemicals with K_{oc} values below 100 are moderately to highly mobile. Thus, K_{oc} values can be useful predictors of the potential leachability of compounds through soil or from aqueous sediments.

Values of K_{oc} can be estimated using correlations with other properties of the chemical: (1) water solubility (WS); (2) n-octanol–water partition coefficient (K_{ow}); and (3) bioconcentration factor for aquatic organisms (BCF). Table 2.3 lists some of the commonly used regression equations for estimating K_{oc}.

Table 2.3 Estimation of K_{oc} from Other Related Parameters

Equation	No.[a]	r²*	Chemical Classes Covered	Ref.
log K_{oc} = –0.55 log WS + 3.64 (S in mg/L)	106	0.71	Wide variety, mostly pesticides	Kenaga and Goring 1980
log K_{oc} = –0.54 log WS + 0.44 (S in mole fraction)	10	0.94	Mostly aromatic or polynuclear aromatics; two chlorinated	Karickoff et al. 1979
log K_{oc} = –0.557 log WS + 4.277 (S in µmol/L)	15	0.99	Chlorinated hydrocarbons	Chou et al. 1979
log K_{oc} = 0.544 log K_{ow} + 1.377	45	0.74	Wide variety, mostly pesticides	Kenaga and Goring 1980
log K_{oc} = 0.937 log K_{ow} –0.006	19	0.95	Aromatics, polynuclear aromatics, triazines, and dinitroaniline herbicides	Lyman et al. 1982
log K_{oc} = 1.00 log K_{ow} –0.21	10	1.00	Mostly aromatic or polynuclear aromatics; two chlorinated	Karickoff et al. 1979
log K_{oc} = 1.029 log K_{ow} –0.18	13	0.91	Variety of insecticides, herbicides, and fungicides	Rao and Davidson 1980
log K_{oc} = 0.524 log K_{ow} + 0.855[b]	30	0.84	Substituted phenylureas and alkyl-*N*-phenylcarbamates	Briggs 1973
log K_{oc} = 0.681 log BCF + 1.963	13	0.76	Wide variety, mostly pesticides	Kenaga and Goring 1980
log K_{oc} = 0.681 log BCF + 1.886	22	0.83	Wide variety, mostly pesticides	Kenaga and Goring 1980

[a] No., number of chemicals used to obtain regression equation.
*r² Correlation coefficient for regression equation.
[b] The relationship $k_{om} = k_{oc}/1.724$ was used to rewrite the equation in terms of k_{oc}.

Reprinted with permission from Lyman et al. 1982, *Handbook of Chemical Property Estimation Methods Environmental Behavior of Organic Compounds*, originally published by McGraw-Hill Book Company, copyright 1982 American Chemical Society.

One or more equations should be chosen based on the availability of data and chemical classes covered by the equation. In addition, the range of the input parameter and k_{oc} should be within the range covered by the dataset from which the equation was derived (Figures 2.5 and 2.6). The following equation 2.16 was developed by Mill (1980) for predicting sorption coefficients of monocyclic aromatics:

$$\log k_{oc} = -0.782 \log [C] - 0.27 \quad (2.16)$$

where [C] = concentration in mol liter.⁻¹

The estimate is reliable to a power of 10 for most nonpolar chemicals, which is sufficiently accurate for screening purposes in most cases.

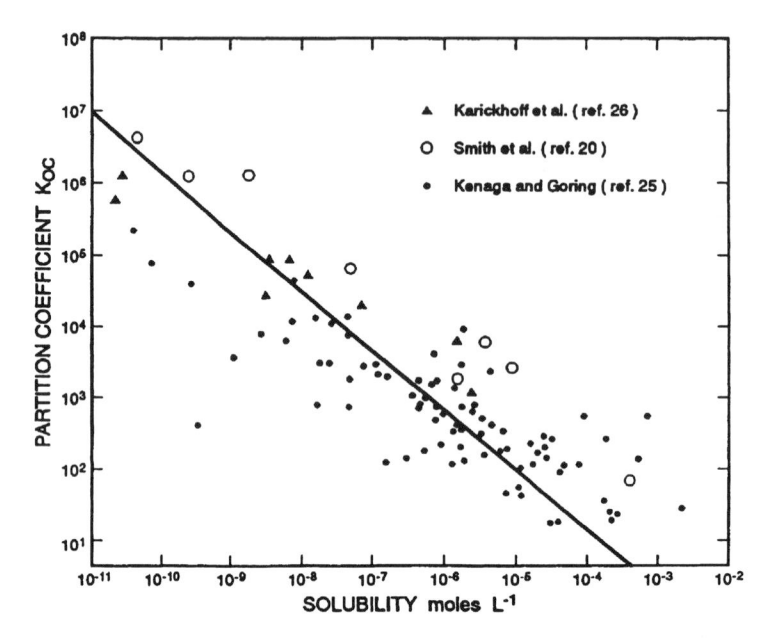

Figure 2.5 Correlation of sorption coefficient normalized to organic carbon (K_{oc}) with WS (water solubility). (Reprinted with permission from Mill 1980, *Dynamics, Exposure and Hazard Assessment of Toxic Chemicals,* Ann Arbor Science, Ann Arbor, MI.)

Bioconcentration

Bioconcentration denotes the concentration of a chemical in an organism or in the tissue of an organism and the bioconcentration factor (BCF) is the ratio of the concentration of the chemical in the organism to its concentration in the surrounding water column. Three commonly used methods for measurement of BCFs are (1) exposure of fish in an aquarium to flowing water spiked with the chemical; (2) a model ecosystem containing plant or animal organism or both in water; and (3) a terrestrial-aquatic model ecosystem containing soil and animal–plant organisms. The BCF obtained is determined by factors such as solubility, residence time in water, hydrophobicity of the chemical and surface-to-volume ratio, and lipid content of the organism. Lieb (1974) measured the lipid content of a 14-week old rainbow trout over a period of 32 weeks. The lipid content doubled in that time period from 4.4% to 8.4%; gill (9.7%), muscle (2.7%), stomach (6.5%), liver (3.5%), and whole fish (8.5%). Concentrations of chemicals that appear safe for organisms can bioconcentrate to levels that are harmful to predators. Reliable bioconcentration values can provide early warning of potential problems in aquatic media without extensive monitoring information. Examples where predators suffered the toxicity effects are with DDD in California in the 1950s–1960s and more recently with organophosphates (Hall and Kolbe 1980). In addition, some stages of the fish life-cycle take up

FATE PROCESSES AND ENVIRONMENTAL MIGRATION

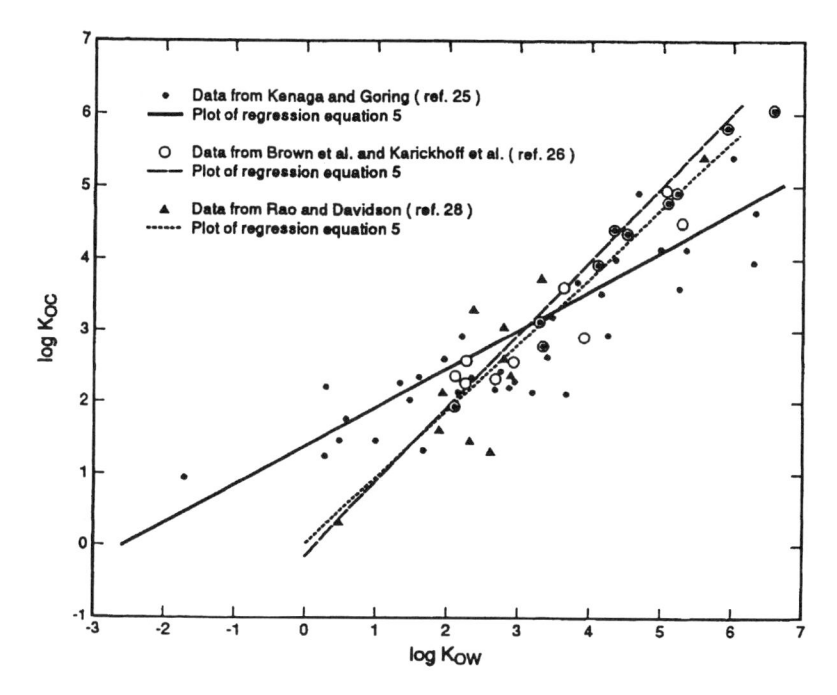

Figure 2.6 Correlation of K_{oc} with log K_{ow}. (Reprinted from Lyman et al. 1982 *Handbook of Chemical Property Estimation Methods Environmental Behavior of Organic Compounds*, originally published by McGraw-Hill Book Company, copyright 1982 American Chemical Society.)

and tolerate chemicals which later either become toxic during periods of stress or are passed on to produce toxic effects in more susceptible stages (Davies and Dobbs 1985).

It is important to measure the uptake and depuration rates of the organisms or alternatively the measurements should be made over a sufficiently long period of time to ensure that equilibrium conditions exist. Flow-through bioassay systems should be used so that chemical concentrations remain relatively constant during the test.

The accumulation of organic chemicals in aquatic organisms can be predicted by several methods. The relationship between bioconcentration and other physicochemical properties e.g., *n*-octanol–water partition coefficient), k_{ow} and aqueous solubility (WS) are used to predict BCFs to provide a preliminary environmental safety assessment of the chemicals. Published correlations of BCF with WS and log k_{ow} are given in Table 2.4. Shortcomings of some of the data used and replacing with data obtained under closely defined and comparable experimental conditions have been discussed in the literature (Davies and Dobbs 1985).

The data on BCF and WS, and log k_{ow} for selected organic chemicals have been plotted in Figures 2.7 and 2.8 using the following regression equations:

$$\log_{10} \text{BCF} = 4.358 - 0.444 \, [\log_{10} \text{WS}(\text{g L}^{-1})]$$
$$r = -0.803, \, n = 29, \text{ and } t = 7.00 \text{ (Figure 2.7)}. \tag{2.31}$$

Table 2.4 Published Regression Equations Between BCF and k_{ow} and WS

Equation Nos.	Equations (log BCF=)	WS Units	BCF vs. WS r	t	n	System	Ref.
2.17	3.995−0.389 (log WS)	ppb	−0.923	7.20	11	Static ecosystem	Hall and Kolble 1980
2.18	3.410−0.508 (log WS)	µmol^{-1}	−0.964	9.04	8	Flow-through	Choi 1977
2.19	2.791−0.564 (log WS)	ppm	−0.72	6.05	36	Flow-through	Kenaga and Goring 1980
2.20	2.183−0.629 (log WS)	ppm	0.66	6.09	50	Static	Kenaga and Goring 1980
2.21	3.710−0.316 (log WS)	ppb	−0.565	3.28	25	Flow-through	Veith et al. 1979, 1980
2.22	5.09−0.85 (log WS)	ppb	0.87	5.29	11	Static	Metcalfe et al. 1973
2.23	2.83−0.55 (log WS)	ppm	—	—	42	Flow-through	Kobayashi 1981
2.24	0.542 (log k_{ow}) +0.124		0.948	7.30	8	Flow-through	Neely et al. 1974
2.25	0.935 (log k_{ow}) −1.495		0.87	8.64	26	Flow-through	Kenaga and Goring 1980
2.26	0.767 (log k_{ow}) −0.973		0.76	6.82	36	Static	Kenaga and Goring 1980
2.27	0.85 (log k_{ow}) −0.70		0.947	21.46	55	Flow-through	Veith et al. 1979, 1980
2.28	0.456 (log k_{ow}) +0.634		0.634	3.93	25	Flow-through	Veith et al. 1979, 1980
2.29	0.634 (log k_{ow}) +0.729		0.788	3.84	11	Static	Lu and Metcalfe 1975
2.30	0.74 (log k_{ow}) −0.77		—	—	40	Flow-through	Kobayashi 1981

Note: r correlation coefficient; n, no. of data points; and t, student's t value for regression.
Reprinted from Davies and Dobbs, 1984. *Water Research*, 18:1256. With kind permission from Elsevier Science Ltd., The Boulevard, Langford Lane, Kidlington, OX5 1GB, U.K.

and
$$\log_{10} BCF = 0.597 (\log_{10} K_{ow}) + 0.188$$
$$r = 0.748, n = 31, \text{ and } t = 6.07 \text{ (Figure 2.8)}. \qquad (2.32)$$

Figure 2.9 presents the correlation of BCF vs. log k_{ow} of hydrocarbons and chlorohydrocarbons that restrict the range but increase the reliability of correlations. Various aspects of BCF measurements such as steady state and kinetic approaches to BCF, experimental factors, relevance of laboratory-measured BCFs to field situations, good laboratory practices, etc. are discussed in detail elsewhere (Davies and Dobbs 1985). However, efforts are underway to make BCF tests to provide results that are consistent and widely applicable, recognizing the differences in fish species and strains, together with appropriate sizes and temperatures to use in BCF tests.

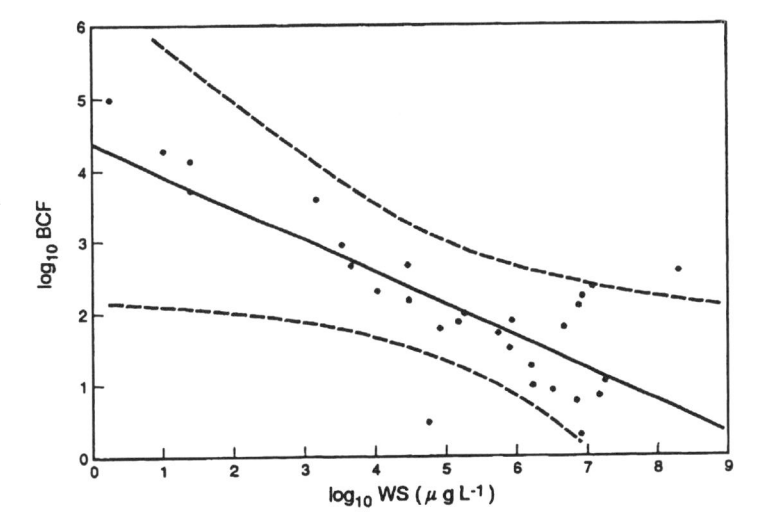

Figure 2.7 Correlation of BCF (bioconcentration factor) with WS (water solubility). (Reprinted from Davies and Dobbs, 1984. *Water Research*, 18:1256. With kind permission from Elsevier Science Ltd., U.K.)

TRANSFORMATION PROCESSES

A chemical in the environment could be altered by transformation processes such as: (1) ionization, (2) hydrolysis, (3) photodegradation, and (4) halogenation–dehalogenation processes. The extent to which a chemical breaks down to simple moities will determine its persistence and toxicity. The transformed derivative could be substantially more hazardous and persistent. Halogenation of aromatic compounds and aliphatic hydrocarbons are environmentally significant. Chlorinated dioxins and furans, and, formation of chloroform in the presence of organic matter are examples of this process. Some of the abiotic transformation processes are briefly described below.

Ionization. An organic acid or base that is extensively ionized in the environment may be significantly different from the corresponding un-ionized neutral molecule

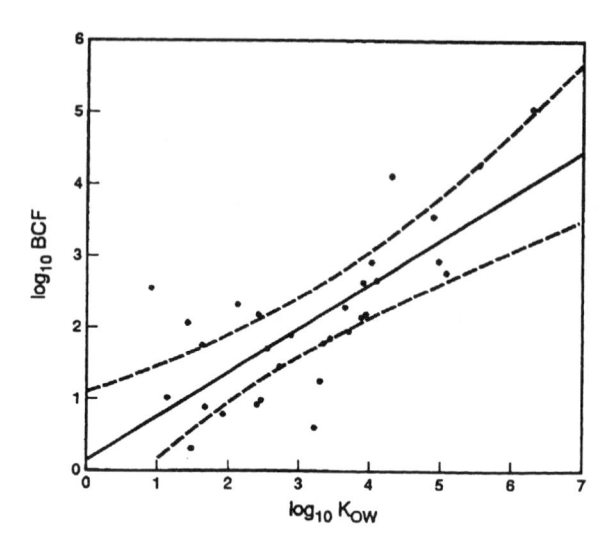

Figure 2.8 Correlation of BCF with log K_{ow}. (Reprinted from Davies and Dobbs, 1984. *Water Research*, 18:1256. With kind permission from Elsevier Science Ltd., U.K.)

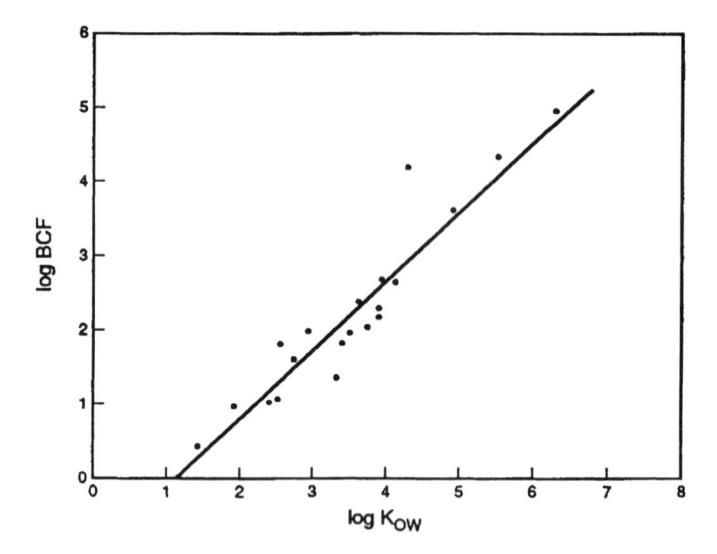

Figure 2.9 Correlation of BCF with log K_{ow} for hydrocarbons and chlorinated hydrocarbons (regression equation, log BCF = −1.30 + 0.98 log K_{ow}). (Reprinted from Davies and Dobbs, 1984. *Water Research*, 18:1256. With kind permission from Elsevier Science Ltd., U.K.)

in solubility, sorption, bioconcentration, and toxic potential. For example, the ionized species have longer residence time in water, less ability to migrate into the organic or lipid part of the abiotic (such as sediment), and the biotic substrates (fish) than the parental neutral un-ionized molecule. The equations for electron and proton changes (ionization and redox conditions, respectively) are given below:

$$pH = \log[H^+]$$
High pH = low H⁺ activity and conversely

$$pH = pK_\alpha + \log[A^-]/[HA]$$

$$pH = pK_\alpha + \text{when } [A^-] = [HA]$$

K_α = acid dissociation constant, $HA \leftrightarrow H^+ + A^-$

$[H^+]$ = Proton concn.

$$p_\varepsilon = -\log[e^-]$$
High p_ε = low ε^- activity and conversely

$$p_\varepsilon = p_\varepsilon^o + \log\frac{[\text{oxidized}]}{[\text{reduced}]}$$

$p_\varepsilon = p_\varepsilon^o$ when [oxidized] = [reduced]

p_ε^o = equilibrium potential

$[e^+]$ = Electron concn.

Hydrolysis. Hydrolysis is likely to be one of the important reactions of organic chemicals in water. A hydrolysis reaction is one where hydrogen, hydroxyl radical or the water molecule interacts with the organic compound depending on the pH and polarity of the site of attack on the molecule.

The rate of hydrolysis of a chemical compound can be calculated using the following equation:

$$-\frac{dc}{dt} = k_h[C] = k_A[H^+][C] + k_B[OH^-][C] + k_N[C] \quad (2.33)$$

where k_h = first order hydrolysis rate constant at a specific pH; k_A and k_B = second order acid and base hydrolysis constants, respectively, and k_N = first order hydrolysis rate constant for pH independent reaction.

Kinetic half-lives of organic compounds. Half-lives of organic compounds are calculated from the respective rate constants. For a first-order kinetic reaction,

$$A \xrightarrow{j} \text{products at a constant volume} \quad (2.34)$$

The rate of disappearance of A is given by:

$$-\frac{dc_A}{dt} = k_j C_A \quad (2.35)$$

where C_A = concentration of A in mol/L
 t = time in appropriate units
 K_j = reaction rate for the process j in units of inverse time,

and

$$-\frac{dc_A}{dt} = \text{rate of change of } C_A \text{ with time}$$

Integrating the equation between the limits of t_o (initial time) and t, yields:

$$k_j = \frac{1}{(t-t_o)} \ln \frac{(^C A_o)}{(C_A)} \tag{2.36}$$

where

$$C_{AO} = \text{initial concentrations of } C_A \text{ at } t_o$$

for

$$C_A = 0.5 \, ^C A_o \text{ the half-life is given by:}$$

$$t_{1/2} = \frac{1}{k_j} \ln \frac{(^{2C} A_o)}{(^C A_o)} \quad \text{or} \quad t_{1/2} = 0.693 \left(\frac{1}{k_j}\right)$$

If all the transformation processes are expressed as a first-order or pseudo first-order kinetic process, the net half-life for the chemical is given by:

$$t_{1/2} = \frac{\ln 2}{\Sigma_j k_j}$$

Photolysis. Structural changes of a molecule induced by electromagnetic impact in the ultraviolet-visible light range (240 to 700 nm) are called photochemical reactions. Rate of disappearance of an organic compound by direct photolysis is calculated by using the following equation:

$$-\frac{dc}{dt} = K_p[C] = k_a \theta[C] \tag{2.37}$$

where k_p = first-order rate constant, θ = reaction quantum yield and k_a = rate constant for light absorption by the chemical that depends on the light intensity, chromaticity of light, and extinction coefficient of the chemical.

Rate of disappearance of an organic compound by indirect photolysis is given by:

$$-\frac{dc}{dt} = K_2[C][X] = k'_p[C] \tag{2.38}$$

where k_2 = second-order constant for the interaction between the chemical and the intermediate, X; for a photosensitized reaction the k_p would be a combined term including the concentration of the excited state species and the quantum yields for the energy transfer to and subsequent reaction of the chemical.

A simple way to determine environmental photolysis of a chemical is to measure θ at an angle wavelength (λ) in the laboratory. Sunlight intensity (I_λ) data as a

FATE PROCESSES AND ENVIRONMENTAL MIGRATION

function of time of day, season, and latitude are available in the literature. The rate constant in sunlight $k_{p(s)}$ is given by the following equation:

$$k_{p(s)} = \theta \Sigma I_\lambda \varepsilon_\lambda \qquad (2.39)$$

and the half-life of the organic chemical in sunlight is calculated by using the equation given below:

$$\left(t_{1/2}\right)_{(s)} = \frac{\ln 2}{k_{p(s)}} \qquad (2.40)$$

Metabolic breakdown. Many organisms and biota, in general, develop resistance to most organic chemicals and transform them to compounds that are not toxic to them, but may be toxic to the total environment. The rate of biotransformation will be a function of the biomass and the concentration of the chemical.

(a) *Rate of substrate utilization*:

$$-\frac{dc}{dt} = \frac{\mu X}{Y} = \frac{(\mu_m)}{(Y)} \cdot \frac{(CX)}{(K_s + C)} = (k_b) \cdot \frac{(CX)}{(K_s + C)} \qquad (2.41)$$

where μ = specific growth rate, X = biomass per unit volume, μ_m = maximum specific growth rate, k_s = concentration of the substrate to support half-maximum specific growth rate (0.5 μ), k_b = substrate utilization constant or biodegradation constant (= μ_m /Y), and Y = biomass produced from a unit amount of the substrate consumed. These constants are characteristics of the microbes, pH, temperature, and media.

(b) *Reduced equation for the rate of substrate utilization:*
When the substrate concentration C >> K_s, equation 2.41 reduces to

$$-\frac{dc}{dt} = k_b X \qquad (2.42)$$

This means that the biodegradation rate is first-order with respect to all biomass concentration and zero order with respect to the chemical concentration.

(c) *Reduced equation for the rate of substrate utilization:*
In actual environmental conditions for many organic compounds, C << K_s, hence equation 2.42 becomes

$$-\frac{dc}{dt} = (k_b)\frac{(CX)}{(K_s)} = k_{b2}[C][X] \qquad (2.43)$$

where k_{b2} is a second-order rate constant.

(d) *Reduced equation for the degradation rate of a chemical:*

When the biomass concentration is relatively large compared to the contaminant concentration, the degradation rate is pseudo-first order rate and given by

$$-\frac{dc}{dt} = k'_b C \qquad (2.44)$$

where k'_b is the pseudo-first order rate constant and dependent on cell concentration (X_o).

The $t^1/_2$ (time required to reduce the chemical to half of its initial concentration) will be given by

$$T_{1/2} = t_o + t_{1/2} \qquad (2.45)$$

where $T^1/_2$ = acclimation time and $t^1/_2$ = half-life of transformation of the chemical.

(e) *Calculation of $t^1/_2$ of a chemical under degradation*:

The half-life of the chemical under degradation ($t^1/_2$ at a given X_o) is given by:

$$t_{1/2} = \frac{\ln 2}{k_{b2} X_o} = \frac{0.693}{k_{b2} X_o} \qquad (2.46)$$

where

$$k_{b2} = \frac{k'_b}{X_o}$$

(k'_b = pseudo-first order rate constant and X_o = cell concentration).

REFERENCES

Banerjee, S., Yalkowsky, S.H., and Valrani, S.C. 1980. Water solubility and octanol–water partition coefficients of organics. *Environ. Sci. Technol.* 14:1227–1229.

Briggs, G.G. 1973. A simple relationship between soil adsorption of organic chemicals and their octanol/water partition coefficient. In: *Proceedings at the 7th British Insecticide and Fungicide Conference Vol. I*, The Boots Company Ltd., Nottingham, Great Britain.

Chiou, C.T., Peters, L.J., and Freed, V.H. 1979. A physical concept of soil-water equilibria for non-ionic organic compounds. *Science*. 206:831–832.

Chiou, C.T. and Freed, V.H. 1977. *Chemodynamics Studies on Benchmark Industrial Chemicals*. MSF/RA-770286. NTIS PB 274263.

Chiou, C.T., Freed, V.H., Schmedding, D.W., and Kohnert, R.L. 1977. Partition coefficient and bioaccumulation of selected organic chemicals. *Environ. Sci. Technol.* 11:475–478.

Cohen, Y., Coccio, W., and Mackay, D. 1978. Laboratory study of liquid-phase controlled volatilization rates in presence of wind waves. *Environ. Sci. Technol.* 12:553–558.

Davies, R.P. and Dobbs, A.J. 1985. The prediction of bioconcentration in fish. *Water Res.* 18:1253–1262.

Dobbs, A.J. and Cull, M.R. 1982. Volatilization of chemicals, relative loss rates, and the estimation of vapor pressures. *Environ. Pollut.* (Series B). 3:289–298.

Dobbs, A.J., Hart, G.F., and Parsons, A.H. 1984. The determination of vapor pressures from relative volatilization rates. *Chemosphere.* 13:687–692.

Dreisbach, R.R. 1952. *Pressure-Volume Relationship of Organic Compounds.* Handbook Publishers, Sandusky, OH. pp. 3–260.

Gowda, T.P.H. and Lock. J.D. 1985. Volatilization rates of organic chemicals of public health concern. *J. Environ. Eng.* 111:755–776.

Greichus, Y.A., Greichus, A., and Emerick, R.J. 1973. Insecticides, polychlorinated biphenyls and mercury in wild cormorants, their eggs, food and environment. *Bull. Environ. Contam. Toxicol.* 9:321–328.

Hall, R.J. and Kolbe, E. 1980. *J. Toxic. Environ. Health.* 6:853–860.

Hamelink, J.L., Waybrant, R.C., and Yant, P.R. 1977. In: *Fate of Pollutants in the Air and Water Environments,* Part 2. R. Suffet (Ed.), John Wiley & Sons, New York, U.S.A.

Hansch, C. and Leo, A.J. 1979. *Substituent Constants for Correlation Analysis in Chemistry and Biology.* John Wiley & Sons, New York.

Haque, R. 1980. *Dynamics, Exposure and Hazard Assessment of Toxic Chemicals.* Ann Arbor Science, Ann Arbor, MI, 496 pp.

Haque, R. and Schemmedding, D. 1975. A method for measuring the water solubility of hydrophobic chemicals; solubility of five polychlorinated biphenyls. *Bull. Environ. Contam. Toxicol.* 14:13–18.

Karickoff, S.W., Brown, D.S., and Scott, T.A. 1979. Sorption of hydrophobic pollutants on natural sediments. *Water Res.* 13:241–248.

Kenaga, E.E. and Goring, C.A.I. 1980. Relationship between water solubility, soil sorption, octanol-water partitioning and bioconcentration of chemicals in biota. In: *Proceedings of the 3rd Symposium on Aquatic Toxicology,* J.G. Eaton, P.R. Parish, and A.C. Hendricks (Eds.), American Society of Testing Materials, Philadelphia, pp.78–115.

Kobayashi, K. 1981. Proceedings of OECD Workshop on the Control of Existing Chemicals, Unwelltrandesant, Berlin. pp.141–163.

Leo, A.C., Hansch, C., and Elkins, D. 1971. Partition coefficients and their uses. *Chem. Rev.* 71, 525–621.

Lieb, A.J., Bills, D.D., and Sinnhuber, R.O. 1974. *J. Agri. Food Chem.* 22:638–642.

Liss, P.S. and Slater, P.G. 1974. Flux of gases across the air-sea interface. *Nature.* 247:181–184.

Lu, P.Y. and Metcalfe, R.L. 1975. Environmental fate and biodegradability of benzene derivatives studied in a model aquatic ecosystem. *Environ. Health Persp.* 10:269–284.

Lyman, W.J., Reehl, W.F., and Rosenblatt, D.H. 1982. *Handbook of Chemical Property Estimation Methods — Environmental Behaviour of Organic Compounds.* McGraw-Hill Book Company, New York.

Mabey, W. and Mill, T. 1978. Critical review of hydrolysis of organic compounds in water under environmental conditions. *J. Phys. Chem. Ref. Data.* 7:383–415.

Mabey, W., Mill, T., and Hendry, D.G. 1979. Test Protocols in Environmental Processes: Direct Photolysis in Water, U.S. EPA Publication No. EPA-68-03-2227.

Mackay, D. and Wolkoff, A.W. 1973. Rate of evaporation of low solubility contaminants from water bodies to atmosphere. *Environ. Sci. Technol.* 7:611–614.

Mackay, D. and Leinonen, P.J. 1975. Rate of evaporation of low solubility contaminants from water bodies to atmosphere. *Environ. Sci. Technol.* 9:1178–1180.

Metcalfe, R.L., Kapoor, I.P., Lu, P.Y., Schuth, C.K., and Sherman, P. 1973. Model ecosystem studies of the environmental fate of six organochlorine pesticides. *Environ. Health Persp.* pp. 35–44.

Mill, T. 1980. In: *Dynamics, Exposure and Hazard Assessment of Toxic Chemicals,* Haque, R. (Ed.), Ann Arbor Science Publishers, Ann Arbor, MI. pp. 297–322.

Moriguchi, I. 1975. Quantitative-structure-activity studies. I. Parameters related to hydrophobicity. *Chem. and Pharma. Bull. (Tokyo).* 23:247–257.

Moore, J.W. and S. Ramamoorthy. 1984. *Organic Chemicals in Natural Waters. Applied Monitoring and Impact Assessment,* Springer-Verlag, New York. 289 pp.

Neely, W.B., Branson, D.R., and Blau, G.E. 1974. Partition coefficients to measure bioconcentration potential of organic chemicals in fish. *Environ. Sci. Technol.* 8:1113–1115.

Quayle, O.R. 1953. The parachors of organic compounds: an interpretation and catalogue. *Chem. Rev.* 53:439–585.

Ramamoorthy, S. and Rust, B.R. 1978. Heavy metal exchange processes in sediment-water systems. *Environ. Geol.* 2:165–172.

Rao, P.S.C. and Davidson, J.M. 1980. In: *Environmental Impact of Non-Point Source Pollution,* M.R. Overcash and J.M. Davidson (Eds.), Ann Arbor Science Publishers, Ann Arbor, MI.

Rathburn, R.E. and Tai, D.Y. 1982. Volatilization of organic compounds from streams. *J. Env.Eng. Div.,* ASCE. 108(EE5):973–989.

Smith, J.H., Bomberger, D.C. Jr., and Haynes, D.L. 1981. Volatilization rates of intermediate and low volatility chemicals from water. *Chemosphere.* 10:281–289.

Smith, J.H., Mabey, W.R., Bohonos, N., et al. 1978. Environmental Pathways of Selected Chemicals in Freshwater Systems, U.S. EPA, Publications Nos. 600/7-77-113 and 600/7-78-074.

Smith, J.H., Bomberger, D.C., and Haynes, D.S. 1980. Prediction of the volatilization rates of high-volatility chemicals from natural water bodies. *Environ. Sci. Technol.* 14:1332–1337.

Veith, G.D., Defoe, D.L., and Bergstedt, B.V. 1979. Measuring and estimating the bioconcentration factor of chemicals in fish. *J. Fish. Res. Bd. Can.* 36:1040–1048.

Veith, G.D., Macek, K.J., Petrocelli, S.R., and Carroll, J. 1980. An evaluation of using partition coefficients and water solubility to estimate bioconcentration factors for organic chemicals in fish. In: *Aquatic Toxicology,* American Society of Testing Materials, Philadelphia, pp. 116–129.

Wallhofer, P.R., Koniger, P.R.N., and Hutzinger, O. 1973. *Analab. Res. Notes.* 13:14.

Weast, R.C. (Ed.). 1974. *CRC Handbook of Chemistry and Physics, 54th edition,* CRC Press, Boca Raton, FL. D162-D188.

CHAPTER 3

Chlorinated Aliphatic Hydrocarbons

Aliphatic hydrocarbons constitute a diverse group of organic compounds characterized by an open-chain structure and a variable number of single, double, and triple bonds. Saturated compounds are known as alkanes and fit the empirical formula C_nH_{2n+2}. Alkanes having four or more carbon atoms can exist in both straight-chain and branched-chain isomers. Compounds having one double bond with a formula C_nH_{2n} are called alkenes. The first two alkenes of the series, ethylene and propene, exist in structural form; whereas, the next higher homolog, C_4H_8, has two straight-chain isomers and one branched-chain isomer. Aliphatic hydrocarbons having two and three double bonds per molecule are called alkadienes and alkatrienes, respectively; those with triple bonds are called alkylenes.

Aliphatic hydrocarbons may undergo chlorination in natural environment to produce chloromethanes. They also undergo industrial chlorination in various processes to yield mono-to-polychlorinated derivatives that have wide use as solvents, degreasers, dry cleaning agents, refrigerants, and as intermediates in organic syntheses.

PROPERTIES

The physical and chemical properties of chlorinated aliphatic hydrocarbons are listed in Table 3.1. In general, the chlorinated alkanes are low-molecular compounds with relatively high vapor pressures, moderate aqueous solubilities and low-octanol–water partition coefficients that do not greatly favor accumulation in biota. The physical and chemical properties determine the transport and distribution of organic compounds in the environment. Chlorinated alkenes are also characterized by relatively high vapor pressures, low to medium aqueous solubilities and low to moderate octanol–water partition coefficients (Table 3.1). Henry's law constants, which provide an estimate of volatilization of the compound, in units of atmosphere·m^3/mol and conversion factors from ppm to mg/m^3, and from mg/m^3 to ppm in air, are also included in Table 3.1.

Table 3.1 Physical and Chemical Properties of Selected Chlorinated Aliphatic Compounds

Properties	Chloro-methane	Dichloro-methane	Chloro-form	Carbontetra-chloride	Chlorodibromo-methane	Bromodichloro-methane	Chloroethane
Molecular weight	50.49	84.93	119.38	153.8	208.29	163.83	64.52
Melting point °C	−97.7	−95.1	−63.2	−23.0	<−20.0	−57.1	−138.3
Boiling point °C	−23.73	40.0	61.3	76.5	119–120.0	90.0	12.4
Solubility Water@ 25°C mg/L	5,325 4,800	20,000 (20°C) 16,700 (25°C)	7,220.0	785 (20°C)	4,000.0	4,500	5,678 (20°C)
Log K_{ow}	0.91	1.3	1.97	2.64	2.24	2.1	1.43
Log K_{oc}	0.7	25	1.65	2.04	1.92	1.8	1.52
Log BCF	0.46	—	—	—	—	—	—
Vapor Press (mm Hg) @ 20°C @ 25°C	3,670 4,310	349 (30°C) 500 (30°C)	159	91.3 (20°C)	76.0	50	1,008
Henry's law constant atm-m³/mol @ 25°C	8.82×10^{-3}	2.03×10^{-3}	3.67×10^{-3} (24.8°C)	2.41×10^{-2}	9.9×10^{-4}	2.41×10^{-3}	1.11×10^{-2} (24.8°C)
Conversion factors ppm (v/v) to mg/m³ in air @ 25°C	2.064	0.28	4.96	6.3	8.5	6.70	2.68
mg/m³ to ppm (v/v) in air at 25°C	0.4845	3.53	0.20	0.16	0.12	0.15	0.373

CHLORINATED ALIPHATIC HYDROCARBONS

	1,2-Dichloro-ethane	1,1-Dichloro-ethane	1,1,1-Trichloro-ethane	1,1-Dichloro-ethene	1,2-Dichloroethene (cis)	1,2-Dichloroethene (Trans)	1,1,2-Trichloro-ethane
Molecular weight	98.97	98.97	133.4	96.95	96.94	96.94	133.41
Melting point °C	−35.3	−96.7	−30.4	−122.5	−80.5	−50.0	−36.53
Boiling point °C	83–84	57.3	74.1	31.7 (1 atm)	60.3	47.5	113.85
Solubility water @ 25°C mg/L	8,690 (20°C)	5,500	1495	2,500	3,500	6,260	4,400 (20°C)
Log K_{ow}	1.48 (1.45)	1.79	2.49	2.13	1.86	2.09	2.42
Log K_{oc}	1.14 (1.28)	1.76	2.03	1.81	1.51–1.69	1.51–1.69	1.06–2.49
Log BCF	—	—	—	—	—	—	—
Vapor press. (mm Hg) @ 20°C	40 (10°C)	182	124	500(20°C)	215 (25°C)	336 (25°C)	22.49 (25°C)
@ 25°C	61 (20°C)	230		591(25°C)			
	105 (30°C)			720(30°C)			
Henry's law constant atm-m³/mol @25°C	4.5×10^{-2}	4.2×10^{-2}	6.3×10^{-3} 17.2×10^{-3}	0.19	4.08×10^{-3}	9.38×10^{-3}	9.1×10^{-4} (25°C) 1.12×10^{-3} (30°C)
Conversion factors ppm (v/v) to mg/m³ in air @ 25°C	—	4.05	5.4	3.97	3.96	3.96	5.55
mg/m³ to ppm (v/v) in air at 25°C	—	0.25	0.185	0.25	0.25	0.25	0.18

Table 3.1 Physical and Chemical Properties of Selected Chlorinated Aliphatic Compounds (Continued)

	Trichloro-ethylene	1,1,2,2-Tetrachloro-ethane	Tetrachloro-ethylene	Hexachloro-ethane	1,2-Dichloro-propane	1,2,3-Trichloro-propane	1,2-Dibromo-3,chloro-propane	Vinyl-chloride
Molecular weight	131.40	167.85	165.83	236.74	112.99	147.43	236.36	62.5
Melting point °C	−87.1	−43.8	−22.7	Sublimes	−100.44	−14.7	6.0	−153.8
Boiling point °C	86.7	145.1	121.2	186.8	96.37	156.8	196.0	−13.4
Solubility water @ 25°C mg/L	1,070 (20°C) 1,366 (25°C)	2,870 (20°C)	150 (25°C)	50 (22°C) 14 (25°C)	2,700 (20°C)	1,750 (20°C)	1,230 (20°C)	2,763 (25°C)
Log K_{ow}	2.42	2.39	2.88	3.82 (3.34)	1.99	1.98	2.26	1.36
Log K_{oc}	—	1.66	—	4.3	1.67	1.99	2.17 (2.11)	1.99
Log BCF	—	—	—	—	—	0.964	1.045	—
Vapor press. (mm Hg) @ 20°C	59 (20°C)		14 (20°)	0.4			0.58	2,530
@ 25°C	74 (25°C)	5.95	24 (30°C)	0.34	49.67	3.1		2,660
Henry's law constant	0.020 (20°C)	4.7×10^{-4}	0.82	2.237×10^{-2}	2.07×10^{-3} (24°C)	3.17×10^{-4} (25°C)	1.47×10^{-4}	1.2 (10°C)
atm-m³/mol @ 25°C	0.011 (25°C)			8.0×10^{-3} 2.8×10^{-3} (20°C)	1.67×10^{-3} (24°C)			
Conversion factors ppm (v/v) to mg/m³ in air @ 25°C	5.46	6.98	6.89	9.68	—	6.03	9.67	2.6
mg/m³ to ppm (v/v) in air at 25°C	0.18	0.14	0.15	0.10	0.21	0.166	0.103	0.39

PRODUCTION AND USE PATTERN

Chlorinated aliphatic hydrocarbons are produced throughout the U.S., Canada, and Europe finding widespread applications in many industrial and commercial processes. Consequently, chlorinated aliphatic hydrocarbons are ubiquitous in the environment, originating from several point and non-point sources. In the United States, production of chlorinated aliphatic hydrocarbons steadily increased during the 1960s, 1970s, and 1980s, except for some chloroaliphatics whose production started to decline in the 1980s (Table 3.2). Large quantities of 1,2-dichloroethane have been produced annually reaching the maximum with the last reported production value in 1988 of 59.22×10^5 metric tons in the U.S. The production of 1,2-dichloromethane started in 1955, when it began to replace more toxic industrial solvents. Although annual production of chloromethane (3.56×10^5 metric tons) and 1,1,1,-trichloroethane (2.95×10^5 metric tons) was moderately high in 1991, chloroethane, 1,1-Dichloroethane, 1,1,1,2-tetrachloroethane, pentachloroethane, hexachloroethane, and trichloroethylene do not appear to be currently produced commercially in the U.S. Direct production information on 1,1,2-trichloroethane and 1,1,2,2-tetrachloroethane is not available. The heavy industrial use of chlorinated ethanes reflect their low cost and properties that make them excellent degreasing agents, cutting fluids and solvents.

Carbon tetrachloride production in the U.S. peaked in 1988 with 3.46×10^5 metric tons and dropped to 1.23×10^5 metric tons in 1990 (Table 3.2). Approximately, 90% of the carbon tetrachloride produced in earlier years was used to manufacture chlorofluorocarbons (CFCs) which were used as aerosol propellants. Carbon tetrachloride was used formerly as a deworming agent, grain fumigant, anaesthetic, and degreaser in the dry-cleaning industry, but has been replaced with other organic compounds because of the toxicity of carbon tetrachloride. Since CFCs are being phased out, under the Montreal Protocol, the demand for carbon tetrachloride has declined dramatically from 3.36×10^5 metric tons in 1988 to 1.23×10^5 metric tons in 1990. Current production figures are not available (Table 3.2). Carbon tetrachloride is used as a solvent and in consumer products such as cleaning fluids and fire extinguishers.

Tetrachloroethylene is used primarily as a solvent in the dry-cleaning industry and to a lesser extent as a degreaser in the metal industries. Its production peaked in 1981 (3.14×10^5 metric tons) in the U.S. and then started to decline. The following years reported an up-and-down trend in its production, with 1.09×10^5 metric tons produced in 1991 (Table 3.2).

Vinyl chloride (chloroethylene) has been used for more than five decades globally in the manufacture of polymers and copolymers containing polyvinyl chloride (PVC) which is the most important feedstock in the production of plastics. Production of vinyl chloride in the U.S. was second only to 1,2-dichloroethane, increasing from an annual average of 6.4×10^5 metric tons in the early 1960s to the latest figure available, 5.3×10^6 metric tons, in 1991 (Table 3.2). Both vinyl chloride and PVC are used in building and construction products, the automotive industry, cables piping, food packaging, and household and commercial implements.

Table 3.2 United States Production (metric tons × 10⁴) of Some Chloroaliphatic Hydrocarbons

Compound	1966–1970*	1971–1975*	1976–1980*	1981	1982	1983	1984
Chloromethane	14.9	20.8	19.3	18.4	16.0	18.6	21.9
Dichloromethane	14.5	19.0	25.2	26.8 [57.0]	23.8	26.5	27.5
Chloroform	9.1	11.6	15.0	18.3	13.5	16.5	18.4
Carbontetrachloride	36.5	45.9	34.7	32.9	26.7	26.0	32.4
Chlorodifluoromethane	—	—	9.0	11.4	7.9	10.7	11.5
Trichlorofluoromethane	10.4	13.6	9.6	7.4	6.4	9.7	8.4
Dichlorodifluoromethane	15.1	19.9	15.1	14.8	11.7	13.2	15.3
Chloroethane	23.1	8.1	25.4	14.7	13.1	12.8	13.2
1,2-Dichloroethane	234.8	379.2	480.2	453.3	346.3	523.0	333.2
1,1,1-Trichloroethane	14.0	21.9	30.1	27.9	25.0	26.6	30.7
Vinyl chloride	142.1	223.1	285.5	312.4	222.8	312.5	276.6
Trichloroethylene	24.5	18.8	13.6	11.7	8.2#	—	—
Tetrachloroethylene	27.0	32.3	29.4	31.4	26.6	24.9	26.1

Compound	1985	1985 9 months	1986 9 months	1987	1988	1989	1990	1991
Chloromethane	18.6 [23.2]	12.8	14.1	16.9	27.1	18.8	22.6	35.6
Dichloromethane	21.2	17.4	18.4	23.5	22.9	21.3	21.3	17.6
Chloroform	12.5	—	14.7	20.9	23.8	26.3	22.5	—
Carbontetrachloride	29.3	24.2	23.5	30.6	34.6	—	12.3	—
Chlorodifluoromethane	10.7	—	—	—	15.1	—	—	—
Trichlorofluoromethane	7.9	16.9	18.8	26.9	11.4	8.8	6.1	4.5
Dichlorodifluoromethane	13.7	—	—	—	18.8	17.8	9.5	6.9
Chloroethane	7.7	—	—	—	6.9	—	—	—
1,2-Dichloroethane	550.0	—	—	—	592.2	—	—	—
1,1,1-Trichloroethane	39.5	20.8	21.8	—	32.9	35.3	35.6	29.5
Vinyl chloride	430.1	261.9	289.8	374.1	411.7	436.2	483.1	530.4
Trichloroethylene	—	—	—	—	—	—	—	—
Tetrachloroethylene	30.8	15.8	13.9	21.3	22.6	21.5	17.4	10.9

Note: * Annual average; # no production, sales figures only; data in [] worldwide production; — No data available.

Sources: U.S. Government Printing Office; Synthetic Organic Chemicals, U.S. Production and Sales; *Chemical Marketing Reporter.*

Canadian production data for chlorinated aliphatic hydrocarbons are scanty. Comprehensive accounting of imports and overall usage of these chemicals are not available. It is known that chloromethanes are used in moderate amounts in Canada, but of course, lower than in the U.S. In 1978, chloromethane production was 12,200 metric tons compared to 10,400 metric tons in 1979. Whereas, chloromethane consumption was 2800 and 10,360 metric tons for the same years (Moore and Ramamoorthy, 1984).

The uses of chlorinated aliphatic hydrocarbons are listed in Table 3.3. The table includes the chemical name, past and present uses, synonyms and trade names, and identification numbers from several registries.

SOURCES IN THE ENVIRONMENT (NATURAL AND ANTHROPOGENIC)

Chlorination of chemicals occurs in nature via enzyme systems in organisms (chloroperoxidase) that produce many chlorinated alkanes, alkenes, and other chlorinated organic compounds in the ambient environment. Chlorinated aliphatics have numerous industrial applications (Table 3.3) and are by-products of chlorination processes in industry (Table 3.4).

Chloromethane is introduced from both anthropogenic and naturally occurring sources. Anthropogenic sources include industrial production, PVC burning, and wood burning. Chloromethane is also produced naturally and exceeds the amount manufactured by at least a factor of 10. Most naturally produced chloromethane comes from the ocean; ($3-5 \times 10^{12}$g/year (Fabian 1986; Rasmussen et al.1980). Other natural sources of chloromethane include biomass burning (forest fires, wood burning, cigarette smoking, burning plastics, and coal burning), which accounts for $0.2-0.4 \times 10^{12}$g/year (Chopra 1972; Fabian 1986; Edgerton et al. 1984, 1986; Kleindienst et al. 1986); and chlorination of drinking-water and wastewater. The total amount of chloromethane from sources other than manufacturing is approximately $3.2-8.2 \times 10^{12}$g/year.

Currently, it is estimated that chloromethane, chloroform, and carbon tetrachloride are produced in quantities by natural processes (Gribble 1992). Natural processes are reported to emit chloromethane at 5×10^6 tons per year into the global environment; whereas the anthropogenic emissions are reported to be 2.6×10^4 tons (Rasmussen et al. 1980). Controlled combustions related to human activities are reported to contribute 2.1×10^5 tons (Rasmussen et al. 1980). Controlled combustions related to human activities are reported to contribute 2.1×10^5 tons of chloromethane per year in the U.S. Chloromethane is known to be produced naturally by several fungal species including *Phellinus pomaceus*.

Dichloromethane primarily comes as an anthropogenic source from uses in aerosols, paint removers, chemical synthesis for metal degreasing, and as a foam-blowing agent; it is also used in electronic industry.

Natural sources of dichloromethane include biosynthesis by barley, incomplete combustion of plant materials and volcanic eruptions (Gribble 1992). The actual amounts of naturally produced chloromethane are not known. Significant amounts of chloroform are released from natural sources such as (1) biosynthesis of marine algae;

Table 3.3 Industrial Uses of Chloroaliphatic Hydrocarbons

Compound (Name and Formulae)	Industrial Uses	Synonyms and Trade Names	Registry	ID No.
Chloromethane CH_3Cl	Manufacture of fumigants, organic chemicals, synthetic rubber, methylcellulose, silicones; as solvent; used in medicine and refrigerants; use as herbicide has been discontinued in many countries.	Methyl chloride Arctic R 40 Freon 40	CAS NIOSH/RTECS EPA Haz. Waste OHM/TADs DOT/UN/IJA/IMCO shipping HSDB NCI	74-87-3 PA6300000 U045 7216794 UN 1063 883 No Data
Dichloromethane CH_2Cl_2	In the manufacture of aerosols, photographic films and synthetic fibers; as solvent, degreasing and cleansing agent; as fumigant and refrigerant; in organic syntheses and pharmaceuticals.	Methylene chloride Methylene dichloride Narkotil Salaesthin Solmethine	CAS NIOSH/RTECS EPA Haz. Waste OHM/TADs DOT/UN/IJA/IMCO shipping HSDB NCI	75-09-2 PA 8050000 U080,F002 7217234 66 — C50102
Chloroform $CHCl_3$	In the manufacture of chlorofluorocarbons and propellants, anaesthetics, and pharmaceuticals; as fumigant, solvent, and as insecticide. Used in consumer products such as toothpaste and liniments.	Trichloromethane Methyltrichloride Freon 20 R 20 R 20 refrigerant	CAS NIOSH/RTECS EPA Haz. Waste OHM/TADs DOT/UN/IJA/IMCO shipping HSDB NCI	67-66-3 FS 9100000 U044 7216639 UN 1888 56 C02686
Carbon tetrachloride CCl_4	In dry-cleaning operations; in the manufacture of fire-extinguishers, refrigerants (chlorofluoromethane), and propellants; and as solvent.	Tetrachloromethane Carbontet Perchloromethane Benzinoform Fasciolin Freon 10 Halon 104 Tetraform Tetrasol	CAS NIOSH/RTECS EPA Haz. Waste OHM/TADs DOT/UN/IJA/IMCO shipping HSDB NCI	56-23-5 FG 4900000 U211 7216634 UN 1846 53 —

CHLORINATED ALIPHATIC HYDROCARBONS

Chlorodibromomethane $CHBr_2Cl$	In the manufacture of fire-extinguishing agents, aerosol propellants, refrigerants, and pesticides.	Dibromochloromethane	CAS NIOSH/RTECS EPA Haz. Waste OHM/TADs DOT/UN/UA/IMCO shipping HSDB NCI	124-48-1 PB 5600000 U225 8100034 UN 2515 IMCO 6.1 2517 C55130
Bromodichloromethane $CHBrCl_2$	In organic synthesis, as standard in drinking-water analyses.	Dichlorobromomethane	CAS NIOSH/RTECS EPA Haz. Waste OHM/TADs DOT/UN/UA/IMCO shipping HSDB NCI	75-27-4 PA 5310000 — — — 4160 C55243
Chloroethane C_2H_5Cl	In the manufacture of tetraethyl lead; polystyrene foam; as alkylating agent in chemical syntheses; in making pharmaceuticals; as local anaesthetic	Ethyl Chloride Aethylis Anodynon Chelen Chloridum Chloryl Kelene	CAS NIOSH/RTECS EPA Haz. Waste OHM/TADs DOT/UN/UA/IMCO shipping HSDB NCI	75-00-3 KH 7525000 C266 7216712 UN 1037; IMCO 2.3 533 C06224
1,2-Dichloroethane $C_2H_4Cl_2$	In cosmetics, as food additive and as fumigant; used in the production of vinyl chloride and tetraethyl lead	Dichloroethylene Ethylenedichloride EDC	CAS NIOSH/RTECS EPA Haz. Waste OHM/TADs DOT/UN/UA/IMCO shipping HSDB NCI	107-06-2 KI 0525000 U077 7216717 1184 65 C00511
1,1-Dichloroethane $C_2H_4Cl_2$	In the manufacture of vinyl chloride, paint and varnish removers, and metal degreasers; in organic syntheses.	Ethylidenechloride Ethylidenedichloride	CAS NIOSH/RTECS EPA Haz. Waste OHM/TADs DOT/UN/UA/IMCO shipping HSDB NCI	75-34-3 KI 0175000 U076 — DOT 2362; UN 2362; IMCO 3.2 64 —

Table 3.3 Industrial Uses of Chloroaliphatic Hydrocarbons (Continued)

Compound (Name and Formulae)	Industrial Uses	Synonyms and Trade Names	Registry	ID No.
1,1,1-Trichloroethane $C_2H_3Cl_3$	As cleaning solvent for metals; in aerosols, cutting oils, coolants, lubricants, solvents, in vapor degreasers, insecticides, and electronic industries.	Methylchloroform Methyltrichloromethane Chloroethene Aeroethene TT Inhibisol	CAS NIOSH/RTECS EPA Haz. Waste OHM/TADs DOT/UN/UA/IMCO shipping HSDB NCI	71-55-6 KJ 2975000 U226 8100101 UN 2831 157 C04626
1,1-Dichloroethene $C_2H_2Cl_2$	As intermediate in organic syntheses; in the production of polyvinylidene chloride copolymers, which are used as flame retardant coatings for fibre and carpet backing; in coating for steel pipes and in adhesive applications.	1,1-Dichloroethylene Vinylidenechloride DCE	CAS NIOSH/RTECS EPA Haz. Waste OHM/TADs DOT/UN/UA/IMCO shipping HSDB NCI	75-34-4 KV 9275000 U078 7216949 UN 1303; IMCO 3.1 1995 C54262
1,2-Dicloroethene $C_2H_2Cl_2$	As a chemical intermediate in the synthesis of chlorinated solvents and compounds; as a low-temperature extraction solvent for dyes, perfumes, lacquers, and thermoplastics.	Acetylene dichloride sym1,2-Dichloroethylene Dioform	CAS NIOSH/RTECS EPA Haz. Waste OHM/TADs DOT/UN/UA/IMCO shipping HSDB NCI	540-59-0 KV 9360000 U079 8300194 UN 1150; IMCO 3.2 149 C 56031
1,1,2-Trichloroethane C_2HCl_3	In the production of 1,1-dichloroethene; as solvent for fats, waxes, and resins; limited use where high solvency is required such as chlorinated rubbers.	Ethane trichloride 1,2,2-Trichloroethane Vinyl trichloride δ-T Cement T-399	CAS NIOSH/RTECS EPA Haz. Waste OHM/TADs DOT/UN/UA/IMCO shipping HSDB NCI	79-00-5 KJ 2975000 U227 8100016 None 1412 C04579

CHLORINATED ALIPHATIC HYDROCARBONS

Trichloroethylene C_2HCl_3	As feedstock in the production of trichloroethylene, tetrachloroethylene, and 1,2-dichloroethylene; as solvent in paint removers, varnishes, and lacquers; in photographic films; as extractant for oils and fats. Use as biocide has been discontinued.	Acetylenetrichloride 1,1,2-Trichloroethylene Algylen Cecolene Dow-TriPhilex Triad Trimar V-strol Westrosol	CAS 79-01-6 NIOSH/RTECS KX 4550000 EPA Haz. Waste U228 OHM/TADs 7216931 DOT/UN/UA/IMCO shipping UN 1710 HSDB 133 NCI C04546
1,1,2,2-Tetrachloro-ethane $C_2H_2Cl_4$		Acetylene tetrachloride sym-Tetra chloroethane Bonoform Cellon Westron	CAS 79-34-5 NIOSH/RTECS KI 8575000 EPA Haz. Waste U209 OHM/TADs 8100014 DOT/UN/UA/IMCO shipping UN 1702; IMCO 6.1 HSDB 123 NCI C03554
Tetrachloroethylene C_2Cl_4	Commercially as a chlorinated hydrocarbon solvent and as a chemical intermediate. The current end-use pattern is estimated as 50% for dry-cleaning and textile processing, 28% as a chemical intermediate (in the synthesis of fluorocarbons, 113, 114, 115, and 116), 9% for industrial metal cleaning, 10% for exports and 3% for other uses. Miscellaneous uses include carrier applications for rubber coatings, solvent soaps, printing inks, adhesives and glues, sealants, polishes, lubricants, and silicones. Additional uses are as nonflammable recyclable dielectric fluid for power transformers, heat transfer medium, and extractant in pharmaceutical industry; pesticide intermediate. Its uses as antihelmintic in the treatment of hook worms and nematodes and consumer uses in aerosols and shoe polish have been replaced or discouraged.	Ethylene tetrachloride PERC Perchlor Perchloroethylene Perk Ankilostin Antisol-1 Dee-solv Dow-per Nema, Perclene Percosolv Tetlen Tetracap Tetravec Tetropil Perawin Tetralex Dow-clene EC	CAS 127-18-4 NIOSH/RTECS KX 3850000 EPA Haz. Waste U210 OHM/TADs 7216847 DOT/UN/UA/IMCO shipping UN 1897; IMCO 6.1 HSDB 49403 55 NCI C04580

Table 3.3 Industrial Uses of Chloroaliphatic Hydrocarbons (Continued)

Compound (Name and Formulae)	Industrial Uses	Synonyms and Trade Names	Registry	ID No.
Hexachloroethane C_2Cl_6	Large amounts in smoke and pyrotechnic devices by the military; as constituent in candles, grenades, pressure lubricants, submarine paints, and fire-extinguishing fluids; in making degassing pellets to force air out of molten ore in aluminum foundries; as moth repellant; plasticizer for cellulose esters; polymer additive; accelerator in rubber; and a retardant in fermentation processes.	Perchloroethane Carbonhexachloride Avlothane Distokal Distopan Egitol Falkitol Fasciolin Phenohep	CAS NIOSH/RTECS EPA Haz. Waste OHM/TADs DOT/UN/UA/IMCO shipping HSDB NCI	67-72-1 KI 4025000 U131 — NA 9037 2033 C04604
Vinyl chloride C_2H_3Cl	In the manufacture of PVC homo and co-polymer resins; end-use products are automotive products and accessories, furnitures, packaging materials, wall coverings, wire coatings, films, and resins; in the manufacture of 1,1,1-trichloroethane. Its uses as refrigerant, extraction solvent for heat-sensitive materials, aerosol propellant have been banned.	Chloroethene Chloroethylene Ethylene monochloride VC, VCM	CAS NIOSH/RTECS EPA Haz. Waste OHM/TADs DOT/UN/UA/IMCO shipping HSDB NCI	75-01-4 KU 9625000 U043 7216947 1086 169 —

Table 3.4 Industrial Products and Processes That Yield Aliphatic Hydrocarbons in Untreated Wastewater

Industrial Product	Industrial Process	Chemical Feedstock	Untreated Wastewater Composition
Carbon tetrachloride	Chlorination	Methane, ethylene dichloride	Chloromethanes, chloroethanes, chloroethylenes
Chloroform	Chlorination	Methane, methyl chloride	Chloromethanes, chloroethane, chloroethylenes
1,2-Dichloroethane	Oxychlorination	Ethylene, HCl	Chloroethanes, chloroethylenes
Methyl chloride	Chlorination, hydrochlorination	Methane, methanol	Chloromethanes, chloroethanes, chloroethylenes
Methylene chloride	Chlorination	Methane, methyl chloride	Chloromethanes, chloroethanes, chloroethylenes
Tetrachloroethylene	Chlorination	1,2-dichloroethane	Chloromethanes, -ethanes, -ethylenes, dichloropropane, dichloropropylene
Trichloroethylene	Chlorination	1,2-dichloroethane	Chloromethanes, chloroethanes, chloroethylenes
Vinyl chloride	Dehydrochlorination	1,2-dichloroethane	Chloromethanes, chloroethanes, chloroethylenes

(2) biosynthesis of plants such as barley, lemon, orange, and moss; (3) incomplete combustion of plant materials; and (4) volcanic eruptions. The amount of chloroform routinely produced from these processes are not quantitated and yet the amount would appear to be substantial based on quantitative data for other chlorinated alkanes such as chloromethane. There is also evidence (Gribble 1992; RTP 1994) for production of dichloromethane by plants and volcanoes and for 1,1,1-trichloroethane released from oceans as well as a number of chlorofluoroalkanes. Several chlorinated alkanes have also been identified as natural products of humic soils.

The environmental fate evaluation (MacKay et al. 1993) concluded that the vapor pressures and volatility of chlorinated alkanes distribute them mostly into air (approximately 98% to 99%) with smaller portions ($\leq 1\%$) in water. The environmental fate characteristics of chlorinated alkanes account for their ambient concentrations in water and air regardless of their anthropogenic or natural source. Chloroalkanes are primarily removed from the water column via vaporization although some biological degradation might occur. In the atmosphere, chlorinated alkanes are degraded primarily via photolytic oxidation reactions with atmospheric half-lives of hundreds of days. The only exception seems to be carbon tetrachloride which resists photochemical oxidation and hence has an atmospheric half-life of anywhere from ten to hundreds of years.

In general, chlorinated alkanes are low molecular weight compounds with medium to high vapor pressures, low to moderate water solubilities, and low $k o_w$ coefficient values. The variety of industrial uses of chlorinated alkanes are responsible for their presence in industrial waste waters. Carbon tetrachloride has been used primarily in the production of refrigerants such as CFC_{11} and CFC_{12} in addition to other industrial uses. Dichloromethane is used in aerosols, chemical syntheses, in paint removers, metal degreasing, electronic industry, and as a foam-blowing agent (such as urethane), and as an extraction solvent in the food industry.

Uses of 1,1,1-trichloroethane include (1) cleaning and vapor degreasing of metal (Carchman et al. 1984), (2) in the manufacture of a variety of products such as aerosols, adhesives, cutting oils, lubricants, coolants, solvents, paints, insecticides, and printed circuit boards. This compound is scheduled to be phased out by January 1996 according to Montreal Protocol due to its ozone-depleting potential and also due to its high-volume use globally. The anthropogenic sources of chloroform include drinking-water chlorination, municipal waste water chlorination, and bleaching of pulp at paper mills. Other anthropogenic sources might include automobile exhaust, chemical and pharmaceutical manufacturing, food processing, pesticides, combustion of plastic and tobacco products treated with chlorinated pesticides, and also from the reaction with humic material in natural water. Most of this chloroform ends up in the atmosphere. Chloroform might also be released to soil from improper land filling of chlorinated waste. Smaller amounts of chloroform may leach from soil and enter groundwater (Syracuse Research Corp. 1989). 1,2-Dichloroethane may be emitted to the environment during its production, during manufacturing of vinyl chloride and when it is used as a fumigant, lead scavenging agent, and during storage, distribution, or disposal (Clement Associates 1989; Moore et al. 1991). Older consumer products that contain 1,2-dichloroethane could also be a likely source in the environment.

Carbon tetrachloride may enter the environment as a degradation product of several other compounds. Other sources of carbon tetrachloride in the environment are from its production or as an intermediate in the manufacturing of other chemicals. Until its prohibition in 1986, its use as a grain fumigant was the largest source of release for this compound.

Natural sources for chlorinated alkanes such as trichloroethylene, tetrachloroethylene, and vinyl chloride include volcanic eruptions and various oceanic sources such as marine algae. CFCs have also been identified from these sources. The naturally formed chloroaliphatics in soils include dichloropropane, several chlorinated alkanes, and chlorinated alkenes. The anthropogenic sources of chlorinated alkenes in the environment include dry cleaning operations and use of products such as fumigants, refrigerants, solvents, metal degreasers, and paint removers. Chlorinated alkenes detected in pulp mill effluents include trichloroethylene, tetrachloroethylene, chloropropenes, chloropropadienes and chlorobutatrienes, chlorobutadienes, and hexachlorohexatrienes. (Kringstad and Lindstrom, 1984; Sunito et al. 1988).

ENVIRONMENTAL RESIDUES AND EXPOSURE ROUTES (AIR, WATER, SOIL, AND OTHER MEDIA)

Chloromethane

Air

The volatile organic carbon (VOC) database containing 706 data points (300 cities from 42 states) reported the following values for chloromethane (Shah and Singh 1988):

Average	740 ppt
Upper Quartile	721 ppt
Median	652 ppt
Lower Quartile	607 ppt

The average value may be skewed from a few high numbers because it is higher than the upper quartile value. The type of air masses and their chloromethane concentrations are given below.

Air Mass	Median (ppt)	Data Points
Remote	713	5
Rural	923	2
Suburban	641	599
Urban	810	100

It appears that anthropogenic contributions do not significantly affect the ambient levels of chloromethane. Average urban levels reported were 660 to 960 ppt and the background levels were 600 to 700 ppt (Shah and Singh 1988).

Water

Chloromethane has been detected in surface-waters, groundwaters, drinking-waters, municipal and hazardous landfill leachates, and industrial effluents (ATSDR 1990b). When detected, concentrations are in the ppb–ppt range, possibly due to rapid volatilization of chloromethane. It is likely formed during the chlorination of drinking-water. Presence of chloromethane in surface-waters may be the result of rainout from the atmosphere as well as from anthropogenic input. Thirty-four species of fungi can produce chloromethane biosynthetically (Harper et al. 1988) and these fungi may be present in many natural surface-waters. Similar explanation is given for the detection of chloromethane in groundwaters.

Soil

Chloromethane has been detected in soil at hazardous waste sites at 5 to 500 ppb (mean) (CLPSD 1987). Presence of chloromethane in ambient soils could be due to production by fungi (Harper et al. 1988).

Other Media

Chloromethane has been detected in wood smoke, coal burning, volcanoes, and burning plastics (ATSDR 1990b).

Primary exposure routes of chloromethane for the general public is via air and probably drinking-water. In urban centers, the general population might be exposed to higher levels of chloromethane.

Dichloromethane (Methylene Dichloride)

Air

Dichloromethane Levels in Air

Location	Concentration ($\mu g/m^3$) Maximum	Mean	References
Background	—	0.17	Singh et al. 1982
Oceanic	—	0.07–0.13	Singh et al. 1982
Rural/suburban U.S.	—	0.18–2.1 (median)	Shah and Heyerdahl 1988
Urban sites	22–200	0.8–6.7	Harkov et al. 1984, 1985
Source-dominated	—	2.0	Shah and Heyerdahl 1988
Hazardous waste sites	10–190	0.3–3.9	Harkov et al. 1985, LaRegina et al. 1986
Indoor (non-residential)	19,000	0.2–19,000 (median)	Otson et al. 1983, Harsch 1977

Note: 1 $\mu g/m^3$ = 0.29 ppb.

Water

Dichloromethane has been detected in surface-water, groundwater, and finished drinking-water. It was detected in 30% of 8,917 surface-water samples at 0.1 µg/L (median) (Staples et al. 1985). A survey in the state of New Jersey found dichloromethane in 45% of 605 surface-water samples with a maximum concentration of 743 µg/L (Page 1981). Dichloromethane has been detected in groundwater for several years in the U.S. at a range of 0–3,600 µg/L (Dyksen and Hess 1982; Kelley 1985). According to CERCLA monitoring records in 1987, dichloromethane was the sixth most frequently detected organic contaminant in groundwater with a detection frequency of 19% (Plumb 1987).

Dichloromethane has been detected in numerous drinking-water supplies in the U.S. (ATSDR 1993c). Reported concentrations are less than 1 µg/L (EPA 1975): water chlorination appears to increase both the concentration and the frequency of occurrence in drinking-water supplies (NAS 1977).

Soil

Dichloromethane residues in soil have not been reported. In 20% of 338 sediment samples, the compound was detected at 13 µg/kg (median) (Staples et al. 1985).

Other Media

Dichloromethane is used in food processing (solvent extraction of coffee, spices, hops) and as a fumigant for strawberries, grain, etc. But very little residue is detected in foods; in decaffeinated coffee beans, 0.32-0.42 mg/kg (IARC 1986) with FDA limits of 10 ppm. Dichloromethane is no longer used as a decaffeinating agent by most coffee decaffeinators (Mansville Chemical Products Corp. 1988). Inhalation of dichloromethane from ambient air is the predominant exposure route for the general population. Other routes are unlikely to be important under usual circumstances (ATSDR 1993c).

Chloroform

Air

A most recent study (1982 to 1985 air samples) reported that the background level of chloroform concentrations over the North Atlantic Oceans range from 2×10^{-5} to 5×10^{-5} ppm (Class and Ballschmidter 1986). This range is very close to that which was reported previously (1976–1979) and that was reported in 1987 (EPA 1988a). The maximum and background levels reported in seven U.S. cities for the period of 1980 to 1981 were 5.1×10^{-3} and 2×10^{-5} ppm, respectively (Singh et al. 1982). The median concentration in 1977 and 1980 was 7.2×10^{-5} ppm and in 1987 was 6×10^{-5} ppm (ATSDR 1993b and references cited therein).

The median value for source-dominated areas in the U.S. was 8.2×10^{-4} ppm in 1977 and 1980 and was 5.1×10^{-4} ppm in 1987 (Brodzinski and Singh 1982; EPA 1988a). Chloroform levels in air can be much higher in areas near hazardous and municipal waste sites and incinerators. Typical median indoor concentrations of chloroform range from approximately 2×10^{-4} to 4×10^{-3} ppm and ratios of indoor to outdoor air concentrations range from <1 to 25 (Pellizarri et al. 1986). One of the significant sources of chloroform is chlorinated tap water and showers are likely to contribute to the indoor levels of chloroform (Wallace 1987).

Water

Recent monitoring data on the presence of chloroform in surface-water, sediment, and groundwater are not available in the literature. The most recent data were on levels in drinking-water in 1988 from 35 sites across the U.S. at 9.5 to 15 µg/L (median) (Krasner et al. 1989). Forty-five percent of sample sites in the National Groundwater Supply Survey indicated median and maximum values of 1.5 and 300 µg/L, respectively. (Westrick et al. 1984). EPA STORET database indicates chloroform detection in 64% of 11,928 surface-water samples at 0.30 µg/L (median) (Staples et al. 1985). Current data on surface-waters are lacking. Low levels of chloroform were detected in sediments. Data from EPA STORET indicate detection of chloroform in 8% of 425 sediment samples, at <5µg/kg (median) (Staples et al. 1985).

Soil

Data from CLPSD (Contract Laboratory Program Statistical Database) indicate detection of chloroform in 9.9% of hazardous waste sites (NPL) soils at 12.5 µg/kg (median) (CLPSD 1989).

Other Media

Chloroform has been detected in various foods at the following concentrations: beverages, 2.7–178 µg/kg; dairy products, 7–1,110 µg/kg; oils and fats, traces -<12 µg/kg; grains and milled grain products, 1.4–3,000 µg/kg (references cited in ATSDR 1993b). The highest amount of chloroform was found in butter, 1,110 µg/kg, mixed cereals, 220 µg/kg; infant/junior foods, 230 µg/kg; and cheddar cheese, 83 µg/kg (Heikes 1987). The exposure routes of chloroform for the general population are through drinking-water, dietary sources, inhaling contaminated air, and dermal contact with chlorinated water (Benoit and Jackson 1987).

Carbon Tetrachloride

Air

Continuous monitoring studies have revealed the increasing concentrations of carbon tetrachloride (CCl_4) in global atmosphere (Simmonds et al. 1988). The increase was about 1.3% per year, reaching 0.12–0.14 ppb by 1985. Average suburban and urban levels were 0.19 and 0.59 ppb, respectively, near point sources of CCl_4 (Brodzinski and Singh 1983).

Carbon tetrachloride is a common contaminant in indoor air, approximately 0.16 ppb, with maximum value of 1.4 ppb (Wallace 1986). Indoor air levels of CCl_4 were usually higher than outdoor levels, indicating building materials or products as possible sources of CCl_4 (Wallace et al. 1987).

Water

Results of several surveys reveal that about 99% of all groundwater supplies and ≈95% of all surface-water supplies contain <0.5 µg/L of CCl_4 (Letkiewicz et al. 1983). Carbon tetrachloride was detected in 3.2% of 945 drinking-water samples at >0.2 µg/L with the highest value being 16 µg/L. Carbon tetrachloride has been detected in groundwater at 5–6% of all NPL sites and other Superfund chemical waste sites, at <50 to 1000 µg/L (ATSDR 1994a).

Soil

No data were located on background levels of CCl_4 in soils.

Other Media

Since CCl_4 is no longer used as a fumigant for grains, exposure from this source is of no concern to the general population. CCl_4 is not reported in other food items (ATSDR 1994a). Likely exposure routes of CCl_4 for the general population are from ambient air and drinking-water.

Bromoform and Chlorodibromomethane

Air

Bromoform and chlorodibromomethane have been detected in ambient air only at very low concentrations. 37 ng/m^3 (3.6 ppt, mean) and 32 ng/m^3 (3.8 ppt, mean) respectively (Brodzinski and Singh 1983). The mean concentration of bromoform in ambient Arctic Circle air samples was 53 ng/m^3 (5.1 ppt) (Berg et al. 1984).

Water

These two compounds are rarely measurable in nonchlorinated waters (Staples et al. 1985). Their mean concentration in drinking-water was <10 µg/L (ATSDR 1990c).

Soil

These two compounds were detected in 2 of 862 hazardous waste sites, with a mean concentration of 17 µg/kg (bromoform) and 15 µg/kg (chlorodibromomethane). Bromoform was not detected in any of the 353 sediment samples analyzed and no data for chlorodibromomethane was available (Staples et al. 1985).

Chloroethane

During the mid-to-late 1970s and the early 1980s, chloroethane was detected in outdoor air. City and suburban air contained an average of 41 to 140 ppt of chloroethane and in rural air, <5 ppt. Current atmospheric levels are expected to be much lower due to a decrease in production and use of chloroethane in North America and concomitant decrease in emissions. The limited amount of information that is available suggests that extremely low levels of chloroethane may be found in drinking-water from chlorination and other contamination of water supplies used for drinking-water. But actual information is not available in the literature. Exposure to the general population may arise from contact with various consumer products including some solvents, paints, foamed plastics, and refrigerants (ATSDR 1989c).

1,1-Dichloroethene

1,1-Dichloroethene has been detected at very low levels in indoor and outdoor air (about<1 ppt). The levels are higher in occupational settings in factories that make or use 1,1-dichloroethene, e.g., in making food-packaging films, adhesives, flame-retardant coatings for fiber and carpet backing, and piping and coating for steel pipes. They are also high in hazardous waste sites. A small portion (3%) of the drinking-water sources in the U.S. contain low levels of 1,1-dichloroethene (average, 0.3 ppb). Groundwater samples from hazardous sites contain 1,1-dichloroethene from 0.001 to 0.09 ppm (ATSDR 1994c).

The amount of 1,1-dichloroethene in food-packaging films range from <0.02 to 1.26 ppm and FDA regulates the use of plastic packaging films (<10 ppm).

1,1-Dichloroethane

The average concentration of 1,1-dichloroethane in ambient air across the U.S. is reported to be 55 ppt and may originate from building materials or chlorinated waters. 1,1-Dichloroethane has been detected in drinking-water at trace levels up to

4.8 ppb. It has not been detected in any surface-water samples from lakes, rivers, or ponds. No information is available on the levels of 1,1-dichloroethane in soil, sediment, or food (ATSDR 1990d).

1,2-Dichloroethane

1,2-Dichloroethane has been detected in urban air samples across the U.S.; the median daily atmospheric concentration in urban sites was 0.012 ppb (1,214 samples) and 0.26 ppb (182 samples) for source-dominated sites; it was not detected in 648 samples from suburban, rural, or remote sites (EPA 1988). 1,2-Dichloroethane has been detected in ambient air collected in the vicinity of hazardous waste disposal sites. Also, it was detected in indoor ambient air samples.

An average concentration of 175 ppb has been detected in 12% of the surface-water and groundwater samples collected from 2,783 hazardous waste sites in the U.S. was also detected in U.S. drinking-water at 0.05 to 19 ppb. 1,2-Dichloroethane was used as a gasoline additive in leaded gasoline. With the discontinuance of leaded gasoline, this exposure route is nonexistent.

1,1,1-Trichloroethane

1,1,1-Trichloroethane has been detected in air samples around the world. The U.S. city air typically contains about 0.1 to 1 ppb of 1,1,1-trichloroethane and in rural air, <0.2 ppb, since it is used in many home and office products. Based on absolute concentrations over a 12-year period, a global atmospheric concentration of 0.157 ppb has been estimated for 1,1,1-trichloroethane in the middle-1990s (Prinn et al. 1992). An estimate of 1.5 ppb has been calculated for the average global concentration in air that might exist in the year 2030 (Ramanathan et al. 1985). The long atmospheric lifetime of 1,1,1-trichloroethane allows itself to travel a considerable distance from its initial point of release (Class and Ballschmidter 1986; DeBortoli et al. 1986; ATSDR 1993a). The concentration in indoor air is variable and seems to depend on the practices of individuals and building air-exchange characteristics (Cohen et al. 1989; Hartwell et al. 1992). 1,1,1,-trichloroethane has been detected at ≤70 ppb in newly constructed buildings (Wallace et al. 1987). Thus, one is likely to be exposed to 1,1,1,-trichloroethane vapors indoors at higher concentrations than outdoor, or near Superfund sites in the U.S.

The compound has been detected in rivers and lakes (up to 10 ppb), in soil (up to 120 ppm), and in drinking-water from underground wells (up to 12 ppm). Ambient surface-water concentrations are usually <1 ppb; groundwater, 0–18 ppb; drinking-water from surface or groundwater sources, 0.01–3.5 ppb.

1,1,1-Trichloroethane has been detected in raw, processed, and prepared food products; in fish and shrimp from the Pacific Ocean, 2.7 and <0.3 ppm (average), respectively (Young et al. 1983); and in clams and oysters, 39–310 ppm (mean) (Ferrario et al. 1985).

Inhalation or ingestion of contaminated food could likely be the exposure routes of 1,1,1-trichloroethane for the general population (ATSDR 1993a).

1.2-Dichloroethene

This compound is also known as 1,2-dichloroethylene or acetylene dichloride. 1,2-Dichloroethene has an odor threshold of 17 ppm in air. There are isomeric forms of the compound, *cis* and *trans*. It has been detected frequently in urban air samples and in landfill gas. 1,2-Dichloroethene has been detected in surface, ground and drinking-waters, as well as in industrial and municipal effluents, urban run-offs, and landfill leachates (ATSDR 1990d). A maximum concentration of 818.6 ppb was detected in groundwater in a New Jersey study in 1977–1979 (Page 1981); and a maximum 2.9 ppb in a Nebraska Study (Goodenkauf and Atkinson 1986). 1,2-dichloroethene was detected in 5 of 26 landfills at 3,900 ppb (maximum) and in leachate from 8 of 26 landfills at 310 ppb (maximum) (Friedman 1988). A U.S. survey detected 1,2-Dichloroethene in drinking-water derived from groundwater, in 16 of 466 randomly selected sites and 38 of 479 intentionally chosen sites; maximum concentration was 2 ppb (random sites) and 120 ppb (nonrandom sites) (Westrick et al. 1984).

1,2-Dichloroethene was detected in the soil at eight hazardous waste sites in the U.S. at a 5 to 4,000 ppb average. Air in suburban and urban areas contained an average of 76 ppt. Little is known about air levels of 1,2-dichloroethene in rural areas.

1,2-Dichloroethene is not an atypical biota contaminant (Staples et al. 1985). In the early 1980s, it was found in 4% of 361 sediments at >5 ppb, wet weight (Staples et al. 1985).

Urban air and drinking-water are the likely exposure routes of 1,2-dichloroethene for the general population.

Levels of 1,2-Dichloroethene in Urban Air and Landfills

Location	Year	Isomer	Conc. (ppb)	Sample Type	References
Cities in U.S.	1980–81	cis	0.013–0.76 (mean)	Urban air	Singh et al. 1983
Niagara Falls, NY	1978	NS	Trace	—	Barkley et al. 1980
Home near Love Canal, NY	1978	NS	0.015	Basement Air	Barkley et al. 1980
Knoxville, TN	Winter 1982	NS	8.1 (mean)	Air	Gupta et al. 1984
Methane vents at 2 sanitary landfills, Long Island, NY	NS	trans	75,600 (max)	Air samples	Lipsky and Jacot 1985
Selected U.S. landfills	NS	NS	70 (mean) 3,600 (max)	Secondary source	Vogt and Walsh 1985

Note: NS = Not Specified

1,1,2-Trichloroethane

Low levels of 1,1,2-trichloroethane may be detected in outdoor air, mainly from industrial sources that use the chemical as a solvent. Since these industries recycle or burn their waste, emission should be minor. The timber industry, plastics and synthetics industries, and laundries discharge 1,1,2-trichloroethane. The median concentration of source-related areas throughout the U.S. was 45 ppt. Twenty five percent of 97 samples in this study exceeded 210 ppt with a maximum measured in

California, at 2,300 ppt (Brodzinsky and Singh 1982). The National Ambient Volatile Organic Compounds Data Base (NAVOCDB) with 886 data points detected 1,1,2-trichloroethane in rural, urban, and suburban areas at 0 ppt (median); source-dominated areas at 2 ppt (median), and urban levels ranged from 10 to 50 ppt (Shah and Heyerdahl 1988). Eleven of sixteen indoor air samples in Knoxville, TN, showed a mean of 14.1 µg/m^3 (2.5 ppb) (Gupta et al. 1984); the origin could be building materials or solvent-containing products. A nationwide survey showed nondetection of 1,1,2-trichloroethane in drinking-water, but maximum levels of 31 ppb in some well waters (ATSDR 1989a). 1,1,2-Trichloroethane has not been detected in food or soil. The general population may be exposed to 1,1,2-trichloroethane in air; drinking-water is not an exposure route.

Trichloroethylene

Air

A U.S. monitoring data compilation, prior to 1981, compiled by Brodzinsky and Singh (1982) comprising of 2,300 points, reports trichloroethylene (TCE) concentrations of 30 ppt in rural and remote areas, 460 ppt in urban and suburban areas, and 1,200 ppt in areas near emission sources of TCE. A similar compilation (EPA 1985), which includes additional U.S. data and worldwide data, indicates that the average background level of TCE is 11 to 30 ppt in the northern hemisphere and <3 ppt in the southern hemisphere. A lower background level of 5–10 ppt was also reported in the northern hemisphere (Class and Ballschmidter 1986; Fabian 1986).

Recent Ambient Air Monitoring Studies in the U.S. Reported the Following TCE Concentrations:

Location	Year	Conc. (ppt)	References
Portland, OR	1984	44–714	Ligocki et al. 1985
Philadelphia, PA	1983–84	290	Sullivan et al. 1985
Three New Jersey cities	1981–82	210–590	Harkov et al. 1984
Seven cities, U.S.	1980–81	96–225	Singh et al. 1982
Arctic	1982–83	8–9	Khalil and Rasmussen 1983
			Hov et al. 1984
Six landfill sites, New Jersey		80–2,430 (av); 12,300 (max)	Harkov et al. 1985

Note: av = average; max = maximum.

Water

TCE has been detected in many drinking-waters across the U.S. Median values were 0.25 and 0.31 ppb from 133 U.S. cities using surface-water supplies and 25 cities using groundwater supplies, respectively (Coniglio et al. 1980). The U.S. EPA groundwater supply survey of 945 water supplies, nationwide, detected TCE in 91 waters; median was about 1 ppb with a single maximum value of 130 ppb (Westrick et al. 1984). It was not detected in 90.4% of 945 groundwater supplies. TCE at concentrations of 0.1–0.25 ppb were detected in tapwater from homes in the vicinity of Love Canal, NY waste sites (Barkley et al. 1980). Canadian survey of drinking-water sources (30) detected TCE at <1–2 ppb (Otson et al. 1982). TCE was detected in 16.43% of all analyzed groundwater samples from both federal and state surveys in the U.S. (Dykson and Hess 1982). In the state of New Jersey, it was found in 388 of 669 groundwaters with a maximum concentration of 635 ppb (Page 1981). Leachates from landfill waste disposal sites had the highest TCE levels (ATSDR 1989a). Analysis of the EPA STORET database reveals that TCE was detected in 28% of 9,295 surface-water reporting stations nationwide (Staples et al. 1985). Surface-waters of New Jersey, contained TCE in 261 of 462 samples, with a maximum value of 32.6 ppb (Page 1981). During 1978–81, Niagara River and Lake Ontario contained TCE at 0.008 to 0.12 ppb (Strachan and Edwards 1984).

TCE has been detected in soils up to 5.6 ppb and in sediments up to 300 ppb (EPA 1977). TCE has been detected in 6% of sediments from 338 monitoring stations at <5 ppb, median (Staples et al. 1985).

Fish

Trichloroethylene was found in fish from the Irish Sea on a dry weight basis in eel, cod, coalfish, dogfish, and bib at a <detection limit to 479 µg/kg (Dickson and Riley 1976). Levels of 0.8 to 56 µg/kg (wet weight) was detected in 15 species of fish from the coast of Great Britain (Pearson and McConnell 1975). Clams and oysters from Louisiana contained TCE at 0.8 to 5.7 µg/kg (Ferrario et al. 1985).

Food Items

Trichloroethylene has been detected in many natural and processed foods (Entz and Hollifield 1982). The food types and levels are given below.

Product	Conc. (µg/kg)	References
Dairy products	0.3–10	McConnell et al. 1975
Meat (English beef and pig's liver)	12,000–22,000	McConnell et al. 1975
Oils and fats	0–19	McConnell et al. 1975
Beverages	0.02–60	McConnell et al. 1975
Fruits and vegetables	1.7–5	McConnell et al. 1975
Fresh bread	7	McConnell et al. 1975
U.S. margarine (several brands)	440–3,600	Entz et al. 1982
Chocolate chip cookies	2.9	Heikes 1987
Plain granola	8.0	Heikes 1987
Cheddar cheese	3.1	Heikes 1987
Butter	12	Heikes 1987
Cooked pork sausage	5.2	Heikes 1987

Source: ATSDR 1989

Inhalation and dietary sources are the primary routes of exposure of trichlorethylene to the general population.

1,1,2,2-Tetrachloroethane

Low levels of 1,1,2,2-tetrachloroethane (1,1,2,2-T4CE) can be present in both indoor and outdoor air. Studies showed that it was detected only in a small number of air samples. Although the most typical ambient air level is about 5 ppt or less, it could be as high as 5 ppb in urban air. Analysis of a database containing 1,011 monitoring records showed that the overall median concentration of 1,1,2,2-T4CE is 0 ppt and the 75th percentile is 8 ppt (Shah and Heyerdahl 1988). Its average concentration in indoor air of several homes was reported to be 1.8 ppb. The likely origin of the concentration is from home consumer products (Bayer et al. 1988).

1,1,2,2-Tetrachlorethane has been detected in water. A comprehensive study of surface-water and groundwater conducted in a highly industrialized state in the U.S. in 1977–1979 detected 1,1,2,2-T4CE in 6% of groundwater samples and 11% of surface-water samples, with the highest level being 2.7 ppb (in groundwater) and 3 ppb (surface-water). The compound has not been detected in food or soil. It is not expected to build up in the food chain. 1,1,2,2-T4CE was detected in 278 of 1,430 hazardous waste sites on the NPL (National Priority List) and in gases released from these sites. However, it has not been found in gases released from refuse landfills. 1,1,2,2-T4CE is reported to convert slowly to other chemicals including vinyl chloride in landfills and groundwater. 1,1,2,2-T4CE in biota has not been reported in the literature.

The general population might be exposed to 1,1,2,2-tetrachloroethane via ambient air, drinking-water, and dermal exposure to contaminated soil.

Tetrachloroethylene

Air

A compilation of U.S. Ambient Air Monitoring data prior to 1981 includes more than 25,000 monitoring points, reporting a mean tetrachloroethylene concentration of 0.16 ppb in rural and remote areas, 0.79 ppb in urban and suburban areas, and 1.3 ppb in areas near emission sources (Brodzinsky and Singh 1982).

Water

Compilations of tetrachloroethylene monitoring data in drinking-water, groundwater and surface-water are available (EPA 1985; HSDB 1990). Median concentrations of tetrachloroethylene in drinking-water samples from 180 cities across the U.S. using surface-water supplies and 36 cities using groundwater supplies were 0.3 ppb and 3.0 ppb, respectively (EPA 1982). U.S. EPA groundwater supply survey of 945 cities nationwide showed tetrachloroethylene in 75 water supplies at 0.75 ppb (median) with a maximum value being 6 ppb (Westrick et al. 1984). Drinking-water from wells in New York, New Jersey and Connecticut were found to contain tetrachloroethylene at 715 to 1,500 ppb and these are in areas of considerable pollution (Burmaster 1982). Tap water samples from homes near the Love Canal, NY showed levels of 0.38 to 2.9 ppb (Barkley et al. 1980). In a comparison of CERCLA and RCRA groundwater data, tetrachloroethylene was detected in 36% of groundwater samples and was designated as the number two priority pollutant in CERCLA databases (Plumb 1987).

TCE Data from Ambient Air Monitoring in the U.S. and Other Countries

Location	Year	Conc (ppb)	References
Portland, OR	1984	0.058–0.31	Ligocki et al. 1985
Philadelphia, PA	1983–84	0.41–0.77	Sullivan et al. 1985
New Jersey cities	1981–82	0.24–0.46	Harkov et al. 1984
Seven large cities	1980–81	0.29–0.59	Singh et al. 1982
Arctic haze (winter)		0.118	Khalil and Rasmussen 1983
Arctic haze (summer)		0.067	Khalil and Rasmussen 1983
3 Industrialized areas (TEAM study)		0.035–1.33	Hartwell et al. 1992
Several locations, in Hamburg, Germany		0.27–10.44	Bruckmann et al. 1988
Finland			
industrial		19.17	Kroneld 1989
suburban		0.012	Kroneld 1989
Seven waste sites in New Jersey		0.12–1.91	Harkov et al. 1985
		7.24 (max)	Harkov et al. 1985
Germany, 1 meter above surface of a landfill		0.7–0.82	Koenig et al. 1987

It has been detected in groundwater leachates from various landfill sites nationwide in the U.S. and sewage effluent disposal sites at maximum levels of 590 ppb and 980 ppb respectively (ATSDR 1993a and references cited therein; Baker and MacKay 1985). A nationwide Japanese study showed tetrachloroethylene in 27% of shallow wells, and 30% of deep wells at concentrations of 0.2–23,000 ppb in shallow wells and 0.2–150 ppb in deep wells (Magara and Furuichi 1986). The U.S. EPA STROET database reveals that tetrachloroethylene was detectable in 38% of 9,323 surface-water monitoring stations nationwide (Staples et al. 1985). An analysis of 1,140 samples from the Ohio River showed a similar percentage at detected values, usually <1 ppb (Ohio River Valley Sanitation Commission 1980). In New Jersey, 154 of 174 surface-water samples contained tetrachloroethylene with a maximum concentration of 4.5 ppb (Page 1981). The Niagara River and Lake Ontario showed levels of 0.003–0.008 ppb during 1978 to 1981 (Strachan and Edwards 1984). In Canada, Lake St. Clair surface-water concentrations ranged from 0 to 473,000 ppb in June of 1984 (Kaiser and Comba 1986). Highest levels were detected where St. Clair River empties into the lake. River levels ranged from 79,000 to 182,000 ppb.

Global Concentration of Tetrachloroethylene in Rainwater, Ocean Water, and Snow

Location	Sample Type	Conc. (ng/L)	References
North Atlantic Ocean	Ocean water	0.12–0.5 (av.)	Pearson and McConnell 1975
Gulf of Mexico	Open waters	<Detection limit (1 ng/L)	Sauer 1981
Oregon, 1984	Rainwater	0.82–9.2	Kawamura and Kaplan 1983
La Jolla, CA	Rainwater	5.7	
Central-Southern California	Snow	1.4–2.3	Su and Goldberg 1976
Industrial cities in England	Rainwater	Up to 150	Pearson and McConnell 1975

av = Average

Soil

Sediments from Liverpool Bay, U.K. were found to contain 0.003 to 6 ppm of tetrachloroethylene with most detection near the lower end (Pearson and McConnell 1975). The U.S. EPA STORET database reveals that tetrachloroethylene was detected in 5% of 359 sediment samples at 5 ng/kg (median) (Staples et al. 1985). No data on soil residues were available.

Other Media

Tetrachloroethylene has been detected (µg/kg) in dairy products, 0.3–13; meat (English beef), 0.9–1.0; oils and fats, 0.01–7; beverages, 2–3; fruits and vegetables, 0.7–2; and fresh bread, 1 (McConnell et al. 1975). In another study, tetrachloroethylene was detected in a variety of foods, at 1–230 ppb (µg/kg), with a mean of 12 ppb (Daft 1989). Margarine from Washington, D.C. contained tetrachloroethylene ≥50 ppm in 10.7% of samples, the highest level recorded in a grocery store near a dry-cleaning shop, 500–5,000 ppb (Entz and Diachenko 1988).

Tetrachloroethylene was detected in fish from the Irish Sea with values ranging from below detection to 43 ng/g, dry weight (Dickson and Riley 1976). The compound was detected in 15 species of fish collected off the coast of the U.K. at levels of 0.3-43 µg/g, wet weight (Pearson and McConnell 1975).

The most important exposure routes of tetrachloroethylene for the general population appear to be inhalation from ambient air, ingestion from drinking-water, and ingestion of contaminated foods.

Hexachloroethane (HCE)

Air

Based on the limited data available, typical background concentrations in the northern hemisphere were reported to range from 5 to 7 ppt (48–68 ng/m^3) (Singh et al. 1979). Ambient air in the U.S. was analyzed to contain HCE at an average concentration of 0.1 ppt (9.7 ng/m^3), based on 69 measurements (Shah and Heyerdahl 1988). Concentrations of HCE were reported to be higher in the rural and remote areas (3.2 ppt (= 31ng/m^3) than in urban and industrial areas (Howard 1989). Northern and southern hemispheres reported HCE concentrations of 0.5 ppt (4.8 ng/m^3) and 0.34 ppt (3.3 ng/m^3), respectively (Class and Ballschmidter 1986).

Water

The U.S.EPA STORET database cites that HCE was detectable in only 0.1% of 882 ambient water samples (Staples et al. 1985). Concentration of all samples was <10 µg/L. Concentration of HCE in Lake Ontario was reported to be at 0.02 ng/L (Oliver and Niimi 1983). HCE was not detected in any of 86 urban run-off samples (Cole et al. 1984). HCE has been rarely detected in drinking-waters in North

America. HCE was reported in drinking-water from Cincinnati, OH and three water supplies in New Orleans, LA at 0.03–4.3 µg/L (Keith et al. 1976).

Hexachloroethane was not detectable in any of 356 sediment samples reported in the STORET database (Staples et al. 1985). The median detection limit was 500 µg/kg (ATSDR 1994).

Hexachloroethane was not detected in 116 fish samples reported in STROET database (Staples et al. 1985) nor in 28 fish samples from 14 Lake Michigan tributaries. However, Lake Ontario rainbow trout showed the presence of 0.03ng of HCE/g (average) in 10 of 10 samples (Oliver and Niimi 1983). HCE has not been reported in foods.

It seems that the exposure of hexachloroethane (HCE) to the general population is low and ambient air is likely an exposure route.

Vinyl Chloride

Air

Ambient air in rural–remote and urban–suburban areas of North America typically contains no detectable amount of vinyl chloride (ATSDR,TP-92/20 and references cited therein). Air near vinyl chloride and PVC manufacturing facilities has been reported to contain vinyl chloride from trace levels to 105 µg/m^3 (0.041 ppm) (EPA 1979), but may exceed 2,600 µg/m^3 (1 ppm) (Fishbein 1979). Elevated levels of vinyl chloride may be found near hazardous waste sites and municipal landfills; concentrations ranging from <detection limit to 5–8 µg/m^3 (0.002–0.003 ppm) have been reported in air above landfills (Baker and Mackay 1985; Stephens et al. 1986). Homes near a hazardous waste site in southern California were found to contain levels up to 1,040 µg/m^3 (0.4 ppm) (Stephens et al. 1986). Vinyl chloride was found in emissions from 85% of landfills tested at <1 ppm in more than half of the landfill emissions, with a range of 0.24 to 44 ppm (Wood and Porter 1987).

Water

Monitoring in nine studies in the U.S. have detected vinyl chloride up to 9.8 µg/L (0.01 ppm) in surface-waters and up to 380 µg/L (0.38 ppm) in groundwater (Coniglio et al. 1980; Dyksen and Hess 1982).

The 1982 EPA Groundwater Supply Survey across the U.S. identified vinyl chloride only in 0.74% of 945 groundwater supplies (detection limit, 0.001 ppm). It was reported that 0.5% of 186 random sample sites and 3.8% of 158 nonrandom sample sites contained maximum concentrations of vinyl chloride at 0.0011 ppm, and 0.0084 ppm, respectively (Westrick et al. 1984). Other studies reported detection of vinyl chloride in groundwater samples collected throughout the U.S. at levels of ≤0.38 ppm (Coniglio et al. 1980; Goodenkauf and Atkinson 1986).

Soil

No data are available in the literature searched.

Other Media

In the past, vinyl chloride used to be present in various foods due to migration from food wrappings and containers (Gilbert et al. 1980). Presently, U.S. FDA and other regulatory agencies in the world regulate the use of PVC polymers in food packaging materials and the amount of residual monomer in polymers. Studies have shown that at present conditions, no migration of vinyl chloride monomer from packaging materials into food occurs (Kontominas et al. 1985).

Leaching of vinyl chloride from PVC pipes into drinking-water has been reported at 0.0014 ppm; whereas, water running through a nine year old PVC pipe system contained 3×10^{-5} to 6×10^{-5} ppm (Dressman and McFaren 1978).

U.S. EPA study on vinyl chloride in indoor air samples from seven 1975 model cars detected levels of 0.4–1.2 ppm. Ventilation of the cars insides dissipated the vinyl chloride. These cars had a high ratio of plastics to inside volume and thus represented a worst-scenario of concentration levels of vinyl chloride. Cigarette and cigar smoke has been reported to contain 5.6–27 ng of vinyl chloride per cigarette (Hoffman et al. 1976).

For the general population, inhalation is the most probable exposure route of vinyl chloride.

TOXICITY PROFILE

The half-life of most chlorinated aliphatic hydrocarbons in fish tissues is short, averaging <1 day (Table 3.5). This factor, coupled with relatively slow uptake rates and moderate aqueous solubilities, accounts for low residues in biota.

Table 3.5 Concentration Factors and Half-Life of Chloroaliphatic Hydrocarbons (CHCs) in Bluegill Sunfish

CHC	Concentration Factor	Half-life in Tissues (Days)
Chloroform	6	<1
Carbon tetrachloride	30	<1
1,2-Dichloroethane	2	1–2
1,1,1-Trichloroethane	9	<1
1,1,2,2-Tetrachloroethane	8	<1
Pentachloroethane	67	<1
Hexachloroethane	139	<1
1,1,2-Trichloroethylene	17	<1
Tetrachloroethylene	49	<1

Sources: Barrows et al. 1980; Veith et al. 1979; EPA 1980

Most chlorinated aliphatic hydrocarbons are slightly to moderately toxic to algae, invertebrates, and fish (Table 3.6). Toxicity increases with the chlorine content of the compounds. A comparable response was reported with increasing substitution of bromine for chlorine. The LC_{50} for carp eggs exposed to dibromochloroethane averaged 34 mg/L, compared with values of 52, 67, and 97 mg/L reported for

Table 3.6 Acute Toxicity (96h, LC$_{50}$, mg/L) of Chlorinated Aliphatic Hydrocarbons (CHCs) to Algae, Invertebrates, and Fish

		Invertebrates		Fish	
CHC	Algae	Shrimp[1] *Mysidopsis bahia*[1]	Cladoceran[2] *Daphnia magna*[2]	Bluegill sunfish[3]*	Sheepshead minnow[4]
Chlorinated Ethanes					
1,2-Dichloroethane	>433	113	220	430	130–230
1,1,1-Trichloroethane	>670	31	>530	72	71
1,1,2-Trichloroethane	60–260	—	18	40	—
1,1,2,2-Tetrachloroethane	6–136	9	9.3	21	12
1,1,1,2-Tetrachloroethane	—	—	24	20	—
Pentachloroethane	58–121	0.4	63	7.2	116
Hexachloroethane	8–90	0.9	8.1	1.0	2.4
Chlorinated Propanes					
1,1-Dichloropropane	—	—	23	98	—
1,2-Dichloropropane	—	—	52	280	139+
1,3-Dichloropropane	1–93	0.8–10.3	280	>520	87
1,3-Dichloropropene	—	—	6.2	6.1	1.8
Chlorinated Ethylenes					
1,1-Dichloroethylene	>712	224	79	74	250
1,2-Dichloroethylene	—	—	220	140	—
Trichloroethylene	8	—	18	45	41+
Tetrachloroethylene	>507	10	18	13	29–52
Others					
Bromoform	12–114	24	46	29	18
Carbon tetrachloride	—	—	35	27	43+
Dichloromethane	>662	256	220	220	331
Chloroform	—	—	29	13–22	16–22

* LC$_{50}$, 48 h; +fathead minnow; Algae, *Selenastrum capricornutum, Skeletonema costatum*.
Sources: [1]EPA 1980; [2]LeBlanc 1980; [3]Buccafusco et al. 1981; [4]Heitmuller et al. 1981.

bromomethane, bromodichloromethane, and chloroform, respectively (Erickson and Hawkins 1980). The different isomers of the same compound seem to differ in their toxicity to fish. The LC$_{50}$ of 1,1-dichloropropane and 1,3-dichloropropane for bluegill sunfish was 98 and >520 mg/L, respectively (Table 3.6).

Reported EC$_{50}$ values of chloroform for reductions in cell counts and cell volumes for *Skeletonema costatum* were 477 and 437 mg/L respectively; for *Selenastrum capricornutum,* values were >1,000 mg/L for both effects (Cowgill et al. 1989). The reported LC$_{50}$ values of trichloroethylene for the alga *Microcystis aeruginosa* was 63 mg/L (WHO 1985); the EC$_{50}$ of trichloroethylene for the unicellular alga *Phaeodactylum tricornutum* was 5 mg/L (Thomas et al. 1981; Pearson and McConnell 1975).

Acute toxicities of chlorinated aliphatic hydrocarbons towards freshwater and marine invertebrates are listed in Table 3.7 and Table 3.8, respectively. The ranges of LC_{50} values for chlorinated alkanes are as follows: chloroform, 290–353 mg/L; dichloromethane, 170–900 mg/L; trichloromethane, 43–320 mg/L; and hexachloroethane, <1–16 mg/L (Tables 3.7 and 3.8). For chlorinated alkenes, such as trichloroethylene and tetrachloroethylene, the reported LC_{50} values for *Daphnia magna* were 85.2 and 18 mg/L, respectively. The toxicity seems to increase with the chlorine content of the chlorinated aliphatic hydrocarbons. This could likely be due to increased residence time in water (lower vapor pressures) and increased lipophilicities of chlorinated aliphatic hydrocarbons.

Acute toxicity values of CHCs to freshwater and marine fish species are listed in Tables 3.9 and 3.10, respectively. As with invertebrates, acute toxicity increased with increasing chlorine content of the CHCs. LC_{50} values for dichloroethane are 225–550 mg/L and those for hexachloroethane are reported to be 0.73–2.94 mg/L. Only marine fish species for which data are available, the sheepshead minnow, seems to be more resistant than freshwater fish species. Hexachloroethane is more toxic to fish species (both freshwater and marine) than to freshwater invertebrates. Bluegill and rainbow trout species were reported to be most sensitive to hexachloroethane, whereas goldfish was least sensitive (Phipps and Holcombe 1985). A study with six freshwater species, which included rainbow trout, bluegill sunfish, fathead minnow, mosquito fish, channel catfish and goldfish (Table 3.9) found bluegill to be the most sensitive species and catfish, the least sensitive, to hexachlorethane (Thurston et al. 1985).

The first set of data in Table 3.11 (under Ref. 1) do not show any clear toxicity trend for the compounds tested. Whereas, data set 2 (under Ref.2) reported that toxicity to *Daphnia magna* increased with the chlorine content of chloroalkanes tested (Table 3.11).

The nonlethal effects of CHCs include growth effects, reproductive effects, and behavioral effects. Data for growth effects on freshwater and marine algae are given in Tables 3.12 and 3.13, respectively.

The NOEC (No Observed Effect Level) and LOEC (Lowest Observed Effect Level) for growth and reproductive effects on *Daphnia magna* are given in Tables 3.14 and 3.15, respectively.

The NOEC and LOEC values for growth and reproductive effects in fish and shellfish are given in Tables 3.16 and 3.17, respectively.

Available laboratory animal toxicity data indicate that the observed increased incidence of tumors occurred largely at extreme exposures of chloroform, carbon tetrachloride, and dichloromethane. At lower exposure levels of these compounds, the risk of tumor development would virtually be zero. Studies show that these compounds are not mutagenic *in vivo*, with a weak mutagenicity shown in bacteria that lack the metabolic and pharmacokinetic factors possessed by mammalian species. Hence, on a weight-of-evidence basis, chloroform, carbontetrachloride, and dichloromethane are considered nongenotoxic. 1,1,1-Trichloroethane was reported to be tumorigenic in animals but not mutagenic. Whereas, 1,1- and 1,2-dichloroethanes were reported to induce tumors in rodents and also genotoxic (RTP, 1994).

CHLORINATED ALIPHATIC HYDROCARBONS

Table 3.7 Acute Toxicity of Chlorinated Aliphatic Hydrocarbons (CHCs) to Freshwater Invertebrates

CHC	Species	Toxic Effect	Conc. (mg/L)	References
Chloroform	*Daphnia magna*	LC_{50}, 48 h	353	Cowgill and Milazzo 1991
	Ceriodaphnia dubia	LC_{50}, 48 h	290	Cowgill and Milazzo 1991
1,2-Dichloroethane	*Gammarus fasciatus* (amphipod)	LC_{50}, 96 h	>100	Johnson and Finlay 1980
	Pteronarcys sp. (stonefly)	LC_{50}, 96 h	>100	Johnson and Finlay 1980
1,1,2-Trichloroethane	Molluscs			
	Dreissena polymorpha	LC_{50}, 96 h	320	Adema and Vink 1981
	Lymnaea stagnalis	LC_{50}, 96 h	170	Adema and Vink 1981
Hexachloroethane	Crustaceans			
	Ceriodaphnia reticulata	EC_{50}, 48 h	6.8 (4.7–8.6)	Elnabarawy et al. 1986
	Ceriodaphnia reticulata	LC_{50}, 48 h	3.2 (2.3–4.7)	Mount and Norberg 1984
	Daphnia magna	EC_{50}, 48 h	1.04–12.0	Elnabarawy et al. 1986
				Richter et al. 1983
				Thurston et al. 1985
	Daphnia magna	LC_{50}, 48 h	2.0–16.0	Mount and Norberg 1984
				Richter et al. 1983
				Le Blanc 1980
	Daphnia pulex	EC_{50}, 48 h	13 (12–15)	Elnabarawy et al. 1986
	Daphnia pulex	LC_{50}, 48 h	>10.0	Mount and Norberg 1984
	Orconectes immunis (cray fish)	LC_{50}, 96 h	2.2 (2.13–2.41)	Richter et al. 1983
	Insects			
	Tanytarsus dissimilis (midge)	LC_{50}, 48 h	1.07–1.70	Richter et al. 1983
	Gastropods			
	Aplexa hypnorum (snail)	LC_{50}, 96 h	>2.10	Phipps and Holcombe 1985
Trichloroethylene	*Daphnia magna*	LC_{50}, 48 h	85.2	EPA 1978
	Barnacle larvae (nauplii)	LC_{50}, 48 h	20	Thomas et al. 1981
Tetrachloroethylene	*Daphnia magna*	LC_{50}, 48 h	18	Thomas et al. 1981
	Daphnia magna	EC_{50}, 48 h	7.5–8.5	Richter et al. 1983
				Richter et al. 1983

Table 3.8 Acute Toxicity of Chlorinated Aliphatic Hydrocarbons (CHCs) to Marine Invertebrates

CHC	Species	Toxicity Test	Conc. (mg/L)	References
1,2-Dichloroethane	*Orphyrotrocha labronica* (marine polychaete)	shock LC$_{50}$, 96 h	400	Rosenberg et al. 1975
1,1,2-Trichloroethane	*O. labronica*	shock LC$_{50}$, 96 h	160	Rosenberg et al. 1975
	Mytilus edulis (mollusc)	LC$_{50}$, 96 h	110	Adema and Vink 1981
	Artemia salina (crustaceans)	LC$_{50}$, 96 h	52	Adema and Vink 1981
	Palaemonotes varians	LC$_{50}$, 6 h	43	Adema and Vink 1981
	Crangon crangon	LC$_{50}$, 6 h	43	Adema and Vink 1981
Hexachlorethane	*Photobacterium phosphoreum* (protozoa)	LC$_{50}$, 15 min	8.3 (7.1–9.7)	Nacci et al. 1986
	Arabica punctulata (sea urchin) embryo	LC$_{50}$, 3 h	7.43–10.6	Jackim and Nacci 1984
	Egg/sperm	LC$_{50}$, 4 h	4.97 (4.04–5.91)	Nacci and Jackim 1985
	Mysidopsis bahia (mysid shrimp)	LC$_{50}$, 96 h	0.94	EPA 1978
Trichlorethylene	*Eliminus modestus* (saltwater crustacean)	LC$_{50}$, 46 h	20	Pearson and McConnell 1975; WHO 1985
Tetrachloroethylene	Mussels	LC$_{50}$	3.5	McConnell et al. 1975

Table 3.9 Acute Toxicity of Chlorinated Aliphatic Hydrocarbons (CHCs) to Freshwater Fish

CHC	Fish	Toxicity Test	Conc. (mg/L)	Reference
1,2-Dichloroethane	Bluegill sunfish	LC$_{50}$, 96 h	550	Dawson et al. 1975
	Rainbow trout	LC$_{50}$, 96 h	225	Johnson and Finlay 1980
1,1,1-Trichloroethane	Fathead minnow	LC$_{50}$, 96 h (S)	105 (91–126)	Alexander et al. 1978
	Fathead minnow	LC$_{50}$, 96 h (F)	52.8 (43.7–77.7)	Alexander et al. 1978
1,1,2-Trichloroethane	Tooth carp (young)	LC$_{50}$, 24 h	70–72	Adema and Vink 1981
	American flagfish	LC$_{50}$, 96 h (S)	26.8 (21.3–33.7)	Smith et al. 1991
	American flagfish	LC$_{50}$, 96 h (F)	18.5 (16.4–20.8)	Smith et al. 1991
1,1,2,2-Tetrachloroethane	American flagfish	LC$_{50}$, 96 h (S)	89.1 (66.6–110)	Smith et al. 1991
	American flagfish	LC$_{50}$, 96 h (F)	45.1 (42–48.5)	Smith et al. 1991
Hexachloroethane	Rainbow trout	LC$_{50}$, 96 h	1.18 (0.727–1.92)	Thurston et al. 1985
		LC$_{50}$, 96 h	0.97 (0.73–1.28)	Phipps and Holcombe 1985
		LC$_{50}$, 96 h	0.98	EPA 1980
Hexachloroethane	Goldfish	LC$_{50}$, 96 h	1.42 (1.03–1.95)	Thurston et al. 1985
			>2.1	Phipps and Holcombe 1985
	Fathead minnow	LC$_{50}$, 96 h	1.10 (0.967–1.25)	Thurston et al. 1985
			1.39 (1.08–1.78)	Thurston et al. 1985
			1.23 (1.08–1.40)	Phipps and Holcombe 1985
			1.53	EPA 1980
			1.51 (1.43–1.58)	Wallbridge et al. 1983
			1.50	Veith et al. 1983
	Channel catfish	LC$_{50}$, 96 h	1.77 (1.18–2.70)	Thurston et al. 1985
			2.36 (1.90–2.94)	Thurston et al. 1985
			1.52 (1.39–1.65)	Phipps and Holcombe 1985
	Mosquito fish	LC$_{50}$, 96 h	1.38 (1.05–1.81)	Thurston et al. 1985
	Bluegill sunfish	LC$_{50}$, 96 h	0.86 (0.712–1.03)	Thurston et al. 1985
			0.97 (0.73–1.28)	Phipps and Holcombe 1985
Trichloroethylene	Fathead minnow	LC$_{50}$, 96 h (S)	66.8	McConnell et al. 1975
	Bluegill sunfish	LC$_{50}$, 96 h (S)	44.7	EPA 1978
				Thomas et al. 1981
	American flagfish	LC$_{50}$, 96 h (S)	3.1	Smith et al. 1991
Tetrachloroethylene	American flagfish	LC$_{50}$, 96 h (S)	4.0	Smith et al. 1991
	Fathead minnow	LC$_{50}$, 96 h (F)	13.4	Walbridge et al. 1983

Note: S = static; F = flow through.

Table 3.10 Acute Toxicity of Chlorinated Aliphatic Hydrocarbons (CHCs) to Marine Fish

CHC	Fish	Tox. Test	Conc. (mg/L)	Reference
1,1,2- Trichloroethane	Gobi	LC_{50}, 24 h	43	Adema and Vink 1981
	Plaice (diff. life stages)	LC_{50}, 48 h	34–125	Adema and Vink 1981
Hexachlorethane	Sheepshead minnow	LC_{50}, 96 h	2.4 (1.9–3.1)	Heitmuller et al. 1981
Trichloroethylene	Dab	LC_{50}, 96 h (F)	16.0	McConnell et al. 1975
Tetrachloroethylene	Sailfin mollies	LC_{50}, 96 h (F)	1.6	Laska et al. 1978

Note: F = flow-through system.

Table 3.11 Acute Toxicity of Chlorinated Alkanes to *Daphnia magna*

	LC_{50}, 48 h (mg/L)	
CHC	Values[1]	Values[2]
1,2-Dichloroethane	218.0	270–320
1,1,2-Trichloroethane	18.0	170–190
1,1,1,2-Tetrachloroethane	23.9	—
1,1,2,2-Tetrachloroethane	9.32	57–62
Pentachloroethane	62.9	7.3–8.0
Hexachloroethane	8.1	2.4–2.9

References: [1]LeBlanc 1980, Davidson et al. 1988. [2]Richter et al. 1983.

Table 3.12 Growth Effects of Chlorinated Aliphatic Hydrocarbons (CHCs) in Freshwater Algal Species

CHC	Species	EC_{50}, 96 h (mg/L)	Reference
1,2-Dichloroethane	*Selenastrum capricornutum*	>433	EPA 1978
1,1,1-Trichloroethane	*S. capricornutum*	>669	EPA 1978
1,1,2-Trichloroethane	*Chlorella* sp.	170–220	Adema and Vink 1981
1,1,2,2-Tetrachloroethane	*Selanastrum capricornutum*	136	EPA 1978
Pentachloroethane	*S. capricornutum*	121	EPA 1978
Hexachlorethane	*S. capricornutum*	87–93	EPA 1978
Tri- and tetrachloroethylenes	*S. subsicatus*	410–450	Geyer et al. 1985

Environmental exposures, based on ambient concentrations of chloroform, carbon tetrachloride, dichloromethane and 1,1,1-trichloroethane, were typically less than their respective RfD exposure limits. Occasionally, carbon tetrachloride and chloroform levels just slightly exceeded their RfD exposure limits. But, exposures of 1,1-dichloroethane and 1,1,1-trichloroethane were consistently lower than their

Table 3.13 Growth Effects of Chlorinated Aliphatic Hydrocarbons (CHCs) to Marine Algal Species

CHC	Species	EC$_{50}$, 96 h (mg/L)	Reference
1,1,2-Trichloroethane	Phaeodactylum tricornutum	50	Adema and Vink 1981
1,1,2-Trichloroethane	Dunaliella sp.	200	Adema and Vink 1981
Hexachloroethane	Skeletonema costatum	7.75–8.57	EPA 1978

Table 3.14 Growth Effects of CHCs in *Daphnia magna*

CHC	NOEC (mg/L)	LOEC (mg/L)	References
1,2-Dichloroethane	42	72	Richter et al. 1983
1,1,2-Trichloroethane	13	26	Richter et al. 1983
Tetrachloroethylene	0.51	1.1	Richter et al. 1983

Note: See glossary for acronyms, NOEC and LOEC.

Table 3.15 Reproductive Effects of Chloroaliphatics in Invertebrates

CHC	Species	NOEL (mg/L)	LOEC (mg/L)	References
1,2-Dichloroethane	Daphnia	11	72	Richter et.al 1983
1,1,1-Trichloroethane	Daphnia	1.3	2.4	Thompson and Carmichael 1989
1,1,2-Trichloroethane	Daphnia	26	42	Richter et.al 1983
	Daphnia	18	32	Adema and Vink 1981
	Artemia	10	15	Adema and Vink 1981
1,1,2,2-Tetrachloroethane	Daphnia	6.9	14	Richter et al. 1983
Tetrachloroethylene	Daphnia magna	0.51	1.1	Richter et al. 1983

Table 3.16 Growth Effects of Chlorinated Aliphatic Hydrocarbons (CHCs) in Fish and Shellfish

CHC	Species	NOEC (mg/L)	LOEC (mg/L)	References
Trichloroethylene	Plaice (marine sp.)	3.0	—	Adema and Vink 1981
Chloroethanes (2)	American flagfish	15.8	—	Smith et al. 1991
Tri- and tetrachloroethylenes	Black molly	—	1.46–8.11	Loekle et al. 1983
Chlorinated ethylenes	American flagfish	9.3, 21	—	Smith et al. 1991

RfD limits. Based on the observation that humans would not continually be exposed to exceeding ranges of these compounds and on the fact that these compounds are on the average not detected in air or water samples, ambient concentrations of chloroform, carbon tetrachloride, dichloromethane, and 1,1,1-trichloroethane would not likely cause any adverse effects on human health (RTP 1994).

Table 3.17 Reproductive Effects of Chlorinated Aliphatic Hydrocarbons (CHCs) in Fish and Shellfish

CHC	Species	NOEC (mg/L)	LOEC (mg/L)	References
Chloroethanes	American flagfish	4.9	10.6 (larval survival)	Smith et al. 1991
Tri- and tetrachloroethylenes	American flagfish	7.8–21.2	4.85 (reduction of larval survival)	Smith et al. 1991

The following findings were reported in the literature:

1. 1,2-Dichloroethane was shown to be genotoxic *in vitro*, but negative or weakly active *in vivo*
2. 1,2-Dichloroethane is noncarcinogenic in animals by inhalation route
3. 1,2-Dichloroethane was shown to be carcinogenic in animals by gavage route at very high doses in corn oil
4. However, the available toxicological data were evaluated to be inadequate to develop a nonthreshold exposure limit (RsD)
5. Hence, the RfD (threshold-based) exposure limit for 1,2-dichloroethane was used to evaluate the potential health effects of 1,1-dichloroethane
6. Based on the limited ambient concentration data available for 1,1-dichloroethane, it was evaluated that human exposure to mean ambient air concentrations would be less than both RsD and RfD exposure limits. Reported occasional high concentrations of 1,2-dichloroethane would lead to exposure ≥RsD at a lifetime risk level of 1 in 10^5
7. Since 1,1- and 1,2-dichloroethanes are infrequently detected in air or water samples, no adverse health effects are likely to occur at ambient exposure levels

1,1-Dichloroethylene, trichloroethylene and tetrachloroethylene were assessed not likely be human carcinogens, based on their experimental animal data, epidemiology, and their mechanism of action. Prior to 1992, the U.S. EPA employed linearized multistage modeling to predict cancer potency of tri- and tetrachloroethylenes which is currently under review by the EPA in light of newly available mechanistic and epidemiological data. Similar argument would apply for EPA's assessment of 1,1-dichloroethylene (RTP 1994). Human exposure to 1,1-dichloroethylene, trichloroethylene, and tetrachloroethylene estimated from their reported ambient concentrations, are below their RfD exposure limits. Hence, ambient exposures to these chloroalkenes are unlikely to cause any adverse effects on human health.

The ambient surface-water concentrations of CHCs would not likely cause any adverse effects on aquatic species. Of all CHCs, carbon tetrachloride and 1,1,1-trichloroethylene have been designated as stratospheric ozone depleting substances under the Montreal Protocol and were scheduled to be phased out by January 1, 1996. Chlorinated aliphatic hydrocarbons are not likely to cause any adverse effects on aquatic and terrestrial wildlife from air and water at current ambient levels of exposure (RTP 1994). However, it is absolutely necessary to maintain adequate controls over point-sources to ensure that ambient levels of CHCs are kept low.

REGULATIONS

Chloromethane

Agency	Description[a]	Levels	References
IARC	Not classifiable as human carcinogen	—	IARC 1987
EPA	Possible human carcinogen	—	EPA 1987
Regulations			
OSHA	TWA	50 ppm (8 h)	OSHA 1989
	STEL	100 ppm	OSHA 1989
Guidelines			
Air			
ACGIH	TLV-TWA	50 ppm	ACGIH 1988
ACGIH	STEL	100 ppm	ACGIH 1980
NIOSH	TWA	100 ppm	NIOSH 1985
	Ceiling	200 ppm	NIOSH 1985
	Maximum	300 ppm (5 min in 3 h)	NIOSH 1985
Acceptable ambient concentrations			
Air			
	8 h	1050–2500 µg/m^3	NATICH 1988
	24 h	1,750 µg/m^3	NATICH 1988
	Annual	1.6–2,520 µg/m^3	NATICH 1988
Water (drinking)	—	0.19–0.50 µg/L	NATICH 1988

[a] See glossary for acronyms TWA, STEL, PEL, MCL, TLV, and IDLH.

Dichloromethane

Agency	Description	Levels	References
IARC	Possible human carcinogen	—	IARC 1987
EPA	Probable human carcinogen	—	IRIS 1990

Regulations

Air
OSHA	PEL-TWA	25 ppm	OSHA 1991

Water
EPA, ODW	MCL	0.005 ppm	EPA 1992
FDA	Ban on use of DCM in cosmetic products and decaffeinated roasted coffee	10 ppm	FDA 1989

Guidelines

Air
ACGIH	TLV-TWA	50 ppm	ACGIH 1990
NIOSH	IDLH	5,000 ppm	NIOSH 1992

Water
EPA	MCL(DW)	zero	EPA 1992
ODW	Health advisories		
	Longer term (child)	2 mg/L	EPA 1987
	Longer term (adult)	6 mg/L	EPA 1987
EPA	Ambient water quality criteria		
OWRS	Ingesting water and organisms	0.00019 mg/L	EPA 1980
	Ingesting organisms	0.0157 mg/L	EPA 1980
	RfD (oral)	0.06 mg/kg/d	IRIS 1990
	Unit risk (oral)	2.1×10^{-7} µg/L	IRIS 1992

Acceptable Ambient Air Concentrations

All States	Range (8 h)	0.0035–0.0087 µg/m^3	NATICH 1992
	Range (24 h)	0.0058–2.4 µg/m^3	NATICH 1992
	Range (annual)	0.2–24 µg/m^3	NATICH 1992

Chloroform

Agency	Description	Levels	Reference
IARC	Possible human carcinogen	—	IARC 1987
EPA	Probable human carcinogen; RfD chronic (oral)	0.01mg/kg/day	IRIS 1992

Regulations
Air

OSHA	PEL-TWA	2 ppm (9.78 mg/m^3)	OSHA 1987

International Workplace Limits, Air, mg/m^3

Country	TWA	STEL	
Australia	49	—	Sittig, 1994
California	9.78	—	
Switzerland	50	100	
Finland	50	100	
France	25	250	
U.K.	9.8	—	
Sweden	10	25	

Water

EPA	Practical quantitation limit (PQL)	0.5 µg/L	EPA 1987

Guidelines
Air

ACGIH	TLV-TWA	10 ppm (49 mg/m^3)	ACGIH 1992
NIOSH	STEL (60 min)	2 ppm (9.78 mg/m^3)	NIOSH 1990

Water

EPA, ODW	Individual lifetime cancer risk (10^{-5})	60 µg/L	IRIS 1992

State Regulations and Guidelines

Acceptable ambient air concentrations

All States	Range (8 h)	250–1,200 µg/m^3	NATICH 1992

Drinking-water quality standards

All States	Range	1–5 µg/L	FSTRAC 1988

Limits in soil (permissible cleanup levels) mg/kg

	Residential	Industrial	
Michigan	62	410	Sittig, 1994
Massachusetts	100	520	
New Jersey	19 (surface)	28	
	1 (subsurface)	—	
Oregon	100	900	
Texas	0.437	—	

Carbon Tetrachloride

Agency	Description	Levels	References
IARC	Possible human carcinogen	—	IARC 1987
EPA	Probable human carcinogen	—	IRIS 1993
	RfD	0.0007 mg/kg/day	IRIS 1993
	Inhalation unit risk	1.5×10^{-5} $[\mu g/m^3]^{-1}$	IRIS 1993

National Regulations
Air

OSHA	PEL-TWA	10ppm (63 mg/m^3)	OSHA 1993
	Ceiling	25 ppm	OSHA 1989
EPA, QAQPS	Hazardous air pollutant	—	EPA 1985
	Class I ozone-depleting substance	Proposed phase-out (1/1/96)	EPA 1993b
		Phased out 1/1/2000	EPA 1992

International Workplace Limits, Air, mg/m^3

Country	TWA	STEL	
Australia	31	—	Sittig, 1994
Belgium	31	—	
California	12.6	—	
Czechoslovakia	10	20	
Switzerland	30	60	
Finland	31	63	
France	12	—	
U.K.	12.6	—	
Japan	63	—	
Sweden	13	63	

Water

EPA, ODW	Maximum concentration limit (MCL)	0.005 mg/L	EPA 1978

Guidelines
Air

ACGIH	TLV-TWA	5 ppm (31 mg/m^3)	ACGIH 1992
NIOSH	REL, ceiling (60 min)	2 ppm (12.6 mg/m^3)	NIOSH 1992

Water

EPA, ODW	MCLG (max. concn. limit guideline)	0	EPA 1985
	Health Advisories		
	long term (child)	0.71 mg/L	EPA 1987
	long term (adult)	0.25 mg/L	EPA 1987
EPA, OWRS	Ambient water quality criteria for human health protection		
	Ingesting water and organisms (10^{-5})	4 µg/L	EPA 1980
	Ingesting organisms only (10^{-5})	69.4 µg/L	EPA 1980

CHLORINATED ALIPHATIC HYDROCARBONS

Carbon Tetrachloride (Continued)

State Guidelines and Regulations

Acceptable ambient air concentrations
All states	Range (8 h)	300–714 µg/m^3	NATICH 1992
	Range (24 h)	85.5–310 µg/m^3	NATICH 1992
	Range (24 h)	0.03–13 µg/m^3	NATICH 1992

Drinking-water quality standards
All states	Range	0.5–5 µg/L	FSTRAC 1990

Limits in soil (permissible cleanup levels) mg/kg

	Residential	Industrial
Michigan	3	19
New Jersey	2 (surface)	—
	1 (subsurface)	—
Oregon	5	40
Texas	0.414	0.513
New York	5.4	—

Chlorodibromomethane and Bromoform

Agency	Description	Levels	Reference
Regulations			
Air			
OSHA	PEL-TWA (bromoform)	0.5 ppm (5 mg/m^3)	OSHA 1989a
Water			
U.S.EPA	ODWMCL (trihalomethanes)	0.10 mg/L	ATSDR 1990e
Guidelines			
Air			
ACGIH	TLV-TWA (bromoform)	0.5 ppm (5 mg/m^3)	ACGIH 1986
Water			
U.S.EPA,OWRS	Ambient water quality criteria (halomethanes)		
	Ingesting organisms–water (10^{-5})	0.0019 mg/L	EPA 1980
	Ingesting organisms only (10^{-5})	0.157 mg/L	EPA 1980
State Regulations			
Acceptable ambient Air concentrations			
All states	Range (8 h)	0.05–0.119 mg/m^3	NATICH 1988
	Range (24 h)	0.08 mg/m^3	NATICH 1988
Drinking-water standards			
All states	—	1 µg/L	FSTRAC 1988

Chloroethane

Agency	Description	Levels	Reference
Regulations			
OSHA	PEL-TWA	1000 ppm	OSHA 1988
Guidelines			
ACGIH	TEL-TWA	1000 ppm	ACGIH 1988
State Regulations and Guidelines			
Acceptable ambient air concentrations			
All states	8 h	23.8 ppm	NATICH 1988
	24 h	0.136 ppm (0.360 mg/m^3)	NATICH 1988
	Annual	19.7 ppm (52 mg/m^3)	NATICH 1988

1,1-Dichloroethane

Agency	Description	Levels	Reference
Regulations			
Air			
OSHA	PEL-TWA	400 mg/m^3	OSHA 1988
Guidelines			
ACGIH	TLV-TWA	810 mg/m^3	ACGIH 1986
	STEL	1,010 mg/m^3	ACGIH 1986
NIOSH	IDLH	4,000 ppm	NIOSH 1985
State Regulations and Guidelines			
Acceptable ambient air concentrations			
All states	Range (8 h)	0.0193–8.1 mg/m^3	NATICH 1987
	Range (24 h)	13.5 mg/m^3	NATICH 1987
Drinking-water guidelines			
All states	Range	1–850 µg/L	FSTRAC 1988

1,2-Dichloroethane

Agency	Description	Levels	Reference
IARC	Possible human carcinogen		IARC 1987
EPA	Probable human carcinogen		IRIS 1992

National Regulations

Air
OSHA	PEL-TWA	1 ppm	OSHA 1989
	STEL	2 ppm	OSHA 1989

Water
EPA, ODW	MCL (drinking-water)	0.005 mg/L	EPA 1987

Guidelines

Air
ACGIH	TLV-TWA	10 ppm	ACGIH 1986
NIOSH	REL-TWA	1 ppm	NIOSH 1992
	STEL	2 ppm	NIOSH 1992

Water
EPA, ODW	MCLG (drinking-water)	0 mg/L	IRIS 1992
	Health advisories		
	Long term (child)	0.74 mg/L	EPA 1987
	Long term (adult)	2.6 mg/L	EPA 1987
EPA, OWRS	Ambient water quality criteria for human health protection		
	Ingesting of organisms and water	0.94 µg/L	EPA 1987
	Ingesting of organisms only	243 µg/L	EPA 1987

State Guidelines/Standards

Acceptable ambient air concentrations
·All states	Range (8 h)	20–952 µg/m^3	NATICH 1993
	Range (24 h)	11–670 µg/m^3	NATICH 1993
	Range (annual)	0.0038–0.2 µg/m^3	NATICH 1993

Drinking-water quality guidelines
·All states	Range	0.5–5 µg/L	FSTRAC 1990

1,1-Dichloroethene

Agency	Description	Levels	Reference
IARC	Not classifiable as human carcinogen	—	IARC 1987
EPA	Possible human carcinogen	—	IRIS 1992
	RfD (oral)	0.009 mg/kg/day	IRIS 1992
	Inhalation unit risk	5×10^{-5} [µg m^3]$^{-1}$	IRIS 1992

International Guidelines

WHO	MCL (drinking-water)	0.32 µg/L	WHO 1984

National Regulations

Air

OSHA	PEL-TWA	1 ppm	OSHA 1989

Water

EPA, ODW	MCL (drinking-water)	0.7 µg/L	EPA 1983
EPA, OWRS	Direct discharge point sources-effluent limitations(BAT and NSPS)		
	Max for 1 day	25 µg/L	EPA 1987
	Max for monthly average (Direct discharge point sources that do not use biological treatment-effluent limitations BAT & NSPS)	16 µg/L	EPA 1987
	One day maximum	60 µg/L	EPA 1987
	Monthly average maximum	22 µg/L	EPA 1987

Guidelines

Air

ACGIH	TLV-TWA	5 ppm	ACGIH 1986
	STEL	20 ppm	ACGIH 1986
NIOSH	REL-TWA	Lowest feasible concn	NIOSH 1992

Water

EPA, ODW	MCLG	7 µg/L	EPA 1985
	Health Advisories		
	Long term (child)	1.0 mg/L	EPA 1987
	Long term (adult)	3.7 mg/L	EPA 1987
	Life term (adult)	7 µg/L	EPA 1987
EPA, OWRS	Ambient water quality for human health protection		
	Ingesting organisms and water (10^{-5})	0.33 µg/L	EPA 1980
	Ingesting organisms only (10^{-5})	18.5 µg/L	EPA 1980

State Regulations and Guidelines

Acceptable ambient air concentrations

All states	Range (8 h)	0.476–400 µg/m^3	NATICH 1993
	Range (24 h)	1.08–330 µg/m^3	NATICH 1993
	Range (annual)	0.02–200 µg/m^3	NATICH 1993

1,2-Dichloroethene

Agency	Description	Levels	Reference
IARC	No classification	—	—
EPA	Not classifiable	—	IRIS 1994

National Regulations

EPA, OW	Ambient water quality criterion for human health protection	0.0332 µg/L	
EPA, ODW	MCL (for organic contaminants)	0.1 mg/L	EPA 1975

Guidelines

Air

ACGIH	CEL,TLV-TWA	200 ppm (790 mg/m^3)	ACGIH 1991
NIOSH	REL, TWA	200 ppm (790 mg/m^3)	NIOSH 1992

Water

		Cis	**Trans**	
EPA	Health advisory (child & adult)			
	1 day	20 mg/L	4 mg/L	EPA 1994
	10 day	2 mg/L	3 mg/L	
	Lifetime	0.1 mg/L	0.07 mg/L	
	Longterm	6 mg/L	11 mg/L	
	MCLG	0.1 mg/L	0.07 mg/L	EPA 1994

State Regulations and Guidelines

Acceptable ambient air guidelines

All states	Range (8 h)	2.9–79 mg/m^3 (*trans*)	NATICH 1992
	Range (24 h)	2.63–79.3 mg/m^3 (*trans*)	NATICH 1992
	Range (annual)	108–790 µg/m^3 (*trans*)	NATICH 1992
	Annual	790 µg/m^3 (*cis*)	NATICH 1992
	24 h	6.3 mg/m^3 (*cis*)	NATICH 1992
	1 h	23.8 mg/m^3 (*cis*)	NATICH 1992
Water	Water quality health criteria		
	Domestic water source	100 µg/L (T); 70 µg/L (C)	CELDS 1992, 1993
	Fish consumption	13–140 mg/L	
	Drinking-water	10 µg/L (T); 6 µg/L (C)	FSTRAC 1990
	Freshwater fish ingestion only	130 mg/L	
	Freshwater fish and water ingestion	0.7 mg/L	
	Water quality, recreational use	2.8 mg/L	
	Groundwater quality ENF. Std.	0.1 mg/L (T); 0.1 (C)	

1,1,1-Trichloroethane

Agency	Description	Levels	Reference
IARC	Not classifiable as human carcinogen		IARC 1987
EPA	Not classifiable as human carcinogen		IRIS 1993

National Regulations
Air

EPA	Hazardous air pollutant		EPA 1990
OSHA	PEL-TWA	350 ppm (1,900 mg/m^3)	IRIS 1993
	STEL	450 ppm (2,450 mg/m^3)	

International Workplace Limits, Air, mg/m^3

Country	TWA	STEL	
Australia	680	—	Sittig, 1994
Belgium	1,910	2,460	
California	1,900	2,450	
Switzerland	1,080	5,400	
Czechoslovakia	500	2,000	
Finland	540	1,400	
France	1,650	2,500	
U.K.	1,900 (max)	2,450 (max)	
Israel	1,910	2,460	
Japan	1,100	—	
Sweden	300	500	

Guidelines
Air

ACGIH	TLV-TWA	350 ppm (1,910 mg/m^3)	ACGIH 1992
	STEL	450 ppm (2,460 mg/m^3)	ACGIH 1992
NIOSH	Ceiling (REL)	350 ppm (1,910 mg/m^3)	HSDB 1992

Water

EPA, OGWDW	MCL in drinking-water	0.2 mg/L	IRIS 1993
EPA, ODW	MCLG	0.2mg/L	IRIS 1993
	Health advisories		
	Long term (child)	40 mg/L	IRIS 1993
	Long term (adult)	100 mg/L	
	Lifetime (adult)	0.2mg/L	

Ambient water quality criteria for protection of human health

EPA, OWRS	Ingesting water and organisms	0.000184 µg/L	IRIS 1993
	Ingesting organisms only	1.03 µg/L	IRIS 1993

State Regulations and Guidelines
Acceptable ambient air concentrations guidelines

All states	Range (8 h)	19.1–45 mg/m^3	NATICH 1992
	Range (24 h)	1.04–32 mg/m^3	NATICH 1992
		191 mg/m^3 (24 h) (Oklahoma)	

Drinking-water quality standards

All states	Range	0.026–0.3 mg/L	CELDS 1992

1,1,1-Trichloroethane (Continued)

Groundwater quality standards			
All states Range		0.06–0.2 mg/L	CELDS 1992
Maximum leachable concentration (hazardous wastes)			
Texas		300 mg/L	CELDS 1992
Limits in soil (permissible cleanup levels), mg/kg			
	Residential	Industrial	
Massachusetts	100	5,000	Sittig, 1994
Michigan	2,200	4,200	
New Jersey	210 (surface)	3,800	
	50 (subsurface)	—	
New York	7,000	—	
Oregon	7,000	9,000	
Tennessee	1	20	
Texas	9,630	14,000	

1,1,2-Trichloroethane

Agency	Description	Levels	Reference
IARC	Not classifiable as human carcinogen	—	IARC 1987
EPA	Possible human carcinogen	—	EPA 1988
	RfD for chronic oral exposure	0.004 mg/kg/day	EPA 1988

National Regulations

Air
OSHA	PEL	10 ppm	OSHA 1989

Water
EPA, OWRS	Ambient water quality criteria		
	Consumption of organisms and water	0.6 µg/L	EPA 1980
	Consumption of organisms only	41.8 µg/L	EPA 1980

National Guidelines

Air
ACGIH	TWA	10 ppm	ACGIH 1988

Water
EPA, ODW	Health advisories		
	Longterm (child)	0.400 mg/L	EPA 1987
	Longterm (adult)	1.0 mg/L	EPA 1987
	Lifetime	0.003 mg/L	EPA 1987

State Guidelines

All states	Acceptable ambient air concn.	4.5 mg/m^3 (8 h)	ATSDR 1989a
	Drinking-water quality criteria	0.001–0.1 mg/L	FSTRAC 1988

1,1,2,2-Tetrachloroethane

Agency	Description	Levels	Reference
IARC	Not classifiable as human carcinogen	—	IARC 1987
EPA	Possible human carcinogen	—	IRIS 1994
NIOSH	Potential occupational carcinogen	—	NIOSH 1992

National Regulations

Air
OSHA	PEL-TWA	1 ppm	OSHA 1989

Water
U.S. EPA, ODW	Ambient water quality criteria	0.17 µg/L	EPA 1980
	Effluent guidelines and standards		
	Electroplating	<0.01 mg/L	EPA 1981
	Metal finishing	<0.01 mg/L	EPA 1983
	Coil coating	<0.01 mg/L	EPA 1982

National Guidelines

Air
ACGIH	CEL(TLV-TWA) (skin)	1 ppm	ACGIH 1986
NIOSH	REL(TWA) (air)	1 ppm	NIOSH 1992
	IDLH	150 ppm	NIOSH 1992

State Regulations and Guidelines

Air
Acceptable ambient concentration		
8 h	0.060–70 µg/m^3	NATICH 1992
24 h	8.8–122 µg/m^3	NATICH 1992
Annual	0.017–24.0 µg/m^3	NATICH 1992

Water
Drinking-water guideline	0.16–2.0 µg/L	FSTRAC 1990
Fish consumption	10.7–13.5 µg/L	CELDS 1992
Nondrinking-water supply	1.80 µg/L	CELDS 1992
All other uses	10.7 µg/L	CELDS 1992
Water quality–aquatic life		
Acute cold water fishery	4.7 mg/L	ATSDR 1995a
Acute warm water fishery	4.7 mg/L	ATSDR 1995a
Chronic cold water fishery	3.2 mg/L	ATSDR 1995a
Chronic warm water fishery	3.2 mg/L	ATSDR 1995a
Acute fresh water	0.923 mg/L	ATSDR 1995a
Acute salt water	3.0 mg/L	ATSDR 1995a
Acute marine water	0.902 mg/L	ATSDR 1995a
Chronic fresh water	0.462-2.4 mg/L	ATSDR 1995a
Trout waters	10.7 µg/L	ATSDR 1995a
Recreational Use		
Full body contact	7 µg/L	ATSDR 1995a
Partial body contact	110–450 µg/L	ATSDR 1995a
Groundwater quality standard	0.17 µg/L	ATSDR 1995a

CHLORINATED ALIPHATIC HYDROCARBONS

Trichloroethylene

Agency	Description	Levels	Reference
IARC	Not classifiable as human carcinogen	—	IARC 1987
EPA	Under review	—	IRIS 1992
WHO	TWA	135 mg/m^3	WHO 1981
	CELS (15 min)	1,000 mg/m^3	WHO 1981
	Drinking-water guideline level	0.030 mg/L	WHO 1981

National Regulations

Air
OSHA	PEL-TWA	50 ppm	OSHA 1989a
	STEL	200 ppm	OSHA 1989b

Water
EPA, ODW	MCL (drinking-water)	0.005 mg/L	EPA 1989

National Guidelines

Air
ACGIH	TLV-TWA	50 ppm (260 mg/m^3)	ACGIH 1990
	STEL	200 ppm (1,070 mg/m^3)	ACGIH 1990
	In urine at end of work week	100 mg/L	ACGIH 1990
NIOSH	REL-TWA (10 h)	25 ppm	NIOSH 1985
	IDLH	1,000 ppm	NIOSH 1985

Water
EPA, ODW	MCLG	0	EPA 1989
NAS	SNARL (suggested no-adverse-response level)		
	24h	105 mg/L	NAS 1980
	7 days	15 mg/L	NAS 1980
EPA, OWRS	Ambient water quality criteria		
	Ingesting organisms–water (10^{-5})	0.027 mg/L	EPA 1980
	Ingesting organisms only (10^{-5})	0.807 mg/L	EPA 1980

State Regulations and Guidelines

Air
Acceptable ambient concentrations
8 h range	1.35–2.70 mg/m^3	NATICH 1990
Nevada	6.43 mg/m^3	NATICH 1990
24 h range	0.0365–6.750 mg/m^3	NATICH 1990
Annual	0.3–0.769 µg/m^3	NATICH 1990
North Carolina	59 µg/m^3	NATICH 1990
70 years	0.05 ppb	

Water
Drinking-water quality standards
All states	2.8–25 µg/L	FSTRAC 1988
New Mexico only	0.1mg/L	FSTRAC 1988
Groundwater quality standard	0.05 mg/L	CELDS 1990
Enforcement standard	0.0018 mg/L	WAC 1988
Aquatic life, waterfowl, shorebirds and water-oriented wildlife	1,000 mg/L	
Raw water for drinking-water supply	3 mg/L	ATSDR 1995d

Tetrachloroethylene

Agency	Description	Levels	Reference
IARC	Possibly carcinogenic to humans		IARC 1987
WHO	DW Guideline based on cancer end-point	0.010 mg/L	WHO 1984
EPA	Under review		EPA 1991

National Regulations

Air

OSHA	PEL-TWA	25 ppm	OSHA 1989
EPA, QAQPS	Hazardous air pollutant		EPA 1985

International Workplace Limits, Air, mg/m³

Country	TWA	STEC	
Australia	340	1,020	Sittig, 1994
Belgium	339	1,368	
California	170	300 (max)	
Switzerland	345	690	
Finland	335	520	
France	335	—	
Japan	340	—	
Israel	339	1,370	

Water

EPA, ODW	MCL (drinking-water)	0.005 mg/L	EPA 1989
EPA, OWRS	Effluent limitations, max for 1 day	0.085 mg/L	FSTRAC 1988
	Monthly average	< 0.034 mg/L	EPA 1978a,b

National Guidelines

Air

ACGIH	TLV-TWA	50 ppm	ACGIH 1990
	STEL	200 ppm	ACGIH 1990
NIOSH	REL, LOQ	0.4 ppm	NIOSH 1992
	Potential occupational carcinogen		NIOSH 1992

Water

EPA, ODW	MCLG	0	EPA 1989
	Health advisories		
	Longterm (child)	1.4 mg/L	IRIS 1990
	Longterm (adult)	5 mg/L	IRIS 1990
	Lifetime (adult)	0.01 mg/L	IRIS 1990
NAS	SNARL		
	24 h	172 mg/L	NAS 1980
	7 day	24.5 mg/L	NAS 1980
EPA, OWRS	Water quality criteria		
	Ingesting organisms and water (10^{-5})	0.008 mg/L	EPA 1982
	Ingesting organisms only (10^{-5})	0.0885 mg/L	EPA 1982

Tetrachloroethylene (Continued)

U.S. States Regulations and Guidelines

Air
  Acceptable ambient concentrations

	8 h	1.67–7.98 mg/m^3	NATICH 1990
	24 h	0.922–5.7 mg/m^3	NATICH 1990
	Annual	0.00167–0.19 mg/m^3	NATICH 1990

Water

	Drinking-water guidelines	0.001–0.020 mg/L	FSTRAC 1988
	Public Health GW standards	—	
	Enforcement standard	0.001 mg/L	WAC 1988
	Preventive action limit	0.0001 mg/L	WAC 1988

Limits in Soil (Permissible Cleanup Levels) mg/kg

	Residential	Industrial	
Massachusetts	17	110	Sittig 1994
Michigan	7.8	49	
New Jersey	0.9 (surface)	37	
	1 (subsurface)	—	
Oregon	0.9	10	
Tennessee	0.5	0.5	
Texas	79.3	207	

Hexachloroethane

Agency	Description	Levels	Reference
IARC	Not classifiable as human carcinogen		IARC 1987
EPA	Possible human carcinogen		IRIS 1993
	RfD (oral)	0.001 mg/kg/day	IRIS 1993

National Regulations

Air

EPA, QAQPS	Hazardous air pollutant		
OSHA	PEL-TWA (skin)	1 ppm (10 mg/m³)	OSHA 1989

National Guidelines

ACGIH	TLV-TWA	1 ppm (9.7 mg/m³)	ACGIH 1992
NIOSH	REL-TWA (occupational carcinogen)	1 ppm	NIOSH 1992

Water

EPA, ODW	Health advisories		
	Longterm (child)	0.1 mg/L	
	Longterm (adult)	0.450 mg/L	
	Lifetime (adult)	0.001 mg/L	
EPA, OWRS	Ingesting organisms and water	0.0019 mg/L	EPA 1980
	Ingesting organisms only	0.00874 mg/L	EPA 1980

State Regulations and Guidelines

Acceptable ambient air concn.

	8 h	0.050–0.097 mg/m³	
	Nevada	2.38 mg/m³	NATICH 1992
	24 h	0.00053–0.200 mg/m³	NATICH 1992
	Annual	0.00025–0.010 mg/m³	NATICH 1992
	Drinking-water guideline range	0.0007–0.0019 mg/L	FSTRAC 1990

CHLORINATED ALIPHATIC HYDROCARBONS

Vinyl Chloride

Agency	Description	Levels	Reference
IARC	Human carcinogen	—	IARC 1979, 1987
EPA	Human carcinogen	—	EPA 1990
	Unit risk (inhalation)	$8.4 \times 10^{-5} [\mu g/m^3]^{-1}$	EPA 1990
	Unit risk (oral)	$5.4 \times 10^{-5} [\mu g/L]^{-1}$	EPA 1990

National Regulations

Air

OSHA	PEL-TWA	1 ppm	EPA 1993
	STEL	5 ppm	EPA 1993

International Workplace Limits, Air, mg/m^3

Country	TWA	STEL	
Australia	13	—	Sittig, 1994
Belgium	13	—	
Brazil	398	—	
California	2.6	13	
Switzerland	5.2	—	
Finland	15	30	
U.K.	14 (max)	—	
Israel	2.6	—	

Water

EPA, ODW	MCL (drinking-water)	0.002 mg/L	EPA 1989

National Guidelines

Air

ACGIH	TLV-TWA	5 ppm (10 mg/m^3)	ACGIH 1988
NIOSH	REL-TWA (15 min.)	2.55 mg/m^3	NIOSH 1985

Water

EPA, ODW	MCLG (drinking-water)	0	EPA 1989
	Health advisories		
	Longterm (child)	0.013 mg/L	EPA 1987
	Longterm (adult)	0.046 mg/L	EPA 1987
EPA, OWRS	Ingesting organisms and water (10^{-5})	0.020 mg/L	EPA 1980
	Ingesting organisms only (10^{-5})	5.246 mg/L	EPA 1980
	Drinking-water(10^{-5})	0.00015 mg/L	EPA 1985
	Drinking-water(10^{-6})	0.015 $\mu g/L$	EPA 1985

State Guidelines and Regulations

Air

Acceptable ambient concentrations

8 h range	50–238 $\mu g/m^3$	NATICH 1990
24 h range	3.47–130 $\mu g/m^3$	NATICH 1990
Annual	0.0–3.85 $\mu g/m^3$	NATICH 1990

Vinyl Chloride (Continued)

All states	Drinking-water guidelines/stds.	0–5 µg/L	FSTRAC 1988
	Effluent stds.; MAC into saturated or unsaturated zones	0.005 mg/L	CELDS 1990
	Enforcement standard	0.015 µg/L	CELDS 1990
	Groundwater quality standard	0.002 mg/L	
	Preventative action limit	0.0015 µg/L	CELDS 1990
	Warm/cold water fisheries	0.15 µg/L	CELDS 1990

Limits in Soil (Permissible Cleanup Levels) mg/kg

	Residential	Industrial	
Massachusetts	0.3	2	Sittig, 1994
Michigan	0.18	1.1	
New Jersey	2 (surface)	7	
	1 (subsurface)	—	
Oregon	0.03	0.05	
Texas	0.0199	0.0241	

REFERENCES

ACGIH. 1986. Documentation of the Threshold Limit Values and Biological Exposure Indices. American Conference of Governmental Industrial Hygienists, Cincinnati, OH. 184.

ACGIH. 1988. TLVs (Threshold Limit Values) and Biological Exposure Indices for 1988-1989. American Conference of Governmental Industrial Hygienists, Cincinnati, OH. 26

ACGIH. 1990. Threshold Limit Values for Chemical Substances and Physical Agents and Biological Exposure Indices for 1990-1991. American Conference of Governmental Industrial Hygienists, Cincinnati, OH.

ACGIH. 1991. Threshold Limit Values for Chemical Substances and Physical Agents and Biological Exposure Indices, 1991-1992. American Conference of Governmental Industrial Hygienists, Cincinnati, OH 45211-4438.

ACGIH. 1992. Threshold Limit Values for Chemical Substances and Physical Agents and Biological Exposure Indices, 1992-1993. American Conference of Governmental Industrial Hygienists, Cincinnati, OH.

Adema, D.M.M., and Vink, G.J. 1981. A comparative study of the toxicity of 1,1,2-trichloroethane, dieldrin, pentachlorophenol and 3,4-dichloroaniline for marine and freshwater organisms. *Chemosphere.* 10:533–554.

Alexander, H.C., McCarty, W.M., and Bartlett, E.A. 1978. Toxicity of perchloroethylene, trichloroethylene, 1,1,1-trichloroethane, and methylene chloride to fathead minnows. *Bull. Environ. Contam. Toxicol.* 20:344–352.

Atkinson, R. 1986. Kinetics and mechanisms of the gas-phase reactions of hydroxyl radical with organic compounds under atmospheric conditions. *Chem. Rev.* 85:69–201.

ATSDR. 1989a.* Agency for Toxic Substances and Disease Registry. Toxicological Profile for 1,1,2-Trichloroethane. Public Health Services, U.S. Department of Health and Human Services, Atlanta, GA.

---

* ATSDR references given are additional sources of information for specific chlorinated aliphatic hydrocarbons and are not cited in the text.

ATSDR. 1989b*. Agency for Toxic Substances and Disease Registry. Toxicological Profile for Vinyl Chloride, U.S. Department of Health and Human Services, Atlanta, GA. ATSDR/TP-88/25, August 1989.

ATSDR. 1989c. Agency for Toxic Substances and Disease Registry. Toxicological Profile for Chloroethane. Public Health Services, U.S. Department of Health and Human Services, Atlanta, GA.

ATSDR. 1989d*. Agency for Toxic Substances and Disease Registry. Toxicological Profile for Bromodichloromethane. Public Health Services, U.S. Department of Health and Human Services, Atlanta, GA.

ATSDR. 1990a*. Agency for Toxic Substances and Disease Registry. Toxicological Profile for cis- and trans- 1,2,Dichloroethene. Public Health Services, U.S. Department of Health and Human Services, Atlanta, GA. TP-90/13.

ATSDR. 1990b. Agency for Toxic Substances and Disease Registry. Toxicological Profile for Chloromethane. Public Health Services, U.S. Department of Health and Human Services, Atlanta, GA. TP-90-07.

ATSDR. 1990c. Agency for Toxic Substances and Disease Registry. Toxicological Profile for Bromoform, Chlorodibromomethane. Public Health Services, U.S. Department of Health and Human Services, Atlanta, GA. TP-90/05.

ATSDR. 1990d. Agency for Toxic Substances and Disease Registry. Toxicological Profile for 1,1,-Dichloroethane. Public Health Services, U.S. Department of Health and Human Services, Atlanta, GA. TP-90/12.

ATSDR. 1990e. Methylene Chloride Toxicity. Case Studies in Environmental Medicine. Agency for Toxic Substances and Disease Registry. U.S. Department of Health and Human Services, Atlanta, GA.

ATSDR. 1993a. Agency for Toxic Substances and Disease Registry. Toxicological Profile for Trichloroethylene. Public Health Services, U.S. Department of Health and Human Services, Atlanta, GA. TP-92/18.

ATSDR. 1993b. Agency for Toxic Substances and Disease Registry. Toxicological Profile for Chloroform. Public Health Services, U.S. Department of Health and Human Services, Atlanta, GA. TP-92/07.

ATSDR. 1993c. Agency for Toxic Substances and Disease Registry. Toxicological Profile for Methylene Chloride. Public Health Services, U.S. Department of Health and Human Services, Atlanta, GA. TP-92/13.

ATSDR. 1994a. Agency for Toxic Substances and Disease Registry. Toxicological Profile for Carbon Tetrachloride. Public Health Services, U.S. Department of Health and Human Services, Atlanta, GA. TP-93/02.

ATSDR. 1994b*. Agency for Toxic Substances and Disease Registry. Toxicological Profile for 1,2,-Dichloroethene. Public Health Services, U.S. Department of Health and Human Services, Atlanta, GA.

ATSDR. 1994c. Agency for Toxic Substances and Disease Registry. Toxicological Profile for 1,1,-Dichloroethene. Public Health Services, U.S. Department of Health and Human Services, Atlanta, GA. TP-93/07.

ATSDR. 1994d*. Agency for Toxic Substances and Disease Registry. Toxicological Profile for 1,2,-Dichloroethane. Public Health Services, U.S. Department of Health and Human Services, Atlanta, GA. TP-93/06.

ATSDR. 1995a. Agency for Toxic Substances and Disease Registry. Toxicological Profile for 1,1,1,-Trichloroethane. Public Health Services, U.S. Department of Health and Human Services, Atlanta, GA.

---

* ATSDR references given are additional sources of information for specific chlorinated aliphatic hydrocarbons and are not cited in the text.

ATSDR. 1995b.* Agency for Toxic Substances and Disease Registry. Toxicological Profile for Tetrachloroethylene. Public Health Services, U.S. Department of Health and Human Services, Atlanta, GA.

ATSDR. 1995c.* Agency for Toxic Substances and Disease Registry. Toxicological Profile for Vinyl Chloride. Public Health Services, U.S. Department of Health and Human Services, Atlanta, GA.

ATSDR. 1995d. Agency for Toxic Substances and Disease Registry. Toxicological Profile for Trichloroethylene. Public Health Services, U.S. Department of Health and Human Services, Atlanta, GA.

Baker, L.W. and MacKay, 1985. Hazardous waste management. Screening models for estimating toxic air pollution near a hazardous waste landfill. *J. Air Pollution Control Assoc.* 35:1190–1195.

Barkley, J., Bunch, J., Bursey, J.T. et al. 1980. Gas chromatography mass spectrometry computer analysis of volatile halogenated hydrocarbons in man and his environment. A multimedia environmental study. *Biomed. Mass Spectrom.* 7:139–147.

Barrows, M.E., Petrocelli, S.R., Macek, K.J., and Carroll, J.J. 1980. Bioconcentration and elimination of selected water pollutants by bluegill sunfish *(Lepomis macrochirus)*. In: *Dynamics, Exposure and Hazard Assessment of Toxic Chemicals*. Ann Arbor Science, Ann Arbor, MI, pp. 379–392.

Bayer, C.W., Black, M.S., and Galloway, L.M. 1988. Sampling and analysis techniques for trace volatile organic emissions from consumer products. *J. Chromatogr. Sci.* 26:168–173.

Benoit, F.M. and Jackson, R. 1987. Trihalomethane formation in whirlpool spas. *Water Res.* 2:353–357.

Berck, B. 1974. Fumigant residues of carbon tetrachloride, ethylene dichloride, and ethylene dibromide in wheat, flour, bran, middlings, and bread. *J. Agric. Food Chem.* 22:977–985.

Berg, W.W., Heidt, L.E., Pollock, W., et al. 1984. Brominated organic species in the Arctic atmosphere. *Geophys. Res. Lett.* 11:429–432.

Brodzinsky, R., and Singh, H.B. 1982. Volatile Organic Chemicals in the Atmosphere: An Assessment of Available Data. Menlo Park, CA: Atmospheric Science Centre, SRI International. Contract 68-02-3452.

Brodzinski, R. and Singh, H.B. 1983. Volatile Organic Chemicals in the Atmosphere: An Assessment of Available Data. Office of Research and Development, U.S. Environmental Protection Agency, EPA-600/3-83-027(A).

Bruckmann, P., Kersten, W., Funcke, W., et al. 1988. The occurrence of chlorinated and other organic trace compounds in urban air. *Chemosphere.* 17:2363–2380.

Buccafusco, R.J., Ells, S.J., and LeBlanc, G.A. 1981. Acute toxicity of priority pollutants to bluegill *(Lepomis machrochirus)*. *Bull. Environ. Contam. Toxicol.* 26:446–452.

Burmaster, D.E. 1982. The new pollution-groundwater contamination. *Environment* 24: 6–13, 33–36.

Carchman, R., Davidson, I.W.F., Greenberg, M.M., Parker, J.C., Benignus, V., Singh, D.L., and Fowle, J.R., III. 1984. Health Assessment Document for 1,1,1-Trichloroethane (Methyl Chloroform): Final Report. U.S. Environmental Protection Agency, Office of Research and Development, Washington, D.C.

CELDS. 1990. Computer Aided Environmental Legislative Data System. University of Illinois, Urbana, IL. May 15, 1990.

CELDS. 1992. Computer Aided Environmental Legislative Data System. University of Illinois, Urbana, IL.

* ATSDR references given are additional sources of information for specific chlorinated aliphatic hydrocarbons and are not cited in the text.

CELDS. 1994. Computer Aided Environmental Legislative Data System. University of Illinois, Urbana, IL.

Chopra, N.M. 1972. Breakdown of chlorinated hydrocarbon pesticides in tobacco smokes: a short review. In: *Proceedings of the 2nd International IUPAC Congress of Pesticide Chemistry,* Vol VI. Gordon and Breach Science Publishers, New York. pp. 245–261.

Class, T. and Ballschmiter, K. 1986. Chemistry of organic traces in air. VI. Distribution of chlorinated Cl-C4-hydrocarbons in air over the northern and southern Atlantic Ocean. *Chemosphere* 15:413–427.

Clement Associates. 1989. Toxicological Profile for Carbon Tetrachloride. Prepared for Agency for Toxic Substances and Disease Registry, Atlanta, GA. NTIS PB90-168196.

CLPSD. 1987. Contract Laboratory Program Statistical Database. U.S. Environmental Protection Agency. April 13, 1987.

CLPSD. 1989. Contract Laboratory Program Statistical Database. U.S. Environmental Protection Agency. July 12, 1989.

Cohen, M.A., Ryan, P.B., Yanagisawa, Y. et al. 1989. Indoor/outdoor measurements of volatile organic compounds in Kanawha Valley of West Virginia. *J. Air Pollution Control Assoc.* 39:1086–1093.

Cole, R.H., Frederick R.E., Healy R.P. et al. 1984. Preliminary findings of the priority pollutant monitoring project of the nationwide urban runoff program. *J. Water Pollut.* 56:898–908.

Coniglio, W.A., Miller, K., and Mackeever, D. 1980. The occurrence of volatile organics in drinking-water. Briefing prepared by the Criteria and Standards Division, Science and Technology Branch, Exposure Assessment Project, U.S. Environmental Protection Agency.

Cowgill, U.M., Milazzo, D.P., and Landenberger, B.D. 1989. Toxicity of nine benchmark chemicals to *Skeletome costatum*, a marine diatom. *Environ. Toxicol. Chem.* 8:451–455.

Cowgill, U.M. and Milazzo, D.P. 1991. The sensitivity of *Ceriodaphnia dubia* and *Daphnia magna* to seven chemicals utilizing the three brood test. *Arch. Environ. Contam. Toxicol.* 20:211.

Daft, J.L. 1989. Determination of fumigants and related chemicals in fatty and nonfatty foods. *J. Agric. Food Chem.* 37:560–564.

Davidson, K.A., Hovatter, P.S., and Ross, R.H. 1988. Water Quality Criteria for Hexachloroethane: Final Report. U.S. Army Medical Research and Development Command. Oak Ridge National Laboratory, Oak Ridge, TN. 37831-6050.

Dawson, G.W., Jennings, A.L., Drozdowski, D., and Rider, E. 1975. The acute toxicity of 47 industrial chemicals to fresh and saltwater fishes. *J. Hazardous Mat.* 1.

DeBortoli, M., Knoeppel, H., Pecchio, E. et al. 1986. Concentrations of selected organic pollutants in indoor and outdoor air in northern Italy. *Environ. Int.* 12:343–350.

Dickson, A.G. and Riley, J.P. 1976. The distribution of short-chain halogenated aliphatic hydrocarbons in some marine organisms. *Mar. Pollut. Bull.* 7:167–169.

Dressman, R.C. and McFarren, E.F. 1978. Determination of vinyl chloride migration from polyvinylchloride pipe into water. *J. Am. Water Works Assoc.* 70:29–30.

Dyksen, J.E., and Hess, A.F. III. 1982. Alternatives for controlling organics in groundwater supplies. *J. Am. Water Works Assoc.* (August):394–403.

Edgerton, S.A., Khalil, M.A.K., and Rasmussen, R.A. 1984. Estimates of air pollution from backyard burning. *J. Air Pollut. Control Fed.* 34:661–664.

Edgerton, S.A., Khalil, M.A.K., and Rasmussen, R.A. 1986. Source emission characterization of residential wood-burning stoves and fireplaces: fine particle/methyl chloride ratios for use in chemical mass balance modeling. *Environ. Sci. Technol.* 20:803–807.

Elnabarawy, M.T., Welter, A.N., and Robideau, R.R. 1986. Relative sensitivity of three daphnid species to selected organic and inorganic chemicals. *Environ. Toxicol. Chem.* 5:393–398.

Entz, R.C., Thomas, K.W., and Diachenko, G.W. 1982. Residues of volatile halocarbons in foods using headspace gas chromatography. *J. Agric. Food Chem.* 30:846–849.

Entz, R.C. and Hollifield, H.C. 1982. Headspace gas chromatographic analysis of foods for volatile halocarbons. *J. Agric. Food Chem.* 30:84–88.

Entz, R.C. and Diachenko, G.W. 1988. Residues of volatile halocarbons in margarines. *Food Additive Contam.* 5:267–276.

EPA. U.S. Environmental Protection Agency. 1975. Region V joint federal/state survey of organics and inorganics in selected drinking-water supplies. U.S. EPA and Illinois Environmental Protection Agency, Chicago, IL.

EPA. U.S. Environmental Protection Agency. 1977. Environmental monitoring near industrial sites, trichloroethylene. Washington, D.C., Office of Toxic Substances, EPA 560-16-77-024.

EPA. U.S. Environmental Protection Agency. 1978. In-depth studies on health and environmental impacts of selected water pollutants. U.S.E.P.A. Contract 68-01-4646.

EPA. U.S. Environmental Protection Agency. 1980. Ambient water quality criteria reports. Office of Water Regulations and Standards (OWRS), Washington, D.C.

EPA. U.S. Environmental Protection Agency. 1982. Coil coating point source category. Total toxic organics. Code of Federal Regulations, 40CFR465.

EPA. U.S. Environmental Protection Agency. 1983. Metal finishing point source category. Total toxic organics. 40CFR433.11.

EPA. U.S. Environmental Protection Agency. 1985. Part III, *Federal Register,* 50:46880-46901.

EPA. U.S. Environmental Protection Agency. 1987. Health effects assessment for chloromethane. EPA/600/8-88-024. Office of Research and Development, Env. Criteria and Assessment, Cincinnati, OH, NTIS PB88-279932.

EPA. U.S. Environmental Protection Agency. 1988a. National Ambient Volatile Organic Compounds (VOCs), Database Update. EPA/600/3-88/010.

EPA. U.S. Environmental Protection Agency. 1988b. Integrated Risk Information Systems (IRIS). Risk estimate for carcinogenicity for 1,1,2-trichloroethane. Office of Health and Environmental Assessment, Environmental Criteria and Assessment Office, Cincinnati, OH.

EPA. U.S. Environmental Protection Agency. 1989a. National drinking water regulations. 40CFR141,142,143.

EPA. U.S. Environmental Protection Agency. 1989b. National primary and secondary drinking water regulations; proposed rule, *Federal Register.* 54-22062-22160.

EPA. U.S. Environmental Protection Agency. 1990. Clean Air Act of 1990. 42USC.7412. Title I, Part A, Sec. 112, 53.

EPA. U.S. Environmental Protection Agency. 1991. Science Advisory Board's review of the office of Research and Development draft document: Response to issues and data submissions on the carcinogenicity of tetrachloroethylene (perchloroethylene) February 1991. EPA-SAB-EHC-91-013.

EPA. U.S. Environmental Protection Agency. 1992. Part III, *Federal Register,* 57:31789-31790.

EPA. U.S. Environmental Protection Agency. 1993a. Emissions standard for vinyl chloride and polyvinyl chloride plants. 40CFR61.63, 61.64.

EPA. U.S. Environmental Protection Agency. 1993b. Proposed phase-out by 01/01/96; 40CFR82.

EPA. U.S. Environmental Protection Agency references are cited in the respective ATSDR Toxicological Profile documents as well as in Regulatory Toxicology and Pharmacology, Vol 20. No.1. August, 1994.

Erickson, W.J. and Hawkins, J. 1980. Effects of halogenated organic compounds on photosynthesis in estuarine phytoplankton. *Bull. Environ. Contam. Toxicol.* 24:910–915.

Fabian, P. 1986. Halogenated hydrocarbons in the atmosphere. In: *The Handbook of Environmental Chemistry,* Vol. 4, Part A. Hutzinger, O., Ed. Springer-Verlag, Berlin, pp. 23–51.

FDA. 1989. Food and Drug Administration. *Federal Register* 54:27328–27342.

Ferrario, J.B., Lawler, G.C., DeLeon, I.R., and Laseter, J.L. 1985. Volatile organic pollutants in biota and sediments of Lake Pontchartrain. *Bull. Environ. Contam. Toxicol.* 34:246–255.

Fishbein, L. 1979. Potential halogenated industrial carcinogenic and mutagenic chemicals. 2. Halogenated saturated hydrocarbons. *Sci. Total Environ.* 11:163–195.

Friedman, M.A. 1988. Volatile organic compounds in groundwater and leachate at Wisconsin landfills. Madison, WI. Wisconsin Department of Natural Resources. PUBL-WK-192 88.

FSTRAC. 1988. Federal-State Toxicology and Regulatory Alliance Committee. Summary of state and federal drinking-water standards and guidelines. Chemical Communication Subcommittee, FSTRAC database, Washington, D.C. March 1988.

FSTRAC. 1990. Federal–State Toxicology and Regulatory Alliance Committee. Summary of state and federal drinking-water standards and guidelines. Chemical Communication Subcommittee, FSTRAC database, Washington, D.C.

Geyer, H., Scheunert, I., and Korte, F. 1985. The effects of organic environmental chemicals on the growth of the alga *Scenedesmus subspicatus*: a contribution to environmental biology. *Chemosphere.* 14:1355–1370.

Gilbert, J., Shepherd, M.J., Startin, J.R. et al. 1980. Gas chromatographic determination of vinylidene chloride monomer in packaging films and in foods. *J. Chromatogr.* 197:71-78.

Goodenkauf, O. and Atkinson, J.C. 1986. Occurrence of volatile organic chemicals in Nebraska groundwater. *Groundwater.* 24:231–233.

Gribble, G.W. 1992. Naturally occurring organohalogen compounds - a survey. *J. Nat. Prod.* 55:1353–1395.

Gupta, K.C., Vlsamer, A.G., and Gammage, R. 1984. Volatile organic compounds in residential air: levels, sources and toxicity. *Proc. APCA. Ann. Meet.* 77:84–93.

Harkov, R., Kebbekus, B., Bozzelli, J.W. et al. 1984. Comparison of selected volatile organic compounds during the summer and winter at urban sites in New Jersey. *Sci. Total Environ.* 38:259–274.

Harkov, R., Giante, S.J. Jr., Bozelli, J.W. et al. 1985. Monitoring volatile organic compounds at hazardous and sanitary landfills in New Jersey. *J. Environ. Sci. Health.* A20:491–501.

Harper, D.B., Kennedy, J.T., and Hamilton, J.T.G. 1988. Chloromethane biosynthesis in poroid fungi. *Phytochemistry.* 27:3147–3153.

Harsch, D. 1977. Study of halocarbon concentrations in indoor environments. Final report. Report to U.S. Environmental Protection Agency, Office of Research and Development, Washington, D.C., by Washington State University, College of Engineering, Pullman, WA. Project 1505. Contract WA-6-99-2922-J.

Hartwell, T.D., Perritt, R.L., and Pellizzari, E.D. 1992. Results from the 1987 Total Exposure Assessment Methodology (TEAM) study in Southern California. *Atmos. Environ.* 26:1519–1527.

Heikes, D.L. 1987. Purge and trap method for determination of volatile halocarbons and carbon disulfide in table-ready foods. *J. Assoc. Off. Anal. Chem.* 70:215–226.

Heitmuller, P.T., Hillister, T.A., and Parrish, P.R. 1981. Acute toxicity of 54 industrial chemicals to sheepshead minnows *(Cyprinodon variegatus). Bull. Environ. Contam. Toxicol.* 27:596–604.

Hoffman, D., Patrianakos, C., and Brunnemann, K.D. 1976. Chromotographic determination of vinyl chloride in tobacco smoke. *Anal. Chem.* 48:47–50.

Hov, O., Penkett, S.A., Isaksen, I.S.A. et al. 1984. Organic gases in the Norwegian arctic. *Geophys Res Lett.* 11:425–428.

Howard, P.H. 1989. *Handbook of Environmental Fate and Exposure Data for Organic Chemicals, Vol. I: Large Production and Priority Pollutants,* CRC/Lewis, Boca Raton, FL.

HSDB. 1990. Hazardous Substances Data Bank. National Library of Medicine, National Toxicology Information Program, Bethesda, MD. July 5, 1990.

HSDB. 1992. Hazardous Substances Data Bank. National Library of Medicine, National Toxicology Information Program, Bethesda, MD. July 1992.

IARC. 1979. International Agency for Research on Cancer. Monographs on the Evaluation of the Carcinogenic Risk of Chemicals to Humans. Vinyl Chloride: Polyvinyl Chloride and Vinyl Chloride—Vinyl Acetate Copolymers. Some Monomers, Plastics and Synthetic Elastomers and Acrolein. Vol. 19. IARC, World Health Organization, Lyon, France.

IARC. 1986. International Agency for Research on Cancer. Monographs on the Evaluation of Carcinogenic Risk of Chemicals to Humans. Some Halogenated Hydrogenated and Pesticide Exposures: Methyl Chloride. Lyon, France, WHO, 41:161–186.

IARC. 1987. International Agency for Research on Cancer. Monographs on the Evaluation of Carcinogenic Risk of Chemicals to Humans. Supplement 7: Overall Evaluations of Carcinogenicity: An Updating of IARC Monograph Volumes 1-42. WHO, Lyon, France.

IRIS. 1990. Integrated Risk Information System. U.S. Environmental Protection Agency, Office of Health and Environmental Assessment, Environmental Criteria and Assessment Office, Cincinnati, OH.

IRIS. 1992. Integrated Risk Information System. U.S. Environmental Protection Agency, Office of Health and Environmental Assessment, Environmental Criteria and Assessment Office. Cincinnati, OH.

IRIS. 1993. International Risk Information System. Trichloroethylene. U.S. Environmental Protection Agency. Office of Health and Environmental Assessment, Environmental Criteria and Assessment Office, Cincinnati, OH.

IRIS. 1994. Integrated Risk Information System. U.S. Environmental Protection Agency, Office of Health and Environmental Assessment, Environmental Criteria and Assessment Office. Cincinnati, OH.

Jackim, E. and Nacci, D. 1984. A rapid aquatic toxicity assay utilizing labeled thymidine incorporation in sea urchin embryos. *Environ. Toxicol. Chem.* 3:631–636.

Johnson, W.W. and Finley, M.T. 1980. Handbook of Acute Toxicity of Chemicals to Fish and Aquatic Invertebrates. Department of Interior, U.S. Fish and Wildlife Services, Washington, D.C.

Kawamura, K. and Kaplan, I.R. 1983. Organic compounds in the rain water of Los Angeles. *Environ. Sci. Technol.* 17:497–501.

Kaiser, K.L.E. and Comba, M.E. 1986. Volatile halocarbon contaminant survey of the St. Clair River. *Water Pollut. Res. J. Can.* 21:323–331.

Keith, L.H., Garrison, A.W., Allen, F.R., et al. 1976. Identification of organic compounds in drinking-water from thirteen United States cities. In: *Identification and Analysis of Organic Pollutants in Water.* Keith, L.H. Ed. Ann Arbor Press, Ann Arbor, MI. 329–373.

Kelley, R.D. 1985. Synthetic organic compound sampling survey of public water supplies. Report to U.S. Environmental Protection Agency, Washington, D.C. by Iowa Department of Water, Air and Waste Management, Des Moines, IA. NTIS No. PB85-214427.

Khalil, M.A.K. and Rasmussen, R.A. 1983. Gaseous tracers of arctic haze. *Environ. Sci. Technol.* 17:157–164.

Kleindienst, T.E., Shepson, P.B., Edney, E.O. et al. 1986. Wood smoke: measurement of the mutagenic activities of its gas and particulate-phase photooxidation products. *Environ. Sci. Technol.* 20:493–501.

Koenig, H.P., Lahl, U., and Kock, H. 1987. Determination of organic volatiles in ambient air in the area of a landfill. *J. Aerosol Sci.* 18:837–840.

Kontominas, M.G., Gupta, C., and Gilbert, S.G. 1985. Interaction of vinyl chloride with polyvinyl chloride in model food systems by HPLC chromatography: migration aspects. Lebensmittel-Wissenschaft und technologie 18:353–356.

Krasner, S.W., McGuire, M.J., and Jacangelo, J.G. 1989. The occurrence of disinfection by-products in U.S. drinking water. *J. Amer. Water Works Assoc.* 81:41–53.

Kringstad, K.P. and Lindstrom, K. 1984. Spent liquors from pulp bleaching. *Environ. Sci. Technol.* 18:236–248.

Kroneld, R. 1989. Volatile pollutants in suburban and industrial air. *Bull. Environ. Contam. Toxicol.* 42:873–877.

LaRegina, J., Bozzelli, J.W., Harkov, R. et al. 1986. Volatile organic compounds at hazardous waste sites and a sanitary landfill in New Jersey. *Environ. Progress.* 5:18–27.

Laska, A.L., Bartell, C.K., Condie, D.B., Brown, J.W., Evans, R.L., and Laseter, J.L. 1978. Acute and chronic effects of hexachlorobenzene and hexachlorobutadiene in red swamp crayfish *(Procambarus clarki)* and selected fish species. *Toxicol. Appl. Pharmacol.* 43:1–12.

LeBlanc, G.A. 1980. Acute toxicity of priority pollutants to water flea *(Daphnia magna)*. *Bull. Environ. Contam. Toxicol.* 24:684–691.

Letkiewicz, F., Johnston, P., Macaluso, C. et al. 1983. Carbon tetrachloride; occurrence in drinking-water, food and air. Office of Drinking-water, U.S. Environmental Protection Agency, Washington, D.C.

Ligocki, M.P., Leuenberger, C., and Pankow, J.F. 1985. Trace organic compounds in rain - II. Gas scavenging of neutral organic compounds. *Atmos. Environ.* 19:1609–1617.

Lipsky, D. and Jacot, B. 1985. Hazardous emissions from sanitary landfills. Proceedings of 78th Annual Meeting of the Air Pollution Control Association, Detroit, MI. June 16-21, 1985.

Loekle, D.M., Schecter, A.J., and Christian, J.J. 1983. Effects of chloroform, tetrachloroethylene, and trichloroethylene on survival, growth, and liver of *Poecillia sphenops*. *Bull. Environ. Contam. Toxicol.* 30:199–205.

Mackay, D., Shiu, W.Y., and Ma, K.C. 1993. *Illustrated Handbook of Physical-Chemical Properties and Environmental Fate of Organic Chemicals, Volume 3. Volatile Organic Chemicals,* CRC/Lewis, Boca Raton, FL.

Magara, Y. and Furuichi, T. 1986. Environmental pollution by trichloroethylene and tetrachloroethylene: A nationwide survey. *Dev. Toxicol. Environ. Sci.* 12:231–243.

Mansville Chemical Products Corp. 1988. Chemical products synopsis: methylene chloride. Mansville Chemical Products Corp., Asbury Park, NJ.

McConnell, G., Ferguson, D.M., and Pearson, C.R. 1975. Chlorinated hydrocarbons and the environment. *Endeavour.* 34:13–18.

Moore, J.W. and Ramamoorthy, S. 1984. *Organic Chemicals in Natural Waters, Applied Monitoring and Impact Assessment.* Springer-Verlag, New York.

Moore, D.R.J., Walker, S.L., and Koniecki, D. 1991. Canadian Water Quality Guidelines for Chlorinated Ethanes. Scientific series 185. Inland Waters Directorate, Water Quality Branch, Environment Canada, Ottawa, ON. ISBN 0-662-19185-4.

Mount, D.I. and Norberg, T.J. 1984. A seven-day life-cycle cladoceran toxicity test. *Environ. Toxicol. Chem.* 3:425–432.

Nacci, D.E., Jackim, E., and Walsh, R. 1986. Comparative evaluation of three rapid marine toxicity tests: sea urchin early embryo growth test, sea urchin sperm cell toxicity test and microtox. *Environ. Toxicol. Chem.* 5:521–525.

Nacci, D.E. and Jackim, E. 1985. Rapid aquatic toxicity assay using incorporation of tritiated-thymidine into sea urchin. *Arbaci punctulata*, embryo: evaluation of toxicant exposure procedures. In: R.C. Bahner and D.J. Hansen, Eds. Aquatic Toxicology and Hazard Assessment, American Society for Testing Materials, Philadelphia, PA.

NAS. 1977. National Academy of Sciences. *Drinking Water and Health.* Safe Drinking-water Committee, NAS, Washington, D.C.

NAS. 1980. National Academy of Sciences. *Drinking Water and Health.* Vol. 3. Washington, D.C., National Academy Press.

NATICH. 1987. National Air Toxics Information Clearinghouse. U.S. Environmental Protection Agency, Office of Air Quality Planning and Standards, Research Triangle Park, NC.

NATICH. 1988. National Air Toxics Information Clearinghouse. U.S. Environmental Protection Agency, Office of Air Quality Planning and Standards, Research Triangle Park, NC.

NATICH. 1990. National Air Toxics Information Clearinghouse. U.S. Environmental Protection Agency, Office of Air Quality Planning and Standards, Research Triangle Park, NC. September 1990.

NATICH. 1992. National Air Toxics Information Clearinghouse. U.S. Environmental Protection Agency, Office of Air Quality Planning and Standards, Research Triangle Park, NC. May 1992.

NATICH. 1993. National Air Toxics Information Clearinghouse. U.S. Environmental Protection Agency, Office of Air Quality Planning and Standards, Research Triangle Park, NC. May 1993.

NIOSH. 1985. Pocket guide to chemical hazards. Cincinnati, OH: US Department of Health and Human Services, National Institute for Occupational Safety and Health, 98-99.

NIOSH. 1990. Pocket guide to chemical hazards. National Institute for Occupational Safety and Health. U.S. Department of Health and Human Services. 148. Cincinnati, OH.

NIOSH. 1992. National Institute for Occupational Safety and Health. Recommendations for occupational safety, health, compendium of policy documents and statements. U.S. Department of Health and Human Services, Cincinnati, OH.

Ohio River Valley Sanitation Commission. 1980. Valley Water Sanitation Commission. Assessment of water quality conditions. Ohio River Mainstem 1978-79. Cincinnati, OH.

Oliver, B.G. and Niimi, A.J. 1983. Bioconcentration of chlorobenzenes from water by rainbow trout: Correlations with partition coefficients and environmental residues. *Environ. Sci. Technol.* 17:287–291.

OSHA. 1987. Occupational Safety and Health Administration. *Federal Register.* 29 CFR 1910.

OSHA. 1988. Occupational Safety and Health Administration. *Federal Register.* 29 CFR 1910.1000.

OSHA. 1989. Occupational Safety and Health Administration. Part III. *Federal Register.* 54:2332–2983.

OSHA. 1989a. Occupational Safety and Health Administration. Air Contaminants. Code of Federal Regulations. 29 CFR 1910.1000.

OSHA. 1989b. Occupational Safety and Health Administration. Air Contaminants Final Rule. *Federal Register.* 54:2332–2960.

OSHA. 1991. Occupational Safety and Health Administration. Part II. *Federal Register.* 56:57036–57139.

OSHA. 1993. Occupational Safety and Health Administration. Air Contaminants Final Rule. *Federal Register.* 58:35338–35351.

Otson, R., Williams, D.T., and Bothwell, P.D. 1982. Volatile organic compounds in water at thirty Canadian potable water treatment facilities. *J. Assoc. Off. Anal. Chem.* 65:1370–1374.

Otson, R., Doyle, E.E., Williams, D.T. et al. 1983. Survey of selected organics in office air. *Bull Environ. Contam. Toxicol.* 31:222–229.

Page, G.W. 1981. Comparison of groundwater and surface-water for patterns and levels of contamination by toxic substances. *Environ. Sci. Technol.* 15:1475–1481.

Pearson, C.R. and McConnell, G. 1975. Chlorinated C1 and C2 hydrocarbons in the marine environment. *Proc. P. Soc. Lond.* B189:305–332.

Pellizzari, E.D., Hartwell, T.D., Perritt, R.L. et al. 1986. Comparison of indoor and outdoor residential levels of volatile organic chemicals in five U.S. geographical areas. *Environ. Intern.* 12:619–623.

Phipps, G.L. and Holcombe, G.W. 1985. A method for aquatic multiple species toxicant testing: Acute toxicity of ten chemicals to five vertebrates and two invertebrates. *Environ. Pollut.* 38:141–157.

Plumb, R.H. Jr. 1987. A comparison of groundwater monitoring data from CERCLA and RCRA sites. *Groundwater Monitoring Review.* (Fall) 94–100.

Prinn, R.G. Cunnold, D., and Simmonds, P. 1992. Global average concentration and trend for hydroxyl radicals deduced from ale gauge trichloroethane (methyl chloroform) data for 1978-1990. *J. Geophys. Res. Atmos.* 97:2445–2461.

Ramanathan, V., Cicerone, R.J., Singh, H.B. et al. 1985. Trace gas trends and their potential role in climate change. J. Geophys. Res. Atmos. 95:5547–5566.

Rasmussen, R.A., Rasmussen, L.E., and Khalil, M.A.K. 1980. Concentration distribution of methyl chloride in the atmosphere. *J. Geophys. Res.* 85:7350–7356.

Richter, J.E., Peterson, S.F., and Kleiner, C.F. 1983. Acute and chronic toxicity of some chlorinated benzenes, chlorinated ethanes, and tetrachloroethylene to *Daphnia magna*. *Arch. Environ. Contam. Toxicol.* 12:679–684.

RIDOH. 1989. Rhode Island Department of Health. Letter with accompanying data from Bela T. Matyas (Chief, Office of Environmental Health Risk Assessment) to James Gibsor (ATSDR), dated March 3, 1989.

Rosenberg, R., Grahn, O., and Johansson, L. 1975. Toxic effects of aliphatic chlorinated by-products from vinyl chloride production on marine animals. *Water Res.* 9:607–612.

RTP. Regulatory Toxicology and Pharmacology. 1994. Expert Panel Report "Interpretive review of: Potential Adverse Effects of Chlorinated Organic Chemicals on Human Health and the Environment". Vol. 20, 1, August, 1994.

Sauer, T.C. Jr. 1981. Volatile organic compounds in open ocean and coastal surface waters. *Org. Geo. Chem.* 3:91–101.

Shah, J. and Singh, H.B. 1988. Distribution of volatile organic chemicals in outdoor and indoor air. *Environ. Sci. Technol.* 22:1381–1388.

Shah, J. and Heyerdahl, E. 1988. National Ambient Volatile Organic Compounds (VOCs) Data Base Update. Prepared for the U.S. EPA. Research Triangle Park, NC. Atmospheric Sciences Research Laboratory, January 1988. Cited in SRC, 1989.

Simmonds, P.G., Alyea, F.N., Cardelino, C.A. et al. 1983. The atmospheric lifetime experiment. 6. Results for carbon tetrachloride based on three years data. *J. Geophys. Res.* 88:8427–8441.

Simmonds, P.G., Cunnold, D.M., Alyea, F.N. et al. 1988. Carbon tetrachloride lifetimes and emissions determined from daily global measurements during 1978-1985. *J. Atmos. Chem.* 7:35–58.

Singh, H.B., Salas, L.J., Shigeishi, H. et al. 1979. Atmospheric halocarbons, hydrocarbons and sulfur hexafluoride: Global distributions, sources and sinks. *Science.* 203:899–903.

Singh, H.B., Salas, L.J., and Stiles, R.E. 1982. Distribution of selected gaseous organic mutagens and suspect carcinogens in ambient air. *Environ. Sci. Technol.* 16:872–880.

Singh, H.B., Salas, L.J., and Stiles, R.E. 1983. Methyl halides in and over the eastern Pacific (40 deg N - 32 deg S). *J. Geophys. Res.* 88:3684–3690.

Sittig, M. 1994. *Worldwide Limits for Toxic and Hazardous Chemicals in Air, Water and Soil.* Joyes Publishing, Park Ridge, NJ. 792 pp.

Smith, A.D., Bharath, A., Mallard, C., Orr, D., Smith, K., Sutton, J.A., Vukamanich, J., McCarty, L.S., and Ozburn, G.W. 1991. The acute and chronic toxicity of ten chlorinated organic compounds to the American flagfish *(Jordanella floridae)*. *Arch. Environ. Contam. Toxicol.* 20:94–102.

Staples, C.A., Werner, A., and T. Hoogheem. 1985. Assessment of priority pollutant concentrations in the United States using STORET database. *Environ. Toxicol. Chem.* 4:131–142.

Stephens, R.D., Ball, N.D., and Mar, D.M. 1986. A multimedia study of hazardous waste land fill gas migration. In: *Pollutants in a Multimedia Envrironment.* Plenum Press, New York, pp. 265–287.

Strachan, W.M.J., and Edwards, C.J. 1984. Organic pollutants in Lake Ontario. In: Toxic *Contaminants in the Great Lakes.* Nriagu, J.O. and Simmons, M.S. Eds. John Wiley & Sons, New York, 239 pp. 264.

Su, C. and Goldberg, G.D. 1976. Environmental concentrations and fluxes of some halocarbons. In: Marine Pollutant Transfer, H.L. Windom and A. Duce, Eds. Lexington Books, Lexington, MA, pp. 353–374.

Sullivan, D.A., Jones, A.D., and Williams, J.G. 1985. Results of the U.S. Environmental Protection Agency's Air Toxics Analysis in Philadelphia. In: Proceedings of the 78th Annual Meeting of the Air Pollution Control Association. Vol. 2:1–15.

Sunito, L.R., Shiu, W.Y., and Mackay, D. 1988. A review of the nature and properties of chemicals present in pulp mill effluents. *Chemosphere.* 17:1249–1290.

Syracuse Research Corporation (SRC). 1989. Toxicological Profile for Chloroform. SRC. U.S. ATSDR, Atlanta, GA. NTIS PB89-160360.

Thomas, R., Byrne, M.,Gilbert, D., Goyer, M., and Moss, K. 1981. Exposure and Risk Assessment for Trichloroethylene. A.D. Little, Ed., MIT Press, Cambridge, MA. Prepared for EPA, Washington, D.C. PB85-221513.

Thompson, R.S. and Carmichael, N.G. 1989. 1,1,1-Trichloroethane: medium-term toxicity to carp, daphnids and higher plants. *Ecotoxicol. Environ. Saf.* 17:172–182.

Thurston, R.V., Gilfoil, T.A., Meyn, E.L. Zajdel, R.K., Aoki, T.I., and Veith, G.D. 1985. Comparative toxicity of ten organic chemicals to ten common aquatic species. *Water Res.* 19: 1145–1155.

TRI88. 1990. Toxic Chemical Release Inventory. National Library of Medicine, National Toxicology Information Program, Bethesda, MD.

TRI90. 1992. Toxic Chemical Release Inventory. National Library of Medicine, National Toxicology Information Program, Bethesda, MD.

UDH. 1989. State of Utah Department of Health. Division Environmental Health Written Communication (July 20) Regarding Utah State Groundwater Quality Standards for 1,1,1-Trichloroethane. Salt Lake City, UT.

Veith, G.D., DeFoe, D.L., and Bergstedt, B.V. 1979. Measuring and estimating the bioconcentration factor of chemicals in fish. *J. Fish. Res. Bd. Can.* 36:1040–1048.

Veith, G.D., Call, D.J., and Brooke, T.L. 1983. Structure-toxicity relationships for fathead minnows, *Pimephales promelas:* Narcotic industrial chemicals. *Can. J. Fish. Aquat. Sci.* 40:743–748.

Vogt, W.G. and Walsh, J.J. 1985. Volatile organic compounds in gases from landfill simulators. In: Proceedings of the APCA. Annual Meeting. 78:17.

WAC. 1988. Wisconsin Administrative Code. Control of hazardous pollutants. Department of Natural Resources.

Walbridge, C.T., Flandt, J.T., Phipps, G.L., et al. 1983. Acute toxicity of ten chlorinated aliphatic hydrocarbons to the fathead minnow (*Pimephales promelas*). *Arch. Environ. Contam. Toxicol.* 12:661–666.

Wallace, L.A. 1986. Personal exposures, indoor and outdoor air concentrations and exhaled breath concentrations of selected volatile organic compounds measured for 600 residents of New Jersey, North Dakota, North Carolina and California. *Toxicol. Environ. Chem.* 12:215–236.

Wallace, L.A. 1987. The total exposure assessment methodology (TEAM) study. U.S. Environmental Protection Agency. EPA/600/56-87/002.

Wallace, L.C., Pellizzari, D., Leaderer, B., et al. 1987. Emissions of volatile organic compounds from building materials and consumer products. *Atmosph. Environ.* 21:385–393.

Westrick, J., Mellow, J.W., and Thomas, R.F. 1984. The groundwater supply survey. *J. Am. Water Works Assoc.* 76:52–59.

WHO. 1981. World Health Organization. Recommended health-based limits in occupational exposure to selected organic solvents. In: WHO Technical Report Series 664. Geneva, Switzerland.

WHO. 1984. World Health Organization. The International Drinking-water Supply and Sanitation Decade: Review of National Baseline Data (as at December 31, 1980). WHO, Geneva, Switzerland. WHO Offset Publication 85.

WHO. 1985. World Health Organization. Environmental Health Criteria 50: Trichloroethylene. WHO, Geneva, Switzerland.

Windholz, M. Ed. 1983. *The Merck Index.* 10th ed. Merck and Co., Inc., Rahway, NJ, p. 291.

Wood, J.A. and Porter, M.L. 1987. Hazardous pollutants in class II landfills. *J. Air Pollut. Control Assoc.* 37:609–615.

Young D.R., Gossett, R.W., and Baird, R.B. 1983. Wastewater inputs and marine bioaccumulation of priority pollutant organics off Southern California. In: *Water Chlorination: Environmental Impact and Health Effects,* R.F. Jolley et al., Eds., Ann Arbor Science, Ann Arbor, MI. pp. 871–884.

CHAPTER 4

Chlorinated Aromatic Hydrocarbons — Monocyclic

PROPERTIES

Monocyclic aromatic hydrocarbons consist of a basic benzene ring with six carbon atoms, six hydrogen atoms, and three double bonds. Substitution of hydrogen atoms by chlorine atoms yields chlorobenzenes, chlorotoluenes, and other derivatives. Although the carbon atom is bonded to three other atoms rather than four, the ring is not considered unsaturated due to its stability that results from a resonating electronic structure. The stability of chlorobenzenes increases with increasing chlorination of the benzene ring.

Increase in the number of chlorine atoms in the benzene molecule, increases the lipophilicity of the chlorobenzenes. The relatively higher solubility in water of mono- and dichlorobenzenes is offset by the relatively higher vapor pressure, tendering them to associate with the organic part of the aqueous system. This could explain the presence of chlorobenzenes in soil organic matter. Chlorination of the benzene ring yields 12 different compounds: 1 mono-, 3 di-, 3 tri-, 3 tetra-, 1 penta-, and 1 hexachlorobenzene. Most of them are colorless liquids with a pleasant odor. Chlorination increases the solvent power, viscosity and chemical reactivity of the benzene ring. All of the chlorobenzenes are thermally stable (Table 4.1).

ANTHROPOGENIC SOURCES

Table 4.3 presents the wide range of uses for chlorobenzenes from biocides, to dielectric fluids, to styrene rubbers. Anthropogenic sources for the release of chlorobenzenes into the environment are during production and transport, during their use in various products (Table 4.2), or from industrial wastes in industries manufacturing chlorobenzene-containing products (EPA 1985). Due to their relatively higher vapor pressure (Table 4.1), mono- and dichlorobenzenes will reach the

Table 4.1 Selected Physical and Chemical Properties of Chlorinated Benzenes

Compound	Mol. Wt.	Vapor Pressure[a] (Pa)	Water Solubility[a] (mg/L)	Log K_{ow}[a]
Chlorobenzene	112.60	1580	484	2.80
1,2-Dichlorobenzene	147.01	196	118	3.40
1,3-Dichlorobenzene	147.01	307	120	3.40
1,4-Dichlorobenzene	147.01	170.19	83	3.40
1,2,3-Trichlorobenzene	181.45	52.83	21	4.10
1,2,4-Trichlorobenzene	181.45	61	40	4.10
1,3,5-Trichlorobenzene	181.45	78.05	5.3	4.10
1,2,3,4-Tetrachlorobenzene	215.90	8.67	7.8	4.50
1,2,3,5-Tetrachlorobenzene	215.90	19.22	3.6	4.50
1,2,4,5-Tetrachlorobenzene	215.90	9.6	1.27	4.50
Pentachlorobenzene	250.3	0.9565	0.65	5.00
Hexachlorobenzene	284.8	0.2447	0.005	5.50

[a] At 25°C

Source: Mackay et al. 1992.

atmosphere than reside in the water column. It has been estimated that 30–50% of monochlorobenzene and 5–10% of dichlorobenzene produced will be released into the atmosphere and about 0.1% will enter water. About 70 to 90% of 1,4-dichlorobenzene reaches the atmosphere because of its use as an air deodorant (EPA 1985). Manufacturing processes of dichlorobenzenes have been estimated to release 2% of the annual production quantity into the environment (EPA 1980a). Although hexachlorobenzene is not produced directly in the U.S., releases of hexachlorobenzene have been reported to arise from the synthesis of several other chlorocarbon compounds where hexachlorobenzene is a by-product. Several high-temperature processes involving chlorine seem to yield hexachlorobenzene especially when proper technologies are not used (Bidleman and Foreman 1987).

PRODUCTION AND USE PATTERN

Table 4.2 U.S. Production Data (in 10^4 Metric Tons) of Selected Chlorobenzenes

Year(s)	Monochlorobenzene	1,2-Dichlorobenzene	1,4-Dichlorobenzene	All Chlorinated Benzenes
1960–1965*	25.0	1.9	3.2	—
1966–1970*	25.0	2.7	2.9	—
1971–1975*	17.2	2.7	2.9	—
1976–1980*	14.1	2.2	2.8	—
1981	12.9	2.3	3.3	—
1983	—	—	—	20,450.0

* Annual average.

Table 4.3 Industrial Uses of Chlorinated Benzenes

Compound	Uses
Monochlorobenzene	Synthesis of nitrochlorobenzenes; solvent for paints; process solvent for methylene diisocyanate, adhesives, polishes, and waxes; phenol production. Use in the production of DDT has been discontinued.
1,2-Dichlorobenzene	Synthesis of toluene diisocyanate, 3,4-dichloroaniline, dyes, herbicides; solvent for waxes, gums, resins, tars, and asphalts; in industrial odor control; as degreasing agent for hides, metals, and wools; component in drain cleaners and metal polishes
1,3-Dichlorobenzene	As a fumigant and fungicide
1,4-Dichlorobenzene	As air deodorant and insecticide; in pharmaceuticals, dyes, and moth repellant
1,2,3-Trichlorobenzene	As coolant and solvent; used in insecticides
1,2,4-Trichlorobenzene	As dye carrier, herbicide intermediate; component in dielectric fluid, synthetic transformer oils and lubricants; as heat transfer medium, solvent, insecticide and degreaser
1,2,3,4-Tetrachlorobenzene	As a component in dielectric fluid; in the production of fungicide
1,2,4,5-Tetrachlorobenzene	Synthesis of the defoliant 2,4,5-trichlorophenoxy acetic acid, 2,4,5-trichlorophenol, fungicide; used in electric insulation and packing protection
Pentachlorobenzene	Intermediate in the manufacture of specialty chemicals; as pesticide in oyster drills
Hexachlorobenzene	Component in fungicide, herbicide (DCPA), and pesticide (PCNB); additive in pyrotechnic compositions; porosity controller in the production of graphite electrodes; used as wood preservative, intermediate in organic synthesis, as fluxing agent in aluminum smelting, as peptizing agent in making nitroso- and styrene rubber for tires

Sources: Merck, 1989; Sittig, 1980.

SOURCES IN THE ENVIRONMENT

Biological organisms produce, via enzyme systems, a variety of halogenated compounds including chlorinated monocyclic compounds (Gribble 1992; Gschwend et al. 1985). Natural processes such as (i) volcanic eruptions, (ii) forest and grass fires, and (iii) atmospheric processes involving chlorides released from earth's surface. Organic-rich soils are shown to contain at least 20 different chloro-organic compounds, but in low concentrations. Chlorobenzenes, chlorinated alkanes, chlorinated alkenes, chlorophenol, and tetrachlorobenzene were shown to be present in soil (de Lijser et al. 1991; Kitagawa et al. 1991).

ENVIRONMENTAL RESIDUES AND EXPOSURE ROUTES

Air

Chlorobenzenes have relatively high vapor pressure (Table 4.1) and low water solubilities; hence, they are distributed predominantly in the atmosphere. Consequently, chlorobenzenes are detected in the atmosphere in locations far removed from point sources (Li et al. 1976). Bidleman and Foreman (1987) reported detecting chlorobenzenes in a rural area, 100 km south of the city of Stockholm, Sweden. Detection of hexachlorobenzene at a remote location on the North Pacific Ocean could only be explained by long-range atmospheric transport and/or possible release from natural agricultural sources (Atlas and Giam 1981).

Laboratory studies indicate that direct photolysis of most chlorobenzenes is not likely to occur unless a substantial shift in the wavelength of incident light below 280 nm is caused by environmental conditions. Only the absorption spectrum of 1,2,4-trichlorobenzene overlaps the region of the solar spectrum. Even then, its maximum rate of photolytic breakdown was calculated to be 0.03%/h (Bunce et al. 1989). Studies showed that photodegradation of hexachlorobenzene is extremely slow with no identifiable breakdown products and the reaction is unlikely to be accelerated by the presence of known chemical photosensitizers (Plimmer and Klingebiel 1976).

Chlorobenzenes could be transferred from atmosphere to surface water and soil by wet and dry deposition processes (U.S. EPA 1985). Average dry deposition rate of 0.011 $\mu g/m^2/day$ has been reported for 1,2,4-trichlorobenzene (Young et al. 1976).

The atmospheric concentrations of some chlorobenzenes are listed in Table 4.4.

Table 4.4 Atmospheric Concentrations of Chlorobenzenes

Compound	Location	Value (Concentration)	Unit	Year	Reference
1,2-Dichlorobenzene	U.S.	ND–12.443	$\mu g/m^3$	1979	Pellizzari et al. 1979
	U.S.	2.096 (max 46.78)	$\mu g/m^3$	1978	EPA 1983
1,3-Dichlorobenzene	U.S.	ND–33.783	$\mu g/m^3$	1979	Pellizzari et al. 1979
1,4-Dichlorobenzene	Tokyo (central and suburb)	1.5–4.2	$\mu g/m^3$	1975	Morita and Ohi, 1975
	U.S.	1.703 (max 93.560)	$\mu g/m^3$	1978	EPA 1983
	Los Angeles	0.58–30	$\mu g/m^3$	1987	Wallace, 1991
	U.S.	ND–7.00	$\mu g/m^3$	1979	Pellizzari et al. 1979
Trichlorobenzene	U.S.	ND–4.346	$\mu g/m^3$	1979	Pellizzari et al. 1979

Water

Elevated residue levels of chlorobenzenes in surface waters are primarily due to site-specific discharges (Table 4.5).

Table 4.5 Levels of Some Chlorobenzenes in Water and Effluent

Compound	Range (μg L^{-1})	Source
Monochlorobenzene	0.1–5.6	River water receiving textile industry waste (U.S.)
	90–530	Effluent from dye manufacturing plant (U.S.)
	Max. 27.0	Municipal effluents, 2 cities (U.S.)
1,2-Dichlorobenzene	<1	Potable water, 4 cities (U.S.)
	1.0	Groundwater (Miami, FL)
	15–690	Effluent from consumption plants
	<0.01–440	Municipal effluent, 6 cities (U.S.)
	0.006–0.022	Municipal effluent, 4 cities (Canada)
1,3-Dichlorobenzene	<0.5	Potable water, 4 cities (U.S.)
	0.5	Groundwater (Miami)
	0.007–0.013	Municipal effluents, 4 cities (Canada)
1,4-Dichlorobenzene	<0.5	Potable water, 4 cities (U.S.)
	0.5	Ground water (Miami)
	0.48–0.92	Municipal effluents, 4 cities (Canada)
	0.5–230	Municipal effluents, 6 cities (U.S.)
	1.5–2.9	Municipal effluents (Zurich, Switzerland)
	90–380	Effluent from dye-manufacturing plant (U.S.)
1,2,3-Trichlorobenzene	21–46	Creek receiving municipal effluent (North Carolina)
	0.002–0.003	Municipal effluents, 4 cities (Canada)
1,2,4-Trichlorobenzene	<0.01–275	Municipal wastewater, 5 cities (U.S.)
	0.005–0.018	Municipal effluent, 4 cities (Canada)
	12–500	Effluent from consumer industries
	0.007	Los Angeles River (U.S.)
1,3,5-Trichlorobenzene	<0.01–0.9	Municipal wastewater, 6 cities (U.S.)
	26.0	Industrial waste discharge
	0.006	Los Angeles River (U.S.)
	<0.0005	Municipal effluents, 4 cities (Canada)
Hexachlorobenzene	0.0015	Lake Ontario/Grand River (Canada)
	0.001–0.002	Municipal effluent

Sources: U.S. EPA 1980b; Brooks et al. 1981; Schwarzenbach et al. 1979; Jungclaus et al. 1976; Games and Hites, 1977; Oliver and Nicol, 1982.

Volatilization is the predominant route of escape for chlorobenzenes from surface waters, especially for lower chlorinated benzenes. Higher chlorinated benzenes migrate to sediments from water column. Table 4.6 lists the volatilizational half-lives of some selected chlorobenzenes.

Oliver and Nicol (1982) demonstrated that the concentrations of chlorobenzenes (Cl^2–Cl^6) in the upper reaches of the Niagara River (Canada and U.S.) ranged from <0.01 to 1.0 ngL^{-1} but increased 1 to 126 ngL^{-1} downstream of the city of Niagara Falls. This was ascribed to the discharge of effluents from chemical plants into the river. Volatilization from surface waters was considered to move 80% of the total chlorobenzenes to the atmosphere in the area; whereas higher chlorinated benzenes partitioned significantly into sediments. Biotransformation and phototransformation were comparatively minimal routes of loss of chlorobenzenes from water (Oliver 1987). Chlorobenzenes with relatively higher k_{ow} values and lower water solubilities tend to sorb strongly to organic-rich particulates either in the water column or in soil (Moore and Ramamoorthy 1984).

Table 4.6 Volatilizational Half-Lives of Some Chlorobenzenes from Surface Waters

Compound	Half-Life ($t_{1/2}$)	Reference
Chlorobenzene	9.0 hours	Versar 1979
	10–11 hours	Garrison and Hill 1972
1,2-Dichlorobenzene	<9.0 hours	Versar 1979
	8–9 hours	Garrison and Hill 1972
1,3-Dichlorobenzene	~10 hours	Versar 1979
1,4-Dichlorobenzene	<9 hours	Versar 1979
	11–12 hours	Garrison and Hill 1972
Hexachlorobenzene	~8 hours	Mackay and Leinonen 1975

Sorption of chloromonocyclics or chlorobenzenes especially can be calculated using the following regression equation developed by Mill (1980):

$$\log k_{oc} = -0.782 \log [C] - 0.27$$

where [C] = concentration in moles liter^{-1}.

This equation was developed using the data of Karickhoff et al. (1979), Kenaga and Goring (1980) and Smith et al. (1978) and rescaling to one coordinate set. Figure 2.5 presents the data (from above sources) plotted as log k_{oc} vs. log solubility (of the chemical) in moles liter^{-1} and regression line. Sorption coefficient k_{oc} (sorption constant normalized to organic content of the substrate) can be estimated from the equation with an accurate solubility value. This estimated value is reliable to power ten for most nonpolar chemicals, which is accurate for screening purposes in most cases.

Sorption of chlorobenzenes to suspended solids would likely reduce the rate of volatilization (EPA 1979). Sorption coefficients of hexachlorobenzene have been reported to range from 140 to 28,000 for bentonite clay and activated sludge, respectively (RTP 1994). Hexachlorobenzene was reported to readily sorb to organic materials, such as peat and suspended particulate matter, than to pure mineral sorbents, such as quartz and kaolinite. Oliver (1987) reported that chlorobenzenes present in bottom sediments were bioavailable for uptake by benthic organisms.

Table 4.7 presents the concentrations of three dichlorobenzenes, 1,2,4-trichlorobenzene, and hexachlorobenzene in surface waters.

Concentrations in the air of di- and trichlorobenzenes range from nondetectable (ND) to as high as 93.56 µg/m^3 (see Table 4.4); levels in drinking water are low (Table 4.8). In Ontario, Canada, most measurements have been close to or less than detection limits. Measurements for dichlorobenzenes in groundwater from California and Nebraska, range from <0.5 to 3.3 µg/L (Baird et al. 1983; Kizer 1986; CSDHS 1990).

The mechanisms of uptake and elimination of chlorobenzenes are believed to be by passive diffusion or convection by natural fluid flow, similar to passive transport across gill membranes (Connell et al. 1988; Gobas et al. 1991). Elimination of

Table 4.7 Levels of Some Chlorobenzenes in Selected Lakes and Rivers

Compound	Surface Water Sites	Concentration Mean	Range	Unit	Year	Reference
1,2-Dichloro-benzene	Lake Superior	0.141	(0.064–0.369)	ng/L	1986	Stevens and Neilson 1989
	Lake Huron	0.141	(0.068–0.238)	ng/L	1986	Stevens and Neilson 1989
	Lake Erie	0.214	(0.035–1.168)	ng/L	1986	Stevens and Neilson 1989
	Lake Ontario	0.786	(0.143–3.730)	ng/L	1986	Stevens and Neilson 1989
	Georgian Bay	0.108	(0.062–0.202)	ng/L	1986	Stevens and Neilson 1989
	Lake Huron	ND	(DL, 0.001)	µg/L	1980	Oliver and Nicol 1982
	Grand River	0.006	(ND–0.031)	µg/L	1980	Oliver and Nicol 1982
	St. Lawrence River	1.03		ng/L	1987	Germain and Langlois 1988
1,3-Dichloro-benzene	Lake Superior	0.108	(0.059–0.202)	ng/L	1986	Stevens and Neilson 1989
	Lake Huron	0.053	(0.026–0.146)	ng/L	1986	Stevens and Neilson 1989
	Georgian Bay	0.118	(0.113–0.131)	ng/L	1986	Stevens and Neilson 1989
	Lake Erie	0.238	(0.061–0.551)	ng/L	1986	Stevens and Neilson 1989
	Lake Ontario	0.317	(0.115–1.176)	ng/L	1986	Stevens and Neilson 1989
	Lake Ontario	ND	(DL, 0.001)	µg/L	1980	Oliver and Nicol 1982
	Lake Huron	ND	(DL, 0.001)	µg/L	1980	Oliver and Nicol 1982
	Grand River	0.001	(ND–0.004)	µg/L	1980	Oliver and Nicol 1982
	St. Lawrence River	0.23		ng/L	1987	Germain and Langlois 1988
1,4-Dichloro-benzene	Lake Superior	0.342	(0.220-0.530)	ng/L	1986	Stevens and Neilson 1989
	Lake Huron	0.290	(0.198–0.396)	ng/L	1986	Stevens and Neilson 1989
	Georgian Bay	0.3888	(0.236–0.631)	ng/L	1986	Stevens and Neilson 1989
	Lake Erie	1.147	(0.313–4.973)	ng/L	1986	Stevens and Neilson 1989
	Lake Ontario	1.595	(0.554–3.784)	ng/L	1986	Stevens and Neilson 1989
	Lake Ontario	0.045	(0.033–0.006)	µg/L	1980	Oliver and Nicol 1982
	Lake Huron	0.004	(0.003–0.006)	µg/L	1980	Oliver and Nicol 1982
	Grand River	0.010	(ND–0.042)	µg/L	1980	Oliver and Nicol 1982
	St. Lawrence River	2.7		ng/L	1987	Germain and Langlois 1988

Table 4.7 Levels of Some Chlorobenzenes in Selected Lakes and Rivers (continued)

Compound	Surface Water Sites	Mean	Range	Unit	Year	Reference
1,2,4-Trichlorobenzene	Lake Superior	0.231	(0.120–0.524)	ng/L	1986	Stevens and Neilson 1989
	Lake Huron	0.179	(0.124–0.256)	ng/L	1986	Stevens and Neilson 1989
	Georgian Bay	0.295	(0.202–0.431)	ng/L	1986	Stevens and Neilson 1989
	Lake Erie	0.360	(0.051–0.713)	ng/L	1986	Stevens and Neilson 1989
	Lake Ontario	0.472	(0.111–1.616)	ng/L	1986	Stevens and Neilson 1989
	Lake Ontario	0.6	(0.3–1.0)	ng/L	1980	Oliver and Nicol 1982
	Lake Huron	0.2	(0.1–0.4)	ng/L	1980	Oliver and Nicol 1982
	Grand River	2.0	(0.4–8.0)	ng/L	1980	Oliver and Nicol 1982
	St. Lawrence River	1.1	—	ng/L	1987	Germain and Langlois 1988
Hexachlorobenzene	Lake Ontario, Toronto	<2.3	—	µg/L	1987	CanTox 1987
	Lake Superior	0.026	(0.018–0.040)	ng/L	1986	Stevens and Neilson 1989
	Lake Huron	0.033	(0.018–0.073)	ng/L	1986	Stevens and Neilson 1989
	Georgian Bay	0.041	(0.032–0.054)	ng/L	1986	Stevens and Neilson 1989
	Lake Erie	0.078	(0.025–0.260)	ng/L	1986	Stevens and Neilson 1989
	Lake Ontario	0.063	(0.020–0.113)	ng/L	1986	Stevens and Neilson 1989
	Lake Ontario	0.06	(0.02–0.1)	ng/L	1980	Oliver and Nicol 1982
	Lake Huron	0.04	(0.02–0.1)	ng/L	1980	Oliver and Nicol 1982
	Grand River	0.06	(0.02–0.1)	ng/L	1980	Oliver and Nicol 1982
	Nile Delta	—	(1.00–77.08)	ng/L	1991	El-Genday et al. 1981
	Canadian Arctic	—	(17–22)	pg/L	1986	Hargrave et al. 1988

Note: ND = nondetectable; DL = detection limit.

chlorobenzenes from fish seem to vary with degree of chlorination; higher chlorinated compounds are eliminated slowly (Konemann and van Leeuwen 1980; McCarty et al. 1984; van Hoogen and Opperhuizen 1988; Smith et al. 1990). Dichlorobenzene has been reported to be totally eliminated from rainbow trout within 1 day, after a 7-day exposure period (Calamari 1982). Whole fish elimination rate constants seem to be similar for a congener group; elimination constants K_2 are 0.45

Table 4.8 Chlorobenzenes in Drinking Water

Compound	Location	Concentration Value	Units	Year Measured
1,2-Dichlorobenzene	Toronto area WTPs	0 DL, 0	µg/L µg/L	1987
1,3-Dichlorobenzene	Toronto area	0	µg/L	1987
1,4-Dichlorobenzene	Toronto area WTPs	0–0.2	µg/L	1987
1,2,4-Trichlorobenzene	Toronto area WTPs	<5–9	ng/L	1987

Note: WTP = Water treatment plant. DL = Detection limit.

Source: Ontario Ministry of the Environment (MOE) 1988; Ontario Water Treatment Plants Annual Reports, 1987; Drinking Water Surveillance Program, October 1988.

and 0.40 for 1,2,3-trichlorobenzene and 1,3,5-trichlorobenzene respectively (Konemann and van Leeuwen 1980).

Higher chlorinated benzenes, in general, are poorly metabolized and metabolism is reported to be similar among aquatic species (McCarty et al. 1984). Chlorophenols have been reported to be the major metabolites for various chlorobenzenes (Lu and Metcalfe 1975; Safe et al. 1976; Kasokat et al. 1989). Hydroxylation seems to be the metabolic pathway for chlorinated benzenes except hexachlorobenzenes which appears to metabolize by dechlorination pathway (Kasokat et al. 1989).

Species seem to differ in their rate of elimination of hexachlorobenzene. Preexposure to chlorobenzene can significantly affect the depuration rate, possibly the cause for the reported differences in elimination rate.

Similar decreasing trends in hexachlorobenzene residues in fish have been reported in Lake Erie, Lake Huron, Lake Superior, and Lake Michigan (Allan et al. 1991; Keuhl et al. 1981; Oliver and Nicol 1982). A steadily decreasing trend in hexachlorobenzene in eggs of fish-eating herring gulls have been reported, thereby confirming a similar declining trend in fish between 1973 and 1988 (McCarty et al. 1984; Allan et al. 1991).

Although vapor pressures of chlorobenzenes are relatively low compared to aliphatic monocyclic hydrocarbons, the low aqueous solubilities of chlorobenzenes render them higher volatilization rates from water (Mackay and Wolkoff 1973; EPA 1985). Field studies of the Niagara River and Lake Ontario show that volatilization from surface water passes 80% total chlorobenzenes into the atmosphere. Higher chlorinated hydrocarbons partition significantly to the sediments. Biotransformation and photolytic changes are considered relatively minimal pathways of loss (Oliver 1987). Studies on Lake Zurich in Switzerland show that the major route of loss for 1,4-dichlorobenzene through volatilization (Schwarzenbach et al. 1979).

Phototransformation studies on chlorobenzenes carried out under laboratory conditions have shown to yield chlorophenols; chlorobenzene → phenol; dichlorobenzenes → chlorophenols and trichlorobenzenes → dichlorophenols, respectively (Boule et al. 1987). Rate of the photochemical process was slower and the quantum yield decreased with an increase in the number of chlorine atoms. Photohydrolysis

CHLORINATED ORGANIC COMPOUNDS IN THE ENVIRONMENT

Table 4.9 Average Concentration (μg kg⁻¹ Wet Weight) of Chlorobenzenes in Fish

	1,2-Dichloro-benzene	1,4-Dichloro-benzene	1,3,5-Trichloro-benzene	1,2,4-Trichloro-benzene	1,2,3-Trichloro-benzene	1,2,4,5-Tetrachloro-benzene	1,2,3,4-Tetrachloro-benzene	Pentachloro-benzene	Hexachloro-benzene
Lake Superior									
Lake trout[1,2,*]	0.3	<1	<1	1.2	<1	<5	<5	<5	6.5
Lake Michigan									
Lake trout[1,*]	—	—	<1	2	<1	<5	<5	<5	6.3
Lake Ontario									
Lake trout[1,2,*]	1	4	<1	2.4	<1	2.9	4.5	9.1	117
Brown trout[1,*]	—	—	<1	2.8	1.1	2.5	5.0	11.0	115
Ashtabula River, Ohio									
Pike[1,*]	—	—	11	7.5	<1	96	20	157	1140
Fields Brook, Ohio									
Carp, Bass[2,*]	—	—	31	14	1.9	193	43	352	2210
Sucker[2,*]	—	—	75	124	20	1150	205	789	1530
Drava River, Yugoslavia nase[3,+]	1140	450	5	5	48	12	2	75	160
Gulf of Triest, Yugoslavia									
Pilchard[3,+]	220	30	15	7	130	18	3	9	26

Note: *Whole fish; +Fat; —, not measured.
Sources: [1]Kuehl et al. 1981; [2]Oliver and Nicol 1982; [3]Jan and Malnersic 1980.

Table 4.10 Concentration Factors and Half-Life of Chloro- and Bromobenzenes in Muscle Tissue of Fish

Compound	Concentration Factor	Half-Life in Tissues (Days)
1,2-Dichlorobenzene[1]	89	<1
1,3-Dichlorobenzene[1]	66	<1
1,4-Dichlorobenzene[1,2,3]	60–220	0.5
1,2,4-Trichlorobenzene[1]	182	1–3
1,2,3,5-Tetrachlorobenzene[1]	1800	2–4
Pentachlorobenzene[1]	3400	>7
Hexachlorobenzene[2,3,4]	7,800–18,500	6–7
1,2,3-Tribromobenzene[5]	1100	4–5
1,3,5-Tribromobenzene[5]	1130	4
1,2,4,5-Tetrabromobenzene[5]	1400	4–5

Sources: [1]Barrows et al. 1980; [2]Neely et al. 1974; [3]Neely 1979; [4]Veith et al. 1979; [5]Zitko 1979; [6]Woodward et al. 1981.

Table 4.11 Half-Lives ($t_{1/2}$) of Hexachlorobenzene in Aquatic Organisms

Compound	Organism	$t_{1/2}$ (day)	Reference
1,4-Dichlorobenzene	Guppy	0.7	Konemann and van Leeuwen, 1980
	Flagfish	0.7 ± 0.03	Smith et al. 1990
	Fish	0.71[a]	Smith et al. 1990
		0.47[b]	Smith et al. 1990
Dichlorobenzene (isomers not specified)	Bluegill	<1	Barrows et al. 1980
1,2,3-Trichlorobenzene	Guppy	1.5	Konemann and van Leeuwen, 1980
1,2,4-Trichlorobenzene	Flagfish	1.21 ± 0.03	Smith et al. 1990
	Guppy	0.7	Konemann and van Leeuwen, 1980
	Bluegill	>1, <3	Barrows et al. 1990
	Fish	1.22[a]	Smith et al. 1990
		1.51[b]	Smith et al. 1990
1,2,4,5-Tetrachlorobenzene	Flagfish	1.72 ± 0.15	Smith et al. 1990
	Fish	1.73[a]	Smith et al. 1990
		0.30[b]	Smith et al. 1990
1,2,3,5-Tetrachlorobenzene	Guppy	2.7	Konemann and van Leeuwen, 1980
1,2,3,4-Tetrachlorobenzene	Bluegill	>1, <4	Barrows et al. 1980

[a] Estimated from bioconcentration data
[b] Estimated from toxicity data

of chlorobenzenes with chlorine atoms in ortho- and para-positions was more difficult than chlorine in the meta-position due to delocalization of polarization of molecules (Boule et al. 1987).

Susceptibility to biotransformation seems to decrease with increase in number of chlorine atoms in the molecule (Table 4.13, Moore and Ramamoorthy 1984; EPA

Table 4.12 Decreasing Trend of Hexachlorobenzene in Selected Fish (Wholebody) from Lake Ontario

Location	Species	Year	Tissue Conc. (μg/kg)	Reference
Hamilton Harbour, Lake Ontario	Brown bullhead	1972	28	Harlow and Hodson, 1988
		1978	2	Harlow and Hodson, 1988
		1981	<0.1	Harlow and Hodson, 1988
Hamilton Harbour, Lake Ontario	White perch	1972	96	Harlow and Hodson, 1988
		1978	2	Harlow and Hodson, 1988
		1981	1.1	Harlow and Hodson, 1988
Hamilton Harbour, Lake Ontario	Carp	1978	6	Harlow and Hodson, 1988
		1981	<0.1	Harlow and Hodson, 1988
Niagara on the Lake	Young of the year and spottail shiners	1980	9(6–12)	Suns et al. 1994
		1981	5(3–7)	Suns et al. 1994
		1982	4(3–5)	Suns et al. 1994
		1983	3(ND–6)	Suns et al. 1994
Oswego River	Young of the year and spottail shiners	1984	ND	Suns et al. 1994

1985). Static culture flask degradability tests have estimated the percent degradation of the following chlorobenzenes:

Compound	Degradation
Chlorobenzene	
(5 mg/L)	89–108%
(10 mg/L)	30–100%
1,2-Dichlorobenzene	16–70%
1,3-Dichlorobenzene	16–70%
1,4-Dichlorobenzene	16–70%
1,2,4-Trichlorobenzene	16–70%
Hexachlorobenzene	~ 0%

Source: Tabak et al. 1981. *J. Water Pollut. Control Fed.*, 53:1503–1518.

Soil

Soil partitioning of chlorobenzenes correlates linearly with soil organic matter and adsorption to soil increases with an increase in chlorine content in the molecule (Paya-Perez et al. 1991). Transport of adsorbed chlorobenzene in soil is dependent on soil type and leaching capacity of the soil. Between 26 and 49% of chlorobenzene,

Table 4.13 Biotransformation of Aromatic Monocyclic Compounds

Compound	Organism	Mechanism and Products
Benzene[1,2,3]	Soil bacteria	Oxidative degradation yielding catechols
	Oil-degrading bacteria (in presence of dodecane or naphthalene and at low concentration)	Utilization as carbon source
	Mammals	Oxidation to arene oxides and transidihydrodiols
Chlorobenzene[4,5]	*Pseudomonas putida* (in presence of another aromatic hydrocarbon source)	Oxidative degradation yielding 3-chlorocatechol
	Mixed cultures of aerobes	Complete volatilization
1,2-Dichlorobenzene[6]	Biological sewage treatment	Nondegradable
1,2-Dichlorobenzene[7]	Industrial and municipal wastewater treatment plants	Transformed in 7 days
1,4-Trichlorobenzene[8]	Mixed cultures	Nondegradable
Hexachlorobenzene[9]	Atlantic salmon	Direct excretion
	Algae (*Oedogonium cardiacum*)	85% of HCB intact
	Snail (*Physa sp.*)	91% intact
	Daphnia magna	87% intact
	Mosquito larva	58% intact
	Mosquitofish	27% intact
	Snail	84% intact
	Cladoceran	67% intact
	Mosquito larva	65% intact
	Fish	64% intact (pentachlorophenol found in alga, mosquito larva, and water)
Hexachlorobenzene[10]	Microbes in natural water + sediment	100% intact
1,2,4-Trichlorobenzene[11] 1,2,4,5-Tetrachlorobenzene	Pseudonomas strains	Deoxygenation yielding of 3,4,6-trichlorophenols

Sources: [1]Gibson 1976; [2]Claus and Walker 1964; [3]Walker and Colwell 1975; [4]Gibson et al. 1968; [5]Garrison and Hill 1972; [6]Thom and Agg 1975; [7]Davis et al. 1981; [8]Ware and West 1977; [9]Zitko 1977; [10]Lee and Ryan 1979; [11]Sander et al. 1991

1,4,-dichlorobenzene and 1,2,4-trichlorobenzene percolated through sandy soil; the rest was accounted for through volatilization, biotransformation and sorption. Mobility of chlorobenzenes in soil decreased with a decrease in their aqueous solubility. Retardation factor (interstitial water velocity/velocity of chlorobenzene) was calculated to be 1.7, 3.4, and 9.4 for chlorobenzene, 1,4-dichlorobenzene, and 1,2,4-trichlorobenzene, respectively (Wilson et al. 1981).

Biotransformation of chlorobenzenes in soil is very slow (Wilson et al. 1981; McNamara et al. 1981) after initial higher rates that yielded corresponding chlorophenols, half-life of 1460 days for hexachlorobenzene in soils has been reported (OHMTADS, 1986).

Hexachlorobenzene and Other Chlorobenzenes

Hexachlorobenzene is not produced in many countries including the U.S. and Canada. Most of the hexachlorobenzene in those countries is produced as a byproduct in the manufacturing of chlorinated solvents and in metallurgical processes, or as an impurity in several pesticides that are currently in use. Waste sites and incinerations where the best available technologies are not used are additional sources of hexachlorobenzenes in the environment. Since total production of chlorinated solvents has been declining since 1970, hexachlorobenzene from that source will continue to decline. Impure pesticides, currently in use and from past use, appear to be the continuing source of hexachlorobenzene exposure to the environment and human health. These pesticides include Picloram, PCNB (pentachloronitrobenzene), chlorothalonil, Dacthal and PCP (pentachlorophenol). Picloram (4-amino-3,5,6-trichloropicolinic acid) is a herbicide used in agriculture and silviculture to control broad-leaves weeds and conifers in grasses. PCNB, chlorothalonil and Dacthal are widely used in agriculture and also in home gardens, lawn care, and other applications around residences and urban areas (*Farm Chemicals Handbook* 1993). PCNB is a fungicide and seed-dressing agent, widely used on turf and ornamental plants. Chlorothalonil (Bravo®) is also a fungicide used as a biocide in paints and wood preservatives. DCPA (dimethyl tetrachloroterephthalate), sold as Dacthal, is a preemergent herbicide widely used on lawns and turf grass. Pentachlorophenol is a fungicide, herbicide, and disinfectant used as an anti-fouling agent ingredient in paint and wood preservatives (ATSDR 1994; EPA 1986). The levels of hexachlorobenzene in pesticides were as much as 10% in the early 1970s, but current levels are much lower in all five pesticides mentioned above. Registration standards specifying maximum allowable concentration of hexachlorobenzene for pesticides have been issued ranging from 0.0075% in pentachlorophenol; 0.3% in Dacthal; 0.05% in chlorothalonil; 0.1% in PCNB; and 0.02% in Picloram.

Hexachlorobenzene is very persistent in the environment and bioaccumulates in animals and humans. Dietary intake is the major route of exposure to humans. Hexachlorobenzene residues were detected in 76% of samples analyzed as part of the National Human Adipose Tissue Survey (FY82) (EPA 1986a). Calculated yearly uptake was 68 µg, 22 µg and 5 µg for adults, toddlers, and infants, respectively (EPA 1986a), most likely from dietary consumption. Exposure via inhalation or through drinking water is considered to be low.

The range of bioconcentration factors (BCF) for chlorobenzenes in whole fish and crustaceans are given in Table 4.14; the BCF in lake trout are given in Table 4.15. The influence of sediment on bioaccumulation of chlorobenzenes is shown in the following:

- Uptake studies revealed that suspended sediment had no effect on the uptake of 1,2,3-trichlorobenzene by guppies, but significantly increased the uptake of hexachlorobenzene compared to fish from a control experiment. Ingestion of suspended sediment and subsequent release of hexachlorobenzene may be the likely explanation. (Schrap and Opperhuizen 1989).

Table 4.14 Range of Bioconcentration Factors (BCF) in Whole Fish and Crustoceans

Compound	Organism	Range in BCF	Reference
1,2-Dichlorobenzene	Fish	89–560	Hesse et al. 1991
		89–550	McCarty et al. 1984
1,3-Dichlorobenzene	Fish	66–740	Hesse et al. 1991
		66–741	McCarty et al. 1984
1,4-Dichlorobenzene	Fish	15–720	Hesse et al. 1991
		60-724	McCarty et al. 1984
1,2,3-Trichlorobenzene	Fish	700–2,600	Hesse et al. 1991
		708-3,236	McCarty et al. 1984
1,2,4-Trichlorobenzene	Fish	183–3,200	Hesse et al. 1991
		182-2,344	McCarty et al. 1984
1,3,5-Trichlorobenzene	Fish	760–4,100	Hesse et al. 1991
	Fish	760-4,073	McCarty et al. 1984
Hexachlorobenzene	Fish	7,880–22,000	Hesse et al. 1991
		16–44,668	McCarty et al. 1984
Hexachlorobenzene	Crustaceans	13,200–75,000	Hesse et al. 1991

Table 4.15 Field BCFs in Wholebody Lake Trout (*Salvelinus namaycush*)

Compound	BCF
1,2-Dichlorobenzene	200
1,4-Dichlorobenzene	45–250
1,2,3-Trichlorobenzene	5,900
1,2,4-Trichlorobenzene	5,000–5,900
1,3,5-Trichlorobenzene	39,800
Hexachlorobenzene	44,700–1,500,000

Note: Average wholebody lipid concentration is 21%; Trout were from Lake Ontario, Lake Huron, and the Niagara River.

Sources: McCarty et al. 1984; Oliver and Nicol, 1982; Environment Canada/Ontario Ministry of Environment, 1981.

- Using a closed-static system, Schuytema et al. 1988 showed that bioaccumulation of hexachlorobenzene by an amphipod (*H. aztec*) was unaffected by the presence of hexachlorobenzene-spiked sediments, compared to control without the spiked sediment. The cause could be the use of aging adults species in this study and their susceptibility to mortality in the test.
- Fathead minnows and worms showed increased bioaccumulation in the presence of spiked sediment. Since the water column concentration also increased, it was difficult to distinguish between direct uptake from water or through ingestion of spiked sediment. But it was suggested that spiked sediment served as a more efficient sink for hexachlorobenzene than organisms based on results of higher tissue concentration in aquarium studies in the absence of sediment.

Studies conducted to evaluate the role of high and low organic sediments on the uptake of chlorobenzenes on larval stages of the midge (*Chironomus decorus*) showed that the uptake was mediated by chemical uptake from the interstitial water (Knezovich and Harrison 1988). Interstitial BCFs were 30 for 1,2-dichlorobenzene, 200 for 1,2,4-trichlorobenzene, and 800 for hexachlorobenzene. Similarly, a study using chlorobenzene containing Lake Ontario sediments and exposing to worms, *Tubifex* and *Limnodrilus hoffmeisteri*, BCFs calculated on biota/sediment basis was very low (<1) for all chlorobenzenes except hexachlorobenzenes which was 3:1. But, when calculated on biota/interstitial water basis, the BCF for hexachlorobenzene was 24,000.

Colonies of the bald eagle (*Haliaetus leucocephalus*) near Lake Superior had higher hexachlorobenzene residues (Table 4.16) than the colonies at inland sites (not detected). The higher residues were explained by the difference in consumption of herring gull (*Larus argentatus*) between the two colonies (lakeside colony bald eagle consumption ±22% vs. 6% by inland colony). Herring gulls were reported to have 0.02 to 0.06 ppm (w/w) of hexachlorobenzene (Kozie and Anderson 1991). Both colonies of bald eagles, however, consumed large proportions of fish in their diets (50 to 84%). In contrast, the hexachlorobenzene residues in aquatic and terrestrial birds sampled in the late 1970s were much higher (Table 4.17).

Table 4.16 Hexachlorobenzene Residues in Piscivorous Wildlife from the Great Lakes

Location	Species	Tissue	Level/Range	Reference
Lake Superior	Bald eagle (*Haliaetus leucocephalus*)	Brain and breast muscle	0.02–0.1 (mg/kg, wet/wt)	Kozie and Anderson 1991
Lake Huron	Tern (*Sterna hirundo*)	Eggs	ND	Weseloh et al. 1989
Lake Superior	Tern (*Sterna hirundo*)	Eggs	0.06 (mg/kg, wet/wt)	Weseloh et al. 1989
Michigan area	Heron species	Carcasses and brains	0.05–1.3 (mg/kg)	Ohlendorf et al. 1981
Great Lakes region	Herring gull and cormorant	Eggs	Trichlorobenzene (ND–0.027 mg/kg)	Hallett et al. 1982
Maritimes	Herring gull and cormorant	Eggs	1,2,4-Trichlorobenzene (ND–0.037 mg/kg)	Matheson et al. 1980

Table 4.17 Hexachlorobenzene Residues in Birds During Late 1970s

Location	Species	Tissue	Tissue Levels (μg/kg)
Great Lakes	Aquatic birds	Eggs	90–3,900
Unspecified	Aquatic birds	—	60–240 wet/wt 1,350–14,700 dry/wt
Maritimes	Aquatic birds	Eggs	0.7–62
Unspecified	Terrestrial birds	—	ND–52,000

Source: McCarty et al. 1984.

In Lakes Superior, Michigan, and Huron, their large sizes, considerable depths, and significant retention times immobilize chlorobenzenes in bottom sediments, as well as reduce their levels in surface waters. In the Upper Great Lakes the great majority of the on-going mass balance inputs seem to relate to atmospheric deposition (ATSDR 1994). Detailed studies in Lake Erie indicate that over half of hexachlorobenzene (270 kg/year) comes from wet or dry atmospheric deposition processes (Kelly et al. 1991). A significant portion (109.6 kg/year) comes from river pathways, such the Detroit River, from surface run-off, and from suspended contaminated sediments through the connecting waters.

The levels of hexachlorobenzene in Lake Ontario sediment is 10^6 times higher than its level in the water of Lake Ontario. Whereas, 1,4-dichlorobenzene is found in Lake Ontario sediment in concentrations similar to those of hexachlorobenzene, but the water column concentration of 1,4,-dichlorobenzene is 500 times higher than the hexachlorobenzene concentration. Hexachlorobenzene residues in fish is 25 times higher than concentrations of 1,4-dichlorobenzene (Oliver and Nicol 1982).

Due to its high sorption characteristic, hexachlorobenzene would not be expected to leach from surface soils into groundwater. However, at waste disposal sites where bioremediation techniques are proposed, the potential exists for accelerated leaching of hexachlorobenzene and the repartition of it to lipid materials in microbial cell structures. This could lead to "facilitated transport" of hydrophobic chloroorganics by biosorption onto bacteria and by moving them into aquifers along with bioremedial microbial cultures (Lindqvist and Enfield 1992). Except at high priority sites, this pathway of groundwater contamination would seem to be remote. Since hexachlorobenzene has been shown to biodegrade under anaerobic conditions to tri- and dichlorobenzenes (Fathepure et al. 1988), hexachlorobenzene in soil would biodegrade faster than leaching and migration in a timeframe of years.

Dietary Exposure of Hexachlorobenzene

Pesticide residue data from the FDA Adult Total Diet Study of 1980–1982 was evaluated by Gartell et al. (1986). Hexachlorobenzene was detected in a wide variety of domestic foods: highest mean concentrations of hexachlorobenzene were detected in oils and fats (0.9 ppb); meat, poultry, and fish (0.2 ppb). Recent FDA Total Diet Study of 1985–1991 was evaluated by Yess et al. (1993) to be: combination meat dinners — pork (0.4 ppb), beef (0.3 ppb), chicken/turkey (0.3 ppb), chicken/turkey/vegetable (0.3 ppb), beef and vegetable (0.1 ppb); vegetables and fruits — pears (1 ppb), apples (0.4 ppb) and carrots (0.2 ppb); milk products — canned evaporated milk (0.5 ppb), whole milk (0.2 ppb), and low-fat (2%) milk (0.1 ppb); and peanut butter (5 ppb).

The frequency of detection of hexachlorobenzene in FDA Total Diet Study decreased over the years: 1982–84, 9%; 1989, 5%; 1990, 4%; and 1991, 2% (Gunderson 1988; FDA 1990; FDA 1991; FDA 1992). Hexachlorobenzene intakes (μg/kg · bw/day) by humans also decreased over the years.

Human body burden for hexachlorobenzene of 0.7 mg was estimated to be derived primarily from intake of fatty foods (0.2 μg/day). Inhalation pathway contributes 100 times less (0.002 μg/day) and consumption of drinking water contributes negligible amounts of hexachlorobenzene (0.06 μg/year) (Burton and Bennett, 1987).

Year	6–11 Month-Old Infants	14–16 Year-Old Males	60–65 Year-Old Females
1982–84	0.0015	0.0020	0.0011
1989	0.0007	0.0009	0.0005
1990	0.0004	0.0005	0.0002
1991	0.0003	0.0004	0.0002

Sources: FDA, 1990, 1991, 1992

TOXICITY PROFILE OF CHLOROBENZENE

Toxicity of aromatic hydrocarbons to aquatic plants is extremely variable, depending on the species, compound, and environmental conditions (Table 4.18).

Unsubstituted monocyclic compounds such as benzene and toluene generally have low toxicity. Chlorobenzene is only slightly toxic but increasing substitution leads to an increase of toxicity to algae. LC_{50} in 96 h of pentachlorobenzene is about 40 to 70 times lower than that of chlorobenzene. The acute toxicity of chlorobenzenes to invertebrates generally parallels their toxicity to aquatic plants (Table 4.19). 1,2,4,5-Tetrachlorobenzene and pentachlorobenzene seem to be very toxic to fish. Chlorobenzenes cause acute lethality through tissue narcosis; hence, they can be considered to belong to a class of compounds termed as *narcotics*.

Chlorobenzenes have been reported to cause DNA reduction in the diatom *Cyclotella meneghiniana*; EC_{50} for dichlorobenzenes, 23.3. to 51.9 ppm; trichlorobenzenes, 0.59 to 6.42 ppm; hexachlorobenzene, 0.002 ppm (Figueroa and Simmons 1991). Saltwater organisms appear to be less sensitive than freshwater species to chlorobenzenes; for example, 96 h, LC_{50} for sheepshead minnow (*Cyprinodon variegatus*) is 21 ppm compared to LC_{50} range of 1.95–6.3 ppm for freshwater fish (Hesse et al. 1991). Another example is 48 h, EC_{50} for freshwater midge *(Tanytarus dissimilis)* from static tests was reported to be 11,760 and 13,000 µg/L for 1,2,-Dichlorobenzene and 1,4-dichlorobenzene, respectively (EPA 1978). Whereas crayfish (*Procambarus clarki*) were not affected by 20-d exposure of hexachlorobenzene at 28–36 µg/L level (Laska et al. 1976).

Several studies have been reported in the literature on fish embryo toxicity of chlorobenzenes (Tables 4.20 to 4.22).

Chlorobenzenes have been reported to affect reproduction in *Daphnia magna* and the effect seems to increase with an increase in chlorine content of the chlorobenzene. Reported NOEC (fertility) values for 1,4-dichlorobenzene seem to range between 0.22 mg/L (Calamari 1982) and 0.3 mg/L (Jori et al. 1982). From a study of 28-d exposure using a static daily replacement exposure method, Calamari et al. (1983) reported 14-d EC_{50} (fertility) values for chlorobenzenes to *D. magna*; 1,2-dichlorobenzene, 0.55 mg/L; 1,4-dichlorobenzene, 0.93 mg/L; 1,2,3-trichlorobenzene, 0.2 mg/L; 1,2,4-trichlorobenzene, 0.45 mg/L and hexachlorobenzene, 0.16 mg/L. Hexachlorobenzene and 1,2,3-trichlorobenzene seem to be the most toxic congeners. NOEC (reproduction) value for the amphipod *H. azteca* to hexachlorobenzene was reported to be 5 µg/L on 30-d exposure (Nebeker et al. 1989). Data on fish reproductive effect by chlorobenzenes were not available in literature.

Table 4.18 Acute Toxicity (96-h LC_{50}, mg L^{-1}) of Monocyclic Aromatic Hydrocarbons to Algae, Invertebrates, and Fish

Compound	Algae Selenastrum capricornutum[1,2]	Algae Skeletonema costatum[1,2]	Invertebrates Cladoceran Daphnia magna[1,3] (48-h LC_{50})	Invertebrates Shrimp Mysidopsis bahia[1,2]	Fish Bluegill[1,4]	Fish Sheepshead minnow[5]
Benzene	525*	—	200	39+	22.5	33**
Toluene	>433	>433	310	56	13	280–480
Chlorobenzene	228	342	86	16.4	16	10
1,2-Dichlorobenzene	95	44	2.4	2.0	5.6	9.7
1,3-Dichlorobenzene	64	52	28	2.9	5.0	7.8
1,4-Dichlorobenzene	97	57	11	2.0	4.3	7.4
1,2,4-Trichlorobenzene	36	8.9	—	0.5	3.4	21
1,2,3,5-Tetrachlorobenzene	17	0.7–7.1	9.7	0.3	6.4	37
1,2,4,5-Tetrachlorobenzene	50	7.3	—	1.5	1.6	0.8
Pentachlorobenzene	6.7	2.1	5.3	0.2	0.25	0.8

Note: *Chlorella vulgaris;* **Fathead minnow; +Grass shrimp *Palaemonetes pugio*; — not measured.

Sources: [1]U.S. Environmental Protection Agency 1980a,b; [2]Kauss and Hutchinson 1975; [3]LeBlanc 1980; [4]Buccafusco et al. 1981; [5]Heitmuller et al. 1981.

Table 4.19 Ranges of LC_{50} and EC_{50} Values of Chlorobenzenes for Various Aquatic Organisms

Compound	Organism	LC_{50} Range (ppm)	EC_{50} Range (ppm)	Reference
1,2,-Dichlorobenzene	Algae	—	2.2–2.0	Calamari et al. 1983; Wong et al. 1984; Canton 1985
	Daphnia magna	2.20	0.74–2.35	Calamari et al. 1983; Canton 1985; Abernathy et al. 1986; Kuhn 1989
	Fish	5.6–10	—	Calamari et al. 1983; Buccafusco et al. 1981
1,3-Dichlorobenzene	Algae	—	22.9–31	Wong et al. 1984; Canton 1985
	Daphnia magna	6.8–7.4	1.2–7	Canton 1985; Kuhn 1989; Richter et al. 1981
	Fish	5–7.8	—	Buccafusco et al. 1981; Veith et al. 1983; Carlson and Kosian 1987
1,4-Dichlorobenzene	Algae	—	5.2–31	Calamari et al. 1983; Wong et al. 1984; Canton 1985
	Daphnia magna	2.2	0.7–3.2	Calamari et al. 1983; Canton 1985; Kuhn 1989
	Fish	1.18–4.3	—	Calamari et al. 1983; Buccafusco et al. 1981; Veith et al. 1983; Carlson and Kosian 1987; Curtis et al. 1979; Curtis and Ward 1981; Hayes et al. 1985
1,2,3-Trichlorobenzene	Algae	—	2.2–5.99	Calamari et al. 1983; Wong et al. 1984
	Daphnia magna	—	0.35–2.72	Calamari et al. 1983; Abernathy et al. 1986; Bobra et al. 1983
	Fish	0.71–3.1	—	Calamari et al. 1983
1,2,4-Trichlorobenzene	Algae	—	3.9–5.99	Calamari et al. 1983; Wong et al. 1984
	Daphnia magna	2.1	1.2	Calamari et al. 1983; Richter et al. 1981
	Fish	1.95–6.30	—	Calamari et al. 1983; Buccafusco et al. 1981; Veith et al. 1983; Carlson and Kosian 1987
1,3,5-Trichlorobenzene	Algae	—	9.07	Wong et al. 1984
Hexachlorobenzene	Algae	—	<0.03–0.0	Calamari et al. 1983
	Daphnia magna	—	0.16	Calamari et al. 1983
	Fish	>0.03	—	Calamari et al. 1983

Source: Hesse et al. 1991.

Table 4.20 Embryo Toxicity of Chlorobenzenes in Fish

Compound	Species	Exposure	Type of Lethality	Concentration	Reference
1,4-Dichlorobenzene	Rainbow trout	14 d	ILL	0.8 mg/L	Calamari 1982
	Fathead minnow	—	Decrease in hatching	8.7 mg/L	Carlson and Kosian 1987
	Surviving larvae	32 d	Total mortality	2–8.7 mg/L	Carlson and Kosian 1987
			LOEL	1.0 mg/L	Carlson and Kosian 1987
			NOEL	0.57 mg/L	Carlson and Kosian 1987
1,3-Dichlorobenzene	Flagfish alevins	28 d post-hatch (7 d)	LC_o	0.349 mg/L	Smith et al. 1991
	Post hatch larvae	32 d	LOEL	3.9 mg/L	Carlson and Kosian 1987
	Fathead minnow	32 d	NOEL	2.4 mg/L	Carlson and Kosian 1987
	Flagfish	10 d	LOEL	0.263 mg/L	Smith et al. 1991
	Larvae		Egg hatchability-NOEL	0.835 mg/L	Smith et al. 1991

Note: ILL, incipient lethal level; LOEL, lowest observed adverse effect level.

Dose-dependant mortality was observed in the offsprings of mink and ferret (RTP 1994). A comparative study of offspring mortality to hexachlorobenzene exposure involving rat, mink, and ferret reported the following: 45.4% preweaning mortality for rat pups nursed by females exposed to 100 ppm of hexachlorobenzene in the diet (Kitchin et al. 1982); Bleavins et al. 1984); 64.3% preweaning mortality for ferret kits nursed by females exposed to 25 ppm of hexachlorobenzene in diets; and 44.1% preweaning mortality for mink kits nursed by females exposed to 1 ppm of hexachlorobenzene in diet (Bleavins et al. 1984). It seems that mink were most sensitive to toxic effects from hexachlorobenzene.

Fetal and postnatal toxicity of hexachlorobenzene was reported in mink administered with 0, 1, and 5 ppm levels in diets. Reported mortality rate of the weanlings was 8.2% (0 ppm), 44.1% (1 ppm), and 77.4% (5 ppm) (Rush et al. 1983). NOEL for one species need not necessarily be protective for other species. For example, 20 ppm of hexachlorobenzene is NOEC reproduction for rat in a natural diet (Bleavins et al. 1984), whereas dietary concentration of 1 ppm of hexachlorobenzene had significant adverse reproductive effects in mink (Bleavins et al. 1984). Dietary exposure of 5 ppm of hexachlorobenzene was reported to significantly increase hepatic cytochrome P450 and ethoxyresorufin-o-deethylase (EROD) levels by 2.0- and 1.5-fold, respectively (Rush et al. 1983). Exposure to hexachlorobenzene (intrapretoneal route) did not induce microsomal MFO (mixed function oxidase) or EROD levels in fish within 13 days following treatment, whereas PCB 77 induced an increase in MFO and EROD levels (Tyle et al. 1991). Hexachlorobenzene levels in herring gull eggs from Lakes Ontario, Erie, Huron, and Superior declined rapidly from 0.6 ppm in 1974 to 0.1 ppm

Table 4.21 Chlorobenzenes Reported to Affect Algal Growth and Cause Growth Reductions in Fish (Table 4.22)

Compound	Species	Exposure	Effect	Concentration mg/L	Reference
1,2-Dichlorobenzene	Freshwater algae (*Selenastrum capricornutum*)	96 h	EC_{50} (Inhibition of growth for 50% of population)	2.2	Smith et al. 1991
1,4-Dichlorobenzene	Freshwater algae (*S. capricornutum*)	96 h	EC_{50} (Inhibition of growth for 50% of population)	1.6	Smith et al. 1991
1,2,3-Trichlorobenzene	Freshwater algae (*S. capricornutum*)	96 h	EC_{50} (Inhibition of growth for 50% of population)	0.9	Smith et al. 1991
1,2,4-Trichlorobenzene	Freshwater algae (*S. capricornutum*)	96 h	EC_{50} (Inhibition of growth for 50% of population)	1.4	Smith et al. 1991
Hexachlorobenzene	Freshwater algae (*S. capricornutum*)	96 h	EC_{50} (Inhibition of growth for 50% of population)	<0.03	Smith et al. 1991
1,2,3-Trichlorobenzene	Freshwater algae (*S. capricornutum*)	96 h	NOEL	0.22	Smith et al. 1991
1,2,4-Trichlorobenzene	Freshwater algae (*S. subspicatus*)	96 h	EC_{50}	8.4	Geyer et al. 1985

Table 4.22 Growth Reduction in Fish and Invertebrates by Chlorobenzenes

Compound	Species	Exposure	Effect	Concentration	Reference
1,4-Dichlorobenzene	Flag fish	28 d	No growth reduction NOEC growth reduction	0.349 mg/L	Smith et al. 1991
	Fathead minnow	32 d	Reduction in growth	1.0 mg/L	Carlson and Kosian 1987
	Fathead minnow		NOEC growth reduction	0.570 mg/L	Carlson and Kosian 1987
1,3-Dichlorobenzene	Fathead minnow		LOEC growth reduction	2.4 mg/L	Carlson and Kosian 1987
	Fathead minnow		NOEC growth reduction	1.0 mg/L	Carlson and Kosian 1987
Hexachlorobenzene	H. azteca	Flow-through	NOEC growth reduction	0.3–0.38 µg/L	Nebeker et al. 1989

in 1989. Similarly, herring gull eggs from the Niagara River showed a steady decline from ~0.17 ppm in 1979 to <0.05 ppm in 1989 (RTP 1994); these decreases are implicated in significant recoveries in reproductive success and increase in the bird populations (Allan et al. 1991).

REGULATIONS

Monochlorobenzene

International Limits in Workplace, Air, mg/m^3			
Country		TWA	STEL
Argentina		350	350
Australia		345	—
California		350	—
Switzerland		230	460
Commonwealth Independent States (Former USSR)		50	100
Finland		230	345
France		350	—
United Kingdom		230	—
Japan		350	—

Limits in Soil (Permissible Cleanup Levels) mg/kg		
State	Residential	Industrial
Arizona	2,300	—
Canada	1	10
Massachusetts	100	5,000
Michigan	1,500	2,800
New Jersey	37 (surface) 1 (subsurface)	690 —
Oregon	5,000	40,000
Texas	256	—

Source: Sittig 1994.

Dichlorobenzene (para)

International Workplace Limits, Air, mg/m³			
Country	TWA	STEL	
Australia	451	661	Sittig, 1994
Belgium	451	661	
California	450	675	
Switzerland	450	900	
France	450	675	
United Kingdom	150	300	
Japan	300	—	

Source: Sittig 1994.

Dichlorobenzene

Limits in Soil (Permissible Cleanup Levels), mg/kg		
State	Residential	Industrial
Canada	1	10
Massachusetts	37	230
Michigan	16	110
New Jersey	280 (surface)	1,200
	100 (subsurface)	—
Texas	86.4	138

Hexachlorobenzene

Limits in Soil (Permissible Cleanup Levels), mg/kg		
Location	Residential	Industrial
Massachusetts	0.7	3
Michigan	0.8	9.4
New Jersey	0.42 (surface)	2
	50 (subsurface)	—
Oregon	0.4	4

International Workplace Limits, Air, mg/m³		
Czechoslovakia	1 (TWA)	2 (STEL)

REFERENCES

Abernathy, D.J. et al. 1986. Acute toxicity of hydrocarbons and chlorinated hydrocarbons to two planktonic crustaceans: The key role of organism–water partitioning. *Aquat. Toxicol.* 8:163–174.

Allan, R.J., Ball, A.J., Cairns, V.W., Fox, G.A., Gilman. A.P., Peakall, D.B., Piekarz, D.A.,Van Oostdam, J.C., Villeneuve, D.C., and Williams, D.T. 1991. Toxic Chemicals in the Great Lakes and Associated Effects. Contaminant Levels and Trends, Vol. I. Environment Canada, Department of Fisheries and Oceans, Health and Welfare Canada, Ottawa, Canada.

Atlas, E. and Giam, C.S. 1981. Global transport of organic pollutants: Ambient concentration in the remote marine atmosphere. *Science.* 211:163–165.

ATSDR. 1994. Agency for Toxic Substances Disease Registry. Toxicological profile for hexachlorobenzene update (draft). August 1994. U.S. Department of Health and Human Services, Atlanta, GA.

Baird, R., Gute, J., Jacks, C., Jenkins, R., Nieisess, L., Scheybeler, B., van Sluis, R., and Yanko, W. 1983. Health effects of water reuse: a combination of toxicological and chemical methods for assessment. In: *Water Chlorination: Environmental Impact and Health Effects,* R. Jolley, W.A. Brungs, J.A. Cotruvo, R.B. Cumming, J.S. Mattice, and V.A. Jacobs, Eds., Vol. 3, CRC/Lewis Publishers, Boca Raton, FL, pp 925–935.

Barrows, M.E., Petrocelli, S.R., Macek, R.J., and Carroll, J.J. 1980. Bioconcentration and elimination of selected water pollutants by bluegill sunfish (*Lepomis machrochirus*). In: *Dynamics, Exposure and Hazard Assessment of Toxic Chemicals,* R. Haque, Ed., Ann Arbor Science Publishers, Ann Arbor, MI, pp. 379–392.

Bidleman, T.F. and Foreman, W.T. 1987. Vapor-particle partitioning of semivolatile organic compounds. In: *Sources and Fates of Aquatic Pollutants,* R.A. Hites and S.J. Eisenrich, Eds. American Chemical Society, Washington, D.C. pp. 27–56.

Bleavins, M.S., Aulerich, R.J., and Ringer, R.K. 1984. Effects of chlorine dietary hexachlorobenzene exposure on the reproductive performance and survivability of mink and European ferrets. *Arch. Environ. Contam. Toxicol.* 13:357–365.

Bobra, A.M., Shiu, W.Y., and Mackay, D. 1983. A predictive correlation for the acute toxicity of hydrocarbons and chlorinated hydrocarbons to the water flea (*Daphnia magna*). *Chemosphere.* 12:1121–1129.

Boule, P., Guyon, C., Tissot, A., and Lemaire, J. 1987. Specific phototransformation of xenobiotic compounds: chlorobenzenes and halophenols. In: *Photochemistry of Environmental Aquatic Systems,* R.G. Zika and W.J. Cooper, Eds. American Chemical Society, Washington, D.C. pp. 10–26.

Brooks, J.M., Wisenburg, D.A., Burke, R.A., Jr. and Kennicut, M.C. 1981. Gaseous and volatile hydrocarbon inputs from a subsurface soil oil spill in the Gulf of Mexico. *Environ. Sci. Technol.* 15:951–959.

Buccafusco, R.J., Ells, S.J., and LeBlanc, G.A. 1981. Acute toxicity of priority pollutants to bluegill (*Lepomis machrochirus*). *Bull. Environ. Contam. Toxicol.* 26:446–452.

Bunce, N.J., Landers, J.P., Langshaw, J., and Nakai, J.S. 1989. An assessment of the importance of direct solar degradation of some simple chlorinated benzenes and biphenyls in the vapor phase. *Environ. Sci. Technol.* 23:213–218.

Burton, M.A. and Bennett, B.G. 1987. Exposure of man to environmental hexachlorobenzene (HCB)- an exposure commitment assessment. *Sci. Total Environ.* 66:137–146.

Calamari, D. 1982. Evaluating the hazard of organic substances on aquatic life: The paradichlorobenzene example. *Ecotoxicol. Environ. Safety.* 6:369–378.

Calamari, D. et al. 1983. Toxicity of selected chlorobenzenes to aquatic organisms. *Chemosphere.* 12: 253–262.

Canton, J.H. et al. 1985. Toxicity, biodegradability and accumulation of a number of Cl/N-containing compounds for classification and establishing water quality criteria. *Reg. Toxicol. Pharmacol.* 5:123–131.

CanTox. 1987. Comparative Evaluation of Drinking Water Sources (unpublished). CanTox, Oakville, ON.

Carlson, A.R. and Kosian, P.A. 1987. Toxicity of chlorinated benzenes to fathead minnows (*Pimephales promelas*). *Arch. Environ. Contam. Toxicol.* 16:129–135.

Claus, D. and Walker, N. 1964. The decomposition of toluene by soil bacteria. *J. General Microbiol.* 36:107–122.

Connell, D.W., Bowman, M., and Hawker, D.W. 1988. Bioconcentration of chlorinated hydrocarbons from sediments by oligochaetes. *Ecotoxicol. Environ. Safety.* 16:293–302.

CSDHS. California State Department of Health Sciences. 1990. Organic Chemical Contamination of Small Public Water Systems in California. Small Water System AB 1803. Final Status Report. Office of Drinking Water.

Curtis, M.W., Copeland, T.L., and Ward, C.H. 1979. Acute toxicity of 12 industrial chemicals to freshwater and saltwater organisms. *Water Res.* 13:137–141.

Curtis, M.W. and Ward, C.H. 1981. Aquatic toxicity of forty industrial chemicals: testing in support of hazardous substance spill prevention regulation. *J. Hydrol.* 51:359–367.

Davis, Murray, H.E., Liehr, J.G., and Powers, E.L. 1981. Basic microbial degradation rates and chemicals by products of related organic compounds. *Water Res.* 15(9), 1125–1127.

de Lijser, H.J.P., Erkelens, C., Knol, A., Pool, W., and De Leer, E.W.B. 1991. Natural organochlorine in humic soils. GC and GC/MS studies of soil pyrolysates. In: *Humic Substances in the Aquatic and Terrestrial Environment,* B. Allard, H. Boren, and A. Grimwall, Eds., Proceedings of an International Symposium, Linkoping, Sweden, August 21–23, 1989, pp. 485–494.

El-Genday, K.S., Abdalla, A.A., Aly, H.A., Tantaway, G., and El-Sebae, A. 1981. Residue levels of chlorinated hydrocarbon compounds in water and sediment samples from Nile branches in the Delta, Egypt. *J. Environ. Sci. Health.* 26:15–36.

Environment Canada/MOE. 1981. Environmental Baseline Report of the Niagara River. November 1981. Update, Canada-Ontario Review Board.

EPA. 1978. U.S. Environmental Protection Agency. Ambient Water Quality Criteria: Dichlorobenzenes.

EPA. 1979. U.S. Environmental Protection Agency. Water Related Environmental Fate of 129 Priority Pollutants, Vol. II. U.S. EPA, Washington, D.C. PB80-204381.

EPA. 1980a. U.S. Environmental Protection Agency. TSCA chemical assessment series: Assessment of Testing Needs: Chlorinated Benzenes. Support document for proposed health effects test rule. Toxic Substances Control Act. Section 4. Office of Toxic Substances, Washington, DC., EPA-560/11- 80-014, pp. 23–32; 142–163.

EPA. 1980b. U.S. Environmental Protection Agency. Ambient Water Quality Criteria Reports, Office of Water Regulations and Standards, Washington, D.C.

EPA. 1983. U.S. Environmental Protection Agency. Volatile Organic Chemicals in the Atmosphere: An Assessment of Available Data. EPA, Cincinnati, OH, EPA600/3-83-027(A).

EPA. 1985. U.S. Environmental Protection Agency. Health Assessment Document for Chlorinated Benzenes. U.S. EPA., PB85-150332.

EPA. 1986. U.S. Environmental Protection Agency. Exposure Assessment for Hexachlorobenzene. U.S.Environmental Protection Agency, Office of Pesticides and Toxic Substances, Washington DC., (author, Carpenter, C. et al.) EPA-560/5-86-019.

EPA. 1986a. U.S. Environmental Protection Agency. Broad Scan Analysis of the FY82 National Adipose Tissue Survey Specimens; Vol.I-Executive summary. U.S. Environmental Protection Agency, Office of Toxic Substances, Washington D.C. (author, Stanley, J.S.), EPA-560/586-035.

Farm Chemicals Handbook. 1993. Meister Publishing Co., Willoughby, OH.

Fathepure, B.Z., Tiedje, J.M., and Boyd, S.A. 1988. Reductive dechlorination of hexachlorobenzene to tri- and di-chlorobenzenes in anaerobic sewage sludge. *Appl. Environ. Microbiol.* 54:327–330.
FDA. 1990. Food and Drug Administration Program. Residues in foods 1989. *J. Assoc. Off. Anal. Chem.* 73:127A–146A.
FDA. 1991. Food and Drug Administration Program. Residues in foods 1990. *J. Assoc. Off. Anal. Chem.* 74:121A–140A.
FDA. 1992. Food and Drug Administration Program. Residues in foods 1991. *J. Assoc. Off. Anal. Chem.* 75:135A–157A.
Figueora, L.D. and Simmons, M.S. 1991. Structure-activity relationships of chlorobenzenes using DNA measurement as a toxicity parameter in algae. *Environ. Toxicol. Pharmacol.* 10:323–329.
Games, L.M. and Hites, R.A. 1977. Composition, treatment efficiency and environmental significance of dye manufacturing plants and effluents. *Anal. Chem.* 49:443–1440.
Garrison, A.W. and Hill, D.W. 1972. Organic pollutants from mill persist in downstream waters. *American Dyestuff Report.* 61:21–25.
Gartrell, M.J., Craun, J.C., and Podrebarac, D.S. 1986. Pesticides, selected elements, and other chemicals in adult total diet samples, October 1980–March 1982. *J. Assoc. Off. Anal. Chem.* 69:146–159.
Germain, A. and Langlois, C. 1988. Contamination des eaus et des sediments en suspension du fleuve Saint-Laurent par les pesticides organochlores, les biphenyles polychlores et d'autres contaminants organiques prioritaires. *Water Pollut. Res. J. Can.* 23:602–614.
Geyer, H., Schunert, I., and Korte, F. 1985. The effects of organic environmental chemicals on the growth of algae *Scenedesmus sabspicatus*. A contribution to environmental biology. *Chemosphere.* 14:1355–1370.
Gibson, D.T., Koch, J.R., and Kallio, R.E. 1968. Oxidative degradation of aromatic hydrocarbons by microorganisms. II. Metabolism of halogenated aromatic hydrocarbons. *Biochemistry.* 7:3795–3802.
Gibson, D.T. 1976. Initial reactions in the bacterial degradation of aromatic hydrocarbons. *Zbl. Bakt. Hyg. I. Abt. Orig., B.* 162:157–168.
Gobas, F.A.P.C., Lovett-Doust, L., and Haffner, G.D. 1991. A comparative study of the bioconcentration and toxicity of chlorinated hydrocarbons in aquatic macrophytes and fish. In: *Plants for Toxicity Assessment,* Gorsuch and Lower Lewis Wang, Eds., Vol. 2, American Society for Testing Materials, Philadelphia, PA, pp. 178–193.
Gribble, G.W. 1992. Naturally occurring organohalogen compounds — a survey. *J. Nat. Proc.* 55:1353–1395.
Gschwend, P.M., MacFarlane, J.K., and Newman, K.A. 1985. Volatile halogenated organic compounds released to seawater from temperate marine macroalgae. *Science.* 227:1033–1035.
Gunderson, E.L. 1988. FDA total diet study, April 1982–April 1984, dietary intakes of pesticides, selected elements and other chemicals. *J. Assoc. Off. Anal. Chem.* 71:1200–1209.
Hallett, D., Norstrom, R.J., Onuska, F.L., and Comba, M.E. 1982. Incidence of chlorinated benzenes and chlorinated ethylenes in Lake Ontario herring gulls. *Chemosphere.* 11:277–286.
Hargrave, B.T., Vass, W.P., Erickson, P.E., and Fowler, B.R. 1988. Atmospheric transport of organochlorines to the Arctic Ocean. *Tellus.* 40B:480–493.
Harlow, H.E. and Hodson, P.V. 1988. Chemical contamination of Hamilton Harbor: A review. *Can. Tech. Rept. Fish. Aquat. Sci.* 1603:91.

Hayes, W., Hanley, T., Gushow, T., Johnson, L., and John, J. 1985. Teratogenic potential of inhaled dichlorobenzenes in rats and rabbits. *Fundam. Appl. Toxicol.* 5:190–202.

Heitmuller, P.T., Hollister, T.A., and Parrish, P.R. 1981. Acute toxicity of 54 industrial chemicals to sheepshead minnows (*Cyprinodon variegatus*). *Bull. Environ. Contam. Toxicol.* 27:596–604.

Hesse, J.M., Spijers, G.J.A., and Taalman, R.D.R.M. 1991. Integrated Criteria Document; Chlorobenzenes Effects. National Technical Information System, Springfield, VA. Appendix, PB92-133057.

Jan, J. and Malnersic, S. 1980. Chlorinated benzene residues in fish in Slovenia (Yugoslavia). *Bull. Environ. Contam. Toxicol.* 24:824–827.

Jori, A., Calamari, D., Cattabeni, F., Galli, A., Gallie, E., and Ramundo, A. 1982. Ecotoxicolgical profile of p-dichlorobenzene (working party on ecotoxicological profiles on chemicals). *Ecotoxicol. Environ. Safety.* 6:413–432.

Jungclaus, G.A., Games, L.M., and Hites, R.A. 1976. Identification of trace organic compounds in tire manufacturing plant waste waters. *Anal. Chem.* 48:1894–1896.

Karickhoff, S.W., Brown, D.S., and Scott, T.A. 1979. Sorption of hydrophobic pollutants on natural sediments. *Water Res.* 13:241–248.

Kasokat, T., Nagel, R., and Urich, K. 1989. Metabolism of chlorobenzene and hexachlorobenzene by the zebra fish, *Brachydanio rerio*, *Bull. Environ. Contam. Toxicol.* 42:254–261.

Kauss, P.B. and Hutchinson, T.A. 1975. The effects of water-soluble petroleum components on the growth of *Chlorella vulgaris* Beijernick. *Environ. Pollut.* 9:157–174.

Kelly, T.J., Czuczwa, J.M., Sticksel, P.R., and Svedrup, G.M. 1991. Atmospheric and tributary inputs to toxic substances to Lake Erie. *J. Great Lakes Res.* 14:504–516.

Kenaga, E.E. and Goring, C.A.I. 1980. Relationship between water solubility, soil sorption, octanol-water partitioning and concentrations of chemicals in biots. In: *Proceedings of the Third Symposium on Aquatic Toxicology,* J.G. Eaton, P.R. Parrish, and A.C. Hendricks, Eds., American Society for Testing and Materials, Philadelphia, 1980, pp. 78–115.

Keuhl, D., Johnson, K., Butterworth, B. Leonard, E., and Verth, G. 1981. Quantification of octachlorostyrene and related compounds in Great Lakes fish by gas chromatography-mass spectrometry. *J. Great Lakes Res.* 7:330-335.

Kitagawa, Y., Fukuda, Y., Taniyama, T., and Yoshikawa, M. 1991. *Chem. Pharmacol. Bull.* 39:1171.

Kitchin, K.T., Linder, R.E., Scotti, T.M., Walsh, D., Curley, A.P., and Svensgaard, D. 1982. Offspring mortality, lung pathology in female rats fed hexachlorobenzene. *Toxicology.* 23:33–39.

Kizer, K. 1986. Final report on a monitoring program for organic chemical contamination of large public water systems in California. Summary version, Department of Health Sciences, California.

Knezovich, J.P. and Harrison, F.L. 1988. The bioavailability of sediment-sorbed chlorobenzenes to larvae of the midge, *Chironomus decorus. Ecotoxicol. Environ. Safety.* 15:226–241.

Konemann, H. and van Leeuwen, K. 1980. Toxicokinetics in fish: accumulation and elimination of six chlorobenzenes by guppies. *Chemosphere.* 9:3–19.

Kozie, K.D. and Anderson, R.K. 1991. Productivity, diet and environmental contaminants in bald eagles nesting near the Wisconsin shoreline of Lake Superior. *Arch. Environ. Contam. Toxicol.* 20:41–48.

Kuehl, D.W., Johnson, K.L., Butterworth, B.C. et al. 1981. Quantification of octachlorostyrene and related compounds in Great Lakes fish by gas chromatography — mass spectrometry. *J. Great Lakes Res.* 7:330–335.

Kuhn, R. 1989. Results of the harmful effects of water pollutants to *Daphnia magna* in the 21-day reproduction test. *Water Res.* 501–510.

Laska, A.L., Bartell, C.K., and Laseter, J.L. 1976. Distribution of hexachlorobenzene and hexachlorobutadiene in water, soil, and selected aquatic organisms along the lower Mississippi River, Louisiana. *Bull. Environ. Contam. Toxicol.* 15:535–542.

LeBlanc, G.A. 1980. Acute toxicity of priority pollutants to water flea (*Daphnia magna*). *Bull. Environ. Contam. Toxicol.* 24:684–691.

Lee, R.F. and Ryan, C. 1979. Microbial Degradation of Organochlorine Compounds in Estuarine Water and Sediments, National Technical Information System, Springfield, VA, PB298254.

Li, R.T., Spigarelli, J.L., and Going, J.E. 1976. Sampling and analysis of selected toxic substances, Ia. Hexachlorobenzene, OTS, (U.S. EPA), Washington, D.C., EPA 560/676/001.

Lindqvist, R. and Enfield, C.G. 1992. Biosorption of dichlorodiphenyl trichloroethane and hexachlorobenzene in groundwater and its implications for facilitated transport. *Applied Environ. Microbiol.* 58:2211–2218.

Lu, P.Y. and Metcalfe, R.L. 1975. Environmental fate and degradability of benzene derivatives as studied in a model ecosystem. *Environ. Health Perspect.* 10:269–284.

Mackay, D. and Wolkoff, A.W. 1973. Rate of evaporation of low solubility contaminants from water bodies to the atmosphere. *Environ. Sci. Technol.* 9:611–614.

Mackay, D. and Leinonen, P.J. 1975. Rate of evaporation of low solubility contaminants from water bodies to the atmosphere. *Environ. Sci. Technol.* 9:1178–1180.

Mackay, D. and Paterson, S. 1981. Calculating fugacity. *Environ. Sci. Technol.* 15:1006–1014.

Mackay, D., Shiu, W.Y., and Ma, K.C. 1992. *Illustrated Handbook of Physical-Chemical Properties and Environmental Fate for Organic Chemicals Monoaromatic Hydrocarbons, Chlorobenzenes, and PCBs,* Vols. 1 and 2. CRC/Lewis, Boca Raton, FL.

Matheson, R., Hamilton, A., Trites, A., and Whitehead, D. 1980. Chlorinated benzenes in herring gull and double-crested cormorant eggs from three locations in the maritime provinces. Environment Protection Service, Atlantic Region, Halifax, NS, Canada. EPS-5-AR-80-1.

McCarty, L.S., Lupp, M., and Shea, M. 1984. Chlorinated Benzenes in the Aquatic Environment. Ontario Ministry of the Environment, Ontario, Canada, Scientific Criteria Document 3-84.

McNamara, P., Byrne, M., Gouer, M., Lucas, P., Scow, K., and Wood, M. 1981. *An Exposure and Risk Assessment for 1,2,4-Trichlorobenzene.* Little Brown, Cambridge, MA.

Merck. 1989. *The Merck Index: An Encyclopedia of Chemicals, Drugs, and Biologicals.* S. Budaveri, Ed., 11th ed., Merck & Co. Rahway, NJ.

Mill, T. 1980. Data needed to predict the environmental fate of organic chemicals. In: *Dynamics, Exposure and Hazard Assessment of Toxic Chemicals,* R. Haque, Ed., Ann Arbor Science, Ann Arbor, MI, pp. 297–322.

Moore, J.W. and Ramamoorthy, S. 1984. *Organic Chemicals in Natural Waters, Applied Monitoring and Impact Assessment.* Springer-Verlag, New York, 289 pp.

Morita, M. and Ohi, G. 1975. Paradichlorobenzene in human tissue and atmosphere in Tokyo metropolitan areas. *Environ. Pollut.* 8:2669–274.

Nebeker, A.V., Griffis, W.L., Wise, C.M., Hopkins, E., and Barbitta, J.A. 1989. Survival, reproduction and bioconcentration in invertebrates and fish exposed to hexachlorobenzene. *Environ. Toxicol. Chem.* 8:601–611.

Neely, W.B., Branson, D.R., and Blau, G.E. 1974. Partition coefficient to measure bioconcentration potential of organic chemicals in fish. *Environ. Sci. Technol.* 8:1113–1115.

Neely, W.B. 1979. A preliminary assessment of the environmental exposure to be expected from the addition of a chemical to a simulated aquatic system, *Int. J. Environ. Studies.* 13:101–108.

Ohlendorf, H.M., Swineford, D.M., and Locke, L.N. 1981. Organochlorine residues and mortality of herons. *Pestic. Monit. J.* 14:125–135.

OHMTADS. Oil and Hazardous Materials Technical Assistance Data System 1986. OHMS/TADS file for hexachlorobenzene.

Oliver, B.G. and Nicol, K.D. 1982. Chlorobenzenes in sediments, water and selected fish from Lake Superior, Huron, Erie and Ontario. *Environ. Sci. Technol.* 16:532–536.

Oliver, B.G. 1987. Fate of some chlorobenzenes from the Niagara River in Lake Ontario. In: *Sources and Fates of Aquatic Pollutants,* R.A. Hites and S.J. Eisenrich, Eds. American Chemical Society, Washington, D.C. pp.10–26.

Paya-Perez, A., Riaz, M., and Larsen, B.R. 1991. Soil sorption of twenty PCB congeners and six chlorobenzenes. *Ecotoxicol. Environ. Safety.* 21:1–17.

Pellizzari, E.D., Little, L.W., Sparocino C., Hughes, T.J., Claxton, L., and Waters, M.D. 1979. Integrating microbiological and chemical testing into the screening of air samples for mutagenicity. *Environ. Sci. Res.* 15:331–351.

Plimmer, J.R. and Klingbiel, U.I. 1976. Photolysis of hexachlorobenzene, *J. Agr. Food. Chem.* 24:721–723.

Richter, E., Renner, G., Bayeri, J., and Wick, M. 1981. Differences in the biotransformation of hexachlorobenzene (HCB) in male and female rats. *Chemosphere.* 10:779–785.

RTP. 1994. Report of Expert Panel: Interpretive review of the potential adverse effects of chlorinated organic chemicals on human health and environment. *Reg. Toxicol. Pharmacol.* 20: August 1994.

Rush, G.F., Smith, J.H., Maita, K., Bleavins, M., Auerich, R.J., Ringer, R.K., and Hook, J.B. 1983. Perinatal hexachlorobenzene toxicity in the mink. *Environ. Res.* 31:116–124.

Safe, S., Jones, D., Kohli, J., Ruzo, L.O., Hutzinger, O., and Sundstrom, G. 1976. The metabolism of chlorinated aromatic pollutants by the frog. *Can. J. Zool.* 54:1818–1823.

Sander, P., Wittich, R.M., Fortnagel, P., Wilkes, H., and Francke, W. 1991. Degradation of 1,2,4-trichloro- and 1,2,4,5-tetrachlorobenzene by *Pseudomonas* strains. *Appl. Environ. Microbiol.* 57:1430–1440.

Schrap, S.M. and Opperhuizen, A. 1989. Preliminary investigations on the influence of suspended sediments on the bioaccumulation of two chlorobenzenes by the guppy (*Poecilia reticulata*). *Hydrobiologica.* 188/189:573–576.

Schuytema, G.S., Krawzyck, D.F., et al. 1988. Comparative uptake of hexachlorobenzene by fathead minnows, amphipods and oligochaete worms from water and sediment. *Environ. Toxicol. Chem.* 7:1035–1045.

Schwarzenbach, R.P., Molnar-Kubica, E., Giger, W., and Wakeham, S.G. 1979. Distribution, residence time and fluxes of tetrachloroethylene and 1,4-dichlorobenzene in Lake Zurich, Switzerland. *Environ. Sci. Technol.* 13:1367–1373.

Sittig, M. 1980. *Prority Toxic Pollutants. Health Impacts and Allowable Limits.* Noyes Publications, Park Ridge, NJ, 370 pp.

Sittig, M. 1994. *Worldwide Limits for Toxic and Hazardous Chemicals in Air, Water and Soil.* Noyes Publications, Park Ridge, NJ, 792 pp.

Smith, J.H., Mabey, W.R., Bohonos, N., Holt, B.R., Lee, S.S., Chou, T.W., Bomberger, D.C., and Mill, T. 1978. Environmental Pathways of Selected Chemicals in Freshwater Systems, Parts I and II, U.S. Environmental Protection Agency, Publication Nos. 600/7-77-113 and 600/7-78-074.

Smith, A.D., Bharath, A., Mallard,C., Orr, D., McCarty, L.S., and Ozburn, G.W. 1990. Bioconcentration kinetics of some chlorinated benzenes and chlorinated phenols in American flag fish, *Jordanella floridae. Chemosphere.* 20:379–386.

Smith, A.D., Bharath, A., Mallard, C., Orr, D., Smith, K., Sutton, J.A.,Vukamanich, J., McCarty, L.S., and Ozburn, G.W. 1991. The acute and chronic toxicity of ten chlorinated organic compounds to the American flagfish (*Jordanella floridae*). *Arch. Environ. Contam. Toxicol.* 20:94–102.

Stevens, R.J. and Neilson, M.A. 1989. Inter- and intralake distributions of trace organic contaminants in surface waters of the Great Lakes. *J. Great Lakes Res.* 15:377–393.

Suns, K., Hitchin, G., and Adamek, E. 1994. Present status and temporal trends of organochlorine contaminants in young-of-the-year spottail shiners (*Notropus Hudsonis*) from Lake Ontario. *Can. J. Fish. Aquat. Sci.* 48:1568.

Tabak, H.H., Quave, S.A., Mashni, C.I., and Barth, E.F. 1981. Biodegradability studies with organic priority pollutant compounds, *J. Water Pollut. Control. Fed.* 53:1503–1518.

Thom, N.S. and Agg, A.R. 1975. The breakdown of synthetic organic compounds in biological processes. Proc. Royal Soc. London, Series B, 189:347–357.

Tyle, H., Egsmose, M., and Harrit, N. 1991. Mixed function oxygenase in juvenile rainbow trout exposed to hexachlorobenzene or 3,3,′4,4′-tetrachlorobiphenyl. *Comp. Biochem. Physiol.* 100(1/2):161–164.

van Hoogen, G. and Opperhuizen, A. 1988. Toxicokinetics of chlorobenzenes in fish. *Environ. Toxicol. Chem.* 7:213–220.

Veith, G.D., Kuehl, D.W., Leonard, E.N., Puglishi, F.A., and Lemke, A.E. 1979. Polychlorinated biphenyls and other organic chemical residues in fish from major watersheds of the United States. 1976. *Pestic. Monit. J.* 13:1–11.

Veith, G.D., Call, D.J., and Brooke, L.T. 1983. Estimating the acute toxicity of narcotic industrial chemicals to fathead minnows. Aquatic toxicology and hazard assessment 6th symposium, *ASTM Spec. Technol. Publ.* 802:90–97.

Versar, 1979. Water Related Environmental Fate of 129 Priority Pollutants. Vol. II. U.S. Environmental Protection Agency, Publication No. EPA-440/4-79-029b, Washington, DC.

Verschuren, K. 1983. *Handbook of Environmental Data on Organic Chemicals.* 2nd ed. Van Nostrand Reinhold, New York.

Walker, J.D. and Colwell, R.R. 1975. Degradation of hydrocarbons and mixed hydrocarbon substrate by microorganisms from Chesapeake Bay. *Progress Water Technol.* 13:227–236.

Wallace, L.A. 1991. Personal exposure to 25 volatile organic compounds. EPA's 1987 TEAM study in Los Angeles, *California. Toxicol. Ind. Health* 7 (5/6):203–208.

Ware, S.A. and West, W.L. 1977. Investigations of selected potential environmental contaminants; halogenated benzenes, U.S. Environmental Protection Agency Publication No. EPA-560/2-77-004, Washington, D.C.

Weseloh, D.V., Custer, T.W., and Braune, B.M. 1989. Organochlorine contaminants in eggs of common terns from the Canadian Great Lakes, 1981. *Environ. Pollut.* 59:141–160.

Wilson, J., Enfield, C.G., Dunlap, W.J., Coasby, R.L., Foster, D.A., and Baskin, L.B. 1981. Transport and fate of selected organic pollutants in a sandy soil. *J. Environ. Qual.* 10:501–506.

Wong, P.T.S. 1984. Relationship between water solubilty of chlorobenzenes and their effects on a freshwater green alga. *Chemosphere.* 13:991–996.

Woodward, D.F., Mehrle, P.M., Jr., and Mauck, W.L. 1981. Accumulation and sublethal effects of a Wyoming crude oil in cutthroat trout. *Trans. Am. Fish. Soc.* 110:437–445.

Yess, N.J., Gunderson, E.L., and Roy, R.R. 1993. U.S. Food and Drug Administration monitoring of pesticide residues in infant foods and adult foods eaten by infants/children. *J. AOAC Int.* 76:492–507

Young, D.R., Hessen, T.C., and McDermott-Ehrlich, D.J. 1976. Synoptic survey of chlorinated hydrocarbon imputs to the Southern California Bight. Southern California Coastal Water Research Project Authority. El Segundo, CA.

Zitko, V. 1977. Uptake and Excretion of Chlorinated and Brominated Hydrocarbons by Fish. Fish. Marine Serv. Tech. Rep. No. 737. Biological Station, St. Andrews, NB, Canada, 14 pp.

Zitko, V. 1979. The fate of highly brominated aromatic hydrocarbons in fish. In: *Pesticide and Xenobiotic Metabolism in Aquatic Organisms*. M.A.Q. Kahn, J.J. Lech, and J.J. Menn, Eds., *Amer. Chem. Soc. Ser.* 99: American Chemical Society, Washington, D.C. pp. 177–182.

CHAPTER 5

Chlorinated Aromatic Hydrocarbons — Polycyclic

Polychlorinated biphenyls (PCBs) will be discussed as typical examples of this category. PCBs are a family of synthetically produced chemicals that contain 209 individual compounds of varying toxicity. PCBs have been produced since 1929 and most information on their manufacture under trade names and general characteristics are available from trade publications. Commercial preparations of PCBs enter the environment as mixtures containing a variety of PCBs and impurities.

PCBs are produced synthetically by chlorinating the biphenyl molecule with anhydrous chlorine in the presence of iron filings or ferric hydroxide as a catalyst. The crude product is diluted with alkali to remove color, traces of HCl, and the catalyst; this results in a mixture of chlorobiphenyls with different number of chlorine atoms per molecule. Chlorobiphenyls could also be synthesized from (1) arylation of aroyl peroxides, diazonium salts, phenyl hydrazines and other compounds; (2) aryl condensation reactions; (3) chlorination of biphenyl systems; (4) decarboxylation; and (5) dechlorination reactions (Moore and Ramamoorthy 1984).

All Aroclor products (Aroclor is the trade name of Monsanto's PCB) are characterized by a four-digit numbering code in which the first two digits (1 and 2) indicate the parent molecule is biphenyl and the last two digits indicate the chlorine content. The first two digits could be 5 and 4 when it refers to chlorinated terphenyls, although some 54 aroclors could be a mixture of ter- and biphenyls. Aroclors 25 and 44 are mixtures of PCBs and polychlorinated terphenyls (75% and 60% PCB, respectively). For example, Aroclor 1242 is a chlorinated biphenyl mixture with an average chlorine content of 42%. Aroclor 1016 is an exception to this designation and contain primarily mono-, di-, and trichloro isomers. Aroclor 1016 has an average chlorine content of 41.5%, which is similar to Aroclor 1242.

Polychlorinated dibenzofurans (PCDFs) are the major toxic contaminants in Aroclors and Japanese Kanechlors; whereas, European PCBs contain PCDFs and hexachloronaphthalenes as contaminants (Vos et al. 1970). Laboratory investigations have shown the possibility of photochemical formations of CDFs and PCDFs as secondary products in commercial PCB mixtures (Roberts et al. 1978). However, the effect of environmental aging of commercial PCBs on the formation and concentrations of PCDFs has not been studied.

PHYSICAL AND CHEMICAL PROPERTIES AND USE PATTERN

Selected physical and chemical properties of the Aroclors are presented in Table 5.1. Pyrolysis of PCBs and formation of PCDFs have been reviewed by EPA (1988a). Pyrolysis products include PCDFs, chlorinated benzenes, naphthalenes, phenylethylenes, biphenylenes, and hydroxy PCBs. Four major pathways for the formation of PCDFs from PCBs have been reported:

1. Loss of two ortho chlorines
2. Loss of ortho hydrogens as well as chlorine
3. Shift of chlorine from 2 to 3 position
4. Loss of two ortho hydrogens (EPA 1988a)

Table 5.1 Selected Physical–Chemical Properties of PCB Congeners

Congener	Mol. Wt.	Vapor Pressure (Pa)	Water Solubility (µg/L)	K_{ow}
Monochlorobiphenyl	188.7	0.9–2.5	1210–5500	4.3–4.6
Dichlorobiphenyl	223.1	0.008–0.60	60–2000	4.9–5.3
Trichlorobiphenyl	257.5	0.003–0.22	15–400	5.5–5.9
Tetrachlorobiphenyl	292.0	0.002	4.3–10	5.6–6.5
Pentachlorobiphenyl	326.4	0.0023–0.051	4–20	6.2–6.5
Hexachlorobiphenyl	360.9	0.0007–0.012	4– 0.70	6.7–7.3
Heptachlorobiphenyl	395.3	0.00025	0.045–0.2	6.7–7.0
Octachlorobiphenyl	429.8	0.0006	0.2–0.31	7.1
Nonachlorobiphenyl	464.2	—	0.18–0.12	7.2–8.16
Decachlorobiphenyl	498.7	0.00003	0.01–0.761	8.26

Note: Ranges in physico-chemical properties for a given congener account for variance in properties of different isomers of a congener group; -mm of Hg (0°C) = 133.322 pascals.

From: Mackay et al. 1992. *Illustrated Handbook of Physical-Chemical Properties and Environmental Fate for Organic Chemicals.* Vol. 1, CRC/Lewis, Boca Raton, FL.

The general structure of chlorinated biphenyls and the position numbering is shown below.

(n and n′ may vary from 0 to 5)

Chemically, there are 209 chlorinated biphenyls, ranging from three monochlorobiphenyls to one fully chlorinated decachlorobiphenyl. The selected properties of all congener groups of PCBs are given in Table 5.1. Tetrachlorobiphenyl is a congener and all isomers of this congener will have four chlorine atoms in the molecule, but in different positions. For example, 2,2′,4,4′-tetrachlorobiphenyl is an isomer.

Table 5.1 illustrates the wide variation in properties between the congeners; vapor pressures, 100,000-fold; solubilities, about 5 millionfold; log K_{ow}, about 10,000 times from 2-chlorobiphenyl to decachlorobiphenyl. The chemical identification and physical–chemical properties of PCBs are listed in Tables 5.2 and 5.3, respectively.

Aroclors possess properties different from their individual chlorobiphenyl compounds. The individual compounds are solids at room temperature, whereas, the individual aroclors are either mobile oils (1221, 1232, 1242, and 1248), viscous liquids (Aroclor 1254), or sticky resins (Aroclor 1260 and 1262). PCBs are sparingly soluble in water, and solubility decreases with increasing chlorine content (Table 5.4). Accordingly, the water solubilities of some aroclors are also variable: 1242, 0.2 mg/L; 1248, 0.1 mg/L; 1254, 0.040 mg/L; and 1260, 0.025 mg/L. Selective solubilization leading to bias in these data should be considered in evaluating environmental behavior of specific PCBs and their toxicity to organisms. The relative pattern of tissue partitioning often reflects the differential aqueous solubilities of the compounds rather than preferential uptake. Aqueous solubility values of chlorobiphenyls are given in Table 5.4.

Table 5.5 provides the percent chlorine content of common aroclor formulations and Table 5.6 lists the PCB component composition of commonly used aroclors.

PRODUCTION AND USE PATTERN

Production

PCBs have been commercially produced in the United States since 1929. Annual production of PCBs peaked in 1970 when 38,630 metric tons were produced, but decreased sharply in 1971 (Table 5.7). About 10% of the production was exported (about 1900 metric tons in 1961 and 1965). Exports followed a trend more or less similar to production data, reaching 6200 metric tons in 1970, but decreased to 4500 metric tons in 1971. Data were not available for the remainder of the 1970s, but all U.S. production of PCBs ceased in October 1977. This was a voluntary restriction reflecting public concern because PCBs accumulated and persisted in the environment and have toxic effects. Japanese annual production peaked at around 11,800 metric tons; PCBs were apparently never produced in Canada. Little information is available on the production figures from other countries.

In 1974, Monsanto produced about 18,200 metric tons of the aroclor mixtures. Production was stopped in 1977. Of the total PCBs sold in the U.S. since 1970, over 98% were Aroclors 1260, 1254, 1248, 1242, 1232, 1221, and 1016 and less than 2% were Aroclor 1268 and 1262 (EPA 1976; IARC 1978).

Table 5.2 Chemical Identification of Aroclors

Identification	Aroclor 1016	Aroclor 1221	Aroclor 1232	Aroclor 1242	Aroclor 1248	Aroclor 1254	Aroclor 1260
Synonyms	PCB-1016 (41.5% Cl)	PCB-1221 (21% Cl)	PCB-1232 (32% Cl)	PCB-1242 (42% Cl)	PCB-1248 (48% Cl)	PCB-1254 (54% Cl)	PCB-1260 (60% Cl)
CAS Registry	12674-11-2	11104-28-2	11141-16-5	53469-21-9	12672-29-6	11097-69-1	11096-82-5
NIOSH/RTECS	TQ1351000	TQ1352000	TQ1354000	TQ1356000	TQ1358000	TQ1360000	TQ1362000
EPA hazardous waste no.	3502	3502	3502	3502	3502	3502	3502
OHM/TADS	8500400	8500401	8500402	8500403	8500404	8500405	8500406
DOT/UN/NA/IMCO	UN 2315	UN 2315	UN 2315	UN 2315	UN 2315	UN 2315	UN 2315

Note: Aroclor is the trade name for chlorinated biphenyls made by Monsanto. EPA hazardous waste no. designated prior to May 19, 1980.

Sources: SANSS 1990; EPA 1980a; EPA-NIH 1990; Stone 1981; HSDB 1995.

Table 5.3 Physical and Chemical Properties of PCBs

	\multicolumn{7}{c}{Aroclor Designation}
	1016	1221	1232	1242	1248	1254	1260
Molecular weight	257.9	200.7	232.2	266.5	299.5	328.4	375.7
Boiling point	325–356	275–320	290–325	325–366	340–375	365–390	385–420
Water solubility, (mg/L)	0.42	0.59 (24°C)	—	0.10 (24°C)	0.06 (24°C)	0.057 (24°C)	0.0027
Log K_{ow}	5.6	4.7	5.1	5.6	6.2	6.5	6.8
Vapor pressure, (mmHg, 25°C)	4×10^{-4}	6.7×10^{-3}	4.06×10^{-3}	4.06×10^{-4}	4.94×10^{-4}	7.71×10^{-5}	4.05×10^{-5}
Henry' law constant (atm-m³/mol at 25°C)	2.9×10^{-4}	3.5×10^{-3}	no data	5.2×10^{-4}	2.8×10^{-3}	2.0×10^{-3}	4.6×10^{-3}
Conversion factors (air, 25°C, 1 mg/m³)	0.095 ppm	0.12 ppm	0.105 ppm	0.092 ppm	0.08 ppm	0.075 ppm	0.065 ppm

Note: Log K_{ow} are average values of the major component of each aroclor (Hancsch and Leo 1985). Henry's law constants are estimated by dividing vapor pressure by the water solubility; hence Henry's law constant is only an average for the entire mixture. The following Henry's law constants were also reported in the literature: Aroclor 1221, 2.28×10^{-4}; 1242, 3.43×10^{-4}; 1248, 4.4×10^{-4}; 1254, 2.83×10^{-4}; 1260, 4.15×10^{-4} (Burkhard et al. 1985). Air conversion factors were calculated using the average molecular mass as given under molecular weight.

Sources: Hutzinger et al. 1974; Monsanto 1974; Paris et al. 1978; Hollifield 1979; EPA 1985; Callahan et al. 1979; IARC 1978; Hubbard 1964.

Table 5.4 Water Solubilities of Chlorobiphenyls (mg/L)

Chlorobiphenyl	Solubility	Chlorobiphenyl	Solubility
Monochlorobiphenyls			
2-	5.9	Pentachloro-	
3-	3.5	biphenyls	
4-	1.19	2,2',3,4,5-	0.022
		2,2',4,5,5'-	0.031
Dichlorobiphenyls			
2,4-	1.40	Hexachlorobiphenyls	
2,2'-	1.50	2,2',4,4',5,5'-	0.0088
2,4'-	1.88		
4,4'-	0.08	Octachlorobiphenyls	
		2,2',3,3',4,4',5,5'	0.0070
Trichlorobiphenyls			
2,4,4'-	0.085	Decachlorobiphenyl	
2',3,4-	0.078		0.015
Tetrachlorobiphenyls			
2,2',5,5'-	0.046		
2,2'-3,3'-	0.034		
2,2',3,5'-	0.170		
2,2',4,4'-	0.068		
2,3',4,4'-	0.058		
2,3',4',5-	0.041		
3,3',4,4'-	0.175		

Sources: Hutzinger et al. 1974.

Table 5.5 Percent Chlorine Content of Aroclor Formulations

Aroclor	% Chlorine	Average Number of Cl/Aroclor Molecule	Average Mol. Wt.
1221	20.5–21.5	1.15	192
1232	31.5–32.5	2.04	221
1242	42	3.10	261
1248	48	3.90	288
1254	54	4.96	327
1260	60	6.30	372

Note: Aroclor 1221 contains a predominance of Cl_2–Cl_4 biphenyls while Aroclor 1254 and 1260 contain Cl_4–Cl_6 biphenyls and Cl_5–Cl_7 biphenyls, respectively. The last two digits of Aroclor formula represent the percent weight of chlorine in the compound.

Source: Hutzinger et al. 1974.

Table 5.6 PCB Composition of Selected Aroclors

PCB Component	Mol. Wt.	% Chlorine	1016	1221	1232	1242	1248	1254	1260
$C_{12}H_{10}$	154.21	0	<0.1	11	<0.1	<0.1	—	<0.1	ND
$C_{12}H_9Cl$	188.65	18.79	1	51	31	1	—	<0.1	ND
$C_{12}H_8Cl_2$	223.10	31.77	20	32	24	16	2	0.5	ND
$C_{12}H_7Cl_3$	257.54	41.30	57	4	28	49	18	1	ND
$C_{12}H_6Cl_4$	291.99	48.56	21	2	12	25	40	21	1
$C_{12}H_5Cl_5$	326.43	54.30	1	<0.5	4	8	36	48	12
$C_{12}H_4Cl_6$	360.88	58.93	<0.1	ND	<0.1	4	4	23	38
$C_{12}H_3Cl_7$	395.32	62.77	ND	ND	ND	<0.1	—	6	41
$C_{12}H_2Cl_8$	429.77	65.98	ND	ND	ND	ND	—	ND	8
$C_{12}HCl_9$	464.21	68.73	—	—	—	—	—	—	—
$C_{12}Cl_{10}$	498.66	71.18	—	—	—	—	—	—	—
Av. mol. mass			257.9	200.9	232.2	266.8	299.5	328.4	375.7

Sources: Hutzinger et al. 1974; ATSDR, 1995

Table 5.7 Total PCB Production (metric tons) in U.S.

Year	Production	Year	Production
1960	18,850	1966	29,900
1961	18,450	1967	34,200
1962	19,050	1968	37,600
1963	20,300	1969	34,600
1964	23,100	1970	38,630
1965	27,400	1971	18,400

Source: *Chemical Engineering News*, 1971.

Imports of PCBs by U.S. have been reported as follows. No data were found to indicate that PCBs have been imported after 1981.

Year	Volume (metric tons)
1981	162.3
1979	219.6
1977	127.7

Source: ATSDR 1989

Use Pattern

Because of their chemical and thermal stability, inertness, and excellent dielectric properties, PCBs have found widespread industrial and commercial applications over the last 50 years. They have been used as dielectric fluids in transformers and capacitors; as plasticizers in paints, plastics, sealants, resins, inks, printing, copy paper, and adhesives; and as components in gas turbines and vacuum pumps. It was reported that PCB alone had low toxicity to houseflies but had a synergistic influence on the toxicity of dieldrin and DDT (Lichtenstein et al. 1969). The effectiveness of these compounds decreased with increasing chlorine content. For example, Aroclor 1221 increased the mortality of fruit flies from 59 to 93%, whereas, Aroclor 1268 increased it only to 77% (Peakall and Lincer 1970). Although some PCBs possess insecticidal and fungistatic activities, they apparently have never been used as biocides. The use of PCBs for various industrial applications was estimated from the domestic sales figures released by Monsanto for the period 1957–1972. In 1960, about 11,300 metric tons of Aroclor were used in transformers and capacitors and about 29% of the total on other uses as plasticizers and in hydraulic fluids. In 1970, the fraction for other uses increased to about 45%. After voluntary restrictions by Monsanto, this percentage dropped to 23% and total PCB sales dropped to 18.4 metric tons.

A thorough review of PCB use in the U.S. is available in the literature (EPA 1976). By 1974, all U.S. use of PCBs was restricted to closed systems to produce capacitors and transformers. By 1976, 70% of Monsanto's domestic sales of Aroclors was used in the production of capacitors and 30% in transformer production. Although Aroclors are no longer used in the production of capacitors and transformers, many old devices with Aroclors are currently still in service. The life expectancy of PCB-containing transformers and capacitors is >30 years and >20 years, respectively, depending upon the electrical application. A large capacitor typically contains over 2 to 3 lbs of PCB, but transformers may contain several times that amount, depending on the size. As of 1976, only 5% of the transformers made in the U.S. contained PCBs, but 95% of the capacitors used PCBs (Durfee 1976). An estimated 131,200 PCB transformers were still in service in 1981, representing about 1% of all operational transformers (Orris et al. 1986). Aroclors 1260 and 1262 are still used occasionally as a slide-mounting medium for microscopes (IARC 1978), since this use has been permitted by federal regulations (ATSDR 1992).

An estimated 2.8 million capacitors were in use in the U.S. in the early 1980s. About 2000 of them rupture every year, leading to spillage in the environment. There are no tools to predict the failure of transformers and capacitors. Portable screening devices such as X-ray fluorescence meters for transformer oil, acoustical detectors to detect the ultrasonic sound from a faltering capacitor, and an infrared scanner to measure the temperature-rise of the faltering capacitor have been studied in the field (Miller 1982).

SOURCES IN THE ENVIRONMENT

Natural Sources

Until recently, it was believed that there are no natural sources of PCBs. However, PCBs have been identified in volcanic ash from the 1980 eruption of Mt. St. Helens. Controlled studies eliminated the possibility of misidentifying anthropogenic PCBs as naturally derived PCBs. In fact, the three PCB isomers identified would not originate from commercial PCB mixtures (Pereira et al. 1980). A recent study identified subunits of PCBs as components of glycopeptides identified from Amycolatopsis (Box et al. 1991).

Anthropogenic Sources

Since 1972, the use of PCBs has been restricted to controllable close systems such as transformers and capacitors (WHO 1976; Environment Canada 1988). Between 1930 and 1972, PCBs were used in open systems without restriction, which allowed easy dissipation into the environment. Products used in open systems (Table 5.8) were lubricating and cutting oils, immersion oils, heat transfer and hydraulic fluids, plastics, paints, waxes, inks, adhesives, sealants, carbonless copying paper, pesticides, antifowling agents, and electrical equipment (IARC 1978; WHO 1976; Fischbein and Rizzo 1987).

At present, the major source of PCB exposure in the environment seems to be environmental recycling of PCBs from past uses and consequent releases into the environment. Prior to 1972, the main sources of PCB release to the environment included leakage and disposal of industrial and municipal fluid wastes, incineration of PCB-containing wastes, disposal in dumps and landfills, accidental spills, vaporization of plasticizers, and release in industrial ventilation and exhaust systems (WHO 1976; Fischbein and Rizzo 1987).

Worldwide PCB production was estimated to be well over 2×10^9 kg (Hutzinger et al. 1974) and 1.2×10^6 tonnes (1.2×10^9 kg) (Tanabe 1988). It has been estimated that due to lack of restrictions on the use, release, and disposal of PCB-containing materials, most of the hundred million kg of PCB produced has been released into the environment (Hutzinger and Veerkamp 1981). Since Aroclors are no longer produced or used in manufacturing industries in the U.S., pointsource discharges no longer occur. Current sources of PCB release to the environment include releases from landfills containing transformers, capacitors, and other wastes; waste incineration of PCB materials; spills; and improper (or illegal) disposal to open areas (Weant and McCormick 1984; Murphy et al. 1985). Also, explosions or overheating of PCB-containing transformers may also release significant amounts

Table 5.8 Some Past Uses of PCBs

Base Material	Aroclor Used (% Cl Content)	End Use
Polyvinyl chloride	1248, 1254, 1260 (7–8%)	Secondary plasticizers to improve flame retardance and chemical resistance
Polyvinyl acetate	1221, 1232, 1242 (11%)	To improve quick-track and fiber-tear properties
Polyester resins	1260 (10–20%)	Strengthen fiberglass; reinforce resins and render fire-retardants economical
Polystyrene	1221 (2%)	Plasticizer
Epoxy resins	1221, 1248 (20%)	Increase resistance to oxidation and chemical attack; improve adhesive properties
Styrene-butadiene co-polymer	1254 (8%)	Better chemical resistance
Neoprene	1268 (40%) (1.5%)	Fire retardant; injection mouldings
Crepe rubber	1262 (5–50%)	Plasticizer in paints
Nitrocellulose lacquers	1262 (7%)	Co-plasticizer
Ethylene vinyl acetate	1254 (41%)	Pressure-sensitive adhesives
Chlorinated rubber	1254 (5–10%)	Enhance resistance, flame retardance, electrical insulation properties
Varnish	1260 (25% of oil)	Improve water and alkali resistance
Wax	1261 (5%)	Improve moisture and flame resistance

Source: Monsanto Technical Bulletin O/PL 306.

of PCBs into the surrounding environment. Landfills are likely to continue releasing PCBs into the atmosphere because gases, such as methane and carbon dioxide that are generated from anaerobic degradation of organic waste, are likely to carry PCBs with them into the atmosphere. Monitoring data has shown that the amount of PCBs released from the above sources (10–100 kg from landfills and 0.25 kg per stack for incinerators) may not be significant when compared with the amounts of PCBs estimated to cycle annually through the atmosphere over the U.S. (900,000 kg/year) (ATSDR 1992). Atmospheric fallout and washout have been identified as nonpoint sources of PCBs in the environment (Weant and McCormick 1984; Swackhamer and Armstrong 1986; Larsson 1985).

Evidence suggests that the environmental recycling process is the major source of PCB release to the environment (Swackhamer and Armstrong, 1986; Larsson 1985; Murphy et al. 1985). This cycling process involves volatilization of PCBs from water bodies and soil surfaces into the atmosphere, and subsequently returned to earth via washout/fallout. The cycle is repeated with revolatilization. Since the volatilization and degradation rates vary among the congeners of PCBs, this continuous cycling process alters the ratio of PCB in water to air, relative to the original source pattern.

ENVIRONMENTAL RESIDUES AND EXPOSURE ROUTES

Air

Ambient atmospheric PCB (Aroclor 1016, 1242, and 1254) concentrations of 7.1 and 4.4 ng m^{-3} were detected in Boston, MA and Columbia, SC, respectively, in the summer of 1978 (Bidleman 1981). Ambient air concentration of PCB in Australia from 1981 to 1982 ranged 0.02 to 0.18 ng m^{-3} (Tanabe et al. 1983). The following atmospheric PCB concentrations were calculated from a large volume of monitoring databases reported in Eisenreich et al. 1981.

Location	Concentration of PCB (Range) ng/m³	Mean Value
Urban	0.5–30	5–10
Rural	0.1–0.2	0.8
Great Lakes	0.4–3	1
Marine	0.05–2	0.5
Remote	0.02–0.5	0.1

PCBs emitted from some typical incinerators are given below.

Incinerators	Range	Ref.
Municipal (5)	0.01–1.5 ppm (fly ash)	Morselli et al. 1985
Municipal refuse and sewage, midwest U.S.	300–3000 ng m^{-3}	Murphy et al. 1985
Municipal waste, Ohio	260 ng m^{-3}	Tiernan et al. 1983
Combustion of coal and refuse, Iowa	2–10 ng m^{-3}	EPA 1988a

Levels of PCBs in the air at different locations are listed below.

Year	Location	Concn (ng m^{-3})	Ref.
1977	Contaminated industrial sites New England	0.7–1.26 × 10^6	Weaver 1984
<1979	Rural South Carolina	0.44	Bidleman and Christensen 1979
<1983	Rural Ontario	0.19	Singer et al. 1983
<1985	Industrial Toronto	0.14–7.6	Canviro 1985
	Urban Toronto	0.13–6.2	Canviro 1985
	Suburban, Toronto	0.54–5.9	Canviro 1985
	Rural, Toronto	0.19–5.4	Canviro 1985
	Toronto area	9.6, 9.63	Canviro 1985
1986	Canadian Arctic	<0.002–0.009	Hargrave et. al. 1988
1987	Urban Germany	0.670	Ballschmiter and Wittlinger 1991

It is safe to conclude that much of transport and deposition of PCBs in North America and Europe is atmospheric. Eisenreich et al. (1981) suggested that atmospheric inputs into Lake Michigan ranged from 2500 to 9000 kg/year, accounting for about 60% of the total deposition. Although this may represent an upper limit for atmospheric deposition in industrial areas, there is little doubt that detectable levels of PCBs found in Arctic lakes and Atlantic Ocean are almost entirely from atmospheric transport and deposition (Doskey and Andren 1981).

Total PCB levels in air and precipitation have generally declined in recent years, reflecting the ban on its production in many countries around the world. Concentrations of PCBs in rainwater in Chicago and rural areas of the U.K. were 119 and 14.9 ng/L, respectively (Murphy and Rzeszutko 1977; Wells and Johnstone 1978). In the continental U.S., vapor concentrations ranged from 1 to 10 ng/L (Richard and Junk 1981); for British coastal waters, the vapor concentration was <0.2–0.08 ng/L (Dawson and Riley 1977). Concentrations of 25 to 3200 mg/L were found in the rainfall at Fort Edward, NY, owing to the disposal and volatilization of PCB-containing wastes (Brinkman et al. 1980). The average adult male inhales approximately 20 m^3 of air per day. Assuming the breathable outdoor air at a typical urban site contains an average PCB concentration of 5 ng m^{-3}, the average intake per day through inhalation would be 100 ng. This estimate relates to background levels of PCBs in outdoor ambient air. PCB levels in certain indoor air may be higher than the outdoor air (ten times higher).

Water

PCBs are only slightly soluble in water, resulting in low dissolved levels in surface waters (Table 5.9). Unfiltered water samples, containing particulate matter, often have higher PCB residue levels; concentrations >100 ng/L have been reported for both marine and fresh waters. The probability of finding PCB in groundwater is comparable to that of surface waters (Page 1981). This poses a new and long-term hazard to potential potable water supplies in industrialized areas.

Levels of PCBs in the open waters of the oceans can reflect the environmental background levels in water. Concentrations reported for various seawaters include 0.04 to 0.59 ng/L in the North Pacific, 0.035 to 0.069 ng/L in the Antarctic, and 0.02 to 0.20 ng/L in the North Atlantic (Tanabe et al. 1983, 1984; Giam et al. 1978). Higher PCB levels have been detected in seawater from the North Sea, 0.3 to 3 ng/L; however, these sites were receiving an anthropogenic input (Boon and Duinker 1986). The low solubility of PCBs in surface waters limits them from reaching higher concentration levels in drinking water supplies (EPA 1988a). Data from the National Organic Monitoring Survey (NOMS, phase I, II, and III) conducted by U.S. EPA during 1975 to 1977 in finished water supplies of 113 cities nationwide, detected a very low detection frequency for PCB at detection limit (0.12 ppb). The more recent data on the residue levels of PCBs in surface waters are given in Table 5.10.

Table 5.9 PCB Levels in Marine and Fresh Waters

Year	Location	Average	Range
	Dissolved (ng/L)		
1974	English Channel[1]	0.19	0.15–0.30
1974	Irish Sea[2]	0.5	<0.2–1.0
1975	Mediterranean Sea[3]	2.0	<0.2–8.6
1975–1976	Grand River, Canada[4]	5.8	ND-100
1976–1977	Grand River, Canada[4]	3.7	ND-100
1975–1976	Saugeen River, Canada[4]	0.2	ND-3
1976–1977	Saugeen River, Canada[4]	0.3	ND-5
	Suspended Solids (µg/kg)		
1971–1972	Inland waters (19 states)[4]	—	ND-4000
1976	Tiber Estuary, Mediterranean[5]	297	9–1000
1977	Tiber Estuary, Mediterranean[5]	135	ND-380
1977	Baltic Sea[6]	5.0	0.3–139
1978	Dority Reservoir, U.S.[7]	99	70–130
—	Brisbane Estuary, Australia[8]	—	ND-50
1979–1980	Niagara River, Canada[9]	961	—

Note: ND, not detected; —, no data.
Sources: [1]Crump-Wiesner et al. 1974; [2]Dawson and Riley 1977; [3]Elder and Villeneuve 1977; [4]Frank 1981; [5]Pucetti and Leoni 1980; [6]Ehrhardt 1981; [7]Brinkman et al. 1980; [8]Shaw and Connell 1980; [9]Warry and Chan 1981.

Precipitation Samples

Based on the available monitoring data from the literature, the following typical ranges of PCBs (ng/L) were reported for various locations (Eisenreich et al. 1981):

Location	PCB Concn (ng/L)
Urban	10–250
Rural	1–50
Remote	1–30
Marine	0.5–10
Great Lakes	10–150
Antarctic (snow)	0.160–1.0 (Tanabe et al. 1983)

Mazurek and Simoneit (1985) have reviewed the PCB monitoring data in precipitation as shown in Table 5.11.

Soils and Sediments

Ninety-nine soil samples collected from rural and urban sites throughout Great Britain, identified PCBs in all samples (Creaser and Fernandes 1986); the range of PCB detected was 2.3–444 ppb (µg/kg). The mean and median values found were 22.8 and 7.2 ppb, respectively. PCB concentrations ranging from 4.5 to 47.7 were detected in soil samples near incineration facilities in South Wales and Scotland

Table 5.10 Levels of PCBs in Surface Waters at Various Locations Around the World

Year	Location	Concn (ng/L)	Reference
1978–1983	Western Lake Superior	0.63–3.3 (mean)	Baker et al. 1985
1974–1976	Lake Michigan	3.0–9.0 (mean)	Rodgers and Swain 1983
1979–1981	Lake Huron	0.49–17.15 (mean)	Rodgers and Swain 1983
1978–1979	Galveston Bay, Texas (8 sites)	3.1 (average)	Murray et al. 1981
1978	Western Lake Superior	0.5–2.0	Eisenreich et al. 1981
1979	Western Lake Superior	3.2–3.4	Eisenreich et al. 1981
1978	Central Lake Superior	0.4–7.6	Eisenreich et al. 1981
1980	Central Lake Superior	0.4–2.1	Eisenreich et al. 1981
1979	Eastern Lake Superior	0.3–6.0, 0.9–8.4	Eisenreich et al. 1981
1980	Eastern Lake Superior	0.3–1.8, 0.4–1.9	Eisenreich et al. 1981
<1983	Great Lakes	1–10	Eisenreich and Johnson, 1983
<1984	Lake Ontario (nearshore)	40–80	Simmons 1984
1984	Lake Ontario		
	dichlorobiphenyl	0.018	Oliver and Niimi 1988
	trichlorobiphenyl	0.150	Oliver and Niimi 1988
	tetrachlorobiphenyl	0.300	Oliver and Niimi 1988
	pentachlorobiphenyl	0.390	Oliver and Niimi 1988
	hexachlorobiphenyl	0.160	Oliver and Niimi 1988
	heptachlorobiphenyl	0.054	Oliver and Niimi 1988
	octachlorobiphenyl	0.010	Oliver and Niimi 1988
	total chlorobiphenyls	1.100	Oliver and Niimi 1988
1986	Lake Superior	0.34 (0.2–0.6)	Stevens and Neilson 1989
1986	Lake Huron	0.63 (0.2–2.3)	Stevens and Neilson 1989
1986	Georgian Bay	0.69 (0.3–1.4)	Stevens and Neilson 1989
1986	Lake Erie	1.38 (0.3–3.5)	Stevens and Neilson 1989
1986	Lake Ontario	1.41 (0.5–2.6)	Stevens and Neilson 1989
1986	Canadian Arctic	0.004–0.016	Hargrave et al. 1988
1988	Grimsby, Ontario, WTP; raw water	<20.0	MOE 1988
<1988	Grimsby, Ontario, WTP; drinking water	<20.0	MOE 1988
<1988	32 out of 163 wells monitored in industrial areas, New Jersey	60–1,270	EPA 1988a
1975–1977	Agricultural watersheds (11), Ontario, Canada	25–38	Frank et al. 1982
1976	New England industrial effluent	73,000–400,000	Weaver 1984
	Nipigon Bay pulp mill effluent	ND–65.0	Kirby 1986
1983–1984	Siskiwit Lake, a remote lake on an island in Lake Superior, Northern Ontario, with no point sources of PCBs	2.3	Swackhamer et al. 1988
<1987	St. Lawrence River	2.46	Germain and Langlois 1988
<1991	Nile Delta, Egypt	8.28–652.9	El-Gendy et al. 1991

Table 5.11 PCB (ng/L) in Precipitation Samples from North America and Europe

Year	Location	Concn (ng/L)	References
1973–1975	North America and Europe	<3–24	Murray and Andren 1992
1974–1976	North America and Europe	ND–230	Murray and Andren 1992
1976	North America and Europe	9–32	Murray and Andren 1992
1976–1977	North America and Europe	11–44	Murray and Andren 1992
1977–1978	North America and Europe	13	Murray and Andren 1992
1979	North America and Europe	<0.6–3.5	Murray and Andren 1992
1979–1980	North America and Europe	12–75	Murray and Andren 1992
1981	North America and Europe	3.65 (1–61)	Murray and Andren 1992
1983	North America and Europe	5.90 (0.6–48)	Murray and Andren 1992
1983–1985	North America and Europe	ND–143	Murray and Andren 1992
1984	North America and Europe	ND–17	Murray and Andren 1992
1986	North America and Europe	1.3–8	Murray and Andren 1992
1976	Great Lakes		
	rain	8–40	Strachan and Huneault 1979
	snow	18–43	Strachan and Huneault 1979
1983	Lake Superior	0.6–48.0	Strachan 1985
1983–1984	Siskiwit Lake, remote island, Lake Superior	13–17	Swackhamer et al. 1988
1984	Lake Superior	0.02–1.76	Strachan 1988
1984	Cree Lake, Saskatchewan, Canada	1.3–5.0	Strachan 1988
1984	Kouchibouguac, New Brunswick, Canada	0.61–17.0	Strachan 1988
<1985	Ontario, Canada, rainwater	10–50	Canviro 1985
1986	Canadian Arctic	0.74–7.3	Gregor and Gummer 1989

during 1984–1985 (Eduljee et al. 1985, 1986). Japanese soils were analyzed to contain levels as high as 100 ppb; however, 40% of the samples had levels <10 ppb (Creaser and Fernandes 1986). PCB concentrations ranging from <1 to 33 ppb were found in the soils of the Everglades in Florida (Requejo et al. 1979). Soils from 37 states in the U.S. were analyzed as part of the National Soils Monitoring Program and reported to contain PCBs in only 2 of 1,483 soil samples (Carey et al. 1979a); the detection limit in this study was only 0.05 to 1 ppm, not low enough to detect the mean and median values reported in the Great Britain study. The former analytical technique detected PCBs in soils from five urban areas (43–156 samples per site) in 1971; positive detections were reported for three areas with PCB levels of 0.02 to 11.94 ppm. The highest concentration reported, 11.94 ppm was detected in 1 of 55 samples from Gadsden, Alabama (Carey et al. 1979b). Soils collected in 1981 from contaminated industrial sites in New England states contained 4,400–99,000 ppm (µg/g) of PCBs (Weaver 1984).

Sediments are the primary sinks for PCBs and will continue to contaminate the food chain for years to come. In the early 1980s, the levels in some major lakes such as St. Clair have started to decline. Yet, the municipal and industrial sources continued to contribute to the total PCB burden in most industrial zone waters. For example, sediment residues of 10 to 20 ppb (dry weight) in Lake Michigan near Chicago, Milwaukee, and Green Bay were reported (Frank 1981). By contrast, the whole lake average was only 9.7 ppb. Similar trends were reported for other major lakes such as, Huron, St. Clair, Erie, and Superior. However, the residues near the waste outfalls ranged from 2000 to 500,000 ppb (Elder et al. 1981). Sediment levels in the California Bight (Los Angeles) generally exceeded 1000 ppb and in some areas reached 10,000 ppb (Young and Heesen 1978). Similar or slightly lower values were reported for the harbors of most major cities and industrial zone estuaries around the world. Off-shore sediments in the Mediterranean and Baltic Seas contained lower PCB residues, <5 ppb (Basturk et al. 1980).

Sediments from four remote high altitude lakes in the Rocky Mountain National Park were found to contain background levels of PCBs: 0.098 to 0.54 ppm from natural deposition (Heit et al. 1984). Sediment core samples from the Milwaukee harbor, which has received industrial effluents containing PCBs, were found to contain levels of 1.03 to 13.4 ppm (Christensen and Lo 1986). Analysis of sediments from 13 selected streams in the Potomac River Basin found (1) a maximum PCB level of 1.2 ppm in one stream (2) seven of the streams contained zero or trace amounts of PCBs and (3) the rest contained 10–80 ppb (Feltz 1980). Upper sediment layers from the Hudson River and New York harbor in 1977 contained Aroclor 1254 (0.56–1.95 ppm) and Aroclor 1242 (3.95–33.3 ppm) (Bopp et al. 1982). Analysis of surficial sediments from the Great Lakes and associated waters found Aroclor 1254 (2.5–251.7 ppb) with the higher values detected in Lake Erie (Thomas and Frank 1981). An average Aroclor 1260 concentration of 120 ppb has been found in sediment samples from eight sites along the coast of Maine (Ray et al. 1983). More sediment PCB data are given in Table 5.12.

Aquatic Plants and Invertebrates

Aquatic plants sorb PCBs from water column. BCFs, though variable, generally fall within a range of 1×10^4 to 5×10^4. Since dissolved PCB levels are low, residues in natural population of algae seldom exceed 1 mg/kg (wet weight). Attached species may mobilize PCBs directly from sediments (Moore and Ramamoorthy 1984). The rate of sorption seems to depend on the degree of chlorination of PCBs (Mrozek and Leidy 1981). These differences are reflected throughout the food chain and may significantly affect the residue levels in fish; for example, BCFs for Aroclors 1016 and 1260 in fathead minnows varied from 4.25×10^4 to 1.94×10^5 (Veith et al. 1979). The bioconcentration potential of PCBs is species dependent. BCF for freshwater plankton was reported to be 1×10^4 (Oliver and Niimi 1988). Movement of PCBs through the food chain is significant and organisms higher in the chain tend to have greater tissue concentration of PCBs.

PCBs are significant and widespread contaminants of both marine and freshwater invertebrates. Concentrations of >5,000 µg/kg (dry weight) for mussels on the Atlantic and Pacific coasts of the U.S. had been reported (Goldberg et al. 1978). Comparable or slightly lower residues had been detected in benthic invertebrates inhabiting coastal waters of many other industrialized countries. In the Medway Estuary, U.K., shrimp *Crangon vulgaris* had whole body burdens ranging from trace to 275 µg/kg (wet weight), whereas residues in several species of polychaetes and crabs from the Brisbane River Estuary(Australia) reached 520 µg/kg (wet weight)

Table 5.12 Levels of PCBs (ng/g) in Sediments from Several Locations Around the World

Year	Location	Concn (ng/g)	References
1904–1922	Lake Michigan	7.6	Swackhamer and Armstrong 1986
1919–1937	Lake Michigan	18.6	Swackhamer and Armstrong 1986
1934–1952	Lake Michigan	42.0	Swackhamer and Armstrong 1986
1949–1967	Lake Michigan	54.8	Swackhamer and Armstrong 1986
1964–1980	Lake Michigan	91.1	Swackhamer and Armstrong 1986
1931–1948	Dart Lake, WI	2.2	Swackhamer and Armstrong 1986
1948–1962	Dart Lake, WI	8.8	Swackhamer and Armstrong 1986
1962–1971	Dart Lake, WI	20.4	Swackhamer and Armstrong 1986
1971–1981	Dart Lake, WI	19.2	Swackhamer and Armstrong 1986
1943	Lake Superior	11.0	Eisenreich et al. 1983
1950	Lake Superior	17.5	Eisenreich et al. 1983
1955	Lake Superior	22.0	Eisenreich et al. 1983
1961	Lake Superior	44.8	Eisenreich et al. 1983
1965	Lake Superior	61.0	Eisenreich et al. 1983
1967	Lake Superior	64.0	Eisenreich et al. 1983
1970	Lake Superior	88.5	Eisenreich et al. 1983
1972	Lake Superior	109.0	Eisenreich et al. 1983
1975	Lake Superior	76.0	Eisenreich et al. 1983
1954–1963	Emerick Lake, WI	89.0	Swackhamer and Armstrong 1986
1963–1971	Emerick Lake, WI	12.8	Swackhamer and Armstrong 1986
1971–1975	Emerick Lake, WI	12.7	Swackhamer and Armstrong 1986
1975–1978	Emerick Lake, WI	26.3	Swackhamer and Armstrong 1986
1965–1974	Little Pine Lake, WI	2.8	Swackhamer and Armstrong 1986
1974–1978	Little Pine Lake, WI	7.8	Swackhamer and Armstrong 1986
1978–1982	Little Pine Lake, WI	2.6	Swackhamer and Armstrong 1986
1984	Lake Ontario		
	dichlorobiphenyl	ND	Oliver and Niimi 1988
	trichlorobiphenyl	22	Oliver and Niimi 1988
	tetrachlorobiphenyl	200	Oliver and Niimi 1988
	pentachlorobiphenyl	180	Oliver and Niimi 1988
	hexachlorobiphenyl	93	Oliver and Niimi 1988
	heptachlorobiphenyl	48	Oliver and Niimi 1988
	octachlorobiphenyl	22	Oliver and Niimi 1988
	total chlorobiphenyls	570	Oliver and Niimi 1988
1984	Lake Ontario, suspended sediments		
	dichlorobiphenyl	5.4	Oliver and Niimi 1988
	trichlorobiphenyl	27	Oliver and Niimi 1988
	tetrachlorobiphenyl	120	Oliver and Niimi 1988
	pentachlorobiphenyl	130	Oliver and Niimi 1988
	hexachlorobiphenyl	81	Oliver and Niimi 1988
	heptachlorobiphenyl	44	Oliver and Niimi 1988
	octachlorobiphenyl	20	Oliver and Niimi 1988
	total chlorobiphenyls	440	Oliver and Niimi 1988
1982–1983	Lake Ontario	1,300–1,900	Oliver et al. 1989
1983–1984	Lake Ontario	350–570	Oliver et al. 1989
1984–1985	Lake Ontario	410–680	Oliver et al. 1989
1981–1986	Lake Ontario	80–290	Oliver et al. 1989
1986	Lake Superior	5.32–11.73	Baker and Eisenreich 1989

Table 5.12 Levels of PCBs (ng/g) in Sediments from Several Locations Around the World (Continued)

Year	Location	Concn (ng/g)	References
1983–1984	Siskiwit Lake, remote island Lake Superior	48	Swackhamer et al. 1988
1980	Lake Huron	34 (12–51)	Oliver and Bourbonniere 1985
	Lake St. Clair	29	Oliver and Bourbonniere 1985
	Western Lake Erie	300 (140–660)	Oliver and Bourbonniere 1985
	Central Lake Erie	131 (38–190)	Oliver and Bourbonniere 1985
	Eastern Lake Erie	91 (37–140)	Oliver and Bourbonniere 1985
1981	Contaminated industrial sites, New England	$190 \times 10^3 - 190{,}000 \times 10^3$	Weaner 1984
1985	St. Lawrence River	10.0–530.0	Kaiser et al. 1990
1979	Thunder Bay	10–600	MOE 1986
1992	Green Bay, WI	710–810	Ankley et al. 1992
1975	Mediterranean surface sediments	0.8–0.9	Fowler 1987
1976	Mediterranean surface sediments	1.5–33	Fowler 1987
1977	Mediterranean surface sediments	0.6–8.9	Fowler 1987
1990	Atlantic, coastal, and nearshore sediments	0.2–480,000	Fowler 1990
1990	Mediterranean coastal and nearshore sediments	0.3–7,420	Fowler 1990
1990	Pacific coastal and nearshore sediments	0.5–2,000	Fowler 1990
1990	North Sea coastal and nearshore sediments	11.5–134	Fowler 1990
1990	Baltic coastal and nearshore sediments	8.4–212	Fowler 1990
<1983	Gulf of Finland	46–60	Perttila 1985
	Gulf of Bothnia	5–15	Perttila 1985
<1991	Nile Delta sites	6.9–3, 154.9	El-Gendy et al. 1991
1976–1979	Eastern Adriatic	28	Picer and Picer 1991
1980–1983	Eastern Adriatic	23	Picer and Picer 1991
1984–1986	Eastern Adriatic	16	Picer and Picer 1991
>1986	Eastern Adriatic	10	Picer and Picer 1991
1988–1989	Finnish Lakes	2.1–379	Paasivirta et al. 1990

(Shaw and Connell 1980; van den Broek 1979). Since PCB levels in sediments are higher compared to those in water, zooplankton are more likely to have lower body burdens than benthic species. Mixed microplankton collections from the Mediterranean provided average residues of 65 µg/kg, dry weight (Fowler and Elder 1980–1981). However, peak concentrations of 3,300 and 720 µg/kg, wet weight, were reported for collections from the Baltic Sea during 1974 and 1976, respectively (Linko et al. 1979). This was probably due to unusually large seasonal lipid levels in plankton tissues.

Fish

PCBs were widespread contaminants of fish tissues and residues continue to decrease with time in recent years. Based on 1977 data, residues in American shad increased from an average of 2.0 to 6.0 mg/kg, wet weight, during the annual migration up the Hudson River (Pastel et al. 1980); whereas, other nonmigratory species from the same river contained PCB residues of 1.8 to 3.8 mg/kg (Skea et al. 1979). Bluefin tuna, a carnivorous highly migratory fish species collected from the North Aegean Sea (1975–1979), had average PCB residues of 2.6 mg/kg compared with 0.5 to 0.7 mg/kg for smaller, more sedentary fish (Kilikidis et al. 1981). The decline in PCB levels has been dramatic in some instances but less in others. For example, levels in brown trout from the River Rhone, Switzerland averaged 7.2 and 0.2 mg/kg in 1972 and 1975, respectively (Schweizer and Tarradellas 1980). Whereas, smallmouth bass from Lake Erie had mean residues of 0.3 and 1.7 mg/kg in 1968 and 1975, respectively (Frank et al. 1978). During 1980–1981, 315 fish samples were collected from 107 stations nationwide as part of the U.S. National Pesticide Monitoring Program; PCB residues were detected in 94% of all whole fish samples with a geometric mean concentration of all Aroclors at 0.53 µg/g, wet weight (Schmitt et al. 1985). Since the samples were whole fish, the concentrations may not represent the actual oral exposure levels. Commercial fish samples (62) collected in 1980, mainly from Lake Ontario, contained PCB levels of 0.11 to 4.90 ppm (Ryan et al. 1984). Seasonal variations in PCB concentrations could be quite marked in fish. A three- to sevenfold seasonal difference in PCB residues in perch and roach (10–30 mg/kg and 10–70 mg/kg, respectively) collected from a bay on the Baltic coast was reported (Edgren et al. 1981). Similarly, PCB residues in whiting from the Medway Estuary, U.K. peaked in April (at 0.16 mg/kg), declining to 0.01 mg/kg by October (van den Broek 1979).

Food is probably a more important route of uptake in fish than water under natural conditions. Retention from the diet is generally high, (>50%), but decreases with the amount of dietary PCB (Leatherland and Sonstegard 1980). Uptake from water has been reported to be rapid under laboratory conditions; concentration factor (CF) for fathead minnows on exposure to contaminated water for 32 days was 194,000 for Aroclor 1260, 100,000 for 1254, 70,500 for 1248, and 42,500 for 1260 (Veith et al. 1979). Absorption efficiencies and elimination half-lives of PCBs in aquatic organisms depends on the type of PCBs and the species. Although elimination constants were relatively small for *Pontoporeia hoyi* (amphipod) and *Mysis relicta* (Mysid), the amphipod was six times more efficient in eliminating PCBs from its body. BCF and $t_{1/2}$ (elimination) for *P. hoyi* were 1.02×10^5 and 45.6 days. For *M. relicta*, the values were 4.4×10^5 and 222 days (Evans and Landrum 1989).

Absorption efficiencies of 31 PCBs of the homologs from dichlorobiphenyls to decachlorobiphenyls ranged from 62 to 85% (Table 5.13), with an average of $75 \pm 6\%$. Absorption does not seem to relate to degree of chlorination; whereas, $t_{1/2}$ (elimination) ranges from 5 to 105 days for whole fish and for muscle tissue, <5 to 127 days. It was reported from QSAR analysis that elimination was fast for lighter PCBs, those PCBs with chlorine atoms on the 0-positions and those PCBs with two adjacent carbons with no substitution (RTP 1994).

Table 5.13 Elimination Half-Lives (t$_{1/2}$) of PCB Homologues from Rainbow Trout Administered with Single Oral Dose

Homolog	Half-Life (t$_{1/2}$) in days	
	Whole Fish	Muscle Tissue
Dichlorobiphenyls	5–85	<5–56
Trichlorobiphenyls	190–196	81–86
Tetrachlorobiphenyls	44–890	29–127
Pentachlorobiphenyls	155–>1,000	62–101
Hexachlorobiphenyls	850–>1,000	77–91
Octachlorobiphenyls	>1,000	78
Nonachlorobiphenyls	>1,000	84
Decachlorobiphenyls	>1,000	122

Source: Niimi and Oliver 1983

Similar study using juvenile sole (*Solea*) exposed to Clopen A40 (technical mixture of PCBs) showed that elimination, on a lipid basis, was highest from muscle tissue and lowest from brain tissue (Boon 1985). This is in agreement with Niimi and Oliver (1983) in that PCB uptake by organisms was nonselective and did not correlate with the type of PCB administered. Metabolism of PCBs in fish is very poor compared to other animal groups. This could be due to relatively lower MFO (Mixed Function Oxidase) activity in fish, which is responsible for the metabolism of xenobiotics. Recent studies, however, observed a significant metabolism of PCBs in fish (Janz and Metcalfe 1991). PCB metabolic pathway in birds and animals is mainly by (1) inserting an oxygen atom(s) between two adjacent carbon atoms of the benzene rings, and (2) forming an unstable arene oxide intermediate by hepatic microsomal oxygenases. Arene oxide is then converted to hydroxylated PCBs and subsequently conjugated with sulphates and glucoronides and excreted via urine or bile, respectively (Sundstrom et al. 1976). PCBs with at least one adjacent pair of meta, para carbon atoms, which are free (nonsubstituted) are rapidly metabolized by allowing space for enzymes. The second pathway for metabolizing PCBs seems to involve insertion of oxygen between ortho- and meta-positions on the benzene rings (Norstrom 1987). Polar bears easily metabolize PCBs with (4-), (3,4-), (2,3,5-)- and (2,3,5,6-) chlorine substitutions on one ring and those with unsubstituted meta-, para-positions. PCBs with free meta positions do not seem to get metabolized.

The metabolism of PCB in animal tissues is characterized by the disappearance or reduction in the concentration of LCBPs (Lower Chlorinated Biphenyls). This has been observed in rats, birds, cows, and carp. Actively metabolizing rats consistently retain a greater proportion of HCBPs (Higher Chlorinated Biphenyls) of the standard Aroclor 1254. Predamaging the rat liver with CCl$_4$ failed to produce any changes in dose PCB composition, thus confirming preferential metabolism of PCB components with liver (or comparable organ) as the primary active site. PCB analyses in environmental samples reveal an isomer pattern similar to that of Aroclor 1254 or 1254/1260 (Task Force on PCB 1976), although North American sales figures indicate the primary input of PCB is LCBP, particularly Aroclor 1242. This indicates that LCBPs are less persistent than HCBPs in the environment and many recent studies confirm this observation. Monsanto Company reports in the 1970s showed the order of microbial degradation in activated sludge test unit was biphenyl>Aroclor 1221> Aroclor 1016>Aroclor 1254. These studies pointed out that mono- to tetrachloroisomers could be degraded. A variety of biota are capable of metabolizing LCBP up to 6 Cl atoms into polar metabolites. Conjugated and free forms of PCB metabolites have been identified in feces and urine of animals. Present evidence indicates that hydroxylation with the formation of arene oxide intermediary is the primary metabolic mechanism for the breakdown of PCBs. However, direct hydroxylation of PCBs yielding a single monohydroxylated product without the arene oxide intermediary has also been reported (Gardner et al. 1976). Studies in the metabolism of a related compound (biphenyl) by liver microsomal preparations from 11 species of animals reported hydroxylation yielding 2- and 4-hydroxy biphenyls. Species differences with different enzyme systems might account for the variation in results.

Certain PCBs accumulation (with 4,4′-chlorine substitutions) by wild guillemot was attributed to the bird's inability to hydroxylate those PCBs (Jansson et al. 1975). Isomerization and dechlorination reactions have been implicated in the metabolism of HCBPs (McKinney 1976; Hutzinger et al. 1974). However, detection of some dibenzofuran (PCDFs) structures in some metabolites combined with lower ratios of PCBs to PCDFs aroused concern regarding possible metabolic formation and accumulation of PCDFs in the liver (Kuratsune et al. 1976).

Bioconcentration factors (BCFs) for PCBs were reported as:

Algae	1×10^4–10^5	Ernst 1984; Eisler 1986
Plankton (freshwater)	1×10^4	Oliver and Niimi 1988
Invertebrate (marine and fresh waters)	60–3.4×10^5	Table 5.14

Table 5.14 Bioconcentration Factors (BCFs) for the Uptake of Aroclor 1254 by Freshwater Aquatic Organisms (Whole Tissue)[a]

Organism	Exposure (d)	BCF
Cladoceran (*Daphnia magna*)	4	4.7×10^4
Amphipod (*Gammarus pseudolimnaeus*)	4–21	2.4–2.7×10^4
Phantom midge (*Chaborus punctipennis*)	4–14	2.3–2.5×10^4
Mosquito larvae (*Cules tarsalis*)	4	1.8×10^4
Grass shrimp (*Palaemonotes kadiakensis*)	21	1.7×10^4
Grass shrimp (*Palaemonotes kadiakensis*)	4	1.2×10^4
Crayfish (*Oronectes nais*)	21	5.1×10^3
Crayfish (*Oronectes nais*)	4	1.7×10^3
Protozoa (*Tetrahymnea pyriformis*)	4	60

[a] Exposure concentration of Aroclor was 1.0–1.6 µg/L.
Sources: NAS 1979; EPA 1980b; Eisler 1986.

Since PCBs are only slightly soluble in water and mostly associated with particulate organic carbon or dissolved organic carbon, the significance of PCBs are difficult to relate to a water column of varying organic carbon content. Organisms

Table 5.15 Bioconcentration Factors (BCFs) for the Uptake of Aroclor 1254 by Marine Organisms

Organism (Genus)	Exposure (d)	Tissue	BCF	References
American oyster (1)	168	Soft tissues	8.5×10^4[a]	Ernst 1984
Rotifer (2)	45	Dry tissue	5.1×10^4[b]	EPA 1980b
Rotifer (2)	45	Lipid	3.4×10^4[b]	EPA 1980b

Note: (1) *Crassostrea virginica*; (2) *Brachionus plaicatilus*
[a] Exposure concentration was 5.0 µg/L of Aroclor 1254.
[b] Exposure concentration was not reported.
Source: Eisler 1986

on lower trophic levels might uptake PCBs directly from water especially when there is less organic carbon in the water column. Invertebrates that feed on detritus in the sediments are reported to accumulate PCBs. The sandworm (*Nereis virens*) compared to the clam (*Mercenaria*) or the grass shrimp (*Palaemonetes pugio*) accumulated more PCBs under identical exposure conditions. The reason is that sandworms live on the sediment surface feeding on detritus and sediments rich in organic matter. Hence, sandworms are continually exposed to multiple exposure routes for PCBs, sediment ingestion, food, and organic carbon in the water column; whereas, clams are water column dwellers for feeding and respiration, and grass shrimp are in close proximity to aquatic plants.

From the data available in the Lake Ontario ecosystem, the biomagnification concentrations were reported as below:

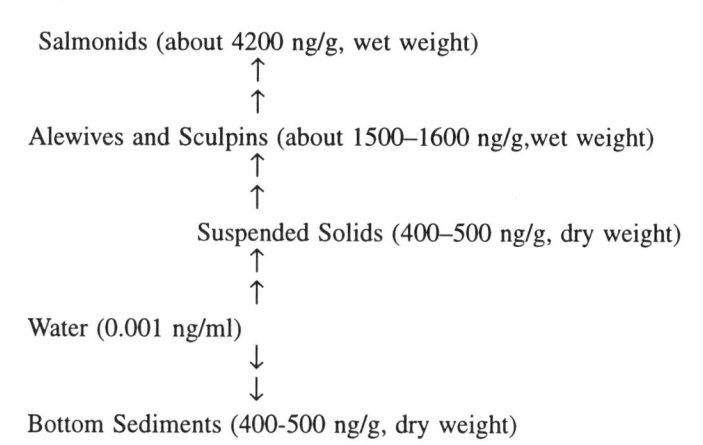

Salmonids (about 4200 ng/g, wet weight)
↑
↑
Alewives and Sculpins (about 1500–1600 ng/g, wet weight)
↑
↑
Suspended Solids (400–500 ng/g, dry weight)
↑
↑
Water (0.001 ng/ml)
↓
↓
Bottom Sediments (400-500 ng/g, dry weight)

The increase in concentrations of PCBs in organisms seems to parallel the lipid content of organisms, which is 0.5% in plankton and 11% in salmonids.

Biomagnification varies among the trophic levels. The reported data from the literature are summarized in Table 5.16.

Generally, higher chlorinated PCBs were found at greater concentrations at higher trophic levels of the ecosystem, as given in Table 5.17.

It has been estimated that PCBs could magnify as high as 25×10^6 from water concentration levels to those residues found in the top predator, namely, the bald

Table 5.16 BCFs for PCBs Between Successive Trophic Levels

Trophic Level	Lake Michigan	Lake Ontario* (West; East)	Mean
Mysids: plankton	1.3	1.3; 1.5	1.4
Amphipods: plankton	3.2	5.4; 8.5	5.7
Fish: plankton	12.9	21.8; 19.1	17.9
Amphipods: mysids	2.4	4.3; 5.9	4.2
Fish: mysids	9.2	17.3; 13.2	13.2
Fish: amphipods	4.0	4.0; 2.3	3.4
Amphipods: sediments	12.2	4.0; 2.3	3.4

Note: *calculated data
Sources: Evans et al. 1991; RTP 1994; Borgmann and Whittle 1991.

Table 5.17 Species with Greater PCB Concentration

PCB Congener	Species with Higher PCB Levels
Tri- and tetrachlorobiphenyls	Water and lower trophic levels (plankton, mysids, and amphipods)
Penta- and hexachlorobiphenyls	Uniformly distributed through trophic levels
Hexa- and heptachlorobiphenyls	Small fish and salmonids

eagle (Norstrom et al. 1978). However, it has to be pointed out the magnification of PCBs at higher trophic levels is via exposure from their food intake. The concentration of PCBs in some adult piscivorous birds are given in Table 5.18.

Table 5.18 Concentrations (μg/g, Wet Weight) of Total PCBs in Adult Piscivorous Birds

Species	Location	PCB Concn.	References
Herring gull	SW shore of Lake Superior	7.4–34	Kozie and Anderson 1991
Black-crowned night heron	Various locations in U.S.	6.3–110	Ohlendorf and Miller 1984
Black-crowned night heron	Lake Michigan	23–127	Heinz et al. 1985
Bald eagle	Lake Superior, Michigan Island	14–40	Kozie and Anderson 1991

Significant differences have been observed in the foodchain movement of PCBs (Table 5.19).

PCB Levels in Food Items

Table 5.20 lists the amounts of PCBs found in raw domestic agricultural commodities during 1970–1976. These were analyzed as part of U.S. Federal Monitoring Programs of U.S. Food and Drug Administration (FDA) and the U.S. Department of Agriculture. It is evident from Table 5.20 that fish are the primary food source containing background PCB levels. Since the early 1960s, the U.S.

Table 5.19 Type of PCBs Found in Highest Concentrations in Foodchain Ladders

PCBs detected	Species in Which Detected
2,2',4,4',5,5'-hexachlorobiphenyl	Fish, piscivorous birds, mammals
2,3,4,4',5'-pentachlorobiphenyl	Alewife-herring gull food chain (Lake Ontario)
2,2'3,4,4',5'-hexachlorobiphenyl	Alewife-herring gull food chain (Lake Ontario)
2,2'3,4,4',5',6'-heptachlorobiphenyl	Alewife-herring gull food chain (Lake Ontario)
2,2'3,3'4,4',5,5'-octachlorobiphenyl	Alewife-herring gull food chain (Lake Ontario)
2,4,4',5'-Tetrachlorobiphenyl	Blue heron and gannet
2,3',4,4'-Tetrachlorbiphenyl	Blue heron and gannet
2,2',4,4',5'-Pentachlorobiphenyl	Blue heron and gannet
2',3,4,4',5-Pentachlorobiphenyl	Blue heron and gannet
3,3',4,4'-Tetrachlorobiphenyl	Spottail shiner to Forster's tern eggs
3,3',4,4',5-Pentachlorobiphenyl	Spottail shiner to Forster's tern eggs
3,3',4,4',5,5'-Hexachlorobiphenyl	Spottail shiner to Forster's tern eggs, BCFs; 0.17,64 and 176, respectively, for PCBs listed
Aroclor 1242:1254:1260	
PCBs 8/5 to PCBs 70/76 of the di- to tetrachlorocongeners	Canadian Arctic cod muscle ratio, 0.6:3:1
Greater chlorinated PCBs	Large marine fish
Lesser chlorinated PCBs	Smaller marine fish
Penta- and hexachlorobiphenyls	Seals
Hexa- and heptachlorobiphenyls	Polar bears in Canadian Arctic
IUPAC #s PCB-99,-153,-138,-180, PCB-170,-194	Liver (99%) and adipose tissue (86%) of polar bears

Sources: Norstrom 1987; Muir et al. 1988; Norheim et al. 1992; Oliver and Niimi 1988; Kubiak et al. 1989; Metcalfe and Metcalfe 1993;

FDA has conducted nationwide Total Diet Studies known as the Market Basket Surveys. These annual surveys analyze ready-to-eat foods collected in markets from a number of cities across the U.S. The estimated dietary intake of PCBs by the average adult (70 kg) is approximately 0.008 µg/g/day; the average daily intake via diet would be 560 ng. Dietary exposure is likely the major source of PCB exposure in humans. Infants and toddlers are not exposed to any detectable PCBs in foods (ATSDR 1992). The primary source of PCBs in the diet has been meat–fish–poultry (Gartrell et al. 1985, 1986); and the source of PCBs in the meat–fish–poultry composite according to U.S. FDA is almost always due to the fish component (Jelinek and Corneliussen 1976). This suggests that persons consuming less amounts of fish will be exposed to lower amounts of PCBs.

The Total Diet Study (Market Basket Survey) conducted by the U.S. FDA monitors the temporal trends of PCBs in food. A total of 234 food items, which represents 100% of the American diet in retail outlets in three cities in each four regions of the U.S. and then analyzed for PCBs and a variety of other contaminants. Analysis of the available data from this survey for the period 1971–1989 shows a dramatic decrease in PCB levels in diet. Today's levels are less than 1/1000th of the levels measured in the mid-1970s and the dietary intake of PCBs has decreased from 6.9 µg/day in 1971 to 0.05 µg/day during 1987–1990. Since 1980 (0.7 µg/day), dietary exposure has been declining steadily to current levels of < 0.05 µg/day (Regulatory Network, Inc. 1992).

Table 5.20 Aroclor Residues in Raw Domestic Agricultural Commodities During 1970–1976 Survey

Commodity	# Samples Analyzed	Percent with Positive Detections	Average Concn (ppm)[a]
Fish	2,901	46.0	0.892
Shellfish	291	18.2	0.056
Eggs	2,303	9.6	0.072
Red meat[b]	15,200	0.4	0.008
Poultry	11,340	0.6	0.006
Fluid milk	4,638	4.1	0.067
Cheese	784	0.9	0.011

[a] Average fall samples, both positive and negative. [b] Fiscal years 1972–1976.
Source: Duggan et al. 1983; ATSDR 1989.

Microbial Transformations

PCBs exist in the environment as complex mixtures of several different congeners with varying degrees of chlorine content. Their biodegradation requires broad-acting enzyme systems. Naturally occurring microorganisms from common soil bacteria to fungi are capable of degrading mixtures of PCBs under laboratory conditions. Most of the bacteria from soils and sediments metabolize only the LCBPs (mono-tetra) (Furakawa and Matsumura 1976; Liu 1982; Hankin and Sawhney 1984), although some strains are capable of degrading HCBPs (penta, hexa, and hepta) (Bopp 1986; Bedard et al. 1987a,b). The study on the effect of the position of chlorine substitution on the rate of biodegradation by *Alcaligenes* sp. and *Acinetobactor* sp. (Furakawa et al. 1978) can be summarized as follows:

- Rate of biodegradation decreased with increase in chlorine content of PCBs
- PCBs with two chlorine atoms in the ortho-positions were resistant to degradation except 2,4,6-trichlorobiphenyl
- PCBs with all chlorine atoms on the same ring were biodegraded faster than PCBs with chlorine atoms on both phenyl rings
- Tetra and penta chlorobiphenyls with chlorine atoms at 2- and 3-positions of one ring were more susceptible for degradation than other tetra- and pentachlorobiphenyls
- Ring cleavage occurs mainly with nonchlorinated or lesser chlorinated biphenyls (Furakawa 1986)

Bedard and Haberl (1990) proposed from a study involving eight bacterial strains that types of breakdown products were dependent on the strain and also on the chlorine atom positioning pattern on the reacting ring, which supported the findings of Furakawa (1986). The major degradation products were chlorobenzoic acids and chloroacetophenols which were further degraded microbially resulting in total mineralization. Other microorganisms that were reported to degrade PCBs were *A. eutrophus H850, Pseudomonas* sp. *LB 400* (Nadim et al. 1987); filamentous fungus, *Aspergillus niger,* the wood-decay white-rot fungus *Phanerochaete chrysosporium* (only low concentrations of PCBs) (Dmochewitz and Ballschmiter 1988);

P. chrysoporium completely degraded HCBPs (Eaton 1985), at low concentrations of about 250 ppb of Aroclor 1254; *LB 400* degraded higher concentrations (1800 ppb) of 2,4,5,2',4',5'-hexachlorobiphenyl and *H 850* degraded 10,000 ppb of Aroclor 1254 (Bedard et al. 1987a). No aerobic microorganisms have been shown to degrade Aroclor 1260 or Clopen A60 (Abramowicz 1990).

Anaerobic microorganisms, present in bottom sediments of rivers were shown to reductively dechlorinate tri-, tetra- and pentachlorophenols (Brown et al. 1984). Dechlorination occurred exclusively from the meta and para positions. Typical pathway was 2,3,4,3',4'-pentachlorobiphenyl → 2,4,3',4'-tetrachlorbiphenyl → 2,4,3'-trichlorobiphenyl → 2,3'-dichlorobiphenyl → 2-monochlorobiphenyl. Anaerobic microorganisms exhibit a broad dechlorination range of activity including Aroclor 1260 (Abramowicz 1990). The removal of chlorine atoms from meta and ortho positions would significantly reduce mammalian toxicity of PCBs (Safe et al. 1985). Microbial degradation has been reported in a number of lakes in the U.S., thus confirming the widespread presence of anaerobic microorganisms (Abramowicz 1990). Environmental fate of PCBs is primarily due to atmospheric transport and sedimentation, with minor roles played by microbial and photolytic degradations.

Fugacity (tendency of a chemical to escape from a given environmental medium), at equilibrium, will be equal in all media/compartments (Murphy et al. 1985). Levels of I, II, and III of fugacity modeling can be used to estimate partitioning of chemicals in environmental compartments (Table 5.21), movement, and behavior of chemicals within different environmental compartments depending on specific physical properties of chemicals related to that compartment.

Level I fugacity calculations indicate LCBPs tend to partition more favorably to soil and sediment than to air and other media (Table 5.21), compared to HCBPs which tend to partition to soil and sediment and fish than other environmental media.

Level II fugacity calculations show that atmospheric removal processes are predominant for LCBPs. Reaction in soil becomes dominant for HCBPs as the major removal process. Removal by advection dominates removal reactions for all PCBs with higher chlorine content. Predicted values from fugacity modeling is about 6 years, compared to actual reported residence times of up to 10 years. The rate of transport between compartments and persistency are calculated by Level III fugacity modeling: (1) LCBPs, rates of transport are greatest to water and to a lesser extent, air, with LCBPs persisting in these compartments; (2) HCBPs, removal from water to sediments with sediments becoming a major sink for HCBPs. Very high concentration of higher homologs are encountered in sediments and fish. If discharged into soil, most PCBs remain in soil with some evaporation. The overall residence time increases with the chlorine content of PCBs to 9 years (calculated $t_{1/2}$ — 6 years). Evaporation is slow because of high PCB sorption to soils. The calculated behavior of all PCBs are consistent with the observed behavior. They accumulate and persist in sediments and soils. They show appreciable atmospheric deposition, sedimentation, bioaccumulation/biomagnification, and long-range transport (Mackay et al. 1992).

Table 5.21 Distribution of PCBs and Exposure Routes in Environment

PCB	Air	Soil	Sediment	Water	Fish	Suspended Sediment
2-Chlorobiphenyl	42.5	53.2	1.2	3.0	0.003	0.037
3-Chlorobiphenyl	29.2	67.4	1.5	1.9	0.004	0.047
4-Chlorobiphenyl	22.5	73.2	1.6	2.6	0.004	0.051
2,2'-Dichlorobiphenyl	14.1	82.9	1.8	1.2	0.005	0.057
2,5-Dichlorobiphenyl	3.4	93.6	2.1	0.8	0.005	0.065
4,4'-Dichlorobiphenyl	1.9	95.3	2.1	0.5	0.005	0.066
2,2',5-Trichlorobiphenyl	4.9	92.7	2.1	0.3	0.005	0.064
2,4,6-Trichlorobiphenyl	3.3	94.1	2.1	0.3	0.005	0.065
2,2',5,5'-Tetrachlorobiphenyl	0.8	96.8	2.1	0.09	0.005	0.067
2,2',4,5,5'-Pentachloro-biphenyl	0.31	97.4	2.16	0.044	0.005	0.068
2,2',3,3',4,4'-Hexachloro-biphenyl	0.026	97.7	2.17	0.011	0.005	0.068
2,2',4,4',6,6'-Hexachloro-biphenyl	0.19	97.55	2.17	0.011	0.005	0.068
2,2',3,3'4,4',6,-Heptachloro-biphenyl	0.02	97.71	2.17	0.02	0.005	0.068
2,2',3,3',5,5',6,6'-Octachlorobiphenyl	0.07	97.68	2.17	0.009	0.005	0.068
2,2',3,3',4,4',5,5',6-Nonachlorobiphenyl	0.001	97.75	2.17	0.007	0.005	0.068
Decachlorobiphenyl	0.003	97.75	2.17	0.006	0.005	0.068

Source: Mackay et al. 1992.

TOXICITY PROFILE

Invertebrates

Members of the insect family, such as dragonfly, damselfly, and stonefly, seem to be relatively less sensitive to PCBs compared to other invertebrates (Tables 5.22a,b). Some invertebrates such as *Daphnia* can accumulate PCBs (tetra-hexa PCBs = 1.81–132 mg/kg) and show no signs of adverse effects (Dillon et al. 1990). Lethality of freshwater and marine fish to PCBs are listed in Table 5.23a,b. Sheepshead minnow seem to be the most sensitive species for acute lethal effects of PCBs. Fry stage is more sensitive than adult stage. Channel catfish seem to be the most resistant species. However, varying test conditions and shorter LC_{50} range for fish cast some doubt on some observations on fish lethality to PCBs.

Early life-stages of fish, such as egg, sac-fry, or fry, appear to be more sensitive than adults to PCBs (Table 5.24).

In laboratory experiments, exposure to PCBs similar to those detected in the Great Lakes, seem to reduce survival, reproduction, and growth of fish (Mayer et al. 1985). However, the poor reproductive performance by salmon in the Great Lakes did not correlate with total organochlorine concentrations; mortality of fry increased from 22% to 92% during 1975–1976 (Berlin et al. 1981) and 1980–1981

Table 5.22a Acute Toxicity of Some Aroclors to Selected Freshwater Invertebrates

Species	Aroclor	Exp (d)	LC$_{50}$ (μg/L)	References
Grass shrimp (*Palaemonotes kadiakensis*)	1254	7	3	NAS 1979
Amphipod (*Gammarus pseudolimnaeus*)	1242	4	10	NAS 1979
Amphipod (*G. pseudolimnaeus*)	1242	10	5	NAS 1979
Amphipod (*G. pseudolimnaeus*)	1248	4	52	NAS 1979
Amphipod (*G. pseudolimnaeus*)	1254	4	2,400	NAS 1979
Crayfish (*Orconectes nais*)	1242	7	30	NAS 1979
Crayfish (*O. nais*)	1254	7	80–100	NAS 1979
Damselfly (*Ischnura vertcalis*)	1242	4	400	Johnson and Finlay 1980
Damselfly (*I. vertcalis*)	1254	4	200	Johnson and Finlay 1980
Dragonfly (*Macromia sp.*)	1242	4	800	Johnson and Finlay 1980
Dragonfly (*M. sp.*)	1254	5	800	Johnson and Finlay 1980
Stonefly (*Pteronarcella badia*)	1016	4	428–878	Johnson and Finlay 1980
Cladoceran (*Daphnia magna*)	1254	14	1.8–24	EPA 1980b
Cladoceran (*D. magna*)	1254	21	1.3	EPA 1980b
Hydra (*Hydra oligactis*)	1016	3	5,000	Adams and Haileselassie 1984
Hydra (*H. oligactis*)	1254	3	10,000	Adams and Haileselassie 1984

Table 5.22b Acute Toxicity of Some Aroclors to Selected Marine Invertebrates

Species	Aroclor	Exp. (d)	LC$_{50}$ (μg/L)	References
Grass shrimp (*Palaemonotes pugio*)	1016	4	12.5	Ernst 1984
Grass shrimp (*P. pugio*)	1254	4	6.1–7.8	Ernst 1984
Brown shrimp (*Penaeus aztecus*)	1016	4	10.5	EPA 1980b
Pink shrimp (*P. duorarum*)	1254	12	1.0	EPA 1980b

(Mac et al. 1985), while total concentrations of PCBs and other organochlorines in fish decreased. It was proposed that these poor correlations may be due to reproductive/growth effects from specific coplanar PCBs (nonortho substituted PCBs) that are believed to act via A$_h$ receptor system, similar to 2378-T$_4$CDD (Kubiak et al. 1989; Safe 1990). Studies were conducted in chinook salmon collected in 1986 from Lake Michigan with mean total PCB concentrations in eggs of 7.02 mg/kg, wet weight, [83.9 mg/kg lipid, or 0.2-12 μg/kg, wet weight, of nonortho chlorosubstituted PCB congeners; or 29 to 514 T$_4$CDD-TEQ, calculated from three different methods (Safe 1990; Newsted 1991; Tillitt et al. 1991)]. Results showed significant differences in mortality of eggs and fry from clutches of different females; however, egg and fry mortality did not correlate with concentrations of either total PCBs or concentrations in T$_4$CDD-TEQ or individual PCBs (IUPAC PCBs 77, 126, 105, and 118). It was concluded that lack of correlation may indicate that factors other than exposure parameters, such as egg ripening, genetics, spawning climate of female fish, and others, could be responsible (Williams and Gisey 1992).

Table 5.23a LC$_{50}$ Values (Acute Toxicity) of Aroclors to Freshwater Fish Species

Species	Aroclor	Exp (d)	LC$_{50}$ (μg/L)	References
Rainbow trout (*Oncorhynchus mykiss*)	1016	4	114–159	Johnson and Finlay 1980
Rainbow trout (*O. mykiss*)	1242	5	67	Johnson and Finlay 1980
Rainbow trout (*O. mykiss*)	1248	5	54	Johnson and Finlay 1980
Rainbow trout (*O. mykiss*)	1254	5	142	Johnson and Finlay 1980
Rainbow trout (*O. mykiss*)	1254	10	8	NAS 1979
Rainbow trout (*O. mykiss*)	1260	20	21	NAS 1979
Bluegill (*Lepomis macrochirus*)	1016	4	390–540	Johnson and Finlay 1980
Bluegill (*L. macrochirus*)	1242	5	125	Johnson and Finlay 1980
Bluegill (*L. macrochirus*)	1242	15	54	NAS 1979
Bluegill (*L. macrochirus*)	1248	20	10	NAS 1979
Bluegill (*L. macrochirus*)	1254	25	54	NAS 1979
Bluegill (*L. macrochirus*)	1260	30	150	NAS 1979
Yellow perch (*Perca flavescens*)	1016	4	240	Johnson and Finlay 1980
Yellow perch (*P. flavescens*)	1242	4	>150	Johnson and Finlay 1980
Yellow perch (*P. flavescens*)	1248	4	>100	Johnson and Finlay 1980
Yellow perch (*P. flavescens*)	1258	4	>150	Johnson and Finlay 1980
Yellow perch (*P. flavescens*)	1260	4	>200	Johnson and Finlay 1980
Salmonoid (four spp.)	1016	4	134–1,154	Johnson and Finlay 1980
Channel catfish (*Ictalurus punctatus*)	1016	4	340–560	Johnson and Finlay 1980
Channel catfish (*I. punctatus*)	1242	15	110	NAS 1979
Channel catfish (*I. punctatus*)	1248	15	130	NAS 1979
Channel catfish (*I. punctatus*)	1254	15	740	NAS 1979
Channel catfish (*I. punctatus*)	1260	30	140	NAS 1979

Source: Eisler 1986.

Table 5.23b LC$_{50}$ (Acute Toxicity) Values of Aroclors to Marine Fish Species

Species	Aroclor	Exp. (d)	LC$_{50}$ (μg/L)	References
Sheepshead minnow (*Cyprinodon variegatus*), adult	1254	21	0.9	EPA 1980
Sheepshead minnow (*Cyprinodon variegatus*), fry	1254	21	0.1–0.32	Ernst 1984

Source: RTP 1994.

Table 5.24 LD$_{50}$ (Acute Toxicity) Values of PCBs to Rainbow Trout Eggs

PCB Congener	LD$_{50}$ (ng/g, egg) Mean	Range	References
3,3',4,4',5-Pentachlorobiphenyl (126)	74.0	44–83	Walker and Peterson 1991
3,3',4,4'-Tetrachlorobiphenyl (77)	1,348	1,064–1,621	Walker and Peterson 1991
2,3,3',4,4'-Pentachlorobiphenyl (105)	>6,970		Walker and Peterson 1991
2,3',4,4',5-Pentachlorobiphenyl (118)	>6,970		Walker and Peterson 1991
2,2',4,4',5,5'-Hexachlorobiphenyl (153)	>6,970		Walker and Peterson 1991

Note: Mortality occurred only from hatch to swim-up stage; mortality included low incidence of half-hatching and sac-fry subcutaneous yolk-sac edema.

Source: Hansen 1987; Walker and Peterson 1991.

Birds

Table 5.25 Acute Toxicity (LD$_{50}$) Values of Aroclors to Selected Bird Species

Species	Aroclor	Exposure Pattern	LD$_{50}$ mg/kg Diet	References
Northern bobwhite (*Colinus virginianus*)	1221	5 d on diet + 3 d unexposed	>6,000	Heath et al. 1972
Northern bobwhite (*C. virginianus*)	1242	5 d on diet + 3 d unexposed	2,098	Heath et al. 1972
Northern bobwhite (*C. virginianus*)	1248	5 d on diet + 3 d unexposed	1,175	Heath et al. 1972
Northern bobwhite (*C. virginianus*)	1254	5 d on diet + 3 d unexposed	604	Heath et al. 1972
Northern bobwhite (*C. virginianus*)	1260	5 d on diet + 3 d unexposed	747	Heath et al. 1972
Ring-necked pheasant (*Phasianus colchicus*)	1221	5 d on diet + 3 d unexposed	>4,000	Heath et al. 1972
Ring-necked pheasant (*P. colchicus*)	1242	5 d on diet + 3 d unexposed	2,078	Heath et al. 1972
Ring-necked pheasant (*P. colchicus*)	1248	5 d on diet + 3 d unexposed	1,312	Heath et al. 1972
Ring-necked pheasant (*P. colchicus*)	1254	5 d on diet +3 d unexposed	1,091	Heath et al. 1972
Ring-necked pheasant (*P. colchicus*)	1260	5 d on diet + 3 d unexposed	1,260	Heath et al. 1972
Mallard (*Anas platyrhynchos*)	1242	5 d on diet + 3 d unexposed	3,182	Heath et al. 1972
Mallard (*A. platyrhynchos*)	1248	5 d on diet + 3 d unexposed	2,798	Heath et al. 1972
Mallard (*A. platyrhynchos*)	1254	5 d on diet + 3 d unexposed	2,699	Heath et al. 1972
Mallard (*A. platyrhynchos*)	1260	5 d on diet + 3 d unexposed	1,975	Heath et al. 1972
Mallard (*A. platyrhynchos*)	1242	Single dose	>2,000	NAS 1979
Mallard (*A. platyrhynchos*)	1254	Single dose	>2,000	NAS 1979
Mallard (*A. platyrhynchos*)	1260	Single dose	>2,000	NAS 1979
Red-winged blackbird (*Agelaius phoeniceus*)	1254	6 d	1,500	Stickel et al. 1984

Source: Eisler 1986.

Mammals

Mink seem to be more sensitive to exposure to PCBs than other wildlife in the field (Wren 1991), whereas guinea pigs are the most sensitive under laboratory conditions (EPA 1989). A total tolerable daily intake limit for mink has been estimated at <1.5 µg of total PCBs/kg body weight. Table 5.26 lists the LD_{50} (acute) values for oral and intraperitoneal routes of exposure to Aroclors.

Table 5.26 LD_{50} (Acute Toxicity) Values of Aroclors to Mink (Single Exposure)

Aroclor	Route of Exposure	LD_{50} (µg/kg · bw)
1221	Oral/intraperitoneal	0.75-1.0/0.5–0.75
1242	Oral/intraperitoneal	3.0/1.0
1254	Oral/intraperitoneal	4.0/1.25–2.25

Sources: Aulerich and Ringer 1977; Ringer 1983.

Table 5.27 LD_{50} (Acute Toxicity) Values of Aroclors to Terrestrial Mammals

Species	Exposure Route	Aroclor	LC_{50} (mg/kg · bw)	References
Raccoon (*Procyon lotor*)	8 d (diet)	1254	>50 in diet	Montz et al. 1982
Cottontail rabbit (*Sylvilagus floridanus*)	12 wk (diet)	1254	>10 in diet	Zepp and Kirkpatrick 1976
Rabbit	Single dose (dermal)	1221	4,000 BW	EPA 1989
Rabbit	Single dose (dermal)	1242	7,700 BW	EPA 1989
Rabbit	Single dose (dermal)	1248	11,000 BW	EPA 1989
Rabbit	Single dose (dermal)	1260	10,000 BW	EPA 1989

Source: Eisler 1986.

Evaluation of toxicity profile of PCBs is complicated by the fact that PCB mixtures are of a variety of congeners and impurities, each with its own characteristics (ATSDR 1989). Toxicological studies used PCB mixtures such as Aroclors, Kanechlors, and Clophens. They are similar to each other, but differ in methods of production, chlorine composition, and PCDF contamination. The reported range of PCDFs in Aroclors is 0 to 2 ppm and in Kanechlors and Clophens, 5 to 20 ppm. The effects produced are generally similar for Aroclors, Kanechlors, and Clophens, provided the chlorine percentages are equivalent. The general population is exposed to PCBs primarily from dietary exposure of PCB-contaminated foods such as fish, dairy products, variety of red-meats, and possibly high-fat plant foods.The dietary exposure decreased by 10-fold over the 1970s (from about 6.9 µg of PCBs/day in 1971 to ~0.7 µg of PCBs/day in 1980), and continued to decrease slowly with time

Table 5.28 Acute Oral Toxicity (LD$_{50}$) of Aroclors to Mammalian Species

Aroclor	Mammal Strain	Sex/Age	LD$_{50}$ (mg/kg)	References
1254	Rat/Wistar	M/30 d	1,300	Grant and Phillipps 1974
1254	Rat/Wistar	F/30 d	1,400	Grant and Phillipps 1974
1254	Rat/Wistar	M/60 d	1,400	Grant and Phillipps 1974
1254	Rat/Wistar	F/60 d	1,400	Grant and Phillipps 1974
1254	Rat/Wistar	M/120 d	2,000	Grant and Phillipps 1974
1254	Rat/Wistar	F/120 d	2,500	Grant and Phillipps 1974
1254	Rat/Sherman	M/weanling	1,295	Linder et al. 1974
1254	Rat/Sherman	NR/adult	4,000–10,000	Linder et al. 1974
1254	Rat/Osborne-Mendel	M/adult	1,010 (single dose)	Garthoff et al. 1981
			1,530 (5 doses over 2.5 wks)	Garthoff et al. 1981
			1,990 (5 doses, 1 day/week)	Garthoff et al. 1981
1254	Mink/pastel	NR/NR	4,000	Aulerich and Ringer 1977
1221	Rat/NR	NR/NR	3,980	Fishbein 1974
1221	Rat/Sherman	F/NR	4,000	Nelson et al. 1972
1221	Mink/pastel	NR/NR	>750–<1,000	Aulerich and Ringer 1977
1260	Rat/Sherman	NR/adult	4,000–10,000	Linder et al. 1974
1260	Rat/Sherman	M/weanling	1,315	Linder et. al. 1974
1232	Rat/NR	NR/NR	4,470	Fishbein 1974
1242	Rat/Sprague-Dawley	M/adult	4,250	Bruckner et al. 1973
1242	Rat/NR	NR/NR	8,650	Fishbein 1974
1242	Mink/pastel	NR/NR	>3,000	Aulerich and Ringer 1977
1248	Rat/NR	NR/NR	11,000	Fishbein 1974

Note: NR, not reported.

down to 0.05 µg of PCBs/day by 1989. Total decline in PCB intake decreased by 140-fold from 1971 to the present. Exposures to media other than diet, such as soil, outdoor-air, and indoor-air, add up to about 0.48 µg of PCBs/day. The total exposure to PCBs dropped from 7.38 µg/day (6.9 µg/day from diet and 0.48 µg/day from other sources) to about 0.53 µg of PCBs/day (0.05 µg from diet plus 0.48 µg from other sources per day). This represents on a body weight basis that the exposure dropped from 0.11 to 0.008 µg/kg · BW/day from 1971 to 1989.

Currently indoor air is a significant source of PCB exposure to humans. Outdoor air is about 10 times lower than indoor air in PCB concentration (~10 ng/m^3 vs. 100 ng/m^3 respectively). Inhalation and dermal exposures are the primary routes of occupational exposure. The bioavailability of PCBs from different exposures are as follows: oral, ~90%, inhalation; 50% (difference between inhalation and exhalation) air-borne particulate; ~13% would be retained by the lung, all of which would likely be bioavailable (Brian and Mosier 1980); and dermal ~50% (only 12% will be bioavailable) (Schmidt et al. 1992).

No pertinent data are available for humans. LD$_{50}$ values, 1300 to 11,300 mg/kg · BW have been reported for male rats (Borlakoglu and Haegle 1991). The range is due to the different Aroclors used in the studies. Species differ in their toxicity to PCBs; the guinea pig is most sensitive followed by rabbit and rat (Drill et al. 1982).

Their chlorine content, and their position on the biphenyl ring markedly affect toxicity with toxic effect being highest for PCB congeners with five or six chlorine atoms on the biphenyl ring (Safe 1989).

Nonlethal Effects

No teratogenic effects have been reported for most mammalian species, but thyroid abnormalities have been reported (Marks et al. 1981). PCBs which are Ah receptor binders, were shown to protect against the teratogenic effects of 2378-T_4CDD (Morrissey et al. 1992). Since some coplanar PCBs are similar to 2378-T_4CDD in many biological and biochemical properties, it is likely that coplanar PCBs could induce teratogenic effects. PCBs have been reported to decrease reproductive ability in most animals under laboratory conditions (decreased fertility and reduced litter sizes). Teratogenic effects were shown in only in-bred populations of mice, which are known to be sensitive to such birth defects. Developmental toxicity such as impaired-learning and decreased reflex development were reported in monkeys and some rodents that were exposed to higher than ambient PCB concentrations. Carcinogenicity of PCB is controversial and interpretation of hyperplastic liver lesions in rodents induced by exposure to certain PCBs had been debatable (RTP 1994). Considering all evidences (laboratory and epidemiological) IARC has classified PCBs as having the potential to cause cancer at higher exposure levels, whereas the U.S. EPA and Health Canada consider that there is inadequate evidence for the carcinogenicity of PCBs. The potential reproductive toxic effects of PCBs as a primary concern has been recognized by these agencies. Genotoxic effects are not expected to occur at exposure to PCBs at the reported ambient environmental levels. Liver is the organ most often implicated in the toxicity of Aroclors in animals. Hepatic effects have been observed in numerous studies involving exposed rats, mice, guinea pigs, rabbits, dogs and monkeys, but rats have been tested most extensively (ATSDR 1992).

Key Toxicity Studies

Inhalation Route

PCB lethality data on humans by inhalation route were not reported in the literature. Exposure to near-saturation vapor concentration of heated Aroclor 1242 at 8.6 mg/m^3 for 7 h/d, 5 d/wk for 24 days was not lethal to cats, rats, mice, rabbits, or guinea pigs (Treon et al. 1956). This represents NOAEL value for intermediate inhalation exposures. Data for acute lethality or decreased longevity of animals due to acute exposure to PCBs were not available in the literature.

It has been established that liver and cutaneous tissues are the primary target organs to PCB exposures. In the only animal inhalation study of PCBs, degenerative liver lesions, a frank effect showed, in cats, rats, mice, rabbits, and guinea pigs that were exposed to 1.5 mg/m^3 of Aroclor 1254 vapor for 7 h/d, 5 d/week for 213 days (Treon et al. 1956). This represents the LOAEL value for liver toxicity on chronic exposure to PCBs. Since LOAEL for Aroclor 1254 was lower than NOAEL for Aroclor 1242, a minimal risk level was not derived for Aroclors as a class.

Data were not available for developmental, reproductive, and genotoxic effects in animals or humans from exposures to PCBs. Occupational studies (Brown 1986; Bertazzi et al. 1987) offer inadequate but suggestive evidence for carcinogenicity of PCBs by inhalation route. Similar data for animals have not been reported.

Oral Route

Quantitative lethality data for oral exposure on humans are not available in the literature. Single dose LC_{50} for oral exposure have been reported for rats and mink. The values were 750 mg/kg for Aroclor 1221 in mink (Aulerich and Ringer 1977) and 1010 mg/kg for Aroclor 1254 in rats (Garthoff et al. 1981). A NOAEL value of 250 mg/kg · day has been reported for mice for short-term (14 d) exposure. Calculated values of NOAEL and LOAEL(FEL) for mice are 32.5 and 130 mg/kg/day for acute oral exposure (EPA 1986a). LD_{50} (dietary) for mink for 28 days continuous exposure was 8.8 mg/kg/day (Hornshaw et al. 1986); LD_{50} (dietary) for 9 months continuous exposure was reported to be 1.25 mg/kg/day. Rats fed diets containing >25 ppm of Aroclor 1254 for 104 weeks showed reduced survival (NCI 1978). Liver and cutaneous tissues are shown to be the primary target organs to PCB toxicity via dietary exposure. Relative liver weights showed significant increase on dietary exposure to 8 ppm of Aroclor 1254 for 4 days and serum cholesterol (HDL) significantly increased at 16 ppm of PCB exposure. For intermediate duration (4 weeks), exposure to diets containing 0.5, 5, or 50 ppm of Aroclor 1242, 1248, 1254 or 1260, hepatic microsomal enzyme activities showed an increase (Litterst et al. 1972). Exposure to 5 ppm of Aroclor 1242 for 2 to 6 months increased the lipid content of liver in rats (Bruckner et al. 1974). Dietary concentrations of 0.5 ppm of Aroclors 1242, 1248, 1254, and 1260 and 5 ppm of Aroclor 1242, therefore represent the highest NOAEL and or lowest LOAEL respectively, for intermediate exposure periods (ATSDR 1992). From data on chronic feeding studies conducted with rats, it was concluded that effect levels for systemic effects were difficult to establish because of the type of liver lesions which were preneoplastic (ATSDR 1992).

Slight developmental toxicity of PCBs observed in infants of mothers who were consumers of PCB-contaminated fish (Jacobson et al. 1985). Reported LOEL for fetotoxicity of PCBs in rats is 50 ppm. Assuming that a rat consumes 5% of its body weight as its daily food consumption, 50 ppm is equivalent to 2.5 mg/kg/day (Collins and Capen 1980). LOAEL for rabbits from a gestational exposure study using Aroclor 1254 by gavage was determined to be 10 mg/kg/day. This represents the LOAEL for developmental effects in rabbits (Villeneuve et al. 1971) LOAEL for mice was determined to be 244 mg/kg/day based on one dose of Aroclor 1254 administration on day 9 of gestation. There are no studies reported on the reproductive toxicity of PCBs in humans. Diets containing >2 ppm of Aroclor 1254 for 4 months prior to mating and during gestation were lethal to fetuses and resulted in reproductive failure in mink (Aulerich and Ringer 1977; Bleavins et al. 1980). Dietary exposures at <5 ppm of Aroclor 1254 in one- and two-generation studies with rats did not reduce litter sizes (Linder et al. 1974). NOAEL concentration for reproductive toxicity of Aroclor 1254 by oral exposure of intermediate duration is 5 ppm or 0.25 mg/kg/day (ATSDR 1992).

PCBs produced generally negative results in *salmonella typhmurium* with or without metabolic activation. Similarly, PCBs produced negative results *in vivo* assays with rats and mice. The study of Norback and Weltman (1985) was used by the U.S. EPA to quantitatively assess the carcinogenicity risk from PCBs. Time-weighted-average (TWA) dosage of 3.45 mg/kg/day was calculated for the rat exposure study of Norback and Weltman, 1985. Incidences of trabecular carcinomas, adeno carcinomas and neoplastic nodules in the liver were combined to produce a total incidence of 45/47 in treated females and 1/49 in controls. Using these data, EPA (1988a) calculated human q^*, of 7.7 mg/kg/day. Dosages corresponding to risk levels of 10^{-4}, 10^{-6}, and 10^{-7} are 1.3×10^{-5}, 1.3×10^{-7}, and 1.3×10^{-8} mg/kg/day, respectively. Since there is no information on PCB mixtures, constituents of Aroclor 1260 is assumed to represent all PCB mixtures (EPA 1988b).

Dermal Route

Inhalation and dermal routes are likely routes under occupational exposure to PCBs.

Human lethality data are not available. Median lethal doses for single dermal applications of PCBs to rabbits ranged from >1,269 mg/kg for Aroclors 1242 and 1248 to < 3,169 mg/kg for Aroclor 1221 (Fishbein 1974). A study involving capacitor workers (Maroni et al. 1981a,b) showed that dermal exposure to PCBs at 2 to 28 µg/cm^2 of skin (on the hands) was not causing liver disease. Some workers might have liver enzyme induction. Dermal application of Aroclor 1260 to rabbits at 5 day/wk at a dose of 118 mg/d for 38 days (27 total applications) resulted in degenerative lesions of liver and kidneys, increased fecal porphyrin elimination, hyperplasia and hyperkeratosis of the follicular, and epidermal epithelium (Vos and Beems 1971). Pertinent data on developmental and reproductive toxicity via dermal exposure of PCBs were not available in the literature. PCBs produced negative genotoxic results *in vivo* and *in vitro* genotoxicity tests. Occupational exposure to PCBs, which includes dermal and inhalation exposures, shows inadequate evidence of carcinogenicity in humans. On dermal exposure to mouse skin, Aroclor 1254 did not produce evidence for promoter or carcinogenic activity (ATSDR 1989).

Populations at High Risk

The adverse effects clearly associated in occupational environment are those on skin, including rashes, burning sensations, and chloracne (Fishbein et al. 1979; Nethercott and Holness 1986; James et al. 1993). Other less frequent effects are erythema, swelling, dryness and thickening, pigmentation, and discoloration of finger nails (Ouw et al. 1976; IDSP 1987). The severity and frequency of observed dermatological effects seems not to correlate with either the duration or the degree of exposure. The correlatable adverse effects include changes in liver function and increase in serum lipids (Reggiani and Bruppacher 1985; Stark et al. 1986; Emmett et al. 1988). Increased xenobiotic metabolism and excretion by the liver as indicated by a decrease in antipyrine clearance half-life were reported in workers occupationally exposed to PCBs. (Fishbein et al. 1979; Alvares et al. 1977). However, lack of statistically significant correlation between antipyrine clearance half-life and either serum or adipose concentrations of PCBs was reported (Emmett et al. 1988). Possibly overlapping exposures to other chemicals might have contributed to contradictory results observed on this aspect of PCB toxicity. Occupational environmental exposure resulting in average blood levels LCBPs in workers in the range of 400, 266, and 1,470 µg/kg of Aroclor 1242 were shown to be associated with alterations in certain serum enzymes related to abnormal liver function (Ouw et al. 1976; Fishbein et al. 1979; Lawton et al. 1985). However, simultaneous exposures to other chemicals or recreational alcohol intake could have contributed to the above observation. A threshold blood level of 0.2 mg/kg assuming continuous exposure has been suggested, below which no adverse liver effects would occur. Improvement in working conditions decreased blood PCB levels from 1,470 µg/kg in 1976 to 277 µg/kg in 1979 (Lawton et al. 1985). Studies have suggested that serum PCBs and serum lipid concentrations appear to be co-variant parameters.

Although malignant melanoma and cancer of brain have been reported (Sinks et al. 1992), no study had statistical evidence to support potential associations between PCBs and human cancers (EPA 1988b; IDSP 1987). To address this problem, an analysis of pooled data from four study populations was conducted (IDSP 1987). It was evident that a statistically significant increase in incidences of cancer of the liver, biliary tract, and gall bladder (combined) was detected in the exposed populations compared to nonexposed groups. These cancers occurred within 2 to 10 years of employment; hence, it was considered unlikely that plant employment was associated with these cancers. Other cancers — rectal, combined kidney and bladder cancers, and leukemia — were not statistically significant. Simultaneous exposures to chemicals other than PCBs and cancer incidence were a relevant concern, considering the evidence that PCBs have substantial cancer-promoting activity. It was estimated that approximately 12,000 U.S. workers were potentially exposed to PCBs annually from 1970 to 1976 (NIOSH 1977a). Currently, PCBs are not manufactured or used industrially in the U.S. The potential for occupational exposure still exists, since PCB-containing transformers and capacitors remain in use. Exposure may occur during repair or accidents with electrical equipment containing PCBs (Wolff 1985). Occupational exposure may also occur during waste-site cleanup of PCB-containing wastes.

Other subpopulations are at high risk from PCBs because they are more sensitive to toxic effects of exposure. Embryos, fetuses, and neonates are potentially susceptible because of physiological differences from adults. They generally lack the hepatic microsomal enzyme systems that facilitate detoxification and excretion of PCBs (Calabrese and Sorenson 1977). Breast-fed infants have additional risks caused by a steroid excreted in human milk, but not cow's milk, that inhibits glucoronyl transferase activity and thus glucoronidation and excretion of PCBs (Gartner and Arias 1966). Other subpopulations that are potentially more sensitive to PCBs include those with incompletely developed glucuronide conjugation mechanisms, such as those with Gilbert's syndrome or Crigler and Najjar syndrome (Lester and Schmid 1964; Calabrese and Sorenson 1977). Persons with hepatic infections may have decreased glucuronide synthesis, rendering them more sensitive to toxicity of PCBs.

The indoor air in seven public buildings (schools, offices) was monitored in Minnesota in 1984 for Aroclors 1242, 1254, and 1260 (Oatman and Roy 1986). The total mean Aroclor concentrations in indoor air of three buildings using PCB transformers was found to be nearly twice as high as that in indoor air in the other four buildings not using PCB transformers (457 ± 223 S.D. vs. 229 ± 106 S.D. ng/m^3). It is also to be noted here that all indoor air levels of PCBs were significantly higher than in typical ambient outdoor air (ATSDR 1989). The indoor air in a number of laboratories, offices, and homes was monitored for various Aroclors. The data revealed that "normal" indoor air concentrations of PCBs were at least one order of magnitude higher than those in the surrounding outdoor air. It was suggested that certain electrical devices and appliances could have been emitting PCB and thus increasing PCBs concentration in indoor air.

PCBs have been shown to decrease the development of liver tumors by hepatocarcinogens (Safe 1989). It was also shown that PCBs inhibit the carcinogenesis of Aflatoxin B$_1$ in rainbow trout and development of skin cancers by PAHs. The cancer-inhibiting mechanism of PCBs was explained to be due to the induction by PCBs of enzyme systems that are responsible for inactivation metabolism of various carcinogens (Safe 1989). Enzyme induction by PCBs may also play a role in their cancer-promoting activity (Swierenga et al. 1990). It has been suggested by weight-of-evidence that Ah receptors mediated response mechanisms for PCDDs, PCDFs, PAHs, and some PCBs. The binding potency of these chemicals and even their isomers to the Ah receptor could be several orders of magnitude. Since all the above-mentioned chemicals affect the same enzyme system(s), their potential to cause adverse effects should be considered additive. This hypothesis formed the basis of extending the Toxic Equivalency Factors (TEFs) developed for PCDDs and PCDFs to include PCBs and maybe other chemicals that follow this mechanism of action (Safe 1990). TEFs for PCBs developed by Safe and co-workers (1990, 1992) should not be considered as "straight forward" as for PCDDs and PCDFs. Instead, TEFs for PCBs pertain only to end-point effects associated with Ah receptor activity: e.g., weight loss, thymic atrophy, immunotoxicity, teratogenicity, and dermatological effects.

Table 5.29 Toxicity Equivalency Factors (TEFs) for PCBs

PCB Isomer	TEF (Safe 1990[a])	TEF (Walker and Peterson 1991[b])
3,3',4,4'-Tetrachlorbiphenyl (77)	0.01	0.00016
3,3',4,4',5-Pentachlorbiphenyl (126)	0.1	0.005
2,3,3',4,4'-Pentachlorobiphenyl (105)	0.001	<0.00007
2,3',4,4',5'-Pentachlorobiphenyl (118)	0.001	<0.00007
3,3',4,4',5,5'-Hexachlorobiphenyl (169)	0.05	—

Note: Both TEF systems are based on 2378-T$_4$CDD as 1.0. It was suggested that mammalian-based TEFs are not good predictors of relative potency for salmonoid (fish) species.

[a] Based on mammalian studies.
[b] Based on fish studies.

Figures 5.1 and 5.2 represent the available information, on a relative basis, on health effects of PCBs on humans and animals.

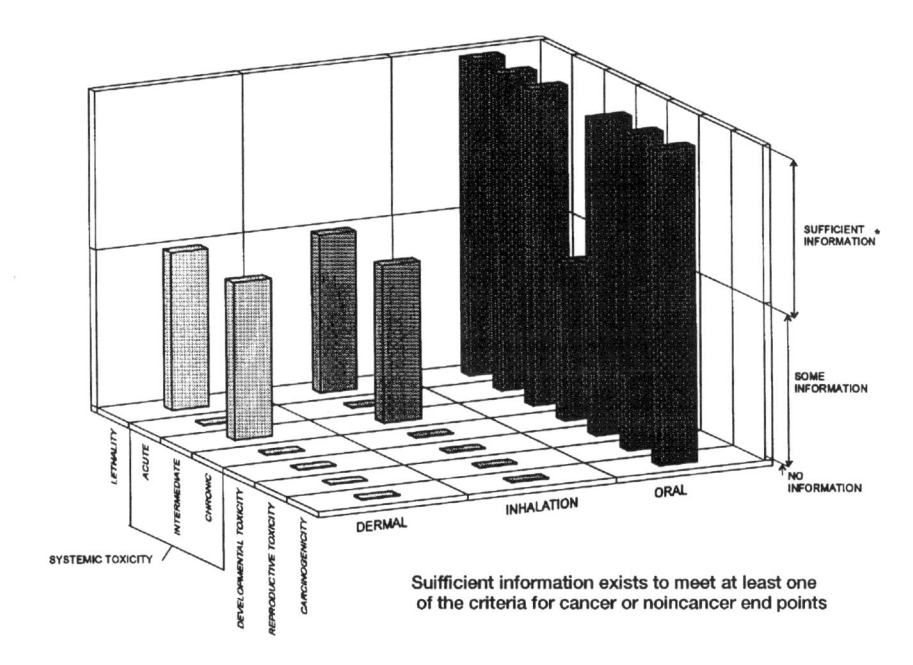

CHLORINATED AROMATIC HYDROCARBONS — POLYCYCLIC 161

Figure 5.1 Available information on health effects of PCBs on animals.

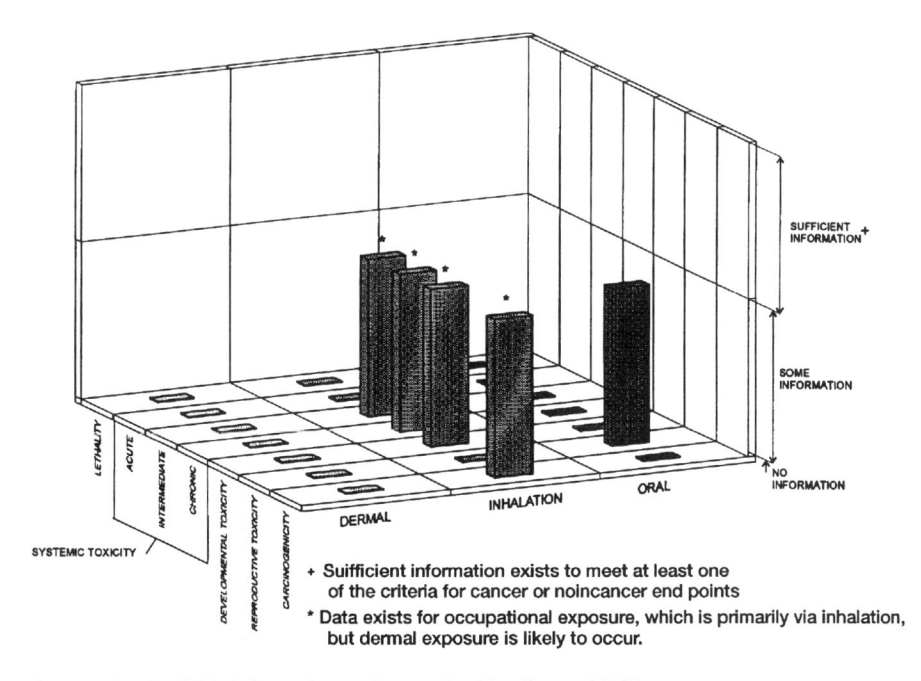

Figure 5.2 Available information on human health effects of PCBs.

REGULATIONS

National Regulations on Air

Agency	Description	Concn	Reference
OSHA	PCB with 42% chlorine		
	TWA–PEL	1.0 mg/m^{-3}	OSHA 1985
	PCB with 54% chlorine		
	TWA–PEL	0.5 mg/m^{-3}	OSHA 1985
U.S. FDA	Food tolerances		
	Foods	0.2–3.0 ppm	FDA 1988
	Food packaging	10.0	FDA 1988
U.S. EPA	Water		
	Following categories are regulated under the Clean Water Act		
	Industrial point sources		EPA 1988c
	electroplating		
	steam electric		
	asbestos manufacturing		
	timber products processing		
	metal finishing		
	paving and roofing		
	paint manufacturing		
	ink formulating		
	gum and wood		
	carbon black		
	aluminum forming		

National Advisory Guidelines

Agency	Description	Concn	Reference
Air			
NIOSH	REL-TWA	1.0 mg m^{-3}	NIOSH 1992
ACGIH	TLV-TWA for 1254	0.5 mg m^{-3}	ACGIH 1992
ACGIH	TLV-TWA for 1242	1.0 mg m^{-3}	ACGIH 1992
Water			
U.S. EPA	Ambient Water Quality Criteria (AWQC) Carcinogenicity risk levels at 10^{-5} to 10^{-7}	0.79–0.0079 ng/L	EPA 1980b
U.S. ODW	Drinking Water Criteria (DWC) carcinogenicity risk levels at 10^{-4} to 10^{-6}	0.5–0.005 µg/L	EPA 1988a
U.S. NAS	SNARL (Suggested No Adverse Response Level)	0.350 mg/L	NAS 1980
U.S. EPA	Health Advisories		
	Longer-term (adult)	0.0035 mg/L	EPA 1988a
	Longer-term (child)	0.001 mg/L	EPA 1988a
U.S. EPA	Permissible PCB soil contamination levels		
	Noncancer 10 d (adult)	0.700 mg/day	EPA 1986b
	Noncancer 10 d (child)	0.100 mg/day	EPA 1986b
	Cancer risk specific doses at risk levels of 10^{-4} to 10^{-7}	1.75–0.00175 µg/d	EPA 1986b

U.S. Regulations and Advisory Guidance

U.S. EPA	Probable human carcinogen			EPA 1988a,b

Permissible Cleanup Levels in Soil (mg/kg)

Location	Residential	Agricultural	Industrial	Reference
Arizona	0.18	—	—	Sittig 1994
Canada	5	0.5	50	
Germany	0.2	—	—	
New Jersey	0.45 (surface)	0.5	5	
	100 (subsurface)	—	—	
Texas	0.0832	—	0.743	

REFERENCES

Abramowicz, D.A. 1990. Aerobic and anaerobic biodegradation of PCBs: a review. *Biotechnology* 10:241–251.

ACGIH. 1992. Threshold limit values for chemical substances and physical agents and biological exposure indices for 1992-1993. Cincinnati, OH.

Adams, J.A. and Haileselassie, H.M. 1984. The effects of polychlorinated biphenyls (Aroclors 1016 and 1254) on mortality, reproduction, and regeneration in *Hydra oligactis. Arch. Environ. Contam. Toxicol.* 13:493–499.

Alvares, A.P., Fischbein, A., and Anderson, K.E. 1977. Alterations in drug metabolism in workers exposed to polychlorinated biphenyls. *Clin. Pharmacol. Ther.* 22:140.

Ankley, G.T. et al. 1992. Integrated assessment of contaminated sediments in the lower Fox Rover and Green Bay, Wisconsin. *Ecotoxicol. Environ. Safety.* 23: 46-63.

ATSDR. 1989. Agency for Toxic Substances and Disease Registry. Decision guide for identifying substance-specific data needs related to toxicological profiles. Division of Toxicology, Atlanta, GA.

ATSDR. 1992. Agency for Toxic Substances and Disease Registry. Draft toxicological profile for chlorinated dibenzofurans (CDFs). Office of External Affairs, Exposure and Disease Registry Branch, Atlanta, GA.

ATSDR. 1993. Agency for Toxic Substances and Disease Registry. Toxicological profile for polychlorinated biphenyls. TP-92/16. April 1993.

ATSDR. 1995. Agency for Toxic Substances and Disease Registry. 1995-Draft. Toxicological profile for polychlorinated biphenyls (Update), August 1995.

Aulerich, R.J. and Ringer, R.K. 1977. Current status of PCB toxicity to mink, and effect on their reproduction. *Arch. Environ. Contam. Toxicol.* 6:279–292.

Bakcr, J.E., Eisenreich, S.J., Johnson, T.C., and Halfman, B.M. 1985. Chlorinated hydrocarbon cycling in the benthic nepreloid layer of Lake Superior. *Environ. Sci. Technol.* 19:854–861.

Baker, J.E. and Eisenreich, S.J. 1989. PCBs and PAHs as tracers of particulate dynamics in large lakes. *J. Great Lakes Res.* 15:84–103.

Ballschmiter, K. and Wittlinger, R. 1991. Interhemisphere exchange of hexachlorocyclohexanes, hexachlorobenzene, polychlorobiphenyls, and 1,1,1-trichloro-2,2-*bis*(=*p*-chlorophenyl)ethane in the lower troposphere. *Environ. Sci. Technol.* 25:1103–1111.

Basturk, O., Dogan, M., Salihoglu, I., and Balkas, T. 1980. DDT, DDE, and PCB residues in fish, crustaceans and sediments from the eastern mediterranean coast of Turkey. *Marine Pollut. Bull.* 11:191–195.

Bedard, D.L., Wagner, R.E., Brennan, M.L., Habert, M.L., and Brown, J.F. 1987a. *Appl. Environ. Microbiol.* 53:1094.

Bedard, D.L., Habert, M.L., May, R.J., and Brennan, M.J. 1987b. *Appl. Environ. Microbiol.* 53:1103.

Bedard, D.L. and Habert, M.L. 1990. Influence of chlorine substitution pattern on the degradation of polychlorinated biphenyls by eight bacterial strains. *Microb. Ecol.* 20:87–192.

Berlin, W.H., Hesselberg, R.J., and Mac, M.J. 1981. Growth and mortality of fry of Lake Michigan lake trout during chronic exposure to PCBs and DDE. In: Technical Papers of the (U.S.) Fish and Wildlife Service. Chlorinated Hydrocarbons as a Factor in the Reproduction and Survival of Lake Trout (*Salvelinus namaycush*) in Lake Michigan, U.S. Department of the Interior, Fish and Wildlife Service, Washington, D.C., pp. 11–22.

Bertazzi, P.A., Riboldi, L., Pesatori, A., Radice, L., and Zocchetti, C. 1987. Cancer mortality of capacitor manufacturing workers. *Am. J. Ind. Med.* 11:165–176.

Bidleman, T.F. and Christensen, E.J. 1979. Atmospheric removal processes for high-molecular weight organochlorines. *J. Geophys. Res.* 84, 7857.

Bidleman, T.F. 1981. Interlaboratory analysis of high molecular weight organochlorines in ambient air. *Atmos. Environ.* 15:619–624.

Bleavins, M.R., Aulerich, R.J., and Ringer, R.K. 1980. Polychlorinated biphenyls (Aroclors 1016 and 1242): effects on survival and reproduction in mink and ferrets. *Arch. Environ. Contam. Toxicol.* 9:627–635.

Boon, J.P. and Duinker, J.C. 1986. Monitoring of cyclic organochlorines in the marine environments. *Environ. Monit. Assess.* 7:189–208.

Boon, J.P. 1985. Uptake, distribution, and elimination of selected PCB components of Clophen A40 in juvenile sole (*Solea*) and effects on growth. In: *Marine Biology of Polar Regions and Effects of Stress on Marine Organisms*. Gray, J.S. and Christensen, M.E., Eds., Wiley, Chichester, pp. 493–512.

Bopp, L.H. 1986. *J. Ind. Microbiol.* 1:23. Cited in Abramowicz, 1990.

Bopp, R.F., Simpson, H.J., Olsen, C.R., Trier, R.M., and Kostyk, N. 1982. Chlorinated hydrocarbons and radionuclide chronologies in sediments of the Hudson River and Estuary, N.Y. *Environ. Sci. Technol.* 16:666.

Borgmann, U. and Whittle, D.M. 1991. Contaminant concentration trends in Lake Ontario lake trout, (*Salvelinus namaycush*): 1977–1988. *J. Great Lakes Res.* 17:368–381.

Borlakoglu, J.T. and Haegle, K.D. 1991. Comparative aspects on the bioaccumulation, metabolism and toxicity of PCBs. *Comp. Biochem. Physiol.* C 100:327–338.

Box, S.J., Coates N.J., Davis C.J., Gilpin M.L., Houge-Frudrych C.S.V., and Milner P.H. 1991. *J. Antibiot.* 44:807.

Brian, D.B. and Mosher, M.J. 1980. Deposition and clearance of grain dusts in the lungs. In: *Occupational Pulmonary Disease: Focus on Grain Dust and Health,* Dossman, J.A. and Cotton, D.J., Eds., pp. 77–94.

Brinkman, M., Fogelman, K., Hoeflein, J. et al. 1980. Distribution of polychlorinated biphenyls in the Fort Edward, New York, water system. *Environ. Manage.* 4:511–520.

Brown, J.F., Wagner, R., Bedard, D.L., Brennan, M.J., Carnahan, J.C., and May, R.J. 1984. *Northeast Environ. Sci.* 3:167.

Brown, D.P. 1986. Mortality of Workers Exposed to Polychlorinated Biphenyls - An Update. Cincinnati, OH: Industry Wide Studies Branch, Div. of Surveillance, Hazard Evaluation and Field Studies, National Institute for Occupational Safety and Health, Centers for Disease Control, U.S. Public Health Service, Dept. of Health and Human Services. NTIS PB86-206000.

Bruckner, J.V., Khanna, K.L., and Cornish, H.H. 1973. Biological responses of the rat to polychlorinated biphenyls. *Toxicol. Appl. Pharmacol.* 24:434–448.

Bruckner, J.V., Khanna, K.L., and Cornish, H.H. 1974. Effect of prolonged ingestion of polychlorinated biphenyls on the rat. *Food Cosmet. Toxicol.* 12:323.

Burkhard, L.P., Armstrong, D.E., and Andren, A.W. 1985. Henry's law constants for the polychlorinated biphenyls. *Environ. Sci. Technol.* 19:590–596.

Calabrese, E.J., and Sorenson, A.J. 1977. The health effects of PCBs with particular emphasis on human high risk groups. *Rev. Environ. Health.* 2:285–304.

Callahan, M.A., Slimak, M.W., Gabel, N.W. et al. 1979. Water-Related Environmental Fate of 129 Priority Pollutants Vol. I. Chap. 36. EPA 440/4-79-029a. Washington, D.C.

Canviro. 1985. Review of PCB Occurrence, Human Exposure and Health Effects. CanViro Consultants Ltd. Prepared for the Ontario Ministry of the Environment, Toronto, ON.

Carey, A.E., Gowen, J.A., Tai, H., Mitchell, W.G., and Wiersma, G.B. 1979a. Pesticide residue levels in soils and crops from 37 states, 1972. National Soils Monitoring Program (IV). *Pestic. Monitor. J.* 12:209–229.

Carey, A.E., Douglas, P., Tai, H., Mitchell, W.G., and Wiersma, G.B. 1979b. Pesticide residue concentrations in soils of five United States cities, 1971. Urban Soils Monitoring Program. *Pestic. Monitor. J.* 13:17–22.

Chemical Engineering News. 1971. Monsanto releases PCB data. 49:15.

Christensen, E.R. and Lo, C.K. 1986. Polychlorinated biphenyls in dated sediments of Milwaukee Harbor, Wisconsin. *Environ. Poll.* 12:217–232.

Collins, W.T. and Capen, C.C. 1980. Fine structural lesions and hormonal alterations in thyroid glands of perinatal rats exposed in utero and by milk to polychlorinated biphenyls. *Am. J. Pathol.* 99:125–142.

Creaser, C.S. and Fernandes, A.R. 1986. Background levels of polychlorinated biphenyls in British soils. *Chemosphere.* 15:499–508.

Crump-Wiesner, H.J., Feltz, H.R., and Yates, M.L. 1974. A study of the distribution of polychlorinated biphenyls in the aquatic environment. *Pest. Monit. J.* 8:157–161.

Dawson, R. and Riley, J.P. 1977. Chlorine-containing pesticides and polychlorinated biphenyls in British coastal waters. *Estuarine Coastal Marine Sci.* 4:55–69.

Dillon, T.M., Benson, W.H., Stackhouse, R.A., and Crider, A.M. 1990. Effects of selected PCB congeners on survival, growth, and reproduction in *Daphnia magna. Environ. Toxicol. Chem.* 9:1317–1326.

Dmochewitz, S. and Ballschmiter, K. 1988. *Chemosphere.* 17:111.

Doskey, P.V. and Andren, A.W. 1981. Modeling the flux of atmospheric PCBs across the air/water interface. *Environ. Sci. Technol.* 15:705.

Drill, Freiss, Hays, Loomis, and Shaffer, Inc. 1982. Potential Health Effects in the Human From Exposure to Polychlorinated Biphenyls (PCBs) and Related Impurities. Report prepared for the National Electric Manufacturers' Association, Washington, D.C.

Duggan, R.E., Corneliussen, P.E., Duggan, M.B., McMahon, B.M., and Martin, R.J. 1983. Pesticide Residue Levels in Foods in the United States from July 1, 1969 to June 30, 1976. Food and Drug Administration, Division of Chemical Technology, Washington, D.C.

Durfee, R.L. 1976. Production and U.S.ge of PCBs in the United States. In: Proceedings of the National Conference on Polychlorinated Biphenyls, Chicago, 1975. EPA-560/6-75-004. Washington, D.C.: U.S. Environmental Protection Agency, pp. 103–107.

Eaton, D.C. 1985. *Enzyme Microbiol. Technol.* 7:194.

Edgren, M., Olsson, M., and Reutergardh, L. 1981. A one year study of the seasonal variations of DDT and PCB levels in fish from heated and unheated areas near a nuclear power plant. *Chemosphere.* 10:447–452.

Eduljee, G., Badsha, K., and Price, L. 1985. Environmental monitoring for PCB and heavy metals in the vicinity of a chemical waste disposal facility-I. *Chemosphere.* 14:1371–1382.

Eduljee, G., Badsha, K., and Scudamore, N. 1986. Environmental monitoring for PCB and trace metals in the vicinity of a chemical waste disposal facility-II. *Chemosphere.* 15:81–93.

Ehrhardt, M. 1981. Organic substances in the Baltic Sea. *Marine Pollut. Bull.* 12:210–213.

Eisenreich, S.J., Looney, B.B., and Thornton, J.D. 1981. Airborne organic contaminants in the Great Lakes ecosystem. *Environ. Sci. Technol.* 15:30–38.

Eisenreich, S.J., Looney, B.B., and Hollod, G.J. 1983. PCBs in the Lake Superior atmosphere 1978–1980. In: Mackay, D. et al., Eds., *Physical Behavior of PCBs in the Great Lakes.* Ann Arbor Science, Ann Arbor, MI.

Eisenreich, S.J. and Johnson, T.C. 1983. *PCBs in the Great Lakes: Sources, Sinks, Burdens.* D'Itri, F.M. and Kamrin, M.A., Eds., Butterworth, Boston, MA, pp. 49–75.

Eisler, R. 1986. Polychlorinated Biphenyl Hazards to Fish, Wildlife and Invertebrates: A Synoptic Review. Fish and Wildlife Service, (U.S.) Department of the Interior, Biological Report 85 (1.7). PB86-170057.

El-Gendy, K.S., Abdalla, A.A., Aly, H.A., Tantawy, G., and El-Sebae, A. 1991. Residue levels of chlorinated hydrocarbon compounds in water and sediment samples from Nile branches in the Delta, Egypt. *J. Environ. Sci. Health.* 26:15–36.

Elder, D.L. and Villeneuve, J.P. 1977. Polychlorinated biphenyls in the Mediterranean Sea. *Marine Pollut. Bull.* 8:19–22.

Elder, V.A., Proctor, B.L., and Hites, R.A. 1981. Organic compounds found near dump sites in Niagara Falls, New York. *Environ. Sci. Technol.* 15:1237–1243.

Emmett, E.A., Maroni, M., Jefferys, J., Schmith, J., Levin, B.K., and Alvares, A. 1988. Studies of transformer repair workers exposed to PCBs : II. Results of clinical laboratory investigations. *Am. J. Ind. Med.* 14:47–62.

Environment Canada 1988. Polychlorinated Biphenyls (PCBs) — Fate and Effects in the Canadian Environment. Report EPS 4/HA/2, p. 69.

EPA. 1976. U.S. Environmental Protection Agency. Industrial uses and environmental distribution. NTIS PB 252-012.

EPA. 1977. U.S. Environmental Protection Agency. Polychlorinated biphenyls (PCBs): Toxic pollutant effluent standards, final rule. *Federal Register.* 42:6531–6555.

EPA. 1979. U.S. Environmental Protection Agency. Polychlorinated biphenyls (PCBs): proposed rulemaking for PCB manufacturing exemptions. *Federal Register.* 44:31564–31567.

EPA. 1980a. Hazard waste generation and commercial hazardous waste management capacity: An assessment, SW-894. Washington, D.C., U.S. EPA, D-4.

EPA. 1980b. U.S. Environmental Protection Agency. Ambient Water Quality Criteria for Polychlorinated Biphenyls (PCBs). EPA, 440/5-80-068, p. 211.

EPA. 1985. U.S. Environmental Protection Agency. Drinking Water Criteria Document for Polychlorinated Biphenyls (PCBs). Draft. Washington, D.C.: Office of Drinking Water. NTIS PB 86-118312/AS.

EPA. 1986a. U.S. Environmental Protection Agency. Broad scan analysis of the FY 82 national human adipose tissue survey specimens. Volume III - Semi-Volatile Organic Compounds. Washington, D.C.: Office of Toxic Substances, EPA-560/5-86-037.

EPA. 1986b. U.S. Environmental Protection Agency. Development of Advisory Levels for Polychlorinated Biphenyls (PCBs) Cleanup. Washington, D.C., EPA/600/6-86-02.

EPA. 1988. U.S. Environmental Protection Agency. Drinking Water Criteria Document for Polychlorinated Biphenyls (PCBs) (Final), April, 1988. Cincinnati, OH. PB89-192256.

EPA. 1988a. U.S. Environmental Protection Agency. Drinking Water Criteria Document for Polychlorinated Biphenyls (PCBs). Final. ECAO-CIN-414. April 1988.

EPA. 1988b. U.S. Environmental Protection Agency. IRIS (Integrated Risk Information System), CRAVE (Carcinogen Risk Assessment Validation Endeavor) for polychlorinated biphenyls. (Verification date: 4/22/87). On-line: input pending. Office of Health and Environmental Assessment, Environmental Criteria and Assessment Office, Cincinnati, OH.

EPA. 1988c. U.S. Environmental Protection Agency. Analysis of Clean Water Act Effluent Guidelines Pollutants. Summary of the Chemicals Regulated by Industrial Point Source Category, 40 CFR Parts 400-475. Draft. Office of Water Regulations and Standards.

EPA. 1989. U.S. Environmental Protection Agency. Exposure Factors Handbook. Office of Health and Environmental Assessment, Washington, D.C. EPA/600/8-89-043.

EPA-NIM. 1990. OHM-TADS (Oil and Hazardous Materials Technical Assistance Data Systems). 1987. U.S. EPA/National Institutes of Health, Washington, D.C.

Ernst, W. 1984. Pesticides and technical organic chemicals. In: *Marine Ecology. Vol. V,* Kinne, O., Ed., John Wiley & Sons, New York, pp. 1617–1709.

Evans, M.S. and Landrum, P.F. 1989. Toxicokinetics of DDE, benzo(a)pyrene, and 2,4,5,2′,4′,5′-hexachlorobiphenyl in *Pontoporeia hoyi* and *Mysis relicta. J. Great Lakes Res.* 15:589–600.

Evans, M.S., Noguchi, G.E., and Rice, C.P. 1991. The biomagnification of polychlorinated biphenyls, toxaphene and DDT compounds in a Lake Michigan offshore food web. *Arch. Environ. Contam. Toxicol.* 20:87–93.

Feltz, H.R. 1980. Significance of bottom material/data in evaluation of water quality. In: *Contaminants and Sediments, Vol. 1, Fate and Transport, Case Studies, Modeling, Toxicity.* Ann Arbor Science, Ann Arbor, MI, 271–287.

FDA. 1988. Food and Drug Administration. Tolerances for unavoidable poisonous or deleterious substances. 21CFR 109.30.

Fischbein, A., Wolff, M.S., Lilis, R., Thornton, J., and Selikoff, I.J. 1979. Clinical findings among PCB-exposed capacitor manufacturing workers. *Ann. NY. Acad. Sci.* 320:703–715.

Fischbein, A. and Rizzo, J.N. 1987. Polychlorinated biphenyls: a review. *Mount Sinai J. Med.* 54:332–336.

Fishbein, L. 1974. Toxicity of chlorinated biphenyls. *Ann. Rev. Pharmacol.* 14:139–156.

Fowler, S.W. 1987. PCBs and the environment: the Mediterranean marine ecosystem. In: *PCBs and the Environment, Vol. III,* Waid, J.S., Ed., CRC Press, Boca Raton, FL, pp. 209–239.

Fowler, S.W. and Elder, D.L. 1980–1981. Chlorinated hydrocarbons in pelagic organisms from the open Mediterranean Sea. *Marine Environ. Res.* 4:87–96.

Fowler, S.W. 1990. Critical review of selected heavy metal and chlorinated hydrocarbon concentrations in the marine environment. *Marine Environ. Res.* 29:1–64.

Frank, R. 1981. Pesticides and PCB in the Grand and Saugeen river basins. *J. Great Lakes Res.* 7:440–454.

Frank, R., Braun, H.E., Holdrinet, M. et al. 1978. Residues of organochlorine insecticides and polychlorinated biphenyls in fish from Lakes Saint Clair and Erie, Canada - 1968–76. *Pestic. Monitor. J.* 12:69–80.

Frank, R., Braun, H.E., and van Hove Holdrinet, M. 1982. Agriculture and water quality in the Canadian Great Lakes Basin: V. Pesticide use in 11 agricultural watersheds and presence in stream water, 1975–77. *J. Environ. Qual.* 11:497–505.

Furukawa, K. and Matsumura, F. 1976. Microbial metabolism of polychlorinated biphenyls, studies on the relative degradability of polychlorinated biphenyl components by *Alkaligenes* sp. *J. Agr. Food Chem.* 24:251.

Furukawa, K. 1986. Modification of PCBs by bacteria and other microorganisms. In: *PCBs and the Environment. Vol. III,* Waid, J.S., Ed., CRC Press, Boca Raton. FL, pp. 89–100.

Furukawa, K., Tonomura, K., and Kamiyashi, A. 1978. Effect of chlorine substitution on the biodegradability of polychlorinated biphenyls. *Appl. Environ. Microbiol.* 35, 301.

Gardner, A.M., Righter, H.R., and Roach, J.A.G. 1976. Excretion of hydroxylate polychlorinated biphenyl metabolites in cow's milk. *J. Assoc. Offi. Anal. Chem.* 59:273–277.

Garthoff, L.H., Cerra F.E., and Marks, E.M. 1981. Blood chemistry alteration in rats after single and multiple gavage administration of polychlorinated biphenyls. *Toxicol. Appl. Pharmacol.* 60:33–44.

Gartner, L.W. and Arias, I.M. 1966. Studies of prolonged neonatal jaundice in the breast-fed infant. *J. Pediatr.* 68:54.

Gartrell, M.J., Craun, J.C., Podrebarac, D.S., and Gunderson, E.L. 1985. Pesticides, selected elements, and other chemicals in adult total diet samples October 1979 – September 1980. *J. Assoc. Off. Anal. Chem.* 68:1184–1197.

Gartrell, M.J., Craun, J.C., Podrebarac, D.S. et al. 1986. Pesticides selected elements and other chemicals in infant and toddler total diet samples. October 1980-March 1982. *J. Assoc. Anal. Chem.* 69:123–145.

Germain, A. and Langlois, C. 1988. Contamination des eaux des sediments en suspension du fleuve Saint-Laurent par les pesticides organochlores, les biphenyles polychlores et d'autres contaminants organiques prioritaires. *Water Pollut. Res. J. Can.* 23:602–614.

Giam, C.S., Chan, H.S., Neff, G.S., and Atlas, E.L. 1978. Phthalate esters plasticizers, a new class of marine pollutant. *Science.* 199:419.

Goldberg, E.D., Bowen, V.T., Farrington, J.W. et al. 1978. The mussel watch. *Environ. Conserv.* 5:101–126.

Grant, D.L. and Philipps, W.E.J. 1974. The effects of age and sex on the toxicity of Aroclor 1254, a polychlorinated biphenyl, in the rat. *Bull. Contam. Toxicol.* 12:145–152.

Gregor, D.J. and Gummer, W.D. 1989. Evidence of atmospheric transport and deposition of organochlorine pesticides and polychlorinated biphenyls on Canadian arctic snow. *Environ. Sci. Technol.* 23:561–565.

HDSB. 1995. Hazardous Substances Data Bank. National Library of Medicine, National Toxicology Program, Bethesda, MD, Jan. 1995.

Hankin, L. and Sawhney, B.L. 1984. Microbial degradation of PCBs in soil. *Soil Sci.* 137:401.

Hansch, C. and Leo, A.J. 1985. Medchem Project. Issue No. 26. Pomona College, Claremont, CA.

Hansen, L.G. 1987. Environment toxicology of polychlorinated biphenyls. In: Polychlorinated Biphenyls (PCB's): Mammalian and Environmental Toxicology, Safe, S. Ed., Springer-Verlag, Berlin, pp. 15–48.

Hargrave, B.T., Vass, W.P., Erickson, P.E., and Fowler, B.R. 1988. Atmospheric transport of organochlorines to the Arctic Ocean. *Tellus.* 40B:480–493.

Heath, R.G., Spann, J.W., Hill, E.F., and Kreitzer, J.F. 1972. Comparative dietary toxicities pesticides to birds. *U.S. Wildlife Service. Spec. Sci. Rep. - Wildlife.* 152, 57.

Heinz, G.A., Erdman, T.C., Haseltine, S.D., and Stafford, C. 1985. Contaminant levels in Colonial Waterbirds from Green Bay and Lake Michigan, 1975–80. *Environ. Monitor. Assess.* 5:223–236.

Heit, M., Klusek, C., and Baron, J. 1984. Evidence of deposition of anthropogenic pollutants in remote Rocky Mountain lakes. *Water Air Soil Pollut.* 22:403–416.

Hollifield, H.C. 1979. Rapid nephelometric estimate of water solubility of highly insoluble organic chemicals of environmental interest. *Bull. Environ. Contam. Toxicol.* 23:579–586.

Hornshaw, T.C., Safronoff, J., Ringer, R.K., and Aulerich, R.J. 1986. LC_{50} test results in polychlorinated biphenyl-fed mink: Age, season and diet comparisons. *Arch. Environ. Contam. Toxicol.* 15(6):717–723.

Hubbard, H.L. 1964. Chlorinated biphenyl and related products. In: *Kirk-Othmer Encyclopedia of Chemical Technology*, 2nd ed., Vol. 5, Standen, A., Ed., John Wiley & Sons, New York, p. 291.

Hutzinger, S., Safe, S., and Zitko, V. 1974. *The Chemistry of PCBs.* CRC Press, Boca Raton, FL.

Hutzinger, O. and Veerkamp, W. 1981. In: *Microbial Degradation of Xenobiotics and Recalcitrant Compounds.* Leisenger, T., Hutter, R., Cook, A., and Nuesch, J., Eds., Academic Press, New York.

IARC. 1978. International Agency for Research on Cancer Monographs on the Evaluation of the Carcinogenic Risk of Chemicals to Humans. Polychlorinated Biphenyls and Polybrominated Biphenyls. Vol. 18. World Health Organization, Lyon, France.

IDSP. 1987. Industrial Disease Standards Panel. Report to the Workers' Compensation Board on Occupational Exposure to PCBs. Toronto, ON. IDSP Report 2.

Jacobson, S.W., Fein, G.G., Jacobson, J.L., Schwartz, P.M., and Dowler, J.K. 1985. The effect of interuterine PCB exposure on visual recognition memory. *Child Dev.* 56:856–860.

James, R., Cusch, H., Tamburro, C., Roberts, S., Schell, J., and Harbison, R. 1993. Polychlorinated biphenyl exposure and human disease. *J. Occup. Med.* 35:136–148.

Jansson, B., Jensen, S., Olsson, M., Renberg, L., Sundstrom, G., and Vaz, R. 1975. Identification by GC-MS of phenolic metabolites of PCB and p,p'-DDE isolated from Baltic guillemot and seal. *Ambio.* 4:93–97.

Janz, D.M. and Metcalfe, C.D. 1991. Relative induction of aryl hydrocarbon hydroxylase by 2,3,7,8-TCDD and two coplanar PCBs in the rainbow trout (*Oncorhynchus mykiss*). *Environ. Toxicol. Chem.* 10:917–923.

Jelinek, C.F. and Corneliussen, P.E. 1976. Levels of PCBs in the U.S. food supply. In: Proceedings of the National Conference on Polychlorinated Biphenyls, Chicago, 1975. EPA-560/6-75-004. U.S. Environmental Protection Agency, Washington, D.C., pp. 147–154.

Johnson, W.W. and Finlay, M.T. 1980. *Handbook of Acute Toxicity of Chemicals to Fish and Aquatic Invertebrates.* U.S. Dept. of Interior, Fish and Wildlife Services, Washington, D.C.

Kaiser, K.L.E., Oliver, B.G., Charlton, M.N., Nichol, K.D., and Comba, M.E. 1990. Polychlorinated biphenyls in St. Lawrence River sediments. *Sci. Total. Environ.* 97/98:495–506.

Kilikidis, S.D., Psomas, J.E., Kamarianos, A.P., and Panetsos, A.G. 1981. Monitoring of DDT, PCBs and other organochlorine compounds in marine organisms from the North Aegean Sea. *Bull. Environ. Contam. Toxicol.* 26:496–501.

Kirby, M.K. 1986. Effect of Waste Discharges on the Water Quality of Nipigon Bay, Lake Superior, 1983. Water Resources Branch, Ontario Ministry of the Environment.

Kubiak, T.J. et al. 1989. Microcontaminants and reproductive impairment of the Forster's tern on Green Bay, Lake Michigan - 1983. *Arch. Environ. Contam. Toxicol.* 18:706–727.

Kozie, K.D. and Anderson, R.K. 1991. Productivity, diet and environmental contaminants in bald eagle nesting near the Wisconsin shoreline of Lake Superior. *Arch. Environ. Contam. Toxicol.* 20:41–48.

Kuratsune, M., Masuda, Y., and Nagayama, J. 1976. Some recent findings concerning Yusho. Proceedings of the National Conference on Polychlorinated Biphenyl, November 19–21, 1975. Chicago, Illinois. U.S. Environmental Protection Agency, Publication No. EPA-560/6-75-004, pp. 14–29.

Larsson, P. 1985. Contaminated sediments of lakes and oceans act as sources of chlorinated hydrocarbons for release to water and atmosphere. *Nature.* 317:347–349.

Lawton, R.W., Ross, M.R., Feingold, J., and Brown, J.F., Jr. 1985. Effects of PCB exposure on biochemical and hematological finding in capacitor workers. *Environ. Health Perspect.* 60:165–184.

Leatherland, L.F. and Sonstegard, R.A. 1980. Effect of dietary Mirex and PCBs in combination with food deprivation and testosterone administration on thyroid activity and bioaccumulation of organochlorines in rainbow trout *Salmo gairdneri* Richardson. *J. Fish Diseases.* 3:115–124.

Lester, R. and Schmid, R. 1964. Bilirubin metabolism. *New Engl. J. Med.* 270:779.

Lichtenstein, E.P., Schulz, K.R., Fuhremann, T.W., and Liang, T.T. 1969. Biological interaction between plasticizers and insecticides. *J. Eco. Entomol.* 62:761–765.

Linder, R.E., Gaines, T.B., and Kimbrough, R.D. 1974. The effect of PCB on rat reproduction. *Food Cosmet. Toxicol.* 12:63.

Linko, R.R., Rantamaki, P., Rainio, K., Urpo, K. 1979. Polychlorinated biphenyls in plankton from the Turku archipelago. *Bull. Environ. Contam. Toxicol.* 23:145–152.

Litterst, C.L., Farber, T.M., Baker, A.M., and van Loon, E.J. 1972. Effect of polychlorinated biphenyls on hepatic microsomal enzymes in the rat. *Toxicol. Appl. Pharmacol.* 23:112–122.

Liu, D. 1982. Assessment of continuous biodegradation of commercial PCB formulations. *Bull. Environ. Contam. Toxicol.* 29:200.

Mac, M.J., Edsall, C.C., and Seelye, J.G. 1985. Survival of lake trout eggs and fry reared in water from the upper Great Lakes. *J. Great Lakes Res.* 11:520–529.

Mackay, D., Shiu, W.Y., and Ma, K.C. 1992. Monoaromatic hydrocarbons, chlorobenzenes, and PCBs. In: *Illustrated Handbook of Physical-Chemical Properties and Environmental Fate for Organic Chemicals,* Vol. 1., Lewis Publishers, Boca Raton, FL.

Marks, T.A., Kimmel, G.L., and Staples, R.E. 1981. Influence of symmetrical polychlorinated biphenyl isomers on embryo and fetal development in mice. *Toxicol. Appl. Pharmacol.* 61:269–276.

Maroni, M., Colombi, A., Arbosti, G., Cantoni, S., and Foa, V. 1981a. Occupational exposure to polychlorinated biphenyls in electrical workers. II. Health effects. *Br. J. Ind. Med.* 38:55–60.

Maroni, M., Colombi, A., Cantoni, S., Ferioli, E., and Foa,V. 1981b. Occupational exposure to polychlorinated biphenyls in electrical workers. I. Environmental and blood polychlorinated biphenyls concentrations. *Br. J. Ind. Med.* 38:49–54.

Mayer, K.S., Mayer, E.L., and Witt, A. 1985. Waste transformer oils and PCB toxicity to rainbow trout. *Tans. Am. Fish. Soc.* 114:869–886.

Mazurek, M.A. and Simonheit, B.R.T. 1985. Organic components in bulk and wet-only precipitation, *CRC Critical Rev. Environ. Control.* 16:41–47.

McKinney, J.D. 1976. Toxicology of selected symmetrical hexachlorobiphenyl isomers; correlating biological effects with chemical structure. Proceedings of the National Conference on Polychlorinated Biphenyl, Nov. 19-21, 1975, Chicago, IL. U.S. Environmental Protection Agency Publication no. EPA-560/6-75-004. pp. 73-76.

Metcalfe, T.L. and Metcalfe, C.D. 1993. The Trophodynamics of Co-Planar PCBs in a Pelagic Food Chain From Lake Ontario. Presented at the 14th Annual Meeting of the Society of Environmental Toxicology and Chemistry, November 14-18, 1993, Houston, TX.

Miller, S. 1982. The persistent PCB problem. *Environ. Sci. Technol.* 16:98A–99A.

MOE. 1986. Ontario Ministry of the Environment. Nearshore Water Quality at Thunder Bay. Lake Superior, 1983. Great Lakes Section, Water Resources Branch.

MOE. 1988. Ontario Ministry of the Environment. Ontario Water Treatment Plants Annual Reports, 1987. Drinking Water Surveillance Program, October 1988.

Monsanto Technical Bulletin O/PL-306., Monsanto Company, St. Louis, Missouri, U.S.A.

Monsanto Industrial Chemical Corporation. 1974. PCBs. Aroclor Technical Bulletin O/PL 306A, St. Louis, MO, 20 pp.

Montz, W.E., Card, W.C., and Kirkpatrick, R.L. 1982. Effects of polychlorinated biphenyls and nutritional restriction on barbituate-induced sleeping times and selected blood characteristics in racoons (*Prociou lotor*). *Bull Environ. Contam. Toxicol.* 28:578–583.

Moore, J.W. and Ramamoorthy, S. 1984. *Organic Chemicals in Natural Waters, Applied Monitoring, and Impact Assessment.* Springer-Verlag, NY.

Morrissey, R.E., Harris, M.W., Diliberto, J.J., and Birnbaum, L.S. 1992. Limited PCB antagonism of TCDD-induced malformations in mice. *Toxicol. Lett.* 60:19–25.

Morselli, L., Brocco, D., and Pirni, A. 1985. The presence of polychlorodibenzo-*p*-dioxins (PCDDs), polychlorodibenzofurans (PCDFs), and incinerators under different technological and working conditions. *Ann. Chem.* 75:59–64.

Mrozek, E., Jr. and Leidy, R.B. 1981. Investigation of selective uptake of polychlorinated biphenyls by *Spartina alterniflora* Loisel. *Bull. Environ. Contam. Toxicol.* 27:481–488.

Muir, D.C.G., Norstrom, R.J., and Simon, M. 1988. Organochlorine contaminants in arctic marine food chains: Accumulation of specific polychlorinated biphenyls and chlordane-related compounds. *Environ. Sci. Technol.* 22:1071–1079.

Murphy, T.J., Formanshi, L.J., Brownwell, B., and Meyer, J.A. 1985. Polychlorinated biphenyl emissions to the atmosphere in the Great Lakes region. Municipal land fills and incinerators. *Environ. Sci. Technol.* 19:924–946.

Murphy, T.J. and Rzeszutko, C.P. 1977. Precipitation inputs of PCBs to Lake Michigan. *J. Great Lakes Res.* 3:305–312.

Murray, H.E., Ray, L.E., and Giam, C.S. 1981. Phthalic acids esters, total DDT and polychlorinated biphenyls in marine samples from Galveston Bay, Texas. *Bull. Environ. Contam. Toxicol.* 26:769–774.

Murray, M.W. and Andren, A.W. 1992. Precipitation scavenging of polychlorinated biphenyl congeners in the Great Lakes Region. *Atmos. Environ.* 26:883–897.

Nadim, L.M., Schocken, M.J., Higson, F.K., Gibson, D.T., Bedard, D.L., Bopp, L.H., and Mondello, F.J. 1987. Remedial Action, Incineration, and Treatment of Hazardous Waste. Proceedings of U.S. Environmental Protection Agency, 13th Annual Research Symposium on Land Disposal, Cincinnati, OH.

NAS. 1979. National Academy of Sciences. Polychlorinated Biphenyls, pp. 182. PCB's in Environment, Environmental Studies Board, Commission on Natural Resources, National Academy of Science, Washington, D.C.

NAS. 1980. National Academy of Sciences. Drinking Water and Health. Vol. 3, Washington, D.C.: National Academy Press, pp. 25–67.

NCI. 1978. National Cancer Institute. Bioassay of Aroclor 1254 for possible carcinogenicity. NCI-GC-TR-38. Bethesda, MD. NTIS PB279624.

Nelson, N.N., Hammon, P.B., Nisbet, I.C.T., Sarofim A.F., and Drury, W.H. 1972. Polychlorinated biphenyls - environmental impact. *Environ. Res.* 5:249–362.

Nethercott, J.R. and Holness, D.L. 1986. A report on an investigation of workers at Ferranti-Packard Ltd., St. Catherines, with reference to PCB exposure. Cited in IDSP, 1987. Industrial Disease Panel (IDSP) 1987. *Report to the Workers' Compensation Board on Occupational Exposure to PCBs.* IDSP, Toronto, ON, IDSP Report 2.

Newsted, J.L. 1991. Biochemical Effects of Planar Halogenated Hydrocarbons on Rainbow Trout *(Oncorhynchus mykiss)*. Ph.D. Dissertation, Michigan State University, East Lansing, MI.

Niimi, A.J. and Oliver, B.G. 1983. Biological half-lives of polychlorinated biphenyl (PCB) congeners in whole fish and muscle of rainbow trout *(Salmo gairdneri)*. *Can. J. Fish. Aquat. Sci.* 40:1388–1394.

NIOSH. 1977a. National Institute for Occupational Safety and Health. Manual of Analytical Methods. 2nd ed., Vol. 1., Taylor, D.G., Ed., U.S. Department of Health and Human Services, Cincinnati, OH., pp. 244–253.

NIOSH. 1977b. National Institute for Occupational Safety and Health. Criteria for a recommended standard. Occupational Exposure to Polychlorinated Biphenyls (PCBs). U.S. Department of Health, Education and Welfare, Rockville, MD., Public Health Service, Centers for Disease Control. NIOSH Publ 77-225.

NIOSH. 1992. NIOSH recommendations for occupational safety and health compendium of policy documents and statements. Cincinnati, OH. U.S. Department of Health. NIOSH. 92–100, B92, 162, 536.

Norback, D.H. and Weltman, R.H. 1985. Polychlorinated biphenyl induction of hepatocellular carcinoma in the Sprague-Dawley rat. *Environ. Health Perspect.* 60:97–195.

Norheim, G., Skaare, J.U., and Wiig, O. 1992. Some heavy metals, essential elements, and chlorinated hydrocarbons in polar bear *(Ursus maritimus)* at Svalbard. *Environ. Pollut.* 77:51–57.

Norstrom, R., Hallett, D., and Sonstegard, R. 1978. Coho salmon *(Oucorhynchus kisutch)* and herring gulls *(Larus agentatus)* as indicators of organochlorine contamination in Lake Ontario. *J. Fish Res. Board Canada.* 35:1401–1409.

Norstrom, R.J. 1987. Bioaccumulation of polychlorinated biphenyls in Canadian wildlife. In: *Hazards, Decontamination, and Replacement of PCB a Comprehensive Guide,* Crine, J.P., Ed., Plenum Press, New York, pp. 85–100.

Oatman, L. and Roy, R. 1986. Surface and indoor air levels of polychlorinated biphenyls in public buildings. *Bull. Environ. Contam. Toxicol.* 37:461–466.

Ohlendorf, H.M. and Miller, M. R. 1984. Organochlorine contaminants in California water fowl. *J. Wildlife Manage.* 48:867–877.

Oliver, B.G. and Bourbonniere, R.A. 1985. Chlorinated contaminants in surficial sediments of Lakes Huron, St. Clair, and Erie: Implications regarding sources along the St. Clair and Detroit Rivers. *J. Great Lakes Res.* 11:366–372.

Oliver, B.G. and Niimi, A.J. 1988. Trophodynamic analysis of polychlorinated biphenyl congeners and other chlorinated hydrocarbons in the Lake Ontario ecosystem. *Environ. Sci. Technol.* 22:388–397.

Oliver, B.G., Charlton, M.N., and Durham, R.W. 1989. Distribution, redistribution, and geochronology of polychlorinated biphenyl congeners and other chlorinated hydrocarbons in Lake Ontario sediments. *Environ. Sci. Technol.* 23:200–208.

Orris, P., Kominsky, J.R., Hryorczyk, D., and Melius, J. 1986. Exposure to polychlorinated biphenyls from an overheated transformer. *Chemosphere.* 15:1305–1311.

OSHA. 1985. Occupational Safety and Health Administration. Code of Federal Regulations. OSHA Occupational Standards. Permissible Limits 29 CFR 1910.1000.

Ouw, H.K., Simpson, G.R., and Siyali, D.S. 1976. Use and health effects of Aroclor 1242, a polychlorinated biphenyl in an electrical industry. *Arch. Environ. Health.* 31:189.

Paasivirta, J., Hakala, H., Knuutinen, J., Otollinen, T., Sarkka, J., Welling, L., Paukku, R., and Lammi, R. 1990. Organic chlorine compounds in lake sediments. III. Chlorohydrocarbons, free and chemically bound chlorophenols. *Chemosphere.* 21:1355–1370.

Page, W.G. 1981. Comparison of ground water and surface water for patterns and levels of contamination by toxic substances. *Env. Sci. Technol.* 15:1475–1481.

Paris, D.F., Steen, W.C., and Baughman, G.L. 1978. Role of the physicochemical properties of Aroclor 1016 and 1242 in determining their fate and transport in aquatic environments. *Chemosphere.* 7(4):319–325.

Pastel, M., Bush, B., Kim, J.S. 1980. Accumulation of polychlorinated biphenyls in American shad during their migration in the Hudson River, spring 1977. *Pestic. Monitor. J.* 14:11–22.

Peakall, D.B. and Lincer, J.J. 1970. Polychlorinated biphenyls. Another long-life widespread chemical in the environment. *Bio-Science.* 20:958–964.

Pereira, W.E., Rostad, C.E., and Taylor, H.E. 1980. *Geophys. Res. Lett.* 7:953.

Perttila, M. 1985. Concentrations of DDT and PCB in sediments in the Finnish sediment areas. *Balt. Sea Environ. Proc.* 14:387–391.

Picer, M. and Picer, N. 1991. Long-term trends of DDTs and PCBs in sediment samples from the Eastern Adriatic coastal waters. *Bull. Environ. Contam. Toxicol.* 47:864–873.

Puccetti, G. and Leoni, V. 1980. PCB and HCB in the sediments and waters of the Tiber estuary. *Marine Pollut. Bull.* 11:22–25.

Ray, L.E., Murray, H.E., and Glam, C.S. 1983. Organic pollutants in marine samples from Portland, Maine. *Chemosphere.* 12:1031–1038.

Reggiani, G. and Bruppacher, R. 1985. Symptoms, signs, and findings in humans exposed to PCBs and their derivatives. *Environ. Health Perspect.* 60:225–232.

Regulatory Network, Inc. 1992. Documenting Temporal Trends of Polychlorinated Biphenyls in the Environment (draft). Regulatory Network, Inc. Prepared for Chemical Manufacturers Association, PCB Panel National Electric Manufacturers' Association.

Requejo, A.G., West, R.H., Hatcher, P.G., and McGillivary, P.A. 1979. Polychlorinated biphenyls and chlorinated pesticides in soils of the Everglades National Park and adjacent agricultural areas. *Environ. Sci. Technol.* 13:931–936.

Richard, J.J. and Junk, G.A. 1981. Polychlorinated biphenyls in effluents from combustion of coal/refuse. *Environ. Sci. Technol.* 15:1095–1100.

Ringer, R.K. 1983. Toxicology of PCB's in mink and ferrets. In: *PCB's: Human and Environmental Hazards,* D'Itri, F.M. and Kamrin, M.A., Eds., Butterworth, Woburn, MA, pp. 227–240.

Roberts, J.R., Rodgers, D.W., Bailey, J.R., and Rorke, M.A. 1978. Polychlorinated biphenyls: biological criteria for an assessment of their effects on environmental quality. Associate Committee on Scientific Criteria for Enviornmental Quality, National Research Council of Canada Publication No. NRCC 16077, 172. pp.

Rodgers, P.W. and Swain, W.R. 1983. Analysis of polychlorinated biphenyl (PCB) loading trends in Lake Michigan. *J. Great Lakes Res.* 9:548–58.

RTP. Regulatory Toxicology and Pharmacology. 1994. Interpretive Review of the Potential Adverse Effects of Chlorinated Organic Chemicals on Human Health and the Environment. Report of an Expert Panel. Coulston, F. and Kolbye, A.C., Jr., Eds. Academic Press, New York, Volume 20, Number 1, August.

Ryan, J.J., Lau, P.Y., Pilon, J.C. Lewis, D., McLeod, H.A., and Gervais, A. 1984. Incidence and levels of 2,3,7,8 - tetrachlorodibenzo-*p*-dioxin in Lake Ontario commercial fish. *Environ. Sci. Technol.* 18:719–721.

SANSS. 1990. Structure and Nomenclature Search System. Chemical Idenification System (CIS), Computer Database.

Safe, S. 1989. Polychlorinated biphenyls (PCBs): Mutagenicity and carcinogenicity. *Mutat. Res.* 220:31–47.

Safe, S. 1990. Polychlorinated biphenyls (PCBs), dibenzo-p-dioxins (PCDD's), dibenzofurans (PCDF's), and related compounds: environmental and mechanistic considerations which support the development of toxic equivalency factors (TEF's). *Crit. Rev. Toxicol.* 21:51–88.

Safe, S. 1992. Development, validation and limitations of toxic equivalency factors. *Chemosphere.* 25:61–64.

Safe, S. et al. 1985. Effects of structure on binding to the 2,3,7,8-TCDD receptor protein and AHH induction-halogenated biphenyls. *Environ. Health Perspect.* 61:21–33.

Schmidt, P., Buhler, F., and Schlatter, CH. 1992. Dermal absorption of PCB in man. *Chemosphere.* 24:1283–1292.

Schmitt, C.J., Zajicek, J.L., and Ribick, M.A. 1985. National pesticide monitoring program. Residues of organochlorine chemicals in freshwater fish, 1980-1981. *Arch. Environ. Contam. Toxicol.* 14:225–260.

Schweizer, C. and Tarradellas, J. 1980. Etat des recherches sur les biphényles polychlorés en Suisse. *Chimia.* 34:507–517.

Shaw, G.R. and Connell, D.W. 1984. Physicochemical properties controlling polychlorinated biphenyl (PCB) concentrations in aquatic organisms. *Environ. Sci. Technol.* 18:18–23.

Simmons, M.S. 1984. PCB contamination in Great Lakes. In: *Toxic Contaminants in the Great Lakes,* Nriagu, J.O. and Simmons, M.S., Eds., John Wiley & Sons, New York, pp. 287–309.

Singer, E., Jarv, T., and Sage, M. 1983. Survey of PCB in Ambient Air across the Province of Ontario.

Sinks, T., Steele, G., Smith, A.B., Watkins, K., and Shuffs, R.A. 1992. Mortality among workers exposed to polychlorinated-biphenyls. *Am. J. Epidemiol.* 136:389–398.

Skea, J.C., Simonin, H.A., Dean, H.J. et al. 1979. Bioaccumulation of Aroclor 1016 in Hudson River fish. *Bull. Environ. Contam. Toxicol.* 22:332–336.

Stark, A.D., Costas, K., Gan Chang, H., and Vallet, H.L. 1986. Health effects of low-level exposure to polychlorinated biphenyls. *Environ. Res.* 41:174–183.

Stevens, R.J. and Neilson, M.A. 1989. Inter- and intralake distributions of trace organic contaminants in surface waters of the Great Lakes. *J. Great Lakes Res.* 15:377–393.

Stickel, W.H., Stickel, L.F., Dyrland, R.A., and Hughes, D.L. 1984. Aroclor 1254 residues in birds: Lethal levels and loss rates. *Arch. Environ. Contam. Toxicol.* 13:7–13.

Stone, P.J., Ed. 1981. *Emergency Handling of Hazardous Materials in Surface Transportation.* Bureau of Explosives, Association of American Railroads, Washington, D.C. 418 p.

Strachan, W.M.J. 1985. Organic substances in the rainfall of Lake Superior. *Environ. Toxicol. Chem.* 4:677–683.

Strachan, W.M.J. 1988. Polychlorinated Biphenyls (PCBs) — Fate and Effects in the Canadian Environment. Environment Canada, Report EPS 4/HA/2.

Strachan, W.M.J. and Huneait, H. 1979. Polychlorinated biphenyls and organochlorine pesticides in Great Lakes precipitation. *J. Great Lakes Res.* 5:61–68.

Strachan, W.M.J. 1985. Organic substances in the rainfall of Lake Superior. *Environ. Toxicol. Chem.* 4:677–683.

Sundstrom, G., Hutzinger, D., and Safe, S. 1976. The metabolism of chlorobiphenyls - a review. *Chemosphere.* 5:267.

Swackhamer, D.L., McVeety, B.D., and Hites, R.A. 1988. Deposition and evaporation polychlorobiphenyl congeners to and from Siskiwit Lake, Isle Royale, Lake Superior. *Environ. Sci. Technol.* 22:664–672.

Swackhamer, W.M. and Armstrong, D.E. 1986. Estimation of the atmospheric and nonatmospheric contributions and losses of polychlorinated biphenyls for Lake Michigan on the basis of sediment records of remote lakes. *Environ. Sci. Technol.* 20:879–883.

Swierenga, S.H.H., Yamasaki, H., Piccoli, C. et al. 1990. Effects on intercellular communication in human keratinocytes and liver-derived cells of polychlorinated biphenyl congeners with differing in vivo promotion activities. *Carcinogenesis.* 11:921–926.

Tanabe, S., Tanaka, H., and Tatsukawa, R. 1984. Polychlorobiphenyls, DDTs and hexachlorocyclohexane isomers in the western North Pacific ecosystem. *Arch. Environ. Contam. Toxicol.* 13:731–738.

Tanabe, S., Hidaka, H., and Tatsukawa, R. 1983. PCBs and chlorinated hydrocarbon pesticides in Antarctic atmosphere and hydrosphere. *Chemosphere.* 12:277–288.

Tanabe, S. 1988. PCB problems in the future: Foresight from current knowledge. *Environ. Pollut.* 50:5–28.

Task Force on PCB. 1976. Background to the regulation of polychlorinated biphenyls (PCB) in Canada. Environment Canada and Health and Welfare Canada, Joint Report No. 76-1, April 1, 1976.

Thomas, R.L. and Frank, R. 1981. PCBs in sediment and fluvial suspended solids in the Great Lakes. In: *Physical Behavior PCDs Great Lakes.* Mackay, D. et. al., Eds., Ann Arbor, MI, Ann Arbor Science, pp. 245–267.

Tiernan, T.O., Taylor, M.L., and Garret, J.H. et al. 1983. PCDDs, PCDFs and related compounds in the effluents from combustion processes. *Chemosphere.* 12:595–606.

Tillitt, D.E., Ankley, G.T., Verbrugge, D.A., and Giesy, J.P. 1991. H411E rat hepatoma cell bioassay-derived 2,3,7,8-tetrachlorodibenzo-p-dioxin equivalents (TCDD-EQ) in colonial fish eating waterbird eggs from the Great Lakes. *Arch. Environ. Contam. Toxicol.* 21:91–101.

Treon, J.F., Cleveland, F.P., Cappel, J.W., and Atchley, R.W. 1956. The toxicity of the vapours of Roclor 1242 and Aroclor 1254. *Am. Ind. Hyg. Assoc. Q* 17:204–213.

van den Broek, W.L.F. 1979. Seasonal levels of chlorinated hydrocarbons and heavy metals in fish and brown shrimps from the Medway estuary, Kent. *Environ. Pollut.* 19:21–38.

Veith, G.D., DeFoe, D.L., and Bergstedt, B.V. 1979. Measuring and estimating the bioconcentration factor of chemicals in fish. *J. Fisheries Res. Bd. Canada.* 36:1040–1048.

Villeneuve, D.C., Grant, D.L., Khera, K., Clegg, D.J., Baer, H., and Phillips, W.E.J. 1971. The fetotoxicity of a polychlorinated biphenyl mixture (Aroclor 1254) in the rabbit and in the rat. *Environ. Physiol.* 1:67–71.

Vos, J.A., Koeman, J.H., Van Der Mass, H.L. et al. 1970. Identification and toxicological evaluation of chlorinated dibenzofuran and chlorinated naphthalenes in two commercial polychlorinated biphenyls. *Food Cosmet. Toxicol.* 8:625–633.

Vos, J.G. and Beems, R.B. 1971. Dermal toxicity studies of technical polychlorinated biphenyls and fractions thereof in rabbits. *Toxicol. Appl. Pharmacol.* 19:317–633.

Walker, M.K. and Peterson, R.E. 1991. Potencies of polychlorinated dibenzo-p-dioxin, dibenzofuran, and biphenyl congeners, relative to 2,3,7,8-tetrachlorodibenzo-p-dioxin, for producing early life stage mortality in rainbow trout *(Oncorhynchus mykiss). Aquat. Toxicol.* 21:219–238.

Warry, N.D. and Chan, C.H. 1981. Organic contaminants in the suspended sediments of the Niagara River. *J. Great. Lakes Res.* 7:394–403.

Weant, G.E. and McCormick, G.S. 1984. Nonindustrial sources of potential toxic substances and their applicability to source apportionment methods. U.S. Environmental Protection Agency, pp. 36, 86.

Weaver, G. 1984. PCB contamination in and around New Bedford, Mass. *Environ. Sci. Technol.* 18:22A-27A.

Wells, D.E. and Johnstone, S.J. 1978. The occurrence of organochlorine residues in rainwater. *Water, Air, and Soil Pollut.* 9:271–280.

WHO. 1976. World Health Organization. Polychlorinated Biphenyls and Terphenyls. WHO, U.K. Environmental Health Criteria 2.

Williams, L.L. and Giesy, J.P. 1992. Relationships among concentrations of individual polychlorinated biphenyl (PCB) congeners, 2,3,7,8-tetrachlorodibenzo-p-dioxins equivalents (TCDD-EQ), and rearing mortality of chinook salmon *(Nocorhynchus tshawtyscha)* eggs from Lake Michigan. *J. Great Lakes Res.* 18:108–124.

Wolff, M.S. 1985. Occupational exposure to polychlorinated biphenyls (PCBs). *Environ. Health Perspect.* 60:133–138.

Wren, C.D. 1991. Cause-effect linkages between chemicals and populations of mink *(Mustela vison)* and otter *(Lutra canadensis)* in the Great Lakes Basin. *J. Toxicol. Environ. Health.* 33:549–585.

Young, D.R. and Heesen, T.C. 1978. DDT, PCB, and chlorinated benzenes in the marine ecosystem off Southern California. In: *Water Chlorination, Environmental Impact and Health Effects,* Vol. 2., Jolley, R.L., Ed., pp. 267–290.

Zepp, R.L. and Kirkpatrick, R.L. 1976. Reproduction in cottontails fed diets containing a PCB. *J. Wildlife Manage.* 4:491–495.

CHAPTER 6

Chlorinated Biocides

Chlorinated biocides is a small but diverse group of synthetically produced chemicals characterized by a cyclic structure and a variable number of chlorine atoms. Pesticides used in the past centuries were derived from natural sources. Natural insecticides and rodenticides were used by Romans. Nicotine was used as an insecticide by the French prior to 1960 and the active ingredient of pyrethrnan powder (insecticide) was extracted from the chrysanthemum plant species as early as 1890 (Green et al. 1987; Mann 1987). Chlorinated pesticides were synthesized first around 1940. DDT was first synthesized in 1874 and its insecticidal properties discovered in 1939. Pesticides developed in the 1940s and 1950s were chlorinated cyclic hydrocarbons. These pesticides were cheap to produce, but effective, with minimal mammalian toxicity. During and after World War II, these pesticides and insecticides were effective in controlling insect-generated and insect-carried disease such as malaria. These chemicals saved thousands of lives by controlling insect-carried diseases. For example, Table 6.1 provides data on cases of malaria in selected countries before and after the use of DDT. It clearly shows the benefit of DDT, although the environmental and health impact of DDT and some other biocides cannot be overlooked and they are well documented. The latter aspect of DDT and other toxic biocides have led to their limited or severely restricted use and, in some countries, total exclusion.

PRODUCTION AND USE PATTERN

In 1942, hexachlorocyclohexane (HCH) was discovered to be an effective and simple insecticide. Of its isomers, the γ-HCH has the greatest insecticidal activity and is therefore widely used. It is marketed as Lindane. The α- and β-isomers of HCH are more toxic than the γ-isomer to mammals. Chlordane was synthesized in 1945 and found to be a highly effective residual insecticide. In 1948, the most active component of chlordane, called heptachlor was synthesized along with two other

Table 6.1 Reported Number of Malaria Cases Before and After the Use of DDT

Country	Year	Number of Cases
Cuba	1962	3,519
	1969	3
Dominican Republic	1950	17,310
	1968	21
Trinidad	1950	5,098
	1969	5
India	1935	100,000,000
	1969	286,962
Sri Lanka	1946	2,800,000
	1961	110
Italy	1945	411,602
	1968	37
Romania	1948	338,198
	1969	4
Taiwan	1945	1,000,000
	1969	9
Turkey	1950	1,118,969
	1969	2,173

Source: Green et al. 1987.

closely related insecticides, known as aldrin and dieldrin. It has been shown that microbial and photochemical transformations of heptachlor in the environment yields heptachlor epoxide and photoheptachlor. The transformed compounds were shown to be equally or more toxic than the parent compound, heptachlor. In the same years, toxaphene was produced by the chlorination of camphene. Toxaphene was shown to contain a considerable number of chlorinated camphenes.

Endrin was marketed in 1951 as a highly effective insecticide, containing at least 95% of the active ingredient. Endrin is not environmentally persistent and it was one of the first generation of degradable, synthetically-made pesticides. Similarly, in 1954, a broad spectrum insecticide, named endosulfan, was introduced to the market and it was also environmentally degradable. Technical endosulfan consists of about four parts of the α-isomer and one part of the β-isomer. The α-isomer, a somewhat more active insecticide, was shown to slowly transform in the environment to the more stable β-isomeric form. Several chlorinated pesticides have appeared on the market from the 1960s to 1980s. Mirex was introduced commercially in 1969 as an insecticide to control fire ants and as a flame-retardant. It is moderately stable in the environment, but slowly degrades into products such as chlordecane and photomirex, which are highly toxic. Methoxychlor did not find widespread application until the ban on DDT in 1969. It is closely related to DDT, containing methoxy ($-OCH_3$) groups on the phenyl groups instead of chlorine. Recent years have seen new chlorinated herbicides, such as Picloram, Triallate, Propachlor, Chloranil, and Dichlobenil. Most of these compounds are toxic and degrade rapidly in the environment, yielding relatively low residue levels.

DDT, DDE, and DDD

DDT (1,1,1-trichloro-2,2-bis)(p-chlorophenyl)ethane), DDE (1,1-dichloro-2,2-bis(p-chlorophenyl ethylene) and DDD (1,1-dichloro-2,2-bis(p-chlorophenyl)ethane) have been found in at least 316 of the sites on the "National Priorities List (NPL)" (ATSDR 1994a). Technical grade DDT is a mixture of three forms, P,P'-DDT (85%), O,P'-DDT (15%) and O,O'-DDT (trace amounts). Technical grade DDT is made from chloral hydrate, chlorobenzene, and sulfuric acid. DDE and DDD sometimes contaminate technical grade DDT. DDD was also used to kill pests; one form of DDD has been used medically to treat cancer of the adrenal gland (ATSDR 1994a). DDT, DDE, and DDD are no longer used in North America, but still used in several other parts of the world. Chemical identity and chemical properties of DDT, DDE, and DDD are given in Table 6.2.

Hexachlorocyclohexane

Hexachlorocyclohexane (HCH), also known as benzene hexachloride (BHC), is a synthetic chemical that exists in eight isomeric forms. The different isomers are named according to the position of the hydrogen atoms on the ring. The gamma γ isomer of HCH was used as an insecticide on fruit, vegetable, and forest crops. It is still used in North America and in other countries as a human medicine to treat head and body lice and scabies, a contagious skin disease caused by mites (ATSDR 1994b). Technical grade HCH, a mixture of several isomers of HCH, was once used as an insecticide and contained about 10 to 15% of γ-HCH as well as α-, β-, δ-, and ε-forms of HCH. Technical grade HCH has not been produced in the U.S. since 1983. Table 6.2 lists the chemical identity and physical and chemical properties of HCH isomers.

Chlordane

Chlordane, a synthetic pesticide, was in use from 1948 to 1988. It is sometimes referred by the trade names Octachlor and Velsicol 1068. Chlordane is not a single compound, but a mixture of about ten major compounds. Some of them include trans-chlordane, cis-chlordane, β-chlordane, heptachlor, and trans-nonachlor (ATSDR 1994c). From 1983 to 1988, chlordane was approved for use to control termites in homes. It was applied underground around the foundation. Until 1978, chlordane was used as a pesticide on agricultural crops, lawns, and gardens and as a fumigant. In 1988, the U.S. EPA banned the use of chlordane to control termites and all other uses (ATSDR 1994d). The Chemical identification and physical and chemical properties are given in Table 6.2.

Aldrin and Dieldrin

Aldrin and dieldrin are common names of two structurally similar synthetic compounds that were used as insecticides. From the 1950s until 1970, aldrin and dieldrin were used extensively as insecticides on crops such as corn and cotton. In

180 CHLORINATED ORGANIC COMPOUNDS IN THE ENVIRONMENT

Table 6.2 Chemical Identities and Properties of Chlorinated Pesticides

Identity/Property	DDT	DDE	DDD	γ-HCH	α-HCH	β-HCH	δ-HCH
CAS Registry	50-29-3	72-55-9	72-54-8	58-89-9	319-84-6	319-85-7	319-86-8
NIOSH/RTECS	KJ3325000	KV9450000	KI0700000	GV4900000	GV3500000	GV4375000	GV4550000
EPA Hazardous Waste	U061	—	U060	U129; D013	—	—	—
OHM/TADS	7216510	—	7215098	7216531	810002	—	—
DOT/UN/NA/IMCO	2761-55; NA2761	—	NA2761; TDE	NA2761 lindane	—	—	—
Shipping	IMCO6.1; UN2761	—		IMCO6.1 lindane UN276	—	—	—
HSDB	200	1625	285	646	6029	6183	6184
NCI	C00465	C00555	C00475	C00204	—	—	—
Chemical formula	$C_{14}H_9Cl_5$	$C_{14}H_8Cl_4$	$C_{14}H_{10}Cl_4$	$C_6H_6Cl_6$	$C_6H_6Cl_6$	$C_6H_6Cl_6$	$C_6H_6Cl_6$
Molecular weight	354.49	318.03	320.05	290.83	290.83	290.83	290.83
Melting point (°C)	108(P,P') 75 (O,P')	88.4–90	109–110	112.5	159–160	314–315	141–142
Boiling point (°C) (~ mm of Hg)	260	—	193.0 (1 mm)	323.4 (760 mm)	288 (760 mm)	60 (0.5 mm)	60 (0.5 mm)
Aq. solubility, mg/L	0.0034	0.12 (25C)	0.16 (24°C)	17.0	10.0	5.0	10.0
Partition coefficient							
Log K_{ow}	6.19	7.00	6.20	3.3; 3.61	3.46; 3.85	4.5; 4.98	2.8; 4.14
Log K_{oc}	5.38	6.64	5.89	3.0; 3.57	3.57	3.57	3.8
Vapor pressure mm of Hg	5.5×10^{-6} (20°C)	6.5×10^{-6} (20°C)	1.02×10^{-6} (30°C)	9.4×10^{-6} (20°C)	—	2.8×10^{-7} (20°C)	1.7×10^{-5} (20°C)
Henry's law constant atm · m³/mol.	5.13×10^{-4}	6.85×10^{-5}	2.16×10^{-5}	7.8×10^{-6} 3.2×10^{-6}	4.8×10^{-6} 6×10^{-6}	4.5×10^{-7}	2.1×10^{-7}
Conversion factors ppm (v/v) to mg/m³ (20°C)	ppm × 14.8	ppm × 13.26	ppm × 13.34	ppm × 4.96	ppm × 4.96	ppm × 4.96	ppm × 4.96
mg/m³ (20°C) to ppm in air (20°C)	mg/m³ × .07	mg/m³ × .08	mg/m³ × .07	mg/m³ × .20	mg/m³ × .20	mg/m³ × .20	mg/m³ × .20

CHLORINATED BIOCIDES 181

Table 6.2 Chemical Identities and Properties of Chlorinated Pesticides (Continued)

Identity/property	Chlordane	Aldrin	Dieldrin	Heptachlor	Heptachlor Epoxide	Toxaphene
CAS Registry	57-74-9	309-00-2	60-57-1	76-44-8	1024-57-3	8001-35-2
NIOSH/RTECS	PB9800000	IO2150000	IO1750000	PC0700000	PB9450000	XW5250000
EPA Hazardous Waste	U036	P004	P037	P059	—	P123
OHM/TADS	7215092	7215090	7216516	7216526	—	7216561
DOT/UN/NA/IMCO	NA2762	IMCO6.1 NA 2762	NA 2761	UN2762,2995 2996; UN2761 NA 3761	—	NA 2761 toxaphene
HSDB	802	199	322	554	6182	1616
NCI	C00099	C00044	C00124	10875-C	—	—
Chemical formula	$C_{10}H_6Cl_8$	$C_{12}H_8Cl_6$	$C_{12}H_8Cl_6O$	$C_{10}H_5Cl_7$	$C_{10}H_5Cl_7O$	$C_{10}H_{10}Cl_8$
Molecular weight	409.76	364.93	380.93	373.5	389.4	414 (AV)
Melting point (°C)	107 (cis) 105 (trans)	104 (pure) 40–60 (tech)	175–176 (pure) 95 (tech)	95–96 (pure) 46–74 (tech)	160–161.5	65–90
Boiling point (°C) (− mm of Hg)	175 (2 mm)	145.0	330.0	135–145 (1 mm)	—	Dechlorination at 155°C
Aq. solubility, mg/L	0.056	0.2	0.18	56.0	0.35	3.0
Partition coefficient						
Log K_{ow}	5.54	3.01	4.55	5.05	4.60	3.3
Log K_{oc}	3.49–4.64	4.69	3.87	—	—	2.47; 5.00
Vapor pressure, mm of Hg	3.9×10^{-6} (20°C)	7.5×10^{-5} (20°C)	1.78×10^{-7} (20°C)	3.0×10^{-4} (25°C)	—	0.4 (20°C)
Henry's law constant atm · m³/ mol.	4.8×10^{-5}	3.2×10^{-4}	1.51×10^{-5}	—	—	0.21
Conversion factors ppm (v/v) to mg/m³ (20°C)	ppm × 16.75	ppm × 14.96	ppm × 15.61	ppm × 65.1	ppm × 62.5	ppm × 16.89
mg/m³ (20°C) to ppm in air (20°C)	mg/m³ × .06	—	—	mg/m³ × 15.3	mg/m³ × 15.9	mg/m³ × 0.059

CHLORINATED ORGANIC COMPOUNDS IN THE ENVIRONMENT

Table 6.2 Chemical Identities and Properties of Chlorinated Pesticides (Continued)

Identity/Property	Endrin	Endosulfan	Mirex	Chlordecone	Methoxychlor	Dichlorvos	Chloro-pyrifos	Chlorfen-vinphos
CAS Registry	72-20-8	115-29-7	2385-85-5	143-50-0	72-43-5	62-73-7	2921-88-2	470-90-6
NIOSH/RTECS	ID1575000	RB9275000	PC8225000	PC8575000	KJ3675000	TC0350000	TF6300000	TB875000
EPA Hazardous Waste	P050; D012	P050	—	U142	U247; D014	—	059101	—
OHM/TADS	7216522	7216559	—	—	7216536	7800015	7800025	810041
DOT/UN/NA/IMCO	UN 2811; NA2761; IMCO6.1	2761	UN 2588	NA 2761; UN 2761;	DOT 2761; UN3018 NA2761; IMCO6.1	NA 2783	NA 2783	UN 2783
HSDB	198	390	1659	1558	1173	319	389	1540
NCI	C00157	C00566	C06428	C00191	C00497	C00113	—	—
Chemical formula	$C_{12}H_8Cl_6O$	$C_9H_6Cl_6O_3S$	$C_{10}Cl_{12}$	$C_{10}Cl_{10}O$	$C_{16}H_{15}Cl_3O_2$	$C_4H_7Cl_2O_4P$	$C_9H_{11}Cl_3NO_3PS$	$C_{12}H_{14}Cl_3O_4P$
Molecular weight	380.9	406.95	545.59	490.68	345.65	220.98	350.57	359.56
Melting point (°C)	235.0	106 (pure)	485.0	350.0	89.0; 77.0	—	41–42	-19 to -23
Boiling point (°C) (— mm of Hg)	245 decomp.	—	—	—	decomposes	140.0 (20 mm)	decomposes (160°C)	167–170 (0.5 mm)
Aq. solubility, mg/L	0.2	0.16	0.6; 0.2	3.0	0.045	16	2.0 (25°C)	145 (23°C)
Partition coefficient								
Log K_{ow}	5.6 (calc)	3.55; 3.62	5.28	4.50	4.68–5.08	1.116; 1.47	5.11; 4.96	3.806
Log K_{oc}	3.2 (calc)	3.5	3.76	3.38–3.41	4.9	1.45	3.73; 4.13	2.45
Vapor pressure (mm of Hg)	2×10^{-7} (25°C)	1×10^{-5} (25°C)	3×10^{-7} (25°C)	$<3 \times 10^{-7}$ (25°C)	1.4×10^{-6} (25°C)	1.88×10^{-6} (20°C)	1.87×10^{-5} (25°C)	1.7×10^{-7} (25°C)
Henry's law constant atm · m³/ mol.	4×10^{-7} (calc)	1×10^{-5} (25°C)	5.16×10^{-4} (25°C)	2.5×10^{-8} (25°C)	1.6×10^{-5} (25°C)	1.88×10^{-6} (25°C)	1.23×10^{-5} (25°C)	2.76×10^{-9} (25°C)
Conversion factors ppm(v/v) to mg/m³ (20°C)	ppm × 15.6	ppm × 16.64	—	—	—	ppm × 0.16	ppm × 0.07	ppm × 0.068
mg/m³(20°C) to ppm in air (20°C)	mg/m³ × .06	mg/m³ × 0.06	mg/m³ × 0.041	mg/m³ × 0.046	—	mg/m³ × 9.03	mg/m³ × 14.3	mg/m³ × 14.7

CAS = Chemical Abstract Service; NIOSH = National Institute for Occupational Safety and Health; RTECS = Registry of Toxic Effects of Chemical Substances; EPA = Environmental Protection Agency; OHMTADS = Oil and Hazardous Materials/Technical Assistance Data System DOT/UN/NA/IMCO = Department of Transportation/United Nations/North America/International Marine Dangerous Goods Code; HSDB = Hazardous Substances Data Bank.

1970, the U.S. Department of Agriculture banned their use. In 1972, however, the U.S. EPA approved aldrin and dieldrin for killing termites. In 1987, the manufacturer voluntarily canceled their registration for use in controlling termites. Aldrin is readily converted to dieldrin in the environment and, thus, they are discussed together in their toxicity profile (ATSDR 1993a). Their chemical identity and physical and chemical properties are listed in Table 6.2.

Heptachlor and Heptachlor Epoxide

Heptachlor was used primarily as insecticide in seed grains and on crops during the 1960s and 1970s. It was also used to terminate termites in homes. Since late 1987, most uses of heptachlor has been phased out and it is not available to the general public. In 1987, the only manufacturer, Velsicol Company, discontinued marketing heptachlor (ATSDR 1989). The transformation product of heptachlor, namely heptachlor epoxide, remains in soil for a long period of time (>=15 years). The chemical identity and physical and chemical properties are given in Table 6.2.

Toxaphene

Toxaphene is a manufactured insecticide containing over 670 chemicals. Until 1982, toxaphene was one of the heavily used insecticides in North America. It was primarily used to control insects, pests on cotton, and other crops. It was also used on insect pests on livestock. In 1988, the U.S. EPA banned all uses of toxaphene and 1993 banned importation of any food containing toxaphene residues into U.S. Table 6.2 lists the chemical identity and properties of toxaphene (ATSDR 1994d).

Endrin

Chemical identity and physical and chemical properties of endrin are listed in Table 6.2. (ATSDR 1994a).

Endosulfan

Endosulfan is a synthetically made insecticide and used to control insects on crops such as grains, tea, fruits, and vegetables, and also on tobacco and cotton. Also used as a wood preservative. Endosulfan is marketed as a mixture of α- and β-isomers. It has a distinct odor similar to turpentine. The chemical identity and properties are given in Table 6.2. (ATSDR 1993b).

Mirex and Chlordecone

Mirex and chlordecone are two separate synthetic insecticides that have similar chemical structures. They are currently produced in the U.S. They were most commonly used in the 1960s and 1970s. Mirex was used to control fire ants and as a flame-retardant in plastics, rubber, paint, paper, and electrical goods from 1959 to 1972, because it does not burn easily. Chlordecone was used to control insects that attached to bananas, citrus trees (with no fruits), tobacco, and ornamental shrubs. It was also used as household ant and roach traps. Chlordecone is also known by its registered trade name, Kepone. All registered products containing mirex and chlordecone were discontinued in 1977 and 1978. Table 6.2 lists the chemical identification as well as properties of mirex and chlordecone.

Chloropyrifos

Chloropyrifos is an organophosphorus pesticide, widely used in the home to control cockroaches, fleas, termites, and in some pet flea and tick collars. It is also used on the farm, as a dip or spray to control ticks on cattle and as dust or spray to control crop pests. The chemical identification and properties of chloropyrifos are given in Table 6.2. (ATSDR 1995b).

Dichlorvos

Dichlorvos is a synthetically produced insecticide. It is marketed under trade names that include Vapona, Nuvan, Atgard, and Task. It also is called DDVP (Dichloro divinyl phosphate). Main uses are for insect control in food storage areas, green houses and barns and for parasite control on live stock. Dichlorvos is not generally used on outdoor crops. It is sometimes used for insect control in workplaces and the home. Table 6.2 lists the chemical identification and properties of dichlorvos (ATSDR 1995c).

Chlorfenvinphos

Chlorfenvinphos is an organophosphorus insecticide used to control insect pests on live stock. It is also used to control household pests such as flies, fleas, and mites. Chemical identification and properties of chlorfenvinphos are given in Table 6.2. (ATSDR 1995d).

ENVIRONMENTAL RESIDUES AND EXPOSURE ROUTES

DDT, DDE, and DDD

Air

During the period when DDT was widely used, a large portion of DDT was released into the air from agricultural or vector control applications. DDT was banned in 1972 and, hence, current release of DDT in North America should be negligible. It is still being used in several areas of the world. DDT, DDE, and DDD in the atmosphere can photodegrade or redeposit by rain or dry deposition. DDT and its breakdown products preferentially bind to soil and sediment, where there may be photodegradation on the surface and biodegradation in the subsurface. DDT may persist for a long time and DDE persists even longer. DDT, DDE, and DDD in water undergo sedimentation, volatilization, photodegradation, and bioconcentration, leading to biomagnification in the food chain. Human exposure in North America is primarily from dietary sources.

Currently, there is no production of DDT in the U.S. (CSCORP 1992) and many other countries. DDT residues in peatlands across the middle latitudes of North America have been determined (Rapaport et al. 1985). These areas receive pollutant input from the atmosphere and thus are important indicators of global levels of certain chemicals. The residue levels in peat cores, snow, and rain samples, indicate that there are continuing sources of direct input of DDT into the environment. However, the levels are relatively low compared to concentrations detected in the 1960s. It is evident that global atmospheric transport is occurring probably from areas where current use of DDT is substantial.

Water

Release of DDT into surface waters occurred when it was used for vector control near open waters. This could still be continuing in countries where DDT is used in insect pest control near open waters.

During 1980 to 1983, data from 3,500–5,700 ambient water samples analyzed for DDT, DDE, and DDD showed that about 45% of the samples contained one of these compounds. The median level reported for DDT and DDE was 1 ppt and for DDD, 0 ppt. Approximately 50 samples of industrial effluents were analyzed to contain 10 ppt of DDT, DDE, and DDD (Staples et al. 1985). A total of 224 surface water samples (unfiltered) taken in 1967-1968 (prior to the ban on DDT) from various sites were analyzed to contain DDT in 27 samples at 0.005-0.316 µg/L; DDE in these samples at 0.02-0.05 µg/L; DDD in six samples at 0.015-0.840 µg/L (Lichtenberg et al. 1970). During 1976 to 1980, U.S. Geological Survey and U.S. EPA collected 2,800 water samples from 177 stations from 150 river sites and analyzed for selected pesticides (Gilliom 1984). Only 2.8, 0.6, and 4.0% were positive for DDT, DDE, and DDD respectively at detection limit of 0.5 µg/L for DDT and DDE and 0.3 µg/L. It was interesting that concentrations of P,P'-DDT equalled or exceeded those of P,P'-DDE, suggesting that P,P'-DDT has an unusually long half-life for DDT in Yakima basin soils (Johnson et al. 1988).

Soil

When it was in use, large quantities of DDT were released to the soil via its direct application or by direct or indirect releases during manufacture, formulation, storage, or disposal. In 1963, approximately 81,000 metric tons of DDT were produced and 27,000 metric tons were utilized in the U.S. alone (WHO 1979). With the banning of DDT in the U.S., some stores of these products were relocated to hazardous waste sites, where they remain as potential sources of release to soil (ATSDR 1994a).

In 1971, the atmospheric levels of DDT and DDE from urban and agricultural areas were 1.4 to 1,560 ng/m^3 and 1.9 to 131 ng/m^3, respectively, and mostly in the particulate phase (Stanley et al. 1971). Air samples taken in 1975 from an extensive use agricultural area showed that the arithmetic mean of DDT and its metabolites decreased 92% in the area sampled since the ban on DDT use in 1972. Samples from the Gulf of Mexico in 1977 had an average DDT level of 0.034 ng/m^3 (Bidleman et al. 1981). Average P,P'-DDT levels in snow samples taken during 1981–1982 were 320 ng/m^3; samples from 1982–1983 were 600 ng/m^3; and samples from 1983–1984 were 180 ng/m^3. Rain samples at these locations contained an average of 300 ng/m^3 of DDT. The results indicate a measurable global contamination in the ppt range (Rapaport et al. 1985).

γ-HCH

HCH can be released into the environment during its formulation and its use as a pesticide. HCH can partition into all media of the environment; in the atmosphere, either as vapor adsorbed to particulates and can undergo photolytic degradation. But it is primarily removed from the atmosphere by rain-out or dry deposition. Biodegradation seems to be the dominant decomposition pathway in soil and water. HCH has been detected in air, surface water, ground water, sediment, soil, fish, and other aquatic organisms, wildlife, food, and humans (ATSDR 1994b). Primary exposure route for humans is from dietary sources. HCH is not a major contaminant in drinking water.

Air

The largest input is from agricultural applications of the pesticide lindane. Other inputs include releases during manufacture and formulation. Since aerial application of lindane (γ-HCH) is now prohibited, releases from this source are not expected. In the vicinity of formulation plants (U.S.), α- and γ-HCH were detected in 1971 in 60 to 90% of the air samples at mean levels of 1–1.3 mg/m^3 (Lewis and Lee 1976).

Use and disposal are likely to introduce α-HCH into air. Examples include, wind erosion of contaminated soil and volatilization from treated agricultural soils and plant foliage (Lewis and Lee 1976). Volatilization from water columns is not considered significant because of the relatively high aqueous solubility of γ-HCH (Mackay and Leinonen 1975). According to the Toxics Release Inventory (TRI), an estimated total of 562 pounds of γ-HCH (and its other isomers), accounting for 99% of the total environmental release of 567 pounds, was discharged from manufacturing and processing facilities in the U.S. in 1991 (TRI91 1993). TRI data should be used with caution since only certain types of facilities are required to report.

Water

Surface run-off (either as dissolved or particulate HCH) could release γ-HCH into surface waters. Also wet deposition of rainfall could contribute to the input (Tanabe et al. 1982). The Great Lakes generally receive from 0.77 to 3.3 metric tons/year of α-HCH and from 3.7 to 15.9 metric tons of γ-HCH from atmospheric deposition (Eisenreich et al. 1981). α- and γ-HCH were detected in urban stormwater run-off in 11–20% of the samples collected. Some monitoring studies suggested that γ-HCH, in spite of its low mobility in soils, does migrate to groundwater.

Levels of DDT, DDE and DDD in Environmental Samples and Food Items

Study	Year	Number of Samples	Levels of DDT, DDE, and DDD	Reference
U.S. Geol. Survey/EPA 150 River Sites	1975–80	1,000 (sediments)	Detectable levels (8%)	Gilliom, 1984
	1980–83	1,100 (sediments)	0.1 and 0.2 ppb (median)	Staples et al. 1985
	1980–83	Biota	14, 26, and 15ppm (median)	Staples et al. 1985
Mississippi River Delta West-central Region	1980–83	Sediments	0.025, 0.07, and 0.03 ppm (AV)	Ford and Hill 1991
Yakima River Basin, Washington, U.S.	1985	River bed Sediments	100–234,000 ppm	Johnson et al. 1988
U.S. National Soil Monitoring Program	1970	1,000 (soil)	0.18 ppm (P,P'-DDT)	Crockett et al. 1974
	1972	1,000 (soil)	0.02 ppm (average)	Carey et al. 1979
Arizona Soil Study	1983	Soil samples	0.39 ppm (average)	Buck et al. 1983
Superfund Site, MA	1985	Soil samples	61, 10, and 70 ppm, respectively	Menzie et al. 1992
Canadian Ice Island Arctic Ocean	1986–88	Plankton	11.8 ppb	Hargrave et al. 1992
		Particles under packed ice	<1-144 ppb	
		Amphipods — open sea	<57-457 ppb	
		fish — abyssal area	819 ppb	
		Amphipod — abyssal area	3,769 ppb	
		Ringed-neck seal	1,482 ppb*	
		Polar bear	266 ppb*	
Market Basket Survey	1969–76	domestic cheese	3 ppb	Duggan 1983
		Ready-to-eat meat, fish, and poultry	5 ppb	
		Eggs	4 ppb	
		Domestic fruits	13 ppb	
		Domestic leaf and stem vegetables	24 ppb	
		Domestic grains	7 ppb	
		Peanut and peanut products	11 ppb	

Levels of DDT, DDE and DDD in Environmental Samples and Food Items (Continued)

Study	Year	Number of Samples	Levels of DDT, DDE, and DDD		Reference
National Contaminant Biomonitoring Study	1976–84	Fish	1976, ppm	1984, ppm	Schmitt et al. 1990
(12 locations across the U.S.)		P,P'-DDT	0.05	0.03	
		P,P'-DDE	0.26	0.19	
		P,P'-DDD	0.08	0.06	

Note: More recent market surveys of pesticides in 13,980 and 13,085 samples taken in 1988 and 1989 from ten states show that P,P'-DDT was detected in 0.028% and 0.122% of samples and P,P'-DDE in 0.178% and 0.252% samples (Minyard and Roberts 1991). Overall, these surveys indicate that DDT and DDE levels are very low in food items, very low ppb levels (Gartrell et al. 1985, 1986). Imported foods may contribute small amounts of DDT and DDE, but no commodities exceeded U.S. EPA tolerance levels for them (Hundley et al. 1988).

* = on lipid weight basis.
Source: ATSDR 1994a

Sediment and Soil

γ-HCH was nondetectable in cropland soils and crops in 37 states monitored in the summer and fall of 1971 as part of the National Soils Monitoring Program (Carey et al. 1978) at the detection limits of 0.002–0.003 ppm. Subsurface soils of Alabama counties contained less than 0.1 ppm in 9.52%; 0.01 ppm in 2.13% of Arkansas soil samples; 0.07 ppm in 3.57% of Georgia soil samples; 0.02 ppm in 0.71% of Illinois soil samples; and at 0.15 ppm in 0.67% of Iowa soil samples (Crockett et al. 1974).

γ-HCH Levels in Sediments and Suspended Sediments in North America

Location	Year	Sample Type	Concentration	Reference
U.S. EPA, STORET	1978–1987	Sediments (596) from across U.S.	<2 µg/kg (median), detected in 0.5% of all samples	Staples et al. 1985
Niagara River	—	Suspended sediments	2 ppb (average) in 33% of samples	Kuntz and Warry 1983
Lake Ontario	1982	Settling particles	2.4 ppb (average)	Oliver and Charlton 1984

γ-HCH was one of the analytes monitored in the National Oceanic and Atmospheric Administration (NOAA) Status and Trends Mussel Watch Program conducted in the Gulf of Mexico. The compound was detected in 19% of the sediment samples collected in 1987 at a mean concentration of 0.07 ng/g (range, <0.02–1.74 ng/g) (Sericano et al. 1990). Sediment samples collected around the Great Lakes in 1989 contained γ-HCH concentration ranging from less than the detection limit (0.10 µg/kg) to 0.99 µg/kg, wet weight (Verbrugge et al. 1991). Sediment samples collected from impoundments along the Indian River Lagoon in Florida contained

γ-HCH (Hexachlorocyclohexane) Levels in Air and Rain Samples Around the World

Location	Year	Sample Type (Number)	Concentration	Reference
Ten states, U.S.	1970–72	Ambient air (2,479 samples)	0.9 ng/m³ (mean) detected in 67.7% of samples, max. 11.7 ng/m3	Kuntz et al. 1976
Nine U.S. locations (urban and agricultural areas)	—	Ambient air		
		Urban sites	0.1–7.0 ng/m³	Stanley et al. 1971
		Rural areas	ND	Stanley et al. 1971
College Station, Texas, U.S.	1979–80	Ground level Ambient air	0.23 (0.01–16) ng/m³	Atlas and Giam 1988
		Rainfall samples	2.81 (0.30–7.8) ng/L	Atlas and Giam 1988
Adirondack mountains New York, U.S.	1985	Troposphere air samples	0.509 ng/m³ (mean)	Knap and Binkley 1991
Remote marine atmosphere	—	Atmosphere (17)	0.015 ng/m³ (mean)	Atlas and Giam 1981
Remote marine locations	—	Rain samples (16)	0.51 ng/L	Atlas and Giam 1981
Germany	1990–91	Rain samples (41)	0.208 µg/L (0.02–0.833) 39 of 41 samples	Scharf et al.1992
West Pacific, Eastern Indian and Atlantic Oceans	—	Air	1.1–2.0 ng/ng/m³	Tanabe et al. 1982
		Water (79)	3.1–7.3 ng/L	Tanabe et al. 1982
Southern Indian Ocean	1986	Lower atmosphere	406 pg/m³	Wittlinger and Ballschmitter 1991
Bermuda	1983–84	Rain samples	0.126 (0.001–0.936) ng/L	Knap and Binkley 1991
	1988	Lower troposphere	0.012 ng/m³	Knap and Binkley 1991
Canadian Arctic	—	Ambient air samples	0.017–0.070 ng/m³	Hargrave et al. 1988
Axel Hieberg Island	1986	Snow and ice samples, respectively	0.211–0.644; 0.186 ng/L	Hargrave et al. 1988
Southern Ontario	1988–1989	Air monitoring samples	α-HCH, 0.145 ng/m³	Hoff et al. 1992
			β-HCH, 0.018 ng/m³	Hoff et al. 1992
			γ-HCH 0.060 ng/m³	Hoff et al. 1992

γ-HCH (Hexachlorocyclohexane) levels in water

Location	Year	Sample Type (Number)	Concentration	Reference
Hampton County, South Carolina, U.S.	—	Surface water	147 (10–319) ppt	Sandhu et al. 1978
		Groundwater	163 ppt	Sandhu et al. 1978
		Drinking water	10 ppt	Sandhu et al. 1978
Washington, D.C., U.S.	—	Drinking water	0.052–0.1 ppb	Cole et al. 1984
Urban and rural Hawaii	1970–71	Groundwater	1.1; 0.5 ppt (mean)	Bevenue et al. 1972
		Drinking water	0.2 ppt	Bevenue et al. 1972
Niagara River near entry to Lake Ontario	1980–81	Drinking water	2.1 ppt (mean) in 99% of samples	Kuntz and Warry 1983
U.S. EPA STORET	—	Surface water samples (4,505)	0.027 µg/L in 27% of samples	Staples et al. 1985
Lake Ontario	1983	Water samples	0.806–1.85 ng/L	Biberhofer and Stevens 1987

γ-HCH concentrations ranging from 34.4 ng/g at the top layer of sediment to 9.4 ng/g in the bottom layer at the same site. Although lindane had been used for mosquito control in the area from the late 1950s to the mid-1960s, the interstitial water samples from the sites did not contain detectable levels of lindane (Wang et al. 1992).

Other Environmental Media

Mean concentrations of γ-HCH in fat from chickens, turkeys, beef, lamb, and pork were reported to range from 0.012 to 0.032 mg/kg; the mean concentration in hen eggs was 0.008 mg/kg (Frank et al. 1990). γ-HCH was detected in 80% of the oyster samples collected from the Gulf of Mexico, in 1987, at 1.74 ng/g (mean) and range of <0.25-9.06 ng/g (Sericano et al. 1990). A monitoring study of fish tissues from 107 fresh water stations in the U.S., reported a decline in tissue occurrence of detectable α- and γ-HCH residues, observed from 1976–1981 (Schmitt et al. 1985). During 1980 to 1981, whole body residues of γ-HCH rarely exceeded 0.01 ppb. Tissue concentrations for α-HCH were 0.03-0.04 ppb and were found in fish from the southwestern and midwestern U.S. An analysis of fish from the Upper Steele Bayou in Mississippi in 1988 indicated that β-HCH concentrations ranged from ND to 0.02 mg/kg, wet weight in fish (Ford and Hill 1991).

Human exposures to γ-HCH include ingestion of plants, animals, animal products, milk, and water containing the pesticide. Farm animals may be exposed to the compound through feed, air, or water, or cutaneous application for protection from ectoparasites. Lipophilic pesticides such as γ-HCH accumulate in adipose tissue. It was reported that γ-HCH levels in the adipose tissues of cattle were 10 times higher than in the feed, 0.002 mg/kg (Clark et al. 1974). The most likely route of nonmedical human exposure to γ-HCH is ingestion of food containing the pesticide. γ-HCH was also detected in 6% of the foods collected in eight market basket surveys from different regions of the U.S. from April, 1982 to April, 1984 (Gunderson 1988). Foods representative of eight infant and adult population groups for consumption prior were analyzed in a revision to the FDA's Total Diet Studies methodology. The estimated mean daily intakes (ng/kg.bw./day) of γ-HCH for these groups in 1982 to 1984 were as follows: (i) 6-11 months-old infants, 1.9; (ii) 2-year-old toddlers, 7.9; (iii) 14–16-year-old females, 3.1; (iv) 14–16-year-old males, 3.4; (v) 25-30-year-old females, 2.0; (vi) 25–30-year-old males, 2.5; (vii) 60–65-year-old females, 1.6; and (viii) 60–65-year-old males, 1.8. γ-HCH intakes (ng/kg.bw./day) for three of these groups in 1988 were estimated in the U.S. FDA's Total Diet Analyses to be: (i) 6-11 month-old infants, 0.8; (ii) 14–16-year-old males, 1.4; and (iii) 60–65-year-old females, 0.9 (FDA 1989).

The Total Diet Studies of the FDA for fiscal year 1981–1982 provided the following daily intake amount of HCH isomers in foods by infants and toddlers (Gartrell et al. 1986a):

HCH isomer	Daily Intake (μg/kg.bw./day)		
	Infants	Toddlers	Adults
α-HCH	0.016	0.019	0.008
β-HCH	ND	<0.001	<0.001
γ-HCH	0.002	0.006	0.002
Total HCH	0.018	0.025	0.010

HCH isomers have been detected in whole milk and other dairy products, meat, fish, and poultry; oils and fats; vegetables; and sugars and adjuncts (Gartrell et al.1986a).

HCH isomers were also detected in adult diet food items, including dairy products; meat, fish, and poultry; garden fruits; oils and fats; leafy and root vegetables; and sugar and adjuncts (Gartrell et al. 1986b). In the Total Diet Study by the U.S. FDA in 1990 on 936 food items, HCH was detected in 23 items, while α-HCH and β-HCH (combined) were detected in 11 food items. Information on the actual levels found were not available (Yess et al. 1991).

Chlordane

Air

Chlordane has been detected in outside and inside urban homes and in rural air. In urban air, the levels ranged from <0.1 ng/m^3 to 58 ng/m^3; rural and background concentrations are 0.01–1.0 ng/m^3 (ATSDR 1994c). Chlordane levels in indoor air are much higher than in either urban or rural air. Since chlordane is poured or injected into soil (to control termites) around the foundations of houses, several studies indicate the levels of chlordane are higher in the basement than in the living area of homes (Anderson and Hites 1989). It was reported that the basement concentrations of chlordane are three to ten times higher than the living areas and two to three orders of magnitude higher than outdoors.

Water

Chlordane has been detected in surface water, groundwater, suspended solids, sediments, bottom detritus, drinking water, sewage sludge, urban run-offs, and in rain (ATSDR 1994c). Recent levels of chlordane in ocean and lake waters are <0.0001 ng/L. Recent studies in the Great Lakes indicate that the levels of the *cis*-chlordane in water is about two to three times that of *trans*-chlordane (Biberhofer and Stevens 1987; Stevens and Neilson 1989). Groundwater monitoring data from 479 disposal sites investigations detected chlordane in 23 samples at ten sites, ranking chlordane the 80th among the 208 RCRA, Appendix IX chemicals investigated (Plumb 1991). From the knowledge of U.S. EPA's STORET database for 1978 to 1987, the west-south-central U.S. had the highest concentrations of chlordane in groundwater and New England states had the lowest (Phillipps and Birchard 1991). Chlordane was used in golf courses from the 1950s to 1970s.

Sediment and Soil

Chlordane was monitored from studies conducted in the late 1960s to the middle 1970s. Chlordane was detected in both rural and urban soils in concentration range of <1 ppb to 141 ppm. The value of 141 ppm was detected in Hartford, CT, where 48 samples were analyzed in 1971. The mean was 4.00 (0.02–141) ppm (Carey et al. 1979). Detections generally reflected use pattern and very few recent general monitoring data are available (ATSDR 1994c); these data show that chlordane is still present in soils, but it is hard to estimate the trends due to insufficient detections. Soil samples around 30 houses in New Orleans treated with chlordane showed that mean residue levels ranged from 22 to 2,540 ppm (Delaplane and LaFarge 1990).

Recent sediment samples collected from Great Lakes harbors had chlordane levels ranging from 1.4 to 14 ppb (Verbrugge et al. 1991). A Missouri study in the 1980s demonstrated that chlordane in sediments of streams were correlated to recent urban development — 1.5 to 310 ppb (Puri et al. 1990). According to average chlordane levels in the U.S. EPA's STORET database for 1978 to 1987, the west-north-central United States had the highest chlordane levels in sediments and New England had the lowest levels of chlordane (Phillipps and Birchard 1991).

Other Environmental Media

Chlordane is widely present in many other environmental media. Tables 6.3, 6.4, and 6.5 list the concentrations of chlordane in food, aquatic, and terrestrial organisms, respectively.

CHLORINATED BIOCIDES

Table 6.3 Concentration of Chlordane in Food Items

Group/Location	Year	Number of Samples	Range ppm	Mean ppm	Percent (%) Occurrence	Reference
Total Adult Food Groups						
U.S. FDA's Total Diet Study	1982–1984	1,872	NS	NS	<1	Gundersen 1988
	1987–1988	1,170	NS	NS	4	FDA 1989
	1989–1990	936	NS	NS	<1	FDA 1990
	1990–1991	936	NS	NS	1	FDA 1991
Ten States Survey	1985–1989	13,085	NS	NS	0.09	Minyard and Robert 1991
	1987–1988	13,980	NS	NS	0.05	Minyard and Robert 1991
United States	1964–1965	216	ND–0.033	NS	0.46	Duggan et al. 1966
	1965–1966	432	ND–0.37	NS	0.23	Duggan et al. 1967
	1966–1967	360	0.005–0.02	0.013	0.56	Martin and Duggan 1968
	1968–1969	360	0.026–0.043	0.035	0.56	Corneliussen 1970
Total Toddler Food Groups						
United States	1975–1976	NS	ND–0.137	NS	NS	Johnson et al.1981
	1977–1978	606	0.010–0.028	0.019	0.33	Podrebarac 1984
	1980–1982	143	ND–0.005	NS	0.7	Gartrell et al. 1986a
Total Infant Food Groups						
United States	1976–1977	445	ND–Trace	NS	0.22	Johnson et al.1984
	1977–1978	417	ND–0.020	NS	0.24	Podrebarac 1984
Dairy Products, Illinois	1972–1981	3,618	ND–>0.3	NS	53.4	Steffey et al. 1984
Animal Products (abdominal fat), Ontario, Canada	1986–1988	539	<0.001	NS	NS	Frank et al. 1990
Hen's eggs, Ontario Canada	1986–1988	63	<0.001	NS	NS	Frank et al. 1990
Seafood, U.S.	1990	172	<0.01–0.13	NS	5	FDA 1991
Produce (fruits and vegetables), U.S.	1989–1991	6,970	ND (DL = 0.625)	ND	0	Schattenberg and Hsu 1992

Table 6.4 Chlordane Residues in Aquatic Organisms

Isomer	Year	Number of Samples	Sample Type	Concentration (ppm) Range	Concentration (ppm) Mean	% Occurrence	References
Major Waterways in U.S. (107 Stations)							
Cis, trans chlordane	1976–77	NS	Whole fish	ND–0.93	0.06	92.5	Schmitt et al. 1985
		NS	Composite	ND–0.32	0.03	84.0	Schmitt et al. 1985
	1978–79	NS	Whole fish	ND–2.53	0.07	94.4	Schmitt et al. 1985
		NS	Composite	ND–0.54	0.02	70.4	Schmitt et al. 1985
	1980–81	315	Whole fish	ND–0.36	0.03	73.8	Schmitt et al. 1985
		315	Composite	ND–0.22	0.02	72.0	Schmitt et al. 1985
Great Lakes and Major Watersheds of the Great Lakes							
Cis, trans chlordane	1979	48	Whole fish	ND–0.61	0.090	79.2	Kuehl et al. 1983
		48	Composite	ND–0.52	0.098	87.5	Kuehl et al. 1983
Chesapeake Bay and Tributaries, Maryland							
Cis, trans chlordane	1978	15	Finfish fillets	ND–0.50	0.013	NS	Eisenberg and Topping 1985
	1979	98	Finfish fillets	ND–0.70	0.08	NS	Eisenberg and Topping 1985
	1980	24	Finfish fillets	0.004–0.31	0.12	NS	Eisenberg and Topping 1985
San Joaquin River and Tributaries, California							
Cis, trans chlordane	July 1981	8	Carp, whole fish composite	0.012–0.273	0.077	100	Saiki and Schmitt 1986
Lake Ontario, Canada, (4 sites)							
Cis, trans chlordane	1977–1988	1,718	Lake trout Whole fish	0.04–0.14	NS	NS	Borgmann and Whittle 1991
Lower Mississippi Basin							
Trans	7-8/1987	17	Whole catfish	0.011–0.170	0.062	100	Leiker et al. 1991
Oysters from Gulf of Mexico (49 sites)							
Cis	1986	147	Composite	0.91–96.3	10.9	100	Sericano et al. 1990
Cis	1987	143	Composite	0.65–292	14.1	100	Sericano et al. 1990
Clams from Portland, Maine							
Cis, trans	November 1980	2	Whole body composite	ND–0.0018	NS	50	Ray et al. 1983

Source: ATSDR,TP-93/03, 1994; ND = Not detected; NS = Not specified.

Table 6.5. Chlordane Residues in Terrestrial Organisms[a]

Location	Year	Number of Samples	Sample Type	Concentration (ppm) Range	Concentration (ppm) Mean	% Occurrence	Reference
Birds							
Herons, Egrets, Kingfishers, and Sandpipers							
Corpus Christie, Texas	1983	10	Eggs	ND–0.88	0.14	NS	White and Krynitsky 1986
Colorado, Wyoming	1979	147	Eggs	0.08–0.23	0.14	5.4	McEwen et al. 1984
United States	1966–NS	105	Carcass	ND–1.8	NS	23.8	Ohlendorf et al. 1981
	1966–NS	105	(brains)	ND–1.4	NS	42.2	Ohlendorf et al. 1981
Brown or White Pelicans							
Louisiana	1971–76	117	Eggs	ND–1.31	0.36	NS	Blus et al. 1979
Klamath Basin California	1969–81	45	Eggs	<0.1–0.12	NS	6.7	Bfoellstorff et al. 1985
Cormorant, Black Skimmer, and Western Grebe							
Galveston Bay, Texas	1980–83	13	Eggs	ND–0.7	NS	<50	King and Krynitsky 1986
South Texas, U.S. and Mexico	1983	30	Carcass	<0.1–<0.3	NS	20	White et al. 1985
Gulls, Common Eider							
Appledore Island, Massachusetts	1977	28	Eggs	0.0–0.43	0.22	>50	King and Krynitsky 1986
Virginia, ME	1977	116	Eggs	0.00–0.50	0.09	>26.7	Szaro et al. 1979
Galveston, Texas	1980–81	10	Eggs	0.1–0.2	0.29	>50	King and Krynitsky 1986
Ducks							
Atlantic flyway	1976–77	NS	Wings	0.01–0.06	NS	57	White et al. 1979
Pacific flyway	1976–77	NS	Wings	0.01–0.02	NS	14	White et al. 1979
Central flyway	1976–77	NS	Wings	0.01–0.02	NS	14	White et al. 1979
Chesapeake Bay, Maryland	1973, 1975	142	Carcass	ND–NS	0.19	6.3	White et al. 1985

Table 6.5. Chlordane Residues in Terrestrial Organisms[a] (Continued)

Location	Year	Number of Samples	Sample Type	Concentration (ppm) Range	Concentration (ppm) Mean	% Occurrence	Reference
Osprey, Eagles, Owls, and Hawks							
United States	1964–1973	26	Carcass Brains	<0.1–1.7	NS	>69.2	Wiemeyer et al. 1980
United States	1969–1979	6	Eggs	ND–3.6	3.9	63.5	Wiemeyer et al. 1984
Earthworms							
Holbrook, MA	NS	29	Whole body	0.8–12.9	NS	45	Callahan et al. 1991
Insects							
Moth							
Washington, D.C.	1977–1979	NS	wholebody	<0.05–0.75	NS	0.22	Beyer and Kaiser 1984
Honey Bees							
Connecticut	1983–85	NS	Whole body 20 brood combs	0.06–0.69	NS	7.5	Anderson and Wojtas 1986

ND = Not detected; NS = Not specified.
[a] No distinction made between *cis* and *trans* isomers of chlordane.

Source: ATSDR TP-93/03, 1994 (for finer details)

Aldrin and Dieldrin

Aldrin readily converts to dieldrin in the environment which is ubiquitous due to its transport and environmental persistence. But it is found only in low levels in all media of exposure.

Air

Analysis of 2,479 air samples from 16 U.S. states from 1970 to 1972 detected the following: aldrin, mean 0.4 ng/m^3 (3 × 10^{-5} ppb), in 13.5% of samples; dieldrin, mean 1.6 ng/m^3 (1 × 10^{-4} ppb) in 94% of samples (Kuntz et al. 1976). After approximately ten years of imposed restriction on their uses, atmospheric concentrations determined in the Great Lakes Basin showed aldrin in 5 of 75 wet precipitation samples with a mean concentration of 0.01 ng/L (1 × 10^{-5} ppb). Dieldrin, conversely, was present widely in more than 60% of the samples at a mean concentration of 0.41–1.81 ng/L (Chan and Perkins 1989). Dieldrin was detected in rainfall during 1984, at mean concentrations of 0.78 ng/L over Lake Superior, 0.27 ng/L in New Brunswick, Canada, and 0.38 ng/L over northern Saskatchewan (Strachan 1988). Dieldrin has been detected indoors in homes 1 to 10 years after they were treated for termites.

Water

In early studies, dieldrin was the most often detected pesticide in surface water samples from all major river basins in the U.S., at a mean concentration of 7.5 ng/L (Weaver et al. 1965). Median values of aldrin and dieldrin from U.S. EPA STORET database are given below:

Medium	Aldrin Median Concentration (ppb)	Aldrin Number of Samples	Aldrin Percent Detection	Dieldrin Median Concentration (ppb)	Dieldrin Number of Samples	Dieldrin Percent Detection
Water	0.001	7,891	40	0.001	7,609	40
Effluent	<0.01	677	3.1	<0.01	676	3.7
Sediment	0.1	2,048	33	0.8	1,812	33
Biota	<0.1	211	0	0.03	530	41

Under the U.S. National Pesticide Monitoring Program conducted during 1975 to 1980, at 160 to 180 stations on major rivers of the United States, aldrin and dieldrin were detected in 0.2% of the 2,946 water samples and in 0.6% and 12% of about 1,016 sediment samples, respectively (Gilliom et al. 1985). Aldrin was detected in 9% of the 422 groundwater samples at a median concentration of 0.01 μg/L from a sandy, alluvial aquifer in Illinois. Water sampling conducted in 1986 during an isothermal period in the Great Lakes did not detect aldrin in any samples. However, dieldrin was measured in all samples at mean concentration of 0.3 ng/L for Lake Superior and 0.402 ng/L for Lake Erie (Stevens and Neilson 1989).

Soil

Dieldrin residues in soil are higher than aldrin as a result of the transformation of aldrin to dieldrin, even though aldrin was applied more frequently to soil. Sediment samples from Lake Ontario, in 1981, showed an increase in dieldrin levels from the 1980 values; from 26 to 48 ng/g, although dieldrin was banned in most of the Great Lakes Basin in the early 1970s (Eisenreich et al. 1989). The National Soils Monitoring Program detected dieldrin in soils in 24 states at a mean concentration ranging from 1 to 49 ppb (Kuntz et al. 1976).

Other Environmental Media

The persistence of dieldrin was shown by a monitoring program conducted in Alabama during 1972 to 1974, where its residues were at 7 to 40 ppb in 50% of soil samples; at <100 ppb in 50% forage samples with levels declining with time; and at an average of 1,490 ppb in 11 of 19 rat tissue samples. This is in spite of the fact that the farmers had not used aldrin or dieldrin for "several years" (Elliot 1975). The half-life of aldrin in crops was estimated to be 1.7 days and that of dieldrin, 2.7 to 6.8 days, depending on the crop (Willis and McDowell 1987).

Levels of Dieldrin in Fish

Fish Species	Location	Year	Mean Dieldrin Concentration (ppb)	References
Carp, white bass	A lake in Kansas	1985	0.069; 0.058	Aruda et al. 1988
Fish	Tributaries around Great Lakes	1980–81	30	DeVault 1988
Fish	Lake Huron	1970–80	10–500	EPA 1985e
Coho salmon	Lake Michigan	1980	60	DeVault et al. 1988
Coho salmon	Lake Michigan	1984	10	DeVault et al. 1988
National Pesticide Monitoring Program	Nationwide (107 Sites)	1978	50 (81% of samples)	Schmitt et al. 1985
National Pesticide Monitoring Program	Nationwide (107 Sites)	1980–81	40 (75% of samples)	Schmitt et al. 1985
Bluefish, raw fillets	Massachusetts	1986	20–40*	Trotter et al. 1989

* Cooking did not degrade dieldrin

In the absence of occupational or domestic use as a termiticide, food is probably the primary source of dieldrin residues in human adipose tissues. Because of the rapid conversion of aldrin to dieldrin in the environment, the average daily intake is <0.001 µg/kg.bw. Dieldrin may be ingested through food items. A 1985 Canadian survey of foods found that, although aldrin was not detected in any of the food samples analyzed, dieldrin was detected in all food composites at 0.00011 µg/g in fruit, 0.0019 µg/g in milk, 0.0031 µg/g in leafy vegetables, eggs, and meat, and at 0.023 µg/g in root vegetables (Davies 1988). Dieldrin residues may persist in foods such as milk, butterfat, and subcutaneous fat in cattle, with an estimated half-life in butterfat of 9 weeks (Dingle et al. 1989). During 1965 to 1970, the total U.S. dietary intake was reported to be 0.05 to 0.08 µg dieldrin/kg.bw./day and 0.0001 to 0.04 µg aldrin/kg.bw./day (IARC 1974b). Since that time, the use of aldrin and dieldrin has been restricted severely and dietary intake has decreased. An FDA Total Diet Study, conducted between 1982 and 1984, found that aldrin intake was less than 0.001 µg/kg.bw./day for all age and sex groups and that toddlers (2 years old) had the highest intake levels for dieldrin (0.016 µg/kg.bw./day), followed by infants, with 0.010 µg/kg.bw./day. Adults had a dieldrin intake of 0.007 µg/kg.bw./day (25 to 30 year-old males) and adolescent males (14 to 16 year olds) had an intake of 0.08 µg/kg.bw./day (Gunderson 1988; Lambardo 1986). Dieldrin was found in 15% of the food samples analyzed. These values represent a decrease from the 1980 Total Diet Study. Between 1980 and 1982 to 1984, daily intakes of dieldrin decreased from 33 ng/kg.bw./day to 10 ng/kg.bw./day for infants, from 46 ng/kg.bw./day to 16 ng/kg.bw./day for toddlers, and from 22 ng/kg.bw./day to 7 to 8 ng/kg.bw./day for adults (Gunderson 1988). Recently, a Total Diet Study conducted by the U.S. FDA, found dieldrin in only 6% of the food items analyzed from 1990 (FDA 1991). Daily intakes of 0.0014, 0.0016, and 0.0016 µ/kg.bw. were estimated for an infant 6 to 11 months old, a 14 to 16 year old male, and a 60 to 65-year-old female, respectively (FDA 1991). Assuming 2 liters of water ingested every day, the average drinking water contribution of dieldrin may range from 0.1 to 0.29 ng/kg/day for a 70-kg adult. These levels are well below the Acceptable Daily Intake (ADI) of 0.1 µg/kg/day recommended by the World Health Organization (WHO) for dieldrin (Geyer et al. 1986).

Heptachlor

Air

During the period of maximum usage, heptachlor was detected in ambient air at 0.5 ng/m^3 (mean). A survey of 16 states, during 1970 to 1972, detected heptachlor in 42% of 2,479 air samples, with a mean concentration of 1 ng/m^3; the maximum concentration recorded was 27.8 ng/m^3 in Tennessee (EPA 1985a). Very little data are available of heptachlor and heptachlor epoxide in air after the suspension of heptachlor use as an agricultural pesticide.

Water

Heptachlor and its epoxide have been detected in ambient waters at 0.001 to 0.5 µg/L (STORET 1987).

Soil

Substantial levels of heptachlor and its epoxide were recorded during heptachlor use; in 1969, it was detected in 68 of 1,729 soil samples at 0.01 to 0.97 mg/kg. In sediments, they were detected at 0.1 to 100 µg/kg (STORET 1987). Recent data are not available.

Other Environmental Media

Data reported in food items are from studies conducted prior to 1980 when heptachlor was still being used as an agricultural pesticide. Recent data are not available.

Toxaphene

Toxaphene has been detected in many environmental media as a result of widespread use in the past as an insecticide and because of its persistence in the environment. Toxaphene may be more a regional than a national problem, due to its selective use in certain areas of the continent or globe. Higher soil concentrations can be found in cotton-growing areas and higher residues are found in fish from the Great Lakes (ATSDR 1994d).

Air

Although detected in ambient air and rainwater samples from a number of sites, current data are lacking. In 1970 to 1972, toxaphene was detected in 3.5% of 2,479 air samples as part of the National Air Pesticide Monitoring Program; mean and maximum concentrations were ND and 8,700 ng/m^3, respectively. Toxaphene was detected in ambient air from remote locations; western and north Atlantic Ocean during 1973 to 1974, <0.04-1.6 ng/m^3; mean concentration in ambient air samples from Bermuda was 0.81 ng/m^3; in rainwater samples from Southern France near the Mediterranean Sea, the mean and range were 7.2 ppt and ND to 53 ppt (Villeneuve and Cattini 1986).

Water

The median ambient surface water concentration of toxaphene in the U.S. was 0.05 ppb (STORET 1987; Staples et al. 1985); in effluents at <0.2 ppb (median concentration for 708 effluent samples and 3.4% was positive). In contrast, toxaphene was not detected in 86 samples of municipal run-off collected from 15 cities in the U.S. in 1982 as part of the Nationwide Urban Run-off Program (Cole et al. 1984). Toxaphene was detected at only 0.2% frequency at the 178 CERCLA sites and 1.1% frequency at the 156 RCRA sites studied. No concentration data were available for these sites.

Sediment and Soil

Toxaphene has been detected in urban and agricultural soils, at concentrations ranging from 0.11 to 52.7 ppm in subsurface soils from 3 to 8 U.S. cities in 1969 (Wiersma et al. 1972). In another study conducted in 1970, toxaphene was detected with a range of 7.73 to 33.4 ppm. Soils monitored in 5 U.S. cities in 1971 show a mean concentration of 0.24 ppm (range, 0.23–4.95 ppm) in 25.6% of samples (Carey et al. 1979). In a recent study, toxaphene was found in 41% of the 56 samples collected at two depths (2.5 to 7.6 cm; 25 to 30.5 cm) along eight different drainages

(Ford and Hill 1991). The geometric mean and maximum wet weight concentrations were 0.12 ppm and 2.80 ppm for upper layer samples, and 0.07 ppm and 4.60 ppm for lower layer samples respectively. Ongoing studies in agricultural soil of the Mississippi Delta indicate the persistence of toxaphene in soils and sediments under a worst case scenario (Cooper 1991). Soils contained the average total toxaphene level of 734 ppb from 69 samples collected during 1983-1984. Sediments in the Moon Lake averaged 12.4 ppb. In a moderately aerobic environment of soils and sediments rich in clay colloids, toxaphene could persist for years. Toxaphene was not detected in core samples from wetland flats with a marked anaerobic environment. It shows that toxaphene biodegrades significantly in such environments.

Other Environmental Media

The U.S. FDA's Total Diet Studies conducted in 1980 to 1982 showed that toxaphene in food groups for infant and toddler diets was at 0.1 to 0.2 ppm. Toxaphene was not detected in drinking water, whole milk, other dairy products, meat, fish, and poultry; grain and cereal products; potatos and vegetables; fruits and fruit juices; and beverages (Gartrell et al. 1986a,b). FDA studies in 1989, 1990, and 1991 clearly suggest that general population intake levels have fallen dramatically over the last decade (ATSDR 1994d).

Toxaphene Levels in Several Fish Species

Type of Study	Fish Species	Year	Number of Samples	Mean (ppm)	Percent Occurrences	Reference
NCBP, U.S. (112 stations including Great Lakes)	Bottom-feeding and predatory fish	1976 1984	— 321	0.34 0.14	— 69	Schmitt et al. 1985, 1990
Lake Michigan area	Rock bass	1983	28	0.04–3.46	—	Camanzo et al. 1987
	Northern pike Smallmouth bass Lake trout Channel catfish Bow fin Largemouth bass Carp		comp. whole fish			

* NCBP = National Contaminant Biomonitoring Program

The occurrence of toxaphene levels above the FDA limit of 5.0 ppm could be a cause for public health concern.

Current human exposure to toxaphene in North America appears to be very limited. The Food Contamination data base for 1988 to 1989 showed no detectable toxaphene residues in food items (Minyard and Roberts 1991). No information was available in the literature regarding the size of human population potentially exposed to toxaphene in the vicinity of hazardous waste sites (ATSDR 1994d). Average daily inhalation exposures are likely to be much less than dietary exposures (HSDB 1994). Toxaphene has not been detected in human adipose tissue (ATSDR 1994d).

Toxaphene Levels in Aquatic Organisms, Yazoo National Wildlife Refuge, Michigan

Species	Number of Samples	Concentration (ppm, wet weight) Mean	Range
Mosquito fish	25	0.25	ND–0.25 (17)
Carp	8	3.06	0.51–6.20 (8)
Smallmouth buffalo	6	5.77	0.75–15.00 (6)
Bowfin	5	2.70	0.27–8.60 (5)
Spotted gar	10	2.71	ND–16.00 (9)
Water snakes	20	0.33	ND–27.00 (17)
Cotton mouths	10	0.03	ND–1.30 (5)

* Numbers in parentheses indicate numbers of samples containing toxaphene.
ND = Not detected (detection limit, 0.01 ppm)
Source: Ford and Hill 1991

Endrin

Air

Levels of Endrin in Ambient Air and Rainfall in North America

Location	Year	Sample Type	Concentration (ng/m^3) Mean	Max.	Reference
U.S.	1970–71	Ambient air	0.2	19.2	Lee 1977
Jackson, MI	1970–71	Ambient air	0.1	—	Bidleman 1981
South Carolina	1970–71	Ambient air	0.2	—	Kuntz et al. 1976
Boston, MA	1980	Ambient air	ND	—	Bidleman 1981
Great Lakes Area	1981	Ambient air	ND	—	Eisenreich et al. 1981
Four stations, Canada	1984	Rainfall	Detectable	—	Strachan 1988

Water

Endrin, unlike other chlorinated pesticides, was never extensively used in urban areas. Results of the EPA's Nationwide Urban Run-off Program showed no detection of Endrin in 86 high-flow-water samples from 51 urbanized watersheds from ten cities (Cole et al. 1984). The U.S. EPA STORET database showed detection in 32% of 8,789 samples close to the detection limit. The national median concentration was reported to be 0.001 ppb (Staples et al. 1985). Endrin is nondetectable to rarely detectable in drinking water (Wnuk et al. 1987). This profile of endrin is similar in groundwaters (Cohen 1986).

Sediment and Soil

Endrin was detected in 10 of 1,483 cropland soil samples in 1972 at concentrations up to 2.13 ppm (Carey et al. 1979). Endrin was not detected in urban soils from 13 of 14 cities included; the exception was Memphis, TN, where a production company was located; the mean concentration for all 28 Memphis sites was <0.01 ppm (Carey et al. 1976). The median endrin concentration for all recorded analyses was 0.001 ppb (Staples et al. 1985). A similar analysis of STORET data indicated that endrin aldehyde was not found in 251 samples of sediment. A recent study on sediments and animals of the Yazoo National Wildlife Refuge in the Mississippi Delta that had heavy usage of pesticides showed no detectable levels of endrin, <DL (0.01 ppm), in sediments (Ford and Hill 1991).

Other Environmental Media

The National Contaminant Biomonitoring Program reported the maximum endrin concentrations in whole fish tissue from around the U.S. for the years 1976 to 1977, 1978 to 1979, 1980 to 1981, and 1984 at 0.40, 0.11, 0.30, and 0.22 ppm, respectively, with a geometric mean of ≤0.01 ppm (Schmitt et al. 1985, 1990). The percentage of stations showing detectable endrin residues declined from 47.2% (1976 to 1977) to 28% (1984). The 1986 EPA National Study of Chemical Residues in Fish reported endrin in fish in 11% of 362 sites surveyed. The maximum, mean and median concentrations of endrin reported were 0.162 ppm, 0.002 ppm, and ND (<0.0025 ppm) respectively (EPA 1992b).

In a recent study of pesticide residues in food conducted during 1988 to 1989 in ten states in the U.S. endrin was not detected in any of the 13,980 samples analyzed. A Canadian study reported endrin detection in composite samples of fresh root vegetables, fruit, leafy and other aboveground vegetables, and cow's milk in a range of 0.27 to 0.37 ppb; endrin was not detected in composite samples of fresh meat and eggs (DL = 0.01 ppb) (Davies 1988). The U.S. FDA Pesticide Residue Monitoring Program for 1989 to 1991 reported the frequency of endrin occurrence was <1% (FDA 1990, 1991, 1992). A food-basket survey, similar to the FDA approach conducted in San Antonio, TX, did not detect endrin (Schattenberg and Hsu 1992). Endrin is no longer used in North America and other countries and the exposure potential for the general population will likely decrease further. The more recent National Human Adipose Tissue Survey did not detect endrin in adipose tissues from the general population (Stanley 1986).

Endosulfan

Air

Endosulfan was not detected in any of the ambient air samples collected from sites in 16 states in 1971 or 1972 (Kuntz et al. 1976). A recent study reported α-endosulfan at a mean concentration of 0.078 ng/m^3 in ambient air samples collected from Columbia, SC, from June to August, 1978 (Bidleman 1981). Rain samples from the Great Lakes area of Canada in 1976 to 1977 contained α-endosulfan at 1 to 2 ng/L (mean) and β-endosulfan at 4 to 5 ng/L (mean) (Strachan et al. 1980). Endosulfan concentrations in snowpack samples from widely distributed sites in the Canadian Arctic were 0.1 to 1.34 ng/L (Gregor and Gummer 1989).

Water

Endosulfan concentrations in surface waters are generally <1 ppb (EPA 1982). Rain samples from the Great Lakes were reported to contain α-endosulfan at 0.1 to 3.8 ng/L and β-endosulfan from 1.0 to 12 ng/L (Strachan and Huneault 1979). Run-off waters from agricultural areas were reported to contain low levels of endosulfan in the aqueous phase and higher concentrations in the particulate phase; <0.002 to 0.18 μg/L in run-off water and 1 to 62 μg/kg in bottom mud. Soils from a farm near the run-off ditch contained 640 μg/kg of endosulfan residues (Miles and Harris 1971).

Soil

According to WHO, surveys of agricultural soils in North America have reported endosulfan residues (α- and β-isomers and endosulfan sulfate) at <1 mg/kg (WHO 1984).

Food Items

U.S. FDA's Total Diet Studies of 1980 to 1982 (survey of 27 cities), reported mean concentrations of endosulfan (α- and β-isomers) and endosulfan sulfate in leafy vegetables, garden fruit, and fruit groups as follows (Gartrell et al. 1986a,b):

Food Groups	α-Endosulfan (ppb) Mean	Range	β-Endosulfan (ppb) Mean	Range	Endosulfan (ppb) Mean	Range
Leafy vegetables	8.3	trace–93	12.3	trace–150	17.9	2–185
Garden Fruits	1.3	trace–7	1.5	trace–10	2.0	trace–16
Fruits	0.3	trace–2	0.6	1–6	0.3	2–20

Endosulfans and residues were not found in dairy products; meat, fish and poultry; grain and cereal products; potatoes; legume vegetables; root vegetables; oils and fats; and sugar or beverages. The main route of exposure to endosulfan for the general population is dietary sources and tobacco products.

Mirex

Air

Rainfall samples collected at several sites around the Great Lakes area contained from >0.2 to <0.5 ng/L of mirex. But mirex was not detected consistently throughout the sampling period; hence, quantitative results were not reported (Strachan 1990). The U.S. EPA reported mirex concentration in wet precipitation over rural areas at <1 ng/L (EPA 1981b). Air samples from Southern Ontario in 1988 detected mirex in 5 of 143 samples at an annual mean concentration of 0.35 pg/m^3 (range, 0.1 to 22 pg/m^3) (Hoff et al. 1992).

Water

Rural drinking water samples reported to contain mirex from ND to 437 ng/L (Sandhu et al. 1978). A survey in 1987, detected mirex in only 5 of 1,147 drinking water samples from Ontario, Canada (Environment Canada 1992). Niagara River samples contained mirex in the aqueous phase at 0.0005 to 0.0075 ng/L (Allan and Ball 1990). During 1981 to 1983, 12% of 104 whole water Niagara River samples detected mirex from <DL (0.06 ng/L) to 2.6 ng/L, with a median concentration of 0.06 ng/L (Oliver and Nicol 1984). Seventeen percent of 42 Niagara River particulate samples contained a mean mirex concentration of 0.022 ng/L (Allan and Ball 1990). During the 1986 spring turnover, mirex was not detectable in any of the Great Lakes (Stevens and Neilson 1989), nor in the St. Lawrence River between 1981 and 1987 (Germain and Langlois 1988). However, 8 of 14 water samples, collected in 1986 at various locations along the St. Lawrence River, contained low levels of mirex with the highest concentration at 0.013 ng/L (Kaiser et al. 1990a). Declining mirex concentration was reported in Lake Ontario from 1.5 ng/L in 1986 to 0.4 in 1988 (Sergeant et al. 1993).

Dissolved average chlordecone concentrations in estuarine water was <10 ng/L in 1977, 12 years after its production and 2 years after cessation of production (Nichols 1990). In 1981, chlordecone water concentrations were from ND to 0.02 µg/L (Lunsford et al. 1987).

Sediment and Soil

Between 1979 and 1981, mirex concentrations in suspended sediments of the Niagara River declined from a mean concentration of 12 ng/l to 1 ng/L; concentrations in bottom sediments were generally low except for hot spots,<1 µg/kg to 890 µg/kg (Allan and Ball 1990). Increasing mirex concentrations, with depth in Lake Ontario sediments, were reported, 1,700 µg/kg at sediment depth of 9 cm (Durham and Oliver 1983). Mirex was nondetectable in run-off waters, but was detected in 10% of 129 run-off sediment samples at 1.3 µg/kg (mean) (Marsalek and Schrocter 1988). With the exception of the James River area, VA, very little information is available on chlordecone residues in soils and sediments. In Hopewell, VA, where a chlordecone manufacturing plant was located, chlordecone was detected at 1 to 2% levels (10,000 to 20,000 mg/kg) and contamination extended to 1000 m at concentrations of 2 to 6 mg/kg (Huggett and Bender 1980). Sediment cores taken from the James River below Hopewell, VA, indicated that chlordecone concentrations were greatest nearest the release site. Down below the Hopewell municipal sewage effluent discharge, sediments contained 2.2 mg/kg (Orndorff and Colwell 1980). Downstream of the river, the highest chlordecone concentration of 0.18 mg/kg from a sediment depth of 46 to 48 cm with a sedimentation rate of 10 cm/year. Immediately below the discharge of effluent, chlordecone was found at sediment depths of 55 to 58 cm at 0.44 to 0.74 mg/kg (Cutshall et al. 1981).

Other Environmental media

Current residues of mirex in various biota are much lower than reported during the early 1970s. Analysis of the food chain in Oxbow Lakes, LA in 1980, showed mirex was detected in secondary consumers (fish and birds that consume invertebrates and insects) and in all tertiary consumers (fish-eating fish, birds, and snakes). Based on fish data, mirex inputs into Lake Ontario appeared to be continuing on an intermittent basis. Very recent analysis of fillet samples, of 11 commercial fish species from the Great Lakes, showed that carp from a closed fishery area contained 120 µg/kg; eel, 56.8 µg/kg; carp from an open fishery area, 5.24 µg/kg; bullhead, 3.63 µg/kg; and trout, 2.38 µg/kg. Mirex concentration in fish decreased from 0.38 µg/kg in 1977 to 0.17 µg/kg in 1988, although there was variability in data (Borgmann and Whittle 1991). There was a distinct east-west gradient across the lake. The western side of the basin recorded the highest mirex residues. Freshwater fish collected during 1980 to 1984 contained mirex in 18% of 1980 samples (maximum concentration, 210 µg/kg and mean concentration, 0.01 µg/kg); in 13% of 1984 samples (maximum concentration, 440 µg/kg, and mean concentration, 10 µg/kg) (Schmitt et al. 1990). A recent U.S. EPA study reported mirex in 38% of 362 sites; mean concentration was 3.86 µg/kg. Food analysis in Canada did not detect any mirex/photomirex in any vegetable, fruit, milk, egg, or meat products (Davies 1988).

Mirex was also not detected in foods domestically produced or imported as part of the U.S. FDA Pesticide Monitoring Study during 1988 to 1989 (FDA 1990), and 1990 to 1991 and 1991 to 1992 (FDA 1992, 1993). Because chlordecone ceased production and use ended almost 20 years ago, current chlordecone levels in various biological organisms are generally lower than reported during its peak production years (1974 to 1975). Chlordecone residues were detected in the FDA Pesticide Residue Monitoring Studies of 49,877 food samples from 1978 to 1982 and of 49,055 food samples from 1982 to 1986; however, the frequency of detection was unspecified but less than 1 and 2%, respectively (Yess et al. 1991a,b). Chlordecone was also detected in 1 of 27,065 food samples collected from ten state laboratories during 1988 and 1989 (Minyard and Roberts 1991). Chlordecone was not detected in any domestically produced or imported foods analyzed as part of the FDA Pesticide Residue Monitoring Studies during 1986 to 1987, 1988 to 1990, 1990 to 1991, and 1991 to 1992 (FDA 1988, 1990, 1991, 1992, 1993).

Mirex has not been produced since 1976 and not used in the U.S. since 1977. The potential for exposure to mirex should decline over time. The general population may be exposed to low concentrations of mirex, through dietary sources such as contaminated fish and shellfish. Drinking water is not a significant exposure pathway.

Methoxychlor

Air

In Canada, the annual mean levels of methoxychlor in ambient air was 1.7 pg/m^3 during 1988 to 1989 (Hoff et al. 1992). Levels in ambient air tended to be higher during insect control periods (up to 27 pg/m^3), whereas, levels were generally below detection limit (0.04 to 0.1 pg/m^3) during non-use periods. A survey of pesticide levels in ambient air in two U.S. cities detected methoxychlor in indoor, outdoor, and personal air samples taken from Jacksonville, FL and were 200 to 300, 0 to 100 and 100 to 600 µg/m^3, respectively (EPA 1990; ATSDR 1994f). Levels of methoxychlor were below detection limit, 36 pg/m^3 in those air samples from Springfield, MA.

Water

Methoxychlor is not commonly detected in surface, ground, or drinking waters. A survey of major rivers in the U.S. did not detect methoxychlor at DL = 100 ng/L in any samples from approximately 180 sites (Gilliom et al. 1985). Methoxychlor was occasionally detected in surface waters from the Great Lakes area at 0.032 to 15.0 ng/L (Biberhofer and Stevens 1987; Kuntz and Warry 1983; Maguire et al. 1983). However, another survey of 783 rural domestic wells and 566 community water systems across the U.S. did not detect methoxychlor, DL = 300 ng/L (EPA 1990f in ATSDR 1994). Similarly, methoxychlor was not detected in 54 wells in California, DL = 5000 ng/L (note the higher detection limit), in groundwater below irrigated farm land in Nebraska, DL = 5-10 ng/L, or in the drinking water in Jacksonville, FL, and Springfield, MA, (Maddy et al. 1982; Spalding et al. 1980; EPA 1990). Methoxychlor has been detected in rain and snow samples in the Great Lakes region at 0.43 to 13.1 ng/L and 0.1 to 5.8 ng/L, respectively (Strachan 1985, 1988; Strachan and Huneault 1979). Methoxychlor was also detected in the Canadian Arctic at 0.2 ng/L (Welch et al. 1991). Methoxychlor intentionally added to control fly larvae in an irrigation canal, was found at peak levels of 1000 ng/L, 75 miles downstream, 45 to 46 hours after application (Stoltz and Pollock 1982). Aerial application of methoxychlor to elm trees, in a short time, was detected at 40 to 160 mg/L, downstream of a nearby river (Wallner et al. 1969), but was not detected 100 feet downstream after 24 hours. A review of groundwater monitoring data from 479 monitoring studies reported methoxychlor in groundwater at 14 (3%) of the sites (Plumb 1991).

Sediment and Soil

Methoxychlor is generally infrequently detected in soils and sediments, but higher levels were detected near release sources. A soil survey in the U.S. detected methoxychlor in 1 of 1,729 cropland soils at 0.28 µg/kg (IARC 1979); in 1 of 45 random soil samples in Alabama at <100 µg/kg (Albright et al. 1974), although detection limit was not reported. Data on methoxychlor levels in selected soils and sediments in North America are given below:

Location	Sample Type	Methoxychlor (µg/kg)	References
Hazardous site, Fresno, CA	Subsurface soil	150–17,000	ATSDR 1989a
Soil beneath elm trees sprayed with 12–16% methoxychlor, U.S.	Soil Soil	2,900–14,600 (after 3 days) 1,000–8,400 (after 130 days)	Wallner et al. 1969
Major rivers, U.S.	Sediments	Only one site detected DL = 1 µg/kg	Gilliom et al. 1985
Niagara River	Sediments	51% of 70 sediments detected, 7 µg/kg (mean) dry weight	Kuntz and Warry 1983

Other Environmental Media

Methoxychlor is either nondetectable or detected at very low concentrations in foods. It was reported in only 5% of 14,000 composite food samples from ten states in 1988 and in 11% of 13,000 samples in 1989 (Minyard and Roberts 1991). The actual concentrations were not reported. A market basket survey in 1980 to 1982 reported methoxychlor in dairy products and cereals/grain products at trace to 4 µg/kg (Gartrell et al. 1986a); it was not detected in vegetables, fruits, meats, or dairy products in Canada (Davies et al. 1988; Frank et al. 1987). Fish from the Great lakes did not contain detectable levels of methoxychlor, except in some odd samples (Camanzo et al. 1987). The most likely exposure route for the general population is dietary sources which contain low levels of methoxychlor.

Chloropyrifos

Air

Chloropyrifos has been detected in both outdoor and indoor air; special concern areas are fogwater, environments receiving broad pesticide application, poorly ventilated environments, and infant breathing zones, 25 cm above the carpet (ATSDR 1995b).

Chloropyrifos Concentration

Location	Fog Water	Air	References
California	320–6,500 ng/L	0.6–14.7 ng/L	Plimmer 1992
Ten U.S. locations		2.1 ng/m^3 (mean)	Carey et al. 1985
1980		100 ng/m^3 (max.)	
		14 of 123 samples	Carey et al. 1985

Substantially higher chloropyrifos concentrations were detected in the infant breathing zone than in the adult breathing zone (Fenske et al. 1990). All concentrations in the infant breathing zone exceeded NAS interim guideline of 10 µg/m^3. Significantly higher levels of chloropyrifos were detected in the air of houses built over sand than those over clay soils, following treatment of crawlspaces. Although residues have been detected, generally ambient levels are not available. Ambient air monitoring at ten U.S. locations in 1980 detected chloropyrifos in 14 of 123 samples with 2.1 ng/m^3 (mean) and 100 ng/m^3 (maximum) (Carey and Kutz 1985).

Water

Chloropyrifos has been rarely detected in surface waters and is generally well below levels of concern; it was detected in 0.2% of 344 groundwater sources used for drinking water supply in Illinois, not detected in 15 Iowa samples, and detected in 3 of 949 surface water samples in Ontario, Canada, collected during 1975 to 1977 (Braun and Frank 1980). No detection was reported in groundwater sampling of 358 wells in Wisconsin, (Krill and Sonzogni 1986). In an intensive monitoring study, chloropyrifos was detected in 0.0 to 1.06% of 750 samples for each of 7 tributaries to Lake Erie from 1983 to 1991 (Richards and Baker 1993). Chloropyrifos was not reported as part of the National Surface Water Monitoring Program for 1976 to 1980 (Carey and Kutz 1985).

Sediment and Soil

Limited data are available on chloropyrifos levels in soil or sediments: (i) no detection at DL = 0.01 mg/kg in sediment samples from Lakes Superior and Huron, including Georgian Bay, in 1974 (ATSDR, 1995b), and (ii) not reported in sediments as part of the National Surface Water Monitoring Program for 1976 to 1980 (Carey and Kutz 1985). One liquid pesticide waste disposal pit detected chloropyrifos to a depth of 67.5 cm (Winterlin et al. 1989).

Other Environmental Media

The U.S. FDA Pesticide Residue Monitoring Program for domestic and imported food items detected chloropyrifos (concentrations not specified) in 33 of 1,044 samples during 1978 to 1982 and in 295 of 3,744 samples during 1982 to 86 (Yess et al. 1991a,b). The same program also detected chloropyrifos during 1988 to 1989, 1989 to 1990, 1990 to 1991, and 1991 to 1992. Chloropyrifos was detected in 121 different domestic foods (0.9% of samples) in 1988 and 128 domestic foods (1% of samples) in 1989 by regulatory agencies (Minyard and Roberts 1991). In a 1989 to 1991 survey in San Antonio, TX, chlorpyrifos was detected in 41 produce samples out of 6,970 produce samples, with DL = 0.25 ppm (Schattenberg and Hsu 1992). A National Study detected chloropyrifos in fish in 26% of 362 sites at 4.09 ng/g (mean) and 344 ng/g (maximum) respectively (EPA 1992a). Chlorpyrifos has rarely been detected in drinking water (ATSDR 1995b).

The primary exposure route for the general population of chloropyrifos is inhalation of indoor air and ingestion of contaminated food items.

Dichlorvos

Air

All reported monitoring in air on dichlorvos was carried out on the indoor locations where it had been used as a fumigant-type pesticide. Product uses of dichlorvos have tapered off dramatically, or ceased altogether, since the 1980s. Most of the existing literature deals with situations when dichlorvos being used to kill insects at homes, commercial aircrafts, hospitals, restaurants (Elgar and Steer 1972; Leary et al. 1974; Deer et al. 1993). Where adequate precautions were taken and barring accidents, levels above 0.25 mg/m^3 (human health safety threshold value) are uncommon (Hayes 1982). A U.S. EPA TEAM study of atmospheric levels of common household pesticides did not commonly detect dichlorvos in the indoor air of sampled homes (Lewis et al. 1988).

Water

The U.S. EPA National Pesticide Survey of drinking water wells included wells from community systems in towns and cities, noncommunity systems such as public places (e.g., truck stops), and private domestic wells. Dichlorvos was not found above the detection limit in any of the drinking water wells sampled. No comprehensive information pertinent to dichlorvos in surface waters or groundwaters was found in the literature. Available data show that dichlorvos was not detected above DL = 0.22 ppb, suggesting that it rapidly dissipates or degrades in surface water (ATSDR 1995c).

Sediment and Soil

Background levels of dichlorvos in soils or sediments have not been reported in the literature. Dichlorvos (i) is relatively volatile, (ii) has low sorption potential to soil or sediments, (iii) lacks bioaccumulation potential, and (iv) rapidly biodegrades through hydrolysis. All these properties indicate that dichlorvos will not likely persist in soil, sediment, biota, and water.

Other Environmental Media

Dichlorvos monitored by the U.S. FDA in food items is reported among those chemicals where some detections were documented, but never at levels of concern (FDA 1988 to 1993). One study reported two detections in 13,085 food items sampled, with none of the detections exceeding any U.S. federal or state tolerance limits (Minyard and Roberts 1991). Similar results were reported of food residue monitoring programs in Canada and the United Kingdom (IARC 1991). Through the late 1980s, the widespread use of dichlorvos in resin strips to control insects created low grade exposure potentials. After 1988, concerns over the carcinogenic potential of dichlorvos led the U.S. EPA to require warning labels on strips and the U.S. FDA to disallow the uses of dichlorvos strips in kitchens, restaurants, and other places where food is prepared. This led to the virtual elimination of dichlorvos in formulation for home use (ATSDR 1995c). Dichlorvos may still be used by professional exterminators and in turf grass treatments. This may create short-term exposures via inhalation, dermal, and oral (by children) intake.

Chlorfenvinphos

Air

Information on ambient air concentrations of chlorfenvinphos was not available in the literature.

Water

Studies on surface water samples during 1975 to 1977 in 11 agricultural watersheds in southern Ontario, Canada, did not detect chlorfenvinphos in any water sample at DL = 1 µg/L (Braun and Frank 1980). A five-year study during 1986 to 1990 on water from the mouths of three major agricultural watersheds, the Grant, the Saugeen, and the Thomas Rivers in Ontario, did not detect chlorfenvinphos in any sample at DL = 1 µg/L (Frank et al. 1991). Possibly due to the fact that chlorfenvinphos was never registered for use as a soil insecticide, no data are available in the U.S.

Sediment and Soil

Monitoring data on chlorfenvinphos in some Canadian soils are given below:

Location	Sample Type	Year	Concn. (ppm) Mean	Range	Reference
Holland March, Ontario, Canada	13 Farm soils	1972–75	1972, 0.12	—	Miles et al.1978
			1973, 0.05	—	Miles et al.1978
			1974, 0.36	—	Miles et al. 1978
			1975, 0.13	—	Miles et al.1978
Fraser Valley, B.C. Canada	Farm soils	July–Dec. 1991	0.031	0.012–0.060	Wan et al. 1994

Other Environmental Media

A monitoring program in San Antonio, TX detected chlorfenvinphos in 1 out 6,970 produce samples at a DL = 0.75 ppm (Schattenberg and Hsu 1992). In a similar study conducted in Canada during 1989 to 1991 on 13,230 domestic and imported food items, chlorfenvinphos was not detected in domestic foods, but was detected in 13 imported food items, including fresh oranges, peppers, pineapples, and spinach (Neidert et al. 1994). Cooking is reported to reduce chlorfenvinphos content in raw foods but to not completely eliminate it (ATSDR 1995d). Japanese tea leaves were reported to contain 3.4 ppm of chlorfenvinphos and more than 12% of it was found to leach into the tea (Nagayama et al. 1989). Chlorfenvinphos detected in anhydrous lanolin samples in 1989, 0.60 to 5.9 mg/kg;those collected in 1991, 0.81 to 10 mg/kg (Heikes and Craun 1992). Chlorfenvinphos was detected in a wide range of pharmaceutical products such as ointments, analgesic balms, creams, etc., at 0.08 to 1.1 mg/kg; in 1992, chlorfenvinphos was detected in antibiotics and cold sore and ophthalmic ointments at trace to 0.32 mg/kg (ATSDR 1995d).

Currently, the primary exposure route of chlorfenvinphos to the general population is via imported fresh fruits and vegetables from countries where it is still used. Additional exposure could be through lanolin-containing products.

TOXICITY PROFILE

The mechanism of higher molecular weight chlorinated hydrocarbon pesticides, having molecular weight >236 g/mol, generally involves the alteration of electrophysiological and enzymatic properties of the nerve cell membranes. Such effects are the results of changes in the kinetics of the Na$^+$ and K$^+$ ion fluxes across the nerve cell membranes and subsequent interference with nerve pulse transmission (Matsumara and Patil 1969; Wooley et al. 1985; Ishikawa et al. 1989). This adverse effect of pesticides and, particularly, this mechanism of toxic effect is nonspecific across the biotic species, affecting both insects (target species) as well nontarget species (mammals). These effects occur at lower exposure concentrations on target species compared to nontarget (mammalian) species. These pesticides include DDT, DDE, methoxychlor, lindane, chlordane, heptachlor, endrin, dieldrin, toxaphene, mirex, and chlordecone (Smith 1991). Toxic effects are mainly manifested in the liver. Hepatocarcinogenic effects with liver toxicity and alterations of various enzyme activities have been observed in laboratory animal studies, particularly at higher exposure levels. Other toxicological effects of pesticides include changes to the structure and function of certain endocrine organs and various hormonal changes. (Singh et al. 1985; Smith 1991).

Chlorinated acyclic or aromatic pesticides are known to bioaccumulate in those organisms higher in the food chain, as well as to persist in the environment. (Gonzales et al. 1991). This accounts for accumulation in aquatic and terrestrial wildlife with resultant chronic toxicity, including reproductive toxic effects. Since the ban on these pesticides, their presence in biological tissues has been steadily decreasing. In 1956, the total DDT in body fat of an average adult human was 15.6 ppm; this level dropped to about 8 ppm in 1974 (2 years after the 1972 ban of DDT) and to approximately 3 ppm in 1980 (Smith 1991). Given the steady disappearance of these pesticides from the environment and from biological organisms since their ban, ecological effects are expected to further decline. Because of the continued use of these pesticides in developing countries, lack of enforcement of pesticide regulations and other field factors, their total environmental and ecological impacts, would not be reduced.

The interpretation of residue data is complicated by factors such as the migratory pattern and dietary habits of avian species, lipid content of the aquatic biota, trophic level, seasonal variations, and transformation of pesticides to more persistent metabolites, such as aldrin to dieldrin.

The role of chlorine in the activity of both natural and synthetic pesticides has been reviewed recently (RTP 1994). The available information reveals that a pesticide, whether synthetic or natural, chlorinated or nonchlorinated, does not solely account for its toxic potential. Low molecular weight chlorinated hydrocarbons that are used as fumigants are not persistent in the environment and have low acute toxicity, whereas higher molecular weight chlorinated hydrocarbons are very toxic and persistent. This demonstrates that the molecular structure and overall physical and chemical properties dictate the behavior of pesticides in the environment. Organophosphorus insecticides, chlorinated and nonchlorinated, have more or less similar toxic potencies regardless of the presence of chlorine (RTP 1994). Many of the nonchlorinated, naturally occurring pesticides induce the same effects such as carcinogenicity, hepatotoxicity, hormone disruption, neurotoxicity, teratogenicity, and reproductive toxicity at low or even lower doses than synthetic chlorinated organic pesticides.

New pesticides are being developed with the goal of narrowing the specificity on target pests, reducing the potential adverse effects on nontarget organisms and greatly reducing environmental persistence by making the molecules rapidly biodegrade after usage. The chlorinated, organo-phosphorus insecticides include chlorpyrifos, dichlorvos, chlorfenvinphos etc. These pesticides, generally inhibit acetylcholine esterase, a neurotransmitter enzyme. This inhibition is reported to lead to the buildup of acetylcholine at nerve and neuromuscular synapses, which could lead to severe muscle contraction and paralysis (Gallo and Lawryk 1991). The range of the neurotoxic potential of the chlorinated and nonchlorinated organophosphorus insecticides is not different (RTP 1994). The causes for the delayed neurotoxicity is not well understood, but seems to relate to poor regenerative abilities of nerve tissues, rather than the prolonged retention of the organo-phosphorus insecticide. These insecticides do not significantly accumulate in the biota nor persist in the environment. Since they are very specific in their activity, they are applied to confined areas that can be well monitored and controlled.

REGULATIONS

DDT, DDE, DDD

Agency	Description	Concentration
IARC	Probable human carcinogen	
WHO	Allowable Daily Intake(ADI)	0.02 mg/kg
OSHA	PEL-TWA (skin) (DDT)	1 mg/m^2
FIFRA	Most uses of DDT and DDD canceled	
Guidelines		
Air		
ACGIH	TLV-TWA(DDT)	1 mg/m^3
NIOSH	REL-TWA(DDT)	0.5 mg/m^3
Water		
U.S. EPA	Ambient water quality criteria for human health protection	
	DDT, DDE, and DDD	
	Ingestion of water and organisms	0.59 ng/L
	Ingestion of organisms only	0.59 ng/L (for 10^{-6} risk level)
U.S. FDA	Recommended action levels	
	Fruits and vegetables	0.1–0.5 ppm
	Eggs	0.5 ppm
	Grain	0.5 ppm
	Milk	0.05 ppm
	Meat	5 ppm

Source: ATSDR TP-93/05, May 1994

	Limits in Soil (Permissible Cleanup Levels), mg/kg					
	DDT		DDE		DDD	
State	Residential	Industrial	Residential	Industrial	Residential	Industrial
Arizona	4	—	4	—	5.7	—
Massachusetts	2	9	2	9	2	13
Michigan	3.8	44	3.8	44	5.4	62
New Jersey	2 (surface)	9	2 (surface)	9	3 (surface)	12
	100 (subsurface)	—	100 (subsurface)	—	100 (subsurface)	—
New York	2.1	—	2.1	—	2.9	—
Oregon	2	20	2	20	3	20
Texas	1.88	16.8	1.88	16.8	2.67	23.8

α-, β-, γ-, and δ-HCH Isomers

Agency	Description	Concentration
IARC	Probable human carcinogen	
WHO	Drinking water guideline	0.003 mg/L
FAO/WHO	Allowable Daily Intake (ADI)	0.–0.1mg/kg.bw.
	Allowable tolerances (γ-isomer)	
	Potatoes	0.05 mg/kg
	Lettuce	2.0 mg/kg
Air		
OSHA	PEL-TWA(skin), γ-HCH	0.5 mg/m³

Lindane

	International Workplace Limits, Air, mg/m³			
Country	TWA	STEL		Reference
Argentina	0.5	1.5		Sittig, 1994
Australia	0.5	—		
California	0.5	—		
United Kingdom	0.5	1.5		
Water				
U.S. EPA, ODW	Drinking water quality guideline		4 µg/L	
Food				
U.S. FDA	Permissible levels in bottled water		4 µg/L	
Guidelines				
Air				
ACGIH	TLV-TWA(skin), γ-HCH		0.5 mg/m³	
NIOSH	REL-TWA(skin), γ-HCH		0.5 mg/m³	
Water				
U.S. EPA, ODW	MCL in drinking water (γ-HCH)		4 µg/L	
U.S. EPA, OWRS	Ambient water quality health criteria			
	Ingestion of water and organisms		0.0186 µg/L (γ-HCH)	
	Ingestion of organisms only		0.0625 µg/L (γ-HCH)	
	Drinking water (γ-HCH)		0.2 µg/L (γ-HCH)	
U.S. EPA	RfD (oral)		3.00×10^{-4} mg/kg.bw./day	
NTP	Reasonably anticipated to be carcinogens			

Source: ATSDR, TP-93/09 May 1994

Limits in soil (Permissible Cleanup Levels), mg/kg

State	Residential	Industrial
Michigan	1	12
New Jersey	0.52 (surface)	2.2
	1.0 (subsurface)	—
Oregon	80	600
Tennessee	0.5	20

Chlordane

IARC	Not classifiable as human carcinogen	
FAO/WHO	Allowable Daily Intake (ADI)	0.0–0.001 mg/kg.bw/day
WHO	Drinking water guideline	0.3 µg/L
FAO/WHO	Residue tolerances for sum of α- and γ- isomers and oxychlordane in food	0.02–0.5 mg/kg
Guidelines Air		
NIOSH	PEL-TWA (skin)	~0.5 mg/m^3
ACGIH	TLV-TWA (8 h workday), skin	0.5 mg/m^3
U.S. EPA	Unit risk (inhalation)	3.7×10^{-4} µg/m^3
OSHA	PEL-TWA (8 h workday, skin)	0.5 mg/m^3

International Limits in Workplace, Air, mg/m^3

Country	TWA	STEL	Reference
Argentina	0.5	2	Sittig, 1994
Australia	0.5	2	
Belgium	0.5	2	
California, U.S.	0.5	—	
France	0.5	—	
United Kingdom	0.5	—	

Water		
U.S. EPA, ODW	Human health guideline	
	Ten day (10 kg child)	0.06 mg/L
	Long-term (child)	0.5 µg/L
	Long-term (70 kg adult)	2.0 µg/L
U.S. FDA	Maximum tolerable levels for all Isomers of chlordane	
	Fruits and vegetables	0.1 ppm
	Animal fat and edible portion of fish	0.3 ppm
U.S. EPA, OWRS	Ambient water quality criteria for lifetime increased cancer risk:	
	10^{-6}, ingestion of water, fish and shellfish	0.58 ng/L
	10^{-6}, ingestion of fish and shellfish	0.59 ng/L
U.S. EPA	RfD (oral)	6×10^{-5} mg/kg.bw/day
	Cancer unit risk (drinking water)	3.7×10^{-5} µg/L
U.S. FDA	Action level for fish	0.3 ppm
	Action level for other foods	0.1–0.3 ppm

Source: ATSDR, TP-93/03, May 1994

Limits in soil (Permissible Cleanup Levels), mg/kg

State	Residential	Industrial	Reference
Arizona	1.0	0.5	Sittig, 1994
Massachusetts	1	3	
Michigan	1	12	
Oregon	0.5	4	
Texas	0.493	4.4	
Tennessee	0.5	3	

Aldrin

IARC	Not classifiable as human carcinogen	
WHO	Allowable Daily Intake (ADI)	0.1 µg/kg.bw/day
Regulations		
Air		
OSHA	PEL-TWA (skin)	0.25 mg/m^3

International Limits in Workplace, Air, mg/m^3

Country	TWA	STEL	Reference
Argentina	0.25	0.75	Sittig, 1994
Australia	0.25		
Belgium	0.25		
California, U.S.	0.25		
Germany	0.25		
United Kingdom	0.25		
France	0.25	0.75	
Israel	0.25		
Finland	0.25	0.75	
Water			
U.S. EPA, OWRS	Ambient water quality criteria (navigable water)	0.003 µg/L	
Guidelines			
Air			
ACGIH	TLV-TWA (skin)	0.25 mg/m^3	
NIOSH	REL-TWA (skin)	0.25 mg/m^3	
Water			
U.S. EPA, OWRS	Ambient water quality criteria for protection of human health		
	10^{-6} risk (ingestion of water and organisms)	0.074 ng/L	
	10^{-6} risk (ingestion of organisms)	0.079 ng/L	
	For aquatic life protection		
	MATC (acute) saltwater	1.3 µg/L	
	MATC (acute) freshwater	3.0 µg/L	
Other			
U.S.EPA	RfD (oral)	3.0×10^{-5} mg/kg.bw/day	
	Unit risk (air)	4.9×10^{-3} µg/m^3	
	Unit risk (water)	4.9×10^{-4} µg/L	

Source: ATSDR, TP-92/01, April 1993

Limits in soil (Permissible Cleanup Levels), mg/kg

State	Residential	Industrial	Reference
Massachusetts	0.03	0.1	Sittig, 1994
Michigan	0.076	0.088	
New Jersey	0.4 (surface)	0.17	
	50 (subsurface)	—	
Texas	0.0377	0.336	

Dieldrin

Carcinogenic classification, regulations and guidelines are very similar to those set for Aldrin.

International Workplace Limits, Air, mg/m³

Country	TWA	STEL	Reference
Australia	0.25	—	Sittig, 1994
California, U.S.	0.25	—	
Finland	0.25	0.75	
United Kingdom	0.25	0.75	

Limits in soil (Permissible Cleanup Levels), mg/kg

State	Residential	Industrial	Reference
Massachusetts	0.03	0.2	Sittig, 1994
Michigan	0.08	0.937	
New Jersey	0.042 (surface)	0.18	
	50 (subsurface)	—	
Texas	0.04	0.357	

Heptachlor

WHO	Allowable Daily Intake (ADI)	0.0–0.5 mg/kg.bw
	Drinking water guideline	0.1 mg/L
Regulations		
OSHA	TWA (skin)	0.5 mg/m³
U.S. EPA	RfD (heptachlor)	0.0005 mg/kg.bw./day
	RfD (heptachlor epoxide)	0.000013 mg/kg.bw./day
IARC	Not classifiable as human carcinogen (Heptachlor)	
	Limited evidence for animal carcinogen (Heptachlor epoxide)	

Source: ATSDR, TP-88/16, April 1989

Toxaphene

IARC	Classifiable as a possible human carcinogen	
Regulations		
Air		
OSHA	PEL-TWA	0.5 mg/m³
Food		
U.S. FDA	Tolerance range for agricultural products	0.1–7 ppm
	In soybean oil	6 ppm
Water		
U.S. EPA	Effluent standard	0–1.5 µg/L/day
U.S. EPA, OSW	Groundwater concentration limit	0.005 mg/L
	Municipal solid waste(MSW) landfills-design criteria-MCL for upper aquifer	0.005 mg/L
	MSW landfills-practical detection limit	2.0 µg/L
Guidelines		
Air		
ACGIH	TLV-TWA (occupational exposure, skin)	0.5 mg/m³
	TLV-STEL	1 mg/m³
NIOSH	Immediately dangerous to life and health (IDLH)	200 mg/m³
U.S. EPA	10 day health advisory	
	child and adult	0.04 mg/L
	MCL	0.003 mg/L
	q*,cancer slope factor (oral)	1.1 mg/kg.bw./day

Source: ATSDR, Toxicity Profile on Toxaphene (Draft), August 1994.

Endrin
IARC   No classification
Regulations
Air
OSHA   PEL-TWA 0.1 mg/m³
                        STEL   0.3 mg/m³ (15 min)
The International Workplace Limits in air (mg/m³) for TWA and STEL are 0.1 and 0.3, respectively for many countries.
Water
U.S. EPA, OW   Ambient water quality criteria for water and fish   1 µg/L; 0.004 µg/L
                        Effluent standards   1.5 µg/L
                        Effluent standards
                        Electroplating   <0.01 mg/L
                        Electroplating and metal finishing   <0.01 mg/L
                        Aluminum forming   <0.01 mg/L
                        MCL goals for organic contaminants   0.2 µg/L
                        Tolerance range for agricultural products   0 ppm
U.S. EPA/OPTS   Pesticides for restricted use   ≤ 2%
U.S. EPA/OSW   Criteria for classification of solid waste (SW)
                          disposal facilities and practices: MCLs   0.2 µg/L
                        Design criteria   0.2 µg/L
                        MSW-landfills   0.1–20 µg/L (PQL
                                                                                  for two methods)
                        MCL of constituents for groundwater protection   0.2 µg/L
                        Standards for owners of hazardous waste treatment
                          facilities, ground water monitoring list   0.1–10 µg/L
                        Standards for owners of hazardous waste treatment
                          facilities; EPA's interim primary drinking water std.   0.1–10 µg/L
Guidelines
Air
ACGIH   Ceiling limit for occupational exposure (TLV-TWA)   0.1 mg/m³
NIOSH   TWA (skin), recommended limit   0.1 mg/m³
Water
U.S. EPA   Health advisories:
                        10 day (child and adult)   0.02 mg/L
                        Lifetime (child and adult)   0.002 mg/L
                        MCL guideline   0.002 mg/L
                        RfD (oral)   0.3 µg/kg.bw./day

Limits in soil (Permissible Cleanup Levels), mg/kg

State	Residential	Industrial	Reference
Massachusetts	6	15	Sittig, 1994
Michigan	44	150	
New Jersey	17 (surface)	310	
	50 (subsurface)	—	
Oregon	80	600	

Source: ATSDR,TP for Endrin and Endrin aldehyde (draft update), August 1994

Endosulfan

FAO/WHO	Allowable Daily Intake (ADI)	0.006 mg/kg
	Residue limits in foods:	
	Fruits and vegetables (except carrots, potatoes, sweet potatoes and onions)	2 mg/kg
	Cottonseed	1 mg/kg
	Milk and milk products (fat basis)	0.5 mg/kg
	Carrots, potatoes, seed potatoes and onions	0.2 mg/kg
	Fat of meat	0.2 mg/kg
	Rice (in husk)	0.1 mg/kg

Regulations
Air

OSHA	PEL-TWA (8h, skin)	0.1 mg/m^3

The International Workplace Limits, in air (mg/m^3) are 0.1 and 0.3 respectively for TWA and STEL for many countries (Argentina, Finland, United Kingdom etc.)

Water

U.S. EPA, OWRS	Maximum discharge in effluent one day maximum	0.01 kg/1000 kg of endosulfan
	Monthly average shall not exceed	0.0018 kg/1000 kg of endosulfan
U.S. EPA	Tolerances in selected food commodities:	
	Cotton seed	1.0 ppm
	Alfalfa (fresh)	0.3 ppm
	Alfalfa (hay)	1.0 ppm
	Meat or meat fat	0.2 ppm
	Milk or milk fat	0.5 ppm
	Produce (range)	0.1–2.0 ppm
U.S. FDA	Dried Tea, tolerance limit	24 ppm

Guidelines
Air

ACGIH	TLV-TWA (skin)	0.1 mg/m^3
NIOSH	REL-TWA	0.1 mg/m^3

Water

U.S. EPA/OWRS	Ambient water quality criteria for protecting human health:	
	Ingestion of organisms and water	75 µg/L
	Ingestion of organisms only	159 µg/L
	Freshwater aquatic life protection	0.056–0.22 µg/L

Limits in soil (Permissible Cleanup Levels), mg/kg

State	Residential	Industrial	Reference
Massachusetts	1	63	Sittig, 1994
Michigan	60	210	
New Jersey	3 (surface)	52	
	50 (subsurface)	—	
Oregon	10	100	

Source: ATSDR, Endosulfan, TP-91/16, April 1993

Mirex

IARC	Possible human carcinogen	
Regulations		
U.S. FDA	Food action level	0.1 ppm
Guidelines		
U.S. EPA	Ambient water quality criteria for protection of aquatic organisms	
	Freshwater	0.001 µg/L
	Marine	0.001 µg/L
	RfD (oral)	0.2 µg/kg.bw./day

Source: ATSDR, Toxicity Profile on Mirex and Chlordecone, August 1995

Methoxychlor

IARC	Not classifiable as human carcinogen	
WHO	Drinking water guideline	0.030mg/L
	Allowable Daily Intake	0.1 mg/kg
Regulations		
Air		
OSHA	PEL-TWA (total dust)	10 mg/m^3

International workplace limits in air (mg/m^3) for TWA and STEL are 10, 10 for many countries, excepting 15 (Denmark) and 10, 20 (Finland).

Water		
U.S. EPA/ODW	MCL	0.040 mg/L
Food		
U.S. FDA	Concentration in bottled water	0.1 mg/L
U.S. EPA	Tolerances in raw agricultural commodities:	
	Forage crops	100 ppm
	Fruits and vegetables	14 ppm
	Sweet potatoes and Yams	7 ppm
	Meat fat	3 ppm
	Milk fat	1.25 ppm
	Potatoes and horseradish	1 ppm
Guidelines		
Air		
ACGIH	TLV-TWA	10 mg/m^3
NIOSH	REL-TWA	0.07 mg/m^3 (LOQ)
Water		
	MCL guidelines	0.04 mg/L
	1 day exposure (child)	6.4 mg/L
	10 day exposure (child)	2.0 mg/L
	Longer-term (child)	0.05 mg/L
	Longer-term (adult)	0.175 mg/L
	Lifetime (adult)	0.04 mg/L
U.S. EPA/OWRS	Ambient water quality criteria for domestic water supply	0.100 mg/L
U.S. EPA	RfD (oral)	5 × 10^{-3} mg/kg.bw./day

Limits in soil (Permissible Cleanup Levels), mg/kg

State	Residential	Industrial	Reference
Michigan	1,300	4,500	Sittig, 1994
New York	80	—	
Washington	400	—	

Source: ATSDR, TP-93/11, May 1994

Chloropyrifos
Air

Country	TWA mg/m³	STEL mg/m³
Australia	0.2	0.6
Belgium	0.2	0.6
Switzerland	0.2	—
Denmark	0.2	—
Finland	0.2	0.6
France	0.2	—
United Kingdom	0.2	0.6

Water

Canada	Domestic drinking water	0.090 mg/L

Regulations

U.S. EPA/OPTS	Tolerance ranges for agricultural Products	0.05–15.0 ppm
	Tolerance in foods:	
	Citrus oil	25 ppm
	Corn oil	3 ppm
	Mint oil	10 ppm
	Peanut oil	1.5 ppm
	Animal feeds	0.5–15 ppm

Guidelines
Air

ACGIH	TLV-TWA (occupational exposure)	0.90 mg/m³ (skin)
NIOSH	REL-TWA	0.2 mg/m³
	STEL	0.6 mg/m³

Water

U.S. EPA/OW	Health advisory	
	10 day (child)	0.03 mg/L
	Lifetime	0.02 mg/L
U.S. EPA	RfD	0.003 mg/kg.bw./day

Limits in soil (Permissible Cleanup Levels), mg/kg

State	Residential	Industrial	Reference
Michigan	780	2,700	Sittig, 1994
Arizona	350	—	
Washington	240	—	

Source: ATSDR, TP (draft) on Chloropyrifos, August 1995.

Dichlorvos
IARC   Probable human carcinogen
Workplace limits:(mg/m³)

Country	TWA	STEL
Argentina	1	3
Australia	0.9	—
Belgium	0.9	—
Switzerland	1	—
CIS	—	0.2
Germany	1	—
Finland	1	3
France	1	3
United Kingdom	1	3
Hungary	—	0.2
Poland	1	—

Regulations
Air
OSHA   PEL-TWA   1 mg/m³
Guidelines
ACGIH   TLV-TWA   0.90 mg/m³
                    (skin)
NIOSH   TWA (skin)   1.0 mg/m³
Water   IDLH   200 mg/m³Water
U.S. EPA   q˙cancer
                    slope
                    factor (oral)   0.029 mg/kg.bw./ day
                    RfD   0.5 µg/kg.bw./day

Limits in soil (Permissible Cleanup Levels), mg/kg

State	Residential	Industrial	Reference
Michigan	4.4	52.0	Sittig, 1994

Source: ATSDR, TP (draft) on Dichlorvos, August 1995

Chlorfenvinphos
IARC   Not classified
U.S. EPA   Not classified
Regulations
Food
U.S. EPA/OPTS   Tolerance range for agricultural   0.005–0.2 ppm
                    products

Source: ATSDR, TP (draft) on Chlorfenvinphos, August 1995

REFERENCES

Albright, R., Johnson, N., Sanderson, T.W. et al. 1974. Pesticide residues in the top soil of five west Alabama counties. *Bull. Environ. Contam. Toxicol.* 12:378–384.

Allan, R.J. and Ball, A.J. 1990. An overview of toxic contaminants in water and sediments of the Great Lakes: Part I. *Water Pollut. Res. J. Canada.* 25:387–505.

Anderson, J.F. and Wojtas, M.A. 1986. Honey bees (*Hymenoptera: Apidae*) contaminated with pesticides and polychlorinated biphenyls. *J. Econ. Entomol.* 79:1200–1205.

Anderson, D.J. and Hites, R.A. 1989. Indoor air: spatial variations of chlorinated pesticides. *Atmos. Environ.* 23, 2063–2066.

Aruda, J.A., Cringan, M.S., Layher, W.G. et al. 1988. Pesticides in fish tissue and water from Turtle Creek Lake, Kansas. *Bull. Environ. Contam. Toxicol.* 41:617–624.

Atlas, E. and Giam, C.S. 1988. Ambient concentrations and precipitation scavenging of atmospheric pollutants. *Water, Air and Soil Pollut.* 38:19–36.

Atlas, E. and Giam, C.S. 1981. Global transport of organic pollutants; ambient concentration in the remote marine atmosphere, *Science.* 211:163–165.

ATSDR. 1989. Agency for Toxic Substances and Disease Registry, U.S. Public Health Service, Department of Health and Human Services, Atlanta, GA, Toxicological Profile for Heptachlor and Heptachlor Epoxide, April 1989.

ATSDR. 1989a. Agency for Toxic Substances and Disease Registry. Health Assessment for FMC Fresno National Priorities List (NPL) site, Fresno, CA, Region 9. CERCLIS No. CAD000629998. Atlanta, GA.

ATSDR. 1993a. Agency for Toxic Substances and Disease Registry, U.S. Public Health Service, Department of Health and Human Services, Atlanta, GA, Toxicological Profile for Aldrin and Dieldrin, TP-92/01, May 1993.

ATSDR. 1993b. Agency for Toxic Substances and Disease Registry, U.S. Public Health Service, Department of Health and Human Services, Atlanta, GA, Toxicological Profile for Endosulfan, TP-91/16, April 1993.

ATSDR. 1994a. Agency for Toxic Substances and Disease Registry, U.S. Public Health Service, Department of Health and Human Services, Atlanta, GA, Toxicological Profile for DDT, DDE, and DDD, TP-93/05, May 1994.

ATSDR. 1994b. Agency for Toxic Substances and Disease Registry, U.S. Public Health Service, Department of Health and Human Services, Atlanta, GA, Toxicological Profile for Hexachlorocyclohexane, TP-93/09, 1994

ATSDR. 1994c. Agency for Toxic Substances and Disease Registry, U.S. Public Health Service, Department of Health and Human Services, Atlanta, GA, Toxicological Profile for Chlordane, TP-93/03, May 1994.

ATSDR. 1994d. Agency for Toxic Substances and Disease Registry, U.S. Public Health Service, Department of Health and Human Services, Atlanta, GA, Toxicological Profile for Toxaphene, Update (draft), August 1994.

ATSDR. 1994e. Agency for Toxic Substances and Disease Registry, U.S. Public Health Service, Department of Health and Human Services, Atlanta, GA, Toxicological Profile for Endrin and Endrin aldehyde, Update (Draft), August 1994.

ATSDR. 1994f. Agency for Toxic Substances and Disease Registry, U.S. Public Health Service, Department of Health and Human Services, Atalnta, GA, Toxicological Profile for Methoxychlor. May 1994. TP-93/a.

ATSDR. 1995a. Agency for Toxic Substances and Disease Registry, U.S. Public Health Service, Department of Health and Human Services, Atlanta, GA, Toxicological Profile for Mirex and Chlordecone, (Draft), August 1995.

ATSDR. 1995b. Agency for Toxic Substances and Disease Registry, U.S. Public Health Service, Department of Health and Human Services, Atlanta, GA, Toxicological Profile for Chlorpyrifos, draft, August 1995.

ATSDR. 1995c. Agency for Toxic Substances and Disease Registry, U.S. Public Health Service, Department of Health and Human Services, Atlanta, GA, Toxicological Profile for Dichlorvos, August 1995.

ATSDR. 1995d. Agency for Toxic Substances and Disease Registry, U.S. Public Health Service, Department of Health and Human Services, Atlanta, GA, Toxicological Profile for Chlorfenvinphos, August 1995.

Bevenue, A., Hylin, J.W., and Kawano, Y. 1972. Pesticides in water. organochlorine pesticide residues in water, sediments, algae and fish: Hawaii; 1970–1971. *Pesticide Monit. J.* 6:56–72.

Beyer, W.N. and Kaiser, T.E. 1984. Organochlorine pesticide residues in moths from the Baltimore, MD-Washington, D.C. area. *Environ. Monitor. Assess.* 4:129–137.

Biberhofer, J. and Stevens, R.J. 1987. Organochlorine contaminants in ambient waters of Lake Ontario, Burlington, Ontario: Inland Waters/Land Directorate Canada, 159, 1–11.

Bidleman, T.F. 1981. Interlaboratory analysis of high molecular weight organochlorines in ambient air. *Atmos. Environ.* 15:619–624.

Bidleman, T., Christensen, E., and Billings, W. 1981. Atmospheric transport of organochlorines in the North Atlantic gyre. *J. Marine Res.* 399:443–464.

Blus, L., Cromartie, E., McNease, L. et al. 1979. Brown pelican: Population status, reproductive success, and organochlorine residues in Louisiana, 1971–1976. *Bull. Environ. Contam. Toxicol.* 22:128–135.

Boellstorff, D.E., Ohlendorf, H.M., Anderson, D.W. et al. 1985. Organochlorine chemical residues in white pelicans and western grebes from the Klamath Basin, California. *Arch. Environ. Contam. Toxicol.* 14:485–493.

Borgmann, V. and Whittle, D.M. 1991. Contaminant concentration trends in Lake Ontario lake trout (*salvelinus namayacush*), 1977–1988, *J. Great Lakes Res.* 17:368–381.

Braun, H.E. and Frank, R. 1980. Organochlorine and organophosphorus insecticides: Their use in eleven agricultural watersheds and their loss to stream waters in southern Ontario, Canada, 1975–1977. *Sci. Total Environ.* 15:169–192.

Buck, N., Estesen, B., and Ware, G. 1983. DDT. Moratorium in Arizona: Residues in soils and alfalfa after 12 years. *Bull. Environ. Contam. Toxicol.* 31:66–72.

Callahan, C.A., Menzie, C.A., Burmaster, D.E. et al. 1991. On-site methods for assessing chemical impact on the soil environment using earthworms: A case study at the Baird and McGuire Superfund site, Holbrook, MA. *Environ. Toxicol. Chem.* 10:817–826.

Camanzo, J., Rice, C.P., Jude, D.J. et al. 1987. Organic priority pollutants in nearshore fish from 14 Lake Michigan tributaries and embayments, 1983. *J. Great Lakes Res.* 13:296–309.

Carey, A.E. and Kutz, F.W. 1985. Trends in ambient concentrations of agrochemicals in humans and the environment of the USA. *Environ. Monit. Assess.* 5:155–164.

Carey, A., Gowen, J., and Tai, H. 1979. Pesticide residue level in soils and crops from 37 states, 1972-national soils monitoring program (IV). *Pestic. Moniot. J.* 12:209–229.

Carey, A.E., Wiersma, G.B., and Tai, H. 1976. Pesticide residues in urban soils from 14 United States cities, 1970. *Pestic. Monit. J.* 10:54–60.

Carey, A.E., Douglas, P., and Tai, H. 1979. Pesticide residue concentrations of five United States cities, 1971- Urban soils monitoring program. *Pestic. Monit. J.* 13:17–22.

Carey, A.E., Gowen, J.A., and Tai, H. 1978. Pesticide residue levels in soils and crops, A 1971 — A National Soils Monitoring Program (III). *Pestic. Monit. J.* 12:117–136.

Carmanzo, J., Rice, C.P., Jude, D.J., et al. 1987. Organic priority pollutants in nearshore fish from 14 Lake Michigan tributaries and embayments. *J. Great Lakes Res.* 13:296–309.

CELDS. 1993. Computer-aided Environmental Legislative Database Systems. University of Illinois at Urbana.

Chan, C.H. and Perkins, L.H. 1989. Monitoring of trace organic contaminants in atmospheric precipitation. *J. Great Lakes Res.* 15:465–475.

Clark, D.E., Smalley, H.E., and Crookshnak, H.R. 1974. Residues in food and feed: chlorinated hydrocarbon insecticides residues in feed and carcasses of feedlot cattle, Texas-1972. *Pestic. Monit. J.* 8:180–183.

Cohen, D.B. 1986. Ground water contamination by toxic substances, California Assessment. Pollutant Investigations Branch, State Water Resources Control Board, Sacramento, CA 95801, American Chemical Society.

Cole, R.H., Frederick, R.E., and Healy, R.P. 1984. Preliminary findings of the priority pollutant monitoring project of the nationwide urban run-off program. *J. Water Pollut. Control Fed.* 56:898–908.

Cooper, C.M., 1991. Persistent organochlorine and current use insecticide concentrations in major watershed components of Moon Lake, Mississippi, USA. *Archiv. Fur Hydrobiologie.* 121:103–113.

Corneliussen, P.E. 1970. Residues in food and feed. Pesticide residues in total diet samples V. *Pestic. Monit. J.* 4:89–105.

Crockett, A., Wiersma, G., and Tai, H. 1974. Pesticides in soils: pesticide residue levels in soils and crops. FY-70-national soils monitoring program (II), *Pestic. Monit. J.* 8:69–76.

CSCORP 1992. STN. International Network, Chemical Abstracts Service, Columbus, OH, April 20, 1992.

Cutshall, N.H., Larsen, I.V., and Nichols, M.M. 1981. Man-made radioneucleotides confirm rapid burial of Kepone in James River sediments. *Science.* 213:440–442.

Davies, K. 1988. Concentrations and dietary intake of selected organochlorines, including PCB's, in fresh food composites grown in Ontario, Canada. *Chemosphere.* 17:263–276.

Deer, H.M., Beck, E.D., and Roe, A.H. 1993. Respiratory Exposure of Museum Personnel to Dichlorvos Insecticide. *Vet. Hum. Toxicol.* 35:226–228.

Delaplane, K.S. and LaFarge, J.P. 1990. Variable chlordane residues in soil surrounding house foundations in Louisiana, USA. *Bull. Environ. Contam. Toxicol.* 45:675–680.

DeVault, D.S., Clark, J.M., Lahvis, G. et al. 1988. Contaminants and trends in fall run coho salmon. *J. Great Lakes Res.* 14:23–33.

Dingle, J.H.P., Palmer, W.A., and Black, R.R. 1989. Residues of DDT and dieldrin in the subcutaneous fat and butterfat of cattle. *Australian J. Exp. Agric.* 29:497–501.

Duggan, R.E., Barry, H.C., and Johnson, L.Y. 1967. Residues in food and feed, Pesticide residues in total diet samples, II. *Pestic. Monit. J.* 1:2–12.

Duggan, R.E., Barry, H.C., and Johnson, L.Y. 1966. Pesticide residues in total-diet samples. *Science.* 151:101–104.

Duggan, R. 1983. Pesticide residue levels in foods in the U.S., from July 1, 1969 to June 30, 1976. *J. Assoc. Off. Anal. Chem.* 66:1534–1535.

Durham, R.W. and Oliver, B.G. 1983. History of Lake Ontario (Canada, U.S.) contamination from the Niagara river by sediment radiodating and chlorinated hydrocarbon analysis. *J. Great Lakes Res.* 9:160–168.

Eisenberg, M. and Topping, J.J. 1985. Organochlorine residues in finfish from Maryland waters, 1976-1980, *J. Environ. Sci. Health. B.* 20:729–742.

Eisenreich, S.J., Looney, B.B., and Thornton, J.D. 1981. Airborne organic contaminants in the Great Lakes ecosystem. *Environ. Sci. Technol.* 15:30–38.

Elgar, K.E. and Steer, B.D. 1972. Dichlorvos concentrations in the air of houses arising from the use of dichlorvos PBC Strips. *Pestic. Sci.* 3:591–600.

Elliott, J. 1975. Monitoring of selected ecological components of the environment in four Alabama counties (1972–1974). In: *Papers of Environ. Chem. Human Animal Health Proc., 4th Annual Conf., Auburn University, Auburn, AL.* Alabama Cooperative Extension Service, 233–279.

Environment Canada. 1992. Toxic chemicals in the Great Lakes and associated effects. Vol. II: Effects. Ottawa, Canada: Environment Canada, Health and Welfare Canada, Department of Fisheries and Oceans. March, 1992.

EPA. 1981. The potential atmospheric impact of chemicals released into the environment: Proceedings of four workshops. U.S. Environmental Protection Agency, Washington, D.C., Document No. PB 82-119447.

EPA. 1990. Non-occupational pesticide exposure study (NOPES), U.S. Environmental Protection Agency, Office of Research and Development, Washington, D.C., EPA/600/3-90/003.

EPA. U.S. Environmental Protection Agency references are cited in the respective ATSDR Toxicological Profile documents as well as in Regulatory Toxicology and Pharmacology, Vol. 20, No. 1, August, 1994.

FDA. 1988. Food and Drug Administration Pesticide Program: Residue in Foods-1987. *J. Assoc. Off. Anal. Chem. Int.* 71:A156–A174.

FDA. 1989. U.S. Food and Drug Administration Program-Residues in Foods-1988. *J. Assoc. Off. Anal. Chem.* 72:33A–152A.

FDA. 1990. U.S. Food and Drug Administration Program-Residues in Foods-1989. *J. Assoc. Off. Anal. Chem.* 73:127A–146A.

FDA. 1991. U.S. Food and Drug Administration Program-Residues in Foods-1990. *J. Assoc. Off. Anal. Chem.* 74:121A–141A.

FDA. 1992. Food and Drug Administration Pesticide Program: Residue Monitoring 1991. *J. Assoc. Off. Anal. Chem. Int.* 75:A135–A157.

FDA. 1993. Food and Drug Administration Pesticide Program: Residue Monitoring 1992. *J. Assoc. Off. Anal. Chem. Int.* 76:A127–A148.

Fenske, R.A., Black, K.G., Elkner, K.P. et al. 1990. Potential exposure and health risks of infants following indoor residential pesticide applications. *Am. J. Public Health.* 80:689–693.

Ford, W.M. and Hill, E.P. 1991. Organochlorine pesticide in soils, sediments and aquatic animals in the upper Stede Bayor Watershed of Mississippi. *Arch. Environ. Contam. Toxicol.* 20:161–167.

Frank, R., Braun, H.E., and Ripley, B.D. 1987. Residues of insecticides, fungicides and herbicides in fruit produced in Ontario, Canada, 1980–1984. *Bull Environ. Contam. Toxicol.* 39:272–279.

Frank, R., Logan, L., and Clegg, B.S. 1991. Pesticide and polychlorinated biphenyl residues in waters at the mouth of the Grand, Saugeen, and Thames rivers, Ontario, Canada, 1986–1990. *Arch. Environ. Contam. Toxicol.* 21:585–595.

Frank, R., Braun, H.E., and Stonefield, K.I. 1990. Organochlorine and organophosphorus residues in the fat of domestic farm animal species, Ontario, Canada, 1956–1988. *Food. Addit. Contam.* 7:629–636.

FSTRAC. 1990. Federal State Toxicology and Regulatory Alliance Committee. Summary of state and federal drinking water standards and guidelines. Chemical Communication Subcommittee, U.S. Environmental Protection Agency, Washington, D.C.

Gallo, M.A. and Lawryk, N.J. 1991. Organic phosphorous pesticides. In: *Handbook of Pesticide Toxicology Classes of Pesticides, Vol. 2*: Hayes, W.J. and E.R. Laws, Eds., Academic Press, San Diego, 917–1123.

Gartrell, M.J., Craun, J.C., Podrebarac, D.S., and Gunderson, E.L. 1986a. Pesticides, selected elements, and other chemicals in infant and toddler total diet samples, October 1980–March 1982. *J. Assoc. Off. Anal. Chem.* 69:123–145.

Gartrell, M.J., Craun, J.C., Podrebarac, D.S. et al. 1986b. Pesticides, selected elements, and other chemicals in adult total diet samples, October 1980–September 1982. *J. Assoc. Off. Anal. Chem.* 69:146–161.

Gartrell, M., Craun, J., and Podrebarac, D. 1985. Pesticides, selected elements and other chemicals in adult total diet samples, October 1979-September 1980, *J. Assoc. Off. Anal. Chem.* 68:1184–1197.

Germain, A. and Langlois, C. 1988. Pollution of the water and suspended sediments of the St. Lawrence River (Ontario, Quebec, Canada) by organochlorine pesticides, polychlorinated biphenyls, and other priority pollutants. *Water Pollut. Res. J. Canada.* 23:602–614.

Geyer, H., Scheunart, I., and Korte, F. 1986. Bioconcentration potential of organic environmental chemicals in humans. *Regul. Toxicol. Pharmacol.* 6:313–347.

Gilliom, R. 1984. Pesticides in the rivers of the United States, National Water Survey 1984 Paper, 2275:85-93.

Gilliom, R.J., Alexander, R.B., and Smith, R.A. 1985. Pesticides in the nations rivers, 1975–1980, and implications for future monitoring. U.S. Geological Survey Water Supply Paper 2271.

Green, M.B., Harteley, G.S., and West, T.F. 1987. *Chemicals for Crop Improvement and Pest Management,* 3rd Ed., Pergamon Press, Ontario, CN.

Gregor, D.J. and Gummer, W.D. 1989. Evidence of atmospheric transport and deposition of organochlorine pesticides and polychlorinated biphenyl in Canadian Arctic snow. *Environ. Sci. Technol.* 23:561–565.

Gunderson, E.L. 1988. FDA total diet study, April 1982–April 1984: dietary intakes of Pesticides, selected elements and other chemicals. *Assoc. Off. Anal. Chem.* 71:1200–1209.

Hargrave, B.T., Vass, W.P., and Ericksen, P.F. 1988. Atmospheric transport of organochlorines to the Arctic Ocean, *Tellus B.* 4:480-493.

Hargrave, B.T., Harding, G.C., and Vass, W.P. 1992. Organochlorine pesticides and polychlorinated biphenyls in the Arctic Ocean foodweb. *Arch. Environ. Contam. Toxicol.* 22:41–54.

Hayes, W.J. Jr. 1982. Organic phosphorus pesticides. In: *Pesticides Studied in Man.* Williams & Wilkins, Baltimore, MD. 343-351.

Heikes, D.L. and Craun, J.C. 1992. Rapid multiresidue procedure for the determination of pesticides in anhydrous lanolin and lanolin-containing pharmaceutical preparations utilizing gel permeation chromatography cleanup with gas chromatographic and mass spectrometric techniques. *J. Agric. Food Chem.* 40:1586–1590.

Hoff, R.M., Muir, D.C.G. and Grift, N.P. 1992. Annual cycle of polychlorinated biphenyls and organohalogen pesticides in air in Southern Ontario, 1. Air concentration data. *Environ. Sci. Technol.* 26:266–275.

HSDB. 1994. Hazardous Substances Data Bank. National Library of Medicine, Bethesda, MD.

Hundley, H.K., Cairns, T., and Luke, M.A. 1988. Pesticide residue findings by the Luke method in domestic and imported foods and animal feeds for fiscal yeras 1982–1986. *J. Assoc. Off. Anal. Chem.* 71:875–892.

IARC. 1974a. International Agency for Research on Cancer. Monographs on the Evaluation of the Carcinogenic Risks to Humans. Dieldrin. World Health Organization, Lyon, France.

IARC. 1974b. International Agency for Research on Cancer. Evaluation of the carcinogenic risks of chemicals to humans. Dieldrin. *IARC Monograph.* 5:125–156.

IARC. 1979. International Agency for Research on Cancer. Monographs on the Evaluation of Carcinogenic Risks to Humans. Vol. 20: Methoxychlor. World Health Organization, Lyon, France.

IARC. 1991. International Agency for Research on Cancer. Monographs on the Evaluation of Carcinogenic Risks to Humans. Occupational Exposures in Insecticide Application, and Some Pesticides; Dichlorvos, Vol. 53. World Health Organization, Geneva, Switzerland. 267–307.

Ishikawa, K., Charlambous, P., and Matsumura, K. 1989. Metabolism of pyrethroids and DDT of phosphrylation activities of rat brain sodium channel. *Biochem. Pharmacol.* 38:2449–2457.

Johnson, R.D., Manske, D.D., and New, D.H. 1984. Pesticide, heavy metal and other chemical residues in infant and toddler total diet samples, III., August 1976–September 1977. *J. Assoc. Off. Anal. Chem.* 67:154–166.

Johnson, R.D., Manske, D.D., and New, D.H. 1981. Food and feed; pesticide, heavy metal and other chemical residues in infant and toddler total diet samples, II. August 1975–July 1976. *Pestic. Monit. J.* 15:39–5.

Johnson, A., Norton, D., and Yake, B. 1988. Persistence of DDT in the Yakima river drainage, Washington. *Arch. Environ. Contam. Toxicol.* 17:289–297.

Kaiser, K.L.E., Lum, K.R., and Camba, M.E. 1990. Organic trace contaminants in St. Lawrence River water and suspended sediments, 1985–1987. *Sci. Total Environ.* 97/98:23–40.

King, K.A. and Krynitsky, A.J. 1986. Population trends, reproductive success and organochlorine chemical contaminants in waterbirds nesting in Galveston Bay, Texas. *Arch. Environ. Contam. Toxicol.* 15:367–376.

Knap, A.H. and Binkley, K.S. 1991. Chlorinated organic compounds in the troposphere over the Western North Atlantic Ocean measured by aircraft. *Atmos. Environ.* 25:1507–1516.

Krill, R.M. and Sonzogni, W.C. 1986. Chemical monitoring of Wisconsin's U.S. groundwater. *Am. Water Works Assoc.* 78:70–75.

Kuehl, D.W., Leonard, E.N., and Butterworth, B.C. 1983. Polychlorinated chemical residues in fish from major watersheds near the Great Lakes, 1979. Environ. Int., 9, 293–299.

Kuntz, K.W. and Warry, N.D. 1983. Chlorinated organic contaminants in water and suspended sediments of the lower Niagara River. *J. Great Lake Res.* 9:241–248.

Kuntz, K.W., Yobs, A.R., and Yang, H.S.C. 1976. National pesticide monitoring programs. In: *Air Pollution from Pesticides and Agricultural Processes,* Lee, R.E. Ed., CRC Press, Boca Raton, FL, 95–136.

Kuntz, K.W. and Wang, N.D. 1983. Chlorinated organic contaminants in water and suspended sediments of the Lower Niagara River, *J. Great lakes Res.* 9:241–248.

Lambardo, P. 1986. The FDA total diet study program. In: *Environmental Epidemiology.* Kopfler, F.C., and Craun, G.F. Eds., Lewis Publishers, Boca Raton, FL, 141–148.

Leary, J.S., Kenae, W.T., Fontenot, C., et al. 1974. Safety evaluation in the home of polyvinyl chloride resin strip containing dichlorvos (DDVP). *Arch. Environ. Health.* 29:308–314.

Lee, R.F. Jr. 1977. Fate of petroleum components in estuarine waters of the southeastern United States. Proceedings of 1977 Oil Spill Conference: Prevention, behavior, control, cleanup, March 8–10, 1977. American Pesticide Institute, pp. 611–616.

Leiker, T.J., Rostad, C.E., and Barnes, C.R. 1991. A reconnaissance study of halogenated organic compounds in catfish from the Lower Mississippi River and its major tributaries. *Chemosphere.* 23:817–830.

Lewis, R.G. and Lee, Jr., R.E. 1976. Air pollution from pesticides: Sources, occurence and dispersion. In: Lee, Jr., R.E., Ed., *Air Pollution from Pesticide and Agricultural Processes.* CRC Press, Boca Raton, FL, 5–51.

Lewis, R.G., Bond, A.E., Johnson, D.E., et al. 1988. Measurement of atmospheric concentrations of common household pesticides: A Pilot Study. *Environ. Monit. Assess.* 10:59–73.

Lichtenberg, J., Eichelberger, J., and Dreeman, R. 1970. Pesticides in surface waters of the United States, A 5-year summary, 1964-1968, *Pestic. Monit. J.* 4:71–81.

Lum, K.R., Kaiser, K.L., and Comba, M.E. 1987. Export of mirex from Lake Ontario to the St. Lawrence estuary. *Sci. Total Environ.* 67:41–51.

Lunsford, C.A., Weinstein, M.P., and Scott, L. 1987. Uptake of Kepone by the estuarine bivalve *Rangia cuneata* during the dredging of contaminated sediments in the James River, VA. *Water Res.* 21:411–416.

Mackay, D. and Leinonen, P.J. 1975. Rate of evaporation of low-solubility contaminants from water bodies to atmosphere. *Environ. Sci. Technol.* 9:1178–1180.

Maddy, K.T., Fong, H.R., Lowe, J.A., et al. 1982. A study of well water in selected California communities for residues of 1,3-dichloropropene, chloroallyl alcohol and 49 organophosphate of chlorinated hydrocarbon pesticides. *Bull. Environ. Contam. Toxicol.* 29:354–359.

Maguire, R.J., Kuntz, K.W., and Hale, E.J. 1983. Chlorinated hydrocarbons in the surface microlayer of the Niagara River. *J. Great Lakes Res.* 9:281–286.

Mann, J. 1987. *Secondary Metabolism*, Oxford University Press. Oxford. Chemistry series, 27.

Marsalek, J. and Schroeter, H. 1988. Annual loadings of toxic contaminants in urban runoff from the Canadian Great Lakes basin. *Water Pollut. Res. J. Canada.* 23:360–378.

Martin, R.J. and Duggan, R.E. 1968. Pesticide residues in total diet samples, *III, Pestic. Monit. J.* 1:11–20.

Matsumura, F. and Patil, K.C. 1969. Adenosine triphosphate sensitive to DDT in synapses of rat brain. *Science.* 166:121–122.

McEwen, L.C., Stafford, C.J., and Hensler, G.L. 1984. Organochlorine residues in eggs of Black-crowned night herons from Colorado and Wyoming. *Environ. Toxicol. Chem.* 3:367–376.

Menzie, C.A., Burmaster, D.E., and Freshman, J.S. 1992. Assessment of methods for estimating ecological risk in the terrestrial component: a case study at the Bacid and McGuire Superfund site in Holbrook, Massachusetts. *Environ. Toxicol. Chem.* 11:245–260.

Miles, J.R.W., Harris, C.R., and Moy, P. 1978. Insecticide residues in organic soil of the Holland Marsh, Ontario, Canada, 1972-1975. *J. Econ. Entomol.* 71:91–101.

Miles, J.R.W. and Harris, C.R. 1971. Insecticide residues in a stream and a controlled drainage system in agricultural areas of southwestern Ontario, 1970. *Pestic. Monit. J.* 5:289–294.

Minyard, J.P. and Roberts, W.E. 1991. Chemical contaminants monitoring: State findings on pesticide residues in foods-1988 and 1989. *J. Assoc. Off. Anal. Chem.* 74:438–452.

Nagayama, T., Maki, T., Kan, K., et al. 1989. Residues of organophosphorus pesticides in commercial tea and their leaching into tea. *J. Pestic. Sci.* 14:39–46.

Neidert, E., Trotman, R.B., and Saschenbrecker, P.W. 1994. Levels and incidences of pesticide residues in selected agricultural food commodities available in Canada. *J. Assoc. Off. Anal. Chem. Int.* 77:18–33.

Nichols, M.M. 1990. Sedimentologic fate and cycling of Kepone in an estuarine system: Example from the James River estuary. *Sci. Total Environ.* 97/98:407–440.

Ohlendorf, H.M., Swineford, D.M., and Locke, L.N. 1981. Organochlorine residues and mortality of herons. *Pestic. Monit. J.* 14:125–135.

Oliver, B.G. and Charlton, M.N. 1984. Chlorinated organic contaminants on settling particulates in the Niagara River vicinity of Lake Ontario. *Environ. Sci. Tecnol.* 18:903–908.

Oliver, B.G. and Nicol, K.D. 1984. Chlorinated contaminants in the Niagara River, 1981–1983. *Sci. Total Environ.* 39:57–70.

Orndorff, S.A. and Colwell, R.R. 1980. Microbial transformation of Kepone. *Appl. Environ. Microbiol.* 39:398–406.

Phillipps, L.J. and Birchard, G.F. 1991. Use of STORET data to evaluate variations in environmental contamination by census division. *Chemosphere.* 22:835–848.

Plimmer, J.R. 1992. Dissipation of Pesticides in the Environment. In: *Fate of Pesticides and Chemicals in the Environment.* Schnoor, J.L., Ed., John Wiley & Sons, New York, 79–90.

Plumb, Jr., R.H. 1991. The occurrence of Appendix IV organic constituents in disposal site ground water. *Ground Water Monit. Rev.* 11:157–164.

Podrebarac, D.S. 1984. Pesticide, heavy metal and other chemical residues in infant and toddler total diet samples, IV., October 1977–September 1988. *J. Assoc. Off. Anal. Chem.* 67:166–175.

Puri, R.K., Orazio, C.E., and Kapila, S. 1990. Studies on the transport and fate of chlordane in the environment, In: Long-Range Transport of Pesticides, Kurtz, D.A., Ed., Lewis Publishers, Boca Raton, FL, 271–280.

Rapaport, R., Urban, N., and Capel, P. 1985. "New" DDT inputs to North America: atmospheric deposition. *Chemosphere.* 14:1167-1174.

Ray, L.E., Murray, H.E., and Giam, C.S. 1983. Organic pollutants in marine samples from Portland, Maine. *Chemosphere.* 12:1031–1038.

Richards, R.P. and Baker, D.B. 1993. Pesticide concentration patterns in agricultural drainage networks in the Lake Erie basin. *Environ. Toxicol. Chem.* 12:13–26.

RTP. 1994. Regulatory Toxicology and Pharmacology. Interpretative review of the potential adverse effects of chlorinated organic chemicals on human health and the environment, Report of the Expert Panel. 20: Part 2, Academic Press, Inc., Harcourt Brace & Company, San Diego.

Saiki, M.K. and Schmitt, C.J. 1986. Organochlorine chemical residues in bluegills and common carp from the irrigated San Joaquin Valley Floor, California. *Arch. Environ. Contam. Toxicol.* 15:357–366.

Sandhu, S.S., Warren, W.J., and Nelson, P. 1978. Pesticide residue in rural potable water. *J. Amer. Water Works Assoc.* 70:41–45.

Scharf, J., Wiesiollek, R., and Balchman, K., 1992. Pesticides in the atmosphere. *Fresenius J. Anal. Chem.* 342:813–816.

Schattenberg, H.J., III and Hsu, J.P. 1992. Pesticide residue survey of produce from 1989–1991. *J. Assoc. Off. Anal. Chem.* 75:925–933.

Schmitt, C.J., Zajicek, J.L., and Ribick, M.A. 1985. National pesticide monitoring program: Residues of organochlorine chemicals in freshwater fish, 1980–81. *Arch. Environ. Contam. Toxicol.* 14:225–260.

Schmitt, C.J., Zajicek, J.L., and Peterman, P.H. 1990. National contaminant biomonitoring program: residues of organochlorine chemicals in U.S. freshwater fish, 1976–1984. *Arch. Environ. Contam. Toxicol.* 19:748–781.

Sericano, J.L., Atlas, E.L., and Wade, T.L. 1990. NOAA's status and trends mussels watch program; chlorinated pesticides and PCBs in oysters (*Crassostrea virginica*) and sediments from the Gulf of Mexico; 1986–1987. *Mar. Environ. Res.* 29:161–203.

Sergeant, D.B., Munaware, M., and Hodson, P.V. 1993. Mirex in North American Great Lakes: new detection and their confirmation. *J. Great Lakes Res.* 19:145–157.

Singh, A., Bhatnagar, M.K., Villeneuve, D.C., and Valli, V.E. 1985. Ultrastructure of the thyroid glands of rats fed photomirex: A 48-week recovery study. *J. Environ. Pathol. Toxicol. Oncol.* 6:115–126.

Sittig, M. 1994. World-Wide Limits for Toxic and Hazardous Chemicals in Air, Water and Soil. Noyes Publications, Park Ridge, NJ, 792 p.

Smith, A.G. 1991. Chlorinated hydrocarbon insecticides. In: *Handbook of Pesticide Toxicology Classes of Pesticides. Vol. 2.* Hayes, W.J. and Laws, E.R., Eds., Academic Press, San Diego: 731–915.

Spalding, R.F., Junk, G.A., and Richard, J.J. 1980. Pesticides in groundwater beneath irrigated farmland in Nebraska, August 1978. *Pestic. Monit. J.* 14:70–73.

Stanley, J.S. 1986. Broad scan analysis of the FY82 national human adipose tissue survey specimens: Vol. I. Executive Summary. U.S. Environmental Protection Agency, Office of Toxic Substances, Washington, D.C.

Stanley, C., Barney, J., and Helton, M. 1971. Measurement of atmospheric levels of pesticides. *Environ. Sci. Technol.* 5:430–435.

Staples, C.A., Werner, A.F., and Hoogheem, T.J. 1985. Assessment of priority pollutant concentrations in the United States using STORET database. *Environ. Toxicol. Chem.* 4:131–142.

Steffey, K.L., Mack, J., and Macmanagle, C.W. 1984. A ten-year study of chlorinated hydrocarbon insecticide residues in bovine milk in Illinois, 1972–1981. *J. Environ. Sci. Health.* B19:49–65.

Stevens, R.J.J. and Neilson, M.A. 1989. Inter- and intralake distributions of trace organic contaminants in surface waters of the Great Lakes. *J. Great Lakes Res.* 15:377–393.

Stoltz, R.L. and Pollock, G.A. 1982. Methoxychlor residues in treated irrigation canal water in south central Idaho. *Bull. Environ. Contami. Toxicol.* 28:473–476.

STORET. 1987. Water Quality Control Information System (STORET) (data base). Office of Water and Hazardous Materials, U.S. Environmental Protection Agency, Washington, D.C.

Strachan, W.M. and Huneault, H. 1979. Polychlorinated biphenyls and organochlorine pesticides in Great Lakes precipitation. *J. Great Lakes Res.* 5:61–68.

Strachan, W.M. 1985. Organic substances in the rainfall of Lake Superior: 1983. *Environ. Toxicol. Chem.* 4:677–683.

Strachan, W.M., Huneault, H., and Schertzer, W.M. et al. 1980. Organochlorines in precipitation in the Great Lakes region. In: *Hydrocarbons and Halogenated Hydrocarbons in the Aquatic Environment.* Afghan, B.K. and McKay, D., Eds., Plenum Press, New York, 387–396.

Strachan, W.M. 1988. Short Communication. Toxic contaminants in rainfall in Canada: 1984 *Environ. Toxicol. Chem.* 7:871–877.

Strachan, W.M. 1990. Atmospheric deposition of selected organochlorine compounds in Canada. In: Kurtz, D.A., Ed., *Long Range Transport of Pesticides.* 195th National Meeting of the American Chemical Society held jointly with the Third Chemical Congress of North American, Toronto, Ontario, Canada, June 1988. American Chemical Society, Washington, D.C. 233–240.

Szaro, R.C., Coon, N.C., and Kolbe, E. 1979. Pesticide and PCB of common eider, herring gull and great black-backed gull eggs. *Bull. Environ. Contam. Toxicol.* 22:394–399.

Tanabe, S., Tatsukawa, R., and Kawano, M. 1982. Global distribution and atmospheric transport of chlorinated hydrocarbons: HCH (BHC) isomers and DDT compounds in the Western Pacific, Eastern Indian and Atlantic Oceans. *J. Oceanogr. Soc. Japan.* 38:137–148.

TRI. 1993. Toxic Chemical Release Inventory. National Library of Medicine. National Toxicology Information Program. Bethesda, MD; TRI 1993. Public Data Release. U.S. EPA. Office of Pollution Prevention and Toxics (7408), Washington, D.C.

Trotter, W.J., Corneliussen, P.E., Laski, R.R. et al. 1989. Levels of polychlorinated biphenyls and pesticides in bluefish before and after cooking. *J. Assoc. Off. Anal. Chem.* 72:501–503.

Verbrugge, D.A., Othondt, R.A., and Grzyb, K.R. 1991. Concentrations of inorganic and organic contaminants in sediments of six harbours on the North American Great Lakes. *Chemosphere.* 22:809–820.

Villeneuve, J.P. and Cattini, C. 1986. Input of chlorinated hydrocarbons through dry and wet deposition to the western Mediterranean. *Chemosphere.* 15:115-120.

Wallner, W.E., Leeling, N.C., and Zabik, M.J. 1969. The fate of methoxychlor applied by helicopter for smaller European elm bark beetle control. *J. Econ. Entomol.* 62:1039–1042.

Wan, M.T., Szeto, S., and Price, P. 1994. Organophosphorus insecticide residues in farm ditches of the Lower Fraser Valley of British Columbia. *J. Environ. Sci. Health.* B29:917–949.

Wang, T.C., Hoffmann, M.E., and David, J. 1992. Chlorinated pesticide residue occurrence and distribution in mosquito control impoundments along the Florida Indian River lagoon. *Bull. Environ. Contam. Toxicol.* 49:217–223.

Weaver, L., Gunnerson, C.G., Briedenbach, A.W. et al. 1965. Chlorinated hydrocarbon pesticides in major U.S. river basins. *Public Health Rep.* 80:481–493.

Welch, H.E., Muir, D.C., and Billeck, B.N. 1991. Brown snow: a long range transport event into the Canadian Arctic. *Environ. Sci. Technol.* 25:280–286.

White, D.H., Mitchell, C.A., and Stafford, C.J. 1985. Organochlorine concentrations, whole body weights, and lipid content of black skimmers wintering in Mexico and in south Texas. *Bull. Environ. Contam. Toxicol.* 34:513–517.

White, D.H. and Krynitsky, A.J. 1986. Wildlife in some areas of New Mexico and Texas accumulate elevated DDE residues. *Environ. Contam. Toxicol.* 15:149–157.

White, D.H., Stendell, R.C., and Mulhern, B.M. 1979. Relations of wintering canvasback to environmental pollutants — Chesapeake Bay, Maryland. *Wilson Bull.* 91:279–287.

WHO. 1984. Endosulfan. International Program on Chemical Safety. Environmental Health Criteria 40. World Health Organization, Geneva, Switzerland. 1–62.

WHO, 1979. DDT and its derivatives. Environmental Health Criteria 9, World Health Organization, Geneva, Switzerland.

Wiemeyer, S.N., Lamont, T.G., and Locke, L.N. 1980. Residues of environmental pollutants and necropsy data for Eastern United States ospreys. *Estuaries.* 3:155–167.

Wiemeyer, S.N., Lamont, T.G., and Bunck, C.M. et al. 1984. Organochlorine pesticide, polychlorobiphenyl, and mercury residues in bald eagle eggs — 1969-79 — and their relationship to shell thinning and reproduction. *Bull. Environ. Contam. Toxicol.* 13:529–549.

Wiersma, G.B., Tai, H., and Sand, P.F. 1972. Pesticide residues in soil from eight cities — 1969. *Pestic. Monit. J.* 6:126–129.

Willis, G.H. and McDowell, L.L. 1987. Pesticide persistence on foliage. *Rev. Environ. Contam. Toxicol.* 100:23–73.

Winterlin, W., Seiber, J.N., Craigmill, A. et al. 1989. Degradation of pesticide waste taken from a highly contaminated soil evaporation pit in California, USA. *Arch. Environ. Contam. Toxicol.* 18:734–747.

Wittlinger, R. and Ballschmitter, K. 1991. Studies of the global baseline pollution, XIII, C_6–C_{14} organohalogens (alpha- and gamma- HCH, HCB, PCB, 4,4'-DDT, 4,4 DDE and trans-chlordane, trans-nonachlor, anisoles) in the lower troposphere of Southern Indian Ocean. *Fresenius J. Anal. Chem.* 336:193–200.

Wnuk, M., Kelley, R., Breuer, G., and Johnson, L. 1987. Pesticides in water samples using surface water sources. Iowa Department of Natural Resources and Iowa University Hygienic Laboratory. PB88-136916. Des Moines, IA.

Wolley, D., Zimmer, L., Dodge, D., and Swanson, K. 1985. Effects of Lindane-type insecticides in mammals: unsolved problems. *Neurotoxicology.* 6:165–192.

Yess, N.J. 1991. Food and Drug Administration Pesticide Program — residues in foods. *J. Assoc. Off. Anal. Chem.* 74:1–20.

Yess, N.J., Houston, M.G., and Gunderson, E.L. 1991a. Food and Drug Administration Pesticide Residue Monitoring of Foods: 1978–1982. *J. Off. Anal. Chem.* 74:265–272.

Yess, N.J., Houston, M.G., and Gunderson, E.L. 1991b. Food and Drug Administration Pesticide Residue Monitoring of Foods: 1983–1986. *J. Off. Anal. Chem.* 74:273–280.

Yin, C. and Hassett, J.P. 1989. Fugacity and phase distribution of mirex in Oswego River and Lake Ontario waters. *Chemosphere.* 19:1289–1296.

CHAPTER 7

Chlorinated Phenols

Phenols are a diverse group of organic compounds with a benzene ring substituted with one or more hydroxyl groups. Phenol, hydroxy benzene (C_6H_6OH) was first isolated from coal tar in 1834. Synthetic production of phenol gradually increased and, by 1930, exceeded production from natural sources. Today, almost all phenol is produced by sulfonation of a benzene ring and hydrolysis of the product, sulfonate. Chlorophenols are produced by replacing one or more hydrogen atoms on the benzene ring with chlorine atom(s).

PROPERTIES

Environmental behavior of individual chlorophenolic compounds is likely related to their physical and chemical properties (Table 7.1). In general, melting and boiling points increase with the chlorine content of the phenolic ring; whereas volatilization and water solubility decrease with increasing chlorination of the phenol ring. Because of the fact that chlorinated phenols are weak acids (low dissociation constants), their aqueous solubility markedly increases above their pK_a values, since the phenolate ion is more water soluble than the parent unionized chlorophenol. These properties are also responsible for reduced sorption to sediments and reduced bioaccumulation at pH values above the PK_a values.

PRODUCTION AND USE PATTERN

The total worldwide production of pentachlorophenol was estimated to be 50×10^6 kg/year (Crosby 1981). Pentachlorophenol is produced by the step-by-step chlorination of phenols in the presence of catalysts, such as anhydrous aluminum chloride or ferric chloride. Outside of the U.S., it is also produced by the alkaline hydrolysis of hexachlorobenzene. Typically, commercial grade pentachlorophenol

Table 7.1 Physical and Chemical Properties of Chlorinated Phenols

Compound	Mol. Wt.[1]	Melting Point[1] °C	Boiling Point °C (kPa)	Vapor Pressure[2] Kpa (°C)
Monochlorophenols				
2-Chlorophenol	128.56	9.0	174.9 (101.3)	0.13 (12.1)
3-Chlorophenol	128.56	33	214 (101.3)	0.13 (44.2)
4-Chlorophenol	128.56	43	219.7 (101.3)	0.13 (49.8)
Dichlorophenols				
2,3-Dichlorophenol	163.00	57–59	206[4] (101.3)	
2,4-Dichlorophenol	163.00	45	210 (101.3)	0.13 (53.0)
2,5-Dichlorophenol	163.00	59	211 (99)	
2,6-Dichlorophenol	163.00	68–69	219–220 (99)	
3,4-Dichlorophenol	163.00	68	253.5 (102)	
3,5-Dichlorophenol	163.00	68	233 (101)	
Trichlorophenols				
2,3,4-Trichlorophenol	197.45	83.5	sublimes	
2,3,5-Trichlorophenol	197.45	62	248–249 (33)	
2,3,6-Trichlorophenol	197.45	58	272 (101.3)[4]	
2,4,5-Trichlorophenol	197.45	68-70.5	sublimes (37)[4]	0.13 (72)
2,4,6-Trichlorophenol	197.45	65.9	246 (101.3)	0.13 (76.5)
3,4,5-Trichlorophenol	197.45	101	271–277 (99)	
Tetrachlorophenols				
2,3,4,5-Tetrachlorophenol	231.98	116–117	sublimes	
2,3,4,6-Tetrachlorophenol	231.98	70	150 (2)	0.13 (100.0)
2,3,5,6-Tetrachlorophenol	231.98	115		
Pentachlorophenol	266.34	191	309–310 (101)	0.0024 (30)

Compound	Aq. Solubility mol/L, pH 5.1 25°C	Dissociation Constant[4] 25°C	Pk_a[3,5,6]	Octanol/Water Partition Coefficient Log K_{ow}
Monochlorophenols				
2-Chlorophenol	2.1	3.2×10^{-9}	8.48; 8.65	2.19[7]
3-Chlorophenol	2.6	1.4×10^{-9}	9.08; 9.12	2.50[7]
4-Chlorophenol	2.1	6.6×10^{-10}	9.42; 9.37	2.44[7]
Dichlorophenols				
2,3-Dichlorophenol	—	3.6×10^{-7}	7.70	
2,4-Dichlorophenol	0.038	2.1×10^{-8}	7.85	2.75
2,5-Dichlorophenol	slight[1]	4.5×10^{-7}	7.51	
2,6-Dichlorophenol		1.6×10^{-7}	6.79; 6.91	2.84
3,4-Dichlorophenol	slight[1]	4.1×10^{-8}	8.59	
3,5-Dichlorophenol		1.2×10^{-7}	8.59	
Trichlorophenols				
2,3,4-Trichlorophenol		2.2×10^{-8}		
2,3,5-Trichlorophenol		4.3×10^{-8}		
2,3,6-Trichlorophenol		7.4×10^{-8}	5.98	
2,4,5-Trichlorophenol	0.0048	3.7×10^{-8}	7.0; 7.07	3.72[9]
2,4,6-Trichlorophenol	0.0022	3.8×10^{-8}	6.1; 6.62	3.38[10]
3,4,5-Trichlorophenol		1.8×10^{-8}	7.83	

Table 7.1 Physical and Chemical Properties of Chlorinated Phenols (continued)

Compound	Aq. Solubility mol/L, pH 5.1 25°C	Dissociation Constant[4] 25°C	Pk$_a$[3,5,6]	Octanol/Water Partition Coefficient Log K$_{ow}$
Tetrachlorophenols				
2,3,4,5-Tetrachlorophenol		1.1×10^{-7}		
2,3,4,6-Tetrachlorophenol	0.00079	4.2×10^{-6}		4.10[8]
2,3,5,6-Tetrachlorophenol		3.3×10^{-6}	5.3	
Pentachlorophenol	0.0024	5.6×10^{-5}	4.8; 5.00	5.01[10]

1. Weast 1974; 2. Sax 1975; 3. Blackman et al. 1955; 4. Doedens 1967; 5. Pearce and Simpkins 1968; 6. Farquaharson et al. 1958; 7. Neeley et al. 1974; 8. Hansch and Leo 1979; 9. Mackay 1982 and 10. Leo et al. 1971.

is 86% pure. It generally contains contaminants such as polychlorinated phenols, PCDDs (polychlorinated dibenzo-*p*-dioxins), and PCDFs (polychlorinated dibenzofurans). Pentachlorophenol has been available in the market as (i) a water soluble Na salt; (ii) a 5% emulsifiable concentrate; and (iii) a 3 to 40% solution in formulation with other chlorophenols, methylene bisthiocyanate, or copper naphthenate (IARC 1979; ATSDR 1994). Production volumes, in 10^6 kg of pentachlorophenol, for 1983 to 1986 were as follows: 20.5 (1983), 19.1 (1984), 17.3 (1985), and 14.5 (1986). Recent production data are not available.

2,4,6-Trichlorophenol is not known to occur naturally (IARC 1979; Scow et al. 1982). It was first synthesized by Laurent in 1836 by chlorination of phenol in the presence of aluminum chloride (Prager et al. 1923), a method that is currently used in the U.S. to produce certain chlorophenols (Herrick et al. 1979; Scow et al. 1982). The products, 2,4,6-trichlorophenol and 2,3,4,6-tetrachlorophenol, are separated by vacuum distillation (Herrick et al. 1979; IARC 1986) or by crystallization (Scow et al. 1982). Yields of this reaction have been as high as 97.5% (Freiter 1979), whereas, 2,4,6-trichlorophenol has been isolated only in small unspecified quantities. In Japan, 2,4,6-trichlorophenol is co-produced with ortho- and para-chlorophenols via chlorination of phenol and has a purity of 97% (IARC 1979).

The production of chlorophenols is known to produce PCDDs as contaminants and these levels are determined by the temperature and the pressure of the reaction (EPA 1980a) and the type of solvents and catalysts used (IARC 1986a,b). 1,3,6,8-T$_4$CDD and 1,3,7,9-T$_4$CDD have been detected upto 10.5 mg, but no 2,3,7,8-T$_4$CDD in a kg of 2,4,6-trichlorophenol was produced as an intermediate in chloranil production. Similarly, 93 ppm and 49 ppm of 2,3,7- and 1,3,6,8-T$_4$CDD, respectively, were found in several samples of 2,4,6-trichlorophenol. Unspecified amounts of tetra-, penta-, and hexa-CDFs were found in one sample (Firestone et al. 1972). It was reported that PCDFs are the major trace contaminants in 2,4,6-trichlorophenol, with hexa-CDFs at highest concentrations and tetra-CDFs in the greatest variety (Rappe et al. 1978a).

Commercial production of 2,4,6-trichlorophenol in the U.S. was first reported in 1950 (U.S. Tariff Commission 1951). The last year of reported production was 1974 (USITC 1975). Production was supposed to have been discontinued after 1974 due to the prohibitive cost of purifying the product from dioxin impurities (NCI 1979). However, it was estimated that up to 16,000 kg were produced in 1977 (Scow et al. 1982). It was also reported in the 1979 TSCA inventory that two U.S. manufacturers of 2,4,6-trichlorophenol had an annual production of 25,000 kg. There were no manufacturers by 1986 and no U.S. manufacturers currently listed for 2,4,6-trichlorophenol. (IARC 1986a,b; CIS 1988). It is produced by three manufacturers in Great Britain, two in Japan, and one each in Germany, Australia, and India. It is also reported to have been produced in Denmark and Sweden (CIS 1988; The Chemical Daily Co. 1984). 2,4,6-trichlorophenol has been produced since 1965, with one company manufacturing 120,000 kg in 1977, none of which was exported. Production volumes for other countries were not available.

Chlorinated phenols are/were used as disinfectants, biocides, preservatives, pesticides, and industrial and medicinal organic chemicals (CCREM 1987; Health Welfare Canada 1980; NRCC 1982). Monochlorophenols are used as antiseptics and as intermediates in the synthesis of higher chlorinated phenols and chlorocresols for biocide production. Ortho-chlorophenol is also used in the production of phenolic resins and also in the extraction of sulphur and nitrogen compounds from coal (U.S. EPA 1980b). Of the ten isomers of dichlorophenol, only 2,4-dichlorophenol has been used as a primary chemical (OME 1984). It was used in the manufacture of some phenoxy herbicides. 2,4-Dichlorophenol was also used as a chemical intermediate in the production of germicides, temporary soil sterilants, plant growth regulators, moth proofing agents, seed disinfectants, miticides, and wood preservatives (U.S. EPA 1980a).

Use of chlorophenols in wood preservatives and stains, for interior home use, was suspended in 1980. Other bans of chlorophenols and their sodium salts included use as herbicides and soil sterilants, as fungicides for mushrooms, as microbiocides in curing hides, and as silmicides in pulp and paper operations (Jones 1984). Some chlorinated phenols detected in the environment are due to their past use as phenoxy herbicides.

Of the six isomers of trichlorophenol, only 2,4,5- and 2,4,6-tri-chlorophenols had commercial uses. 2,4,5-Trichlorophenol was used to manufacture the insecticide *Ronnel*, which was used on livestock. It was also used in the production of hexachlorophene, a compound used in disinfectants and sanitation products for domestic, hospital, and veterinary use. The registration of all trichlorophenols and their salts for pesticidal or antimicrobial uses was discontinued in 1985 (Agriculture Canada 1987). In the U.S. and Canada, only tetrachlorophenol and pentachlorophenol remain in extensive use. Of the three tetrachlorophenol isomers, only 2,3,4,6-tetrachlorophenol is used commercially, usually with pentachlorophenol in the production of wood preservatives. Commercial preparations contain 3 to 10% of tetrachlorophenol and are used in agriculture to prevent wood decay. Pentachlorophenol is the most widely used commercial chlorophenol. It is available as pentachlorophenol or as the sodium salt.

Pentachlorophenol is used primarily (80%) in wood preservation, specifically to treat wood for utility poles (CMR 1987). It was also registered for use by the U.S. EPA as a termicide, fungicide, herbicide, molluscide, algicide, disinfectant, and as an ingredient in antifowling paint (Cirelli 1978). But pentachlorophenol is now a restricted-use pesticide (EPA 1984). It is used in the formulation of fungicidal and insecticidal solutions and for incorporation into other pesticide products. In recent years, nonwood uses account for only 2% of the U.S. pentachlorophenol consumption (Mannsville 1987). Pentachlorophenol is no longer used in wood-preserving solutions or insecticides and herbicides available for home and garden use. The solubility of pentachlorophenol in organic solvents and the sodium salt in water account for their wide spectrum use.

In 1980, Agriculture Canada suspended many of the domestic and commercial uses of wood-preservatives containing pentachlorophenol, including wood-preservatives on food containers and on lumber used for seed flats, stakes, and green houses, and wood preservatives on aboveground interior woodwork of farm buildings (Jones 1981). In the past, pentachlorophenol was used in a variety of applications for its strong biocidal activity. Many of these applications are currently prohibited or obsolete. Presently, pentachlorophenol is mainly used as fungicide in the wood industry and as slimicide in the paper industry (Seiler 1991).

SOURCES IN THE ENVIRONMENT

Natural Sources

Mono-, di-, and tri-chlorophenols may be formed in the aquatic environment through interaction of aqueous chlorine with certain organic molecules, such as natural phenols, humic, and fulvic acids, during the chlorination of drinking water and wastewaters (RTP 1994). Chlorination of domestic sewages has been reported to produce several chlorophenols; 40% of the product was 2,4,6-trichlorophenol (Jolley et al. 1978; Burtshell et al. 1959). A literature survey shows that several chlorophenols are naturally produced by a variety of organisms. A soil penicillium produced 2,4-dichlorophenol and complex phenolic compounds were produced by *Helichrysum* sp. (Siuda and DeBernardis 1973; Engvild 1986). Chloroperoxidase, a fungal enzyme, catalyzes the chlorination of various phenolic substrates to yield mono- to penta-chlorophenols (Wannstedt et al. 1990). This study indicated that enzymatic chlorination of naturally occurring organic matter could likely introduce various chlorophenols to the environment, independent of anthropogenic sources. Further, it was shown that chloroperoxidase-mediated chlorination increases the adsorbable organic halogen (AOX) concentration. By this process, phenol was shown to yield 2- and 4-chlorophenols as reaction products. Burning of untreated wood in an open fire was shown to produce chlorinated phenols (Ahling and Lindskog 1982). Burning of wood was shown to produce tetra- and penta-chlorophenols ranging in concentration from 29 to 70 and 19 to 70 $\mu g/m^3$, respectively; burning of fresh wood was reported to yield 270 and 180 $\mu g/m^3$ of tetra- and penta-chlorophenols, respectively.

Detection of pentachlorophenol in lake sediments, dated to be several hundred years old, adds evidence to its origin from wood or forest fires (Salkinoja-Salonen et al. 1984; Paasivirta et al. 1988).

Anthropogenic Sources

Chlorinated phenols are released into the aquatic environment from sources such as wood-processing plants, wood- and wood-preservation plants, kraft pulp mills, sewage treatment plants, pesticide manufacturing, formulating plants, and leaching of agricultural products (NRCC 1982). The most obvious routes of entry for chlorophenols into the environment are the wood-preserving plant sites, which are usually located close to natural waters.

Primary Routes

1. Wood-Preserving Plants and their Treatment Systems

Pentachlorophenol and tetrachlorophenol are primarily used to protect wood from fungal attack which causes wood decay and discoloration (Agriculture Canada 1987). Hence, wood-treatment operations/plants, saw mills, and planar mills with wood-treatment facilities are the major sources of release of chlorophenols into the environment, through their effluents and emissions. Rainfalls are reported to leach chlorophenols from recently treated wood into the environment.

2. Wood-Protection Facilities

Wood-protection (surface treatment of wood for sap stain protection and mold control) is carried out with water-soluble salt formulations of chlorophenols at saw mills and lumber export terminals. Application of chemicals is made by either dipping or spraying, with dipping having been the most common method in 1980. Dipping systems have been identified by Environment Canada as a serious environmental hazard (EPS 1981).

3. Other Routes

In-Service Treatment with Preservatives — When chlorophenol wood preservatives are applied on site to wood products already in service, there is always a potential for environmental contamination. When a hydro pole was treated in place with a PCP preservative, a nearby stream received a shock load of PCP which resulted in fish kill for a distance of 800 km downstream (EPS 1981).

Petrochemical Drilling Fluids — Disposal from Sumps — Petrochemical fluids include bactericides to prevent fermentation of polysaccharides, starch, and XC polymers (Land 1974; Falk and Lawrence 1973). When sodium pentachlorophenate is/was used for this purpose, it is/was maintained at a concentration of 700 to 1400 ppm in the drilling fluid. The used drilling fluids and associated wastes are contained in large excavations or sumps. Sumps vary in size according to effluent storage needs. The sumps are often subject to flooding and washing-out with concomitant release of toxic materials to nearby surface waters.

Incineration — From a study of release of organic pollutants from waste incinerators, it was reported that the most abundant chlorine-containing compounds in condensate from flue gas were the di-, tri-, and tetra-chlorophenols (Olie et al. 1977). Although it was unable to be quantified, the authors concluded that "municipal incinerators and other combustion processes may be a source of some of the organochlorine compounds."

The use of trichlorophenols and their salts for pesticidal and/or antimicrobial purposes was discontinued in 1985 (Agriculture Canada 1987). In the past, chlorinated phenols and their derivatives had been used as wood-preservatives, insecticides, fungicides, herbicides, and general biocides (Crosby 1981; Chaudhry and Chapalmadugu 1991). Most of the compounds and their uses have been banned. 2,4-Dichlorophenol and 2,4,5-trichlorophenol were used extensively in producing 2,4-D and 2,4,5-T, respectively (Jones 1981, 1984). Only tetrachlorophenol and pentachlorophenol remain in extensive use in North America. Environmental residues of chlorinated phenols detected are due to the past uses of phenoxy herbicides. Pentachlorophenol was also used widely used as preservative for cellulosic products, textiles, and leather goods (Kaufmann 1978); many applications/uses of pentachlorophenol are prohibited. Currently, pentachlorophenol is used as a fungicide and as a slimicide in the wood and paper industries, respectively (Seiler 1991). Chlorinated phenols have been reported to form in the environment, during chlorination of drinking water, wastewater (Health and Welfare Canada 1988) and chlorination of domestic sewage (Jolley et al. 1975). Pentachlorophenol maybe formed from hexachlorobenzene and lindane via biotransformation (Pignatello et al. 1983). Pentachlorophenol may degrade microbially to yield tetra- and trichlorophenols (Boyle et al. 1980). Chlorine bleaching of pulp-in-pulp and paper mills has been reported to form a variety of chlorophenols, from di- to penta-chlorophenols, di- to tetra-chloroguaiacols, and tri- and tetra-chlorocatechols. These compounds have been detected in samples of chlorination-stage and extraction-stage process streams of bleach kraft pulp mills (Paasivirta et al. 1992).

ENVIRONMENTAL RESIDUES AND EXPOSURE ROUTES

In the 1970s, chlorinated phenols were widely distributed in surface waters and occurred at a highly variable concentration, depending on the source point. Monochlorophenol concentrations in rivers and coastal waters of The Netherlands ranged from 3 to 20 µg/L, compared to 0.01 to 1.5 and 0.003 to 0.1 µg/L for dichlorophenols and trichlorophenols, respectively (Piet and DeGrunt 1975). Jones (1981), after having reviewed much of the Canadian monitoring data, reported that the discharge of pulp and paper mill effluents resulted in average concentrations of dichloro- and tri-chlorophenols of 4 to 13 µg/L, respectively, in the coastal waters of Lake Superior. Water samples collected from stream mouths and inshore areas of Lake Ontario contained pentachlorophenol ranging from <0.005 to 1.4 µg/L, with transient highs up to 23 µg/L being recorded after periods of heavy rainfall. Only eight out of 85 samples had no detectable pentachlorophenol concentration.

Snowpack samples collected during the winter of 1977 to 1978, from 19 locations in Ontario, were analyzed to contain <0.001 µg/L to 0.003 µg/L in 8 samples (EPS 1981). In British Columbia in 1972, fish were killed in the Little Campbell River after a misapplication of pentachlorophenol in oil to a hydro pole; water samples taken two days after the incident contained 53.7 ppm of pentachlorophenol. The level dropped to 80 ppb of pentachlorophenol at the same site (EPS 1981). Chlorinated phenols have been detected in municipal, industrial, pulp, and paper mills and wood-preservation plants wastewaters. Table 7.2 lists the concentration ranges of several chlorophenols in municipal and industrial wastewaters in the 1970s. Chemical manufacturing plants seem to contain relatively high levels of chlorophenols in their effluents.

Table 7.2 Concentration Ranges (µg/L) of Various Chlorophenols in Municipal and Industrial Effluents in North America

Chlorophenol	Concentration Range	Industry Type
2,4-Dichlorophenol	51–330	Chemical plant
	<0.1	Wood preservation plant
	<0.1	Landfill leachate
2,6-Dichlorophenol	220	Chemical plant
	2.4	Wood preservation plant
	1.2–5.6	Landfill leachate
2,4,5-Trichlorophenol	<0.05	Treated sewage, four plants
	0.5–2,400	Two chemical plants
	<0.05	Wood preservation plant
	0.05–2	Landfill leachate
2,4,6-Trichlorophenol	<0.05–1	Treated sewage, four plants
	<0.05–3,120	Three chemical plants
	0.4–1.0	Landfill leachate
	25–115	Pulpmill effluent
	0.5–1	Wood preservation plant
2,3,4,6-Tetrachlorophenol	0.6–28	Treated sewage, four plants
	1.2–8, 270	Four wood preservation plants
	0.3–166	Three chemical plants
	0.2–0.8	Landfill leachate
Pentachlorophenol	0.5–4.7	Treated sewage, 4 plants
	0.25–1.3	Treated sewage, 6 cities
	0.05–2,760	Wood preservation plants
	5,400,000	Chemical plant
	0.6–42	Landfill leachate

Sources: Jones 1981; Buikema et al. 1979; Waggot and Wheatland 1978.

Chlorophenols are detected in drinking water supplies, but at low concentrations (ppt = ng/L, range). Dichlorophenols are commonly detected at relatively high concentrations (Paasivirta et al. 1985) (Table 7.3).

Table 7.3 Concentration Ranges of Chlorophenols Detected in Drinking Waters

Chlorophenol	Year	Water Type	Concentration	Units	Reference
2,4-Dichlorophenol	1983	Tap water, Finland	53–93	ng/L	Paasivirta et al. 1985
	1986	Drinking water, Canada	<0.02–0.072	µg/L	Sithole and Williams 1986
2,6-Dichlorophenol	1983	Tap water, Finland	62–272	ng/L	Paasivirta et al. 1985
2,4,5-Trichlorophenol	1983	Tap water, Finland	35–59	ng/L	Paasivirta et al. 1985
	1987	WTPs in Toronto area	<50.0	ng/L	Ministry of Environment (MOE) 1988
2,4,6-Trichlorophenol	1983	Tapwater, Finland	14–30	ng/L	Paasivirta et al. 1985
	1986	Drinking water, Canada	<0.008–0.719	µg/L	Sithole and Williams 1986
	1987	WTPs, Toronto area	<50.0	ng/L	Ministry of Environment (MOE) 1988
	1988	Drinking water, U.S.	<0.3–0.4	µg/L	Krasner et al. 1989
	1988	Drinking water survey, 25 cities all seasons, U.S., median value	0.4	µg/L	McGuire et al. 1989
2,3,4,6-Tetrachlorophenol	1983	Tapwater, Finland	9–16	ng/L	Paasivirta et al. 1985
Pentachlorophenol	1983	Tapwater, Finland	5–23	ng/L	Paasivirta et al. 1985
	1986	Drinking water, Canada	<0.004–0.034	µg/L	Sithole and Williams 1986
	1987	WTPs, Toronto area	<50.0	ng/L	Ministry of Environment (MOE) 1988

WTPs = water treatment plants.

Chlorophenols have been detected in pulpmill waste streams by many studies. Spent liquors have been reported to contain maximum concentrations of chlorophenols, chloroguaiacols and chloro-catechols; it has been estimated to be about <10g/ton of pulp produced. Concentration ranges for chlorophenols, chlorocatechols and chloroguaiacols in pulp and paper mill effluents were reported as ND to 66, 11 to 87 and 0.5 to 27 µg/L respectively (Paasivirta et al. 1985; MOE 1986; Alberta Environment 1989). Based on the actual concentrations, effluent flow rates and mixing rate in receiving waters, concentrations of several chlorophenolic compounds were estimated (Table 7.4 and 7.5). Secondary treated effluent concentrations, where available, were used to estimate the concentrations in the receiving waters.

Table 7.4 Concentrations of Chlorophenolic Compounds in Pulp and Paper Mill Effluents and their Estimated Concentrations in Receiving Waters[a]

Chlorophenolic Compound	Effluent (g/tonne)	Receiving Water (µg/L)	References
Dichlorophenol[b]	5.1	7.85	NCASI 1981
Trichlorophenol[b]	0.106	0.16	Lindstrom and Mohamed 1983
	6.1	9.38	Voss et al. 1980; Kovacs et al. 1984
Pentachlorophenol[b]	0.078	0.12	Lee et al. 1989
Dichlorocatechols[b]	0.065	0.1	Lee et al. 1989
Trichlorocatechols	0.83	1.3	Lee et al. 1989
	28	43.08	Voss and Yunker 1983; Kovacs et al. 1984
Tetrachlorocatechols[b]	0.22	0.34	Lindstrom and Mohamed 1988
	24	36.92	Voss and Yunker 1983; Kovacs et al. 1984
Dichloroguaiacols[c]	2.52	2.33	Lee et al.1989
	16.6	15.32	Axegard 1986
Trichloroguiaacols[b]	0.08	0.12	Lindstrom and Mohamed 1988
	1.72	2.65	Lindstrom and Mohamed 1988
Tetrachloroguaiacols[b]	0.72	1.1	Lee et al. 1989
	22	33.85	Voss and Yunker 1983; Kovacs et al. 1984

[a] Estimated using one or more of the following assumptions.
[b] Final effluent.
[c] Untreated final effluent.

(1) Average mill effluents of 130 m³/tonne (1 m³ = 1000 L); (2) Three-fold dilutions of effluents at C and E stage effluents; (3) Forty % degradation in secondary treatment, and (4) Five-fold dilution of final effluent.

Chlorinated phenolic compounds have been detected in surface waters, particularly downstream of pulp and paper mills (Table 7.6). Compounds include chlorophenols, chlorocatechols, and chloroguaiacols. The current levels of chlorinated phenolic compounds in surface waters are greatly reduced compared to the late 1970s and 1980s, due to improved effluent treatment systems.

Air

In ground level air samples collected in Portland, OR, during seven rainstorms, average trichlorophenol (combined 2,4,6-trichlorophenol and 2,4,5-trichlorophenol) concentration was 1.2 ppt (range, ND to 2.5 ppt). In rain samples, the average

Table 7.5 Levels of Chlorinated Phenolic Compounds in Pulp Mill Effluent Streams

Chlorophenol	Year	Stream/Location	Concentration Range	Unit	References
2,4-Dichlorophenol	1979	Spent bleach liquor, Finland	ND–52	g/ton	Salkinoja-Salonen et al. 1981
	1983	Pulp mill wastewater, Finland	2–11	µg/L	Paasivirta et al. 1985
2,4,6-Trichlorophenol	1979	Spent bleach liquor, Finland	1–3	g/ton	Salkinoja-Salonen et al. 1981
	1982	Pulpmill effluent, Lake Superior, Canada	<0.05–16.2	µg/L	MOE 1986
	1983	Pulpmill effluent, Nipigon Bay, Lake Superior	0.2–6.6	µg/L	Kirby 1986
	1983	Pulpmill effluent, Finland	15–28	µg/L	Paasivirta et al. 1985
Pentachlorophenol	1979	Spent bleach liquor, Finland	0.02–0.1	g/ton	Salkinoja-Salonen et al. 1981
	1982	Pulpmill effluent, Lake Superior, Canada	<0.05–2.25	µg/L	MOE 1986
	1983	Pulpmill effluent, Finland	1.0	µg/L	Paasivirta et al. 1985
	1984	Pulpmill effluent, Ontario, Canada	0.244 (ND–2.2)	µg/L	MOE 1986

2,4,6-trichlorophenol concentration was 1.4 ppt (range, ND to 1.9 ppt), and was found exclusively in the dissolved, rather than particulate, form in the rain. The study concluded that 2,4,6-trichlorophenol is primarily present as a gas in the atmosphere and that rain is a major route for its removal (Leuenberger et al. 1985; ATSDR 1990).

Water

In the late 1970s and early 1980s, 241 New Jersey water samples were analyzed and trichlorophenol was detected in 7.5% of the samples, in 9.6% of the finished water samples, and 4.3% of the delivered water samples. Concentrations were between 0 and 1 ppb with the highest value being 7.3 ppb (NJDEP 1989). In a survey of raw (untreated) water from 40 Canadian drinking water treatment plants, 2,4,6-trichlorophenol was detected in three plants; 1.1 ppt (mean) and 23 ppt (maximum) (Sithole and Williams 1986). Treated water from these 40 plants was also sampled on three occasions; 2,4,6-trichlorophenol was found at 8 to 11 plants at values of 61 to 719 ppt (maximum) and 8.8 to 40 ppt (mean), respectively.

Based on limited surface water sampling data from the EPA STORET database, it seems that higher levels were detected in surface waters in the Pacific Northwest (Scow et al. 1982). Twenty-nine percent of the observations were in the 1.1 to 10 ppb range, while 71% were in the 10.1 to 100 ppb range. Nationally in the U.S., 6% were at 0.1 to 1 ppb range, 57% at 1.1 to 10 ppb range, and 37% were at 10.1 to 100 ppb.

Table 7.6 Concentration Ranges of Chlorinated Phenolic Compounds in Surface Waters Downstream of Pulp and Paper Mills

Location	Compound	Concentration (μg/L) 1977	1978	1980	References
Finland, pulpmill receiving waters	2,4-Dichlorophenol	2–1700	250–1,030	89	Salkinoja-Salonen et al. 1981
	2,4,6-Trichlorophenol	0.1–3.6	0.2–1.6	0.01–13.1	Salkinoja-Salonen et al. 1981
	2,3,4,6-Tetrachlorophenol	0.1–0.3	0.2–0.7	0.01–1.2	Salkinoja-Salonen et al. 1981
	Pentachlorophenol	0.03–0.8	0.2–2.2	0.05–2.2	Salkinoja-Salonen et al. 1981
	Chlorocatechols (3,4-di-, 3,4,5-tri-and tetra-)	0–197	0.1–46	0–172	Salkinoja-Salonen et al. 1981
	Trichloroguaiacol	0.1–1.3	0.1–1.3	0.1–3.1	Salkinoja-Salonen et al. 1981
Finland, pulpmill downstream waters	2,4-Dichlorophenol (S)			10	Salkinoja-Salonen et al. 1984
	2,4-Dichlorophenol (B)			2	Salkinoja-Salonen et al. 1984
	2,6-Dichlorophenol (S)			1.2	Salkinoja-Salonen et al. 1984
Surface waters (S)	2,6-Dichlorophenol (B)			1.6	Salkinoja-Salonen et al. 1984
Bottom waters (B)	2,4,6-Trichlorophenol (S)			1.2	Salkinoja-Salonen et al. 1984
	2,4,6-Trichlorophenol (B)			1.6	Salkinoja-Salonen et al. 1984
	2,3,4,6-Tetrachlorophenol (S)			0.1	Salkinoja-Salonen et al. 1984
	2,3,4,6-Tetrachlorophenol (B)			0.2	Salkinoja-Salonen et al. 1984
	Pentachlorophenol (S)			0.03	Salkinoja-Salonen et al. 1984
	Pentachlorophenol (B)			2.2	Salkinoja-Salonen et al. 1984
Surface of receiving lake, Finland	Chlorinated catechols			0.60	Salkinoja-Salonen et al. 1984
	Trichloroguaiacol (S)			0.4	Salkinoja-Salonen et al. 1984
	Trichloroguaiacol (B)			0.8	Salkinoja-Salonen et al. 1984
Finland, within 40 km of pulp mills	2,4-Dichlorophenol			0–0.014	Paasivirta et al. 1985
	2,6-Dichlorophenol			0–0.073	Paasivirta et al. 1985
	2,4,5-Trichlorophenol			0–0.019	Paasivirta et al. 1985
	2,3,4,6-Tetrachlorophenol			0–0.090	Paasivirta et al. 1985
	Pentachlorophenol			0–0.110	Paasivirta et al. 1985
	3,4-Dichlorocatechol			0–1.02	Paasivirta et al. 1985
	3,4,5-Trichlorocatechol			0–0.690	Paasivirta et al. 1985
	Tetrachlorocatechol			0–0.410	Paasivirta et al. 1985
	4,5-Dichloroguaiacol			0–0.027	Paasivirta et al. 1985
	3,4,5-Trichloroguaiacol			0.013–0.280	Paasivirta et al. 1985
	4,5,6-Trichloroguaiacol			0–0.280	Paasivirta et al. 1985
	Tetrachloroguaiacol			0–0.090	Paasivirta et al. 1985

S = surface water; B = bottom water

Other Media

Effluents from industrial processes and wastewater treatment plants were analyzed for chlorophenols. The average 2,4,6-trichlorophenol concentration was 3,000 ppb in timber-barking effluents, 2,400 ppb in paint and ink effluents, and 526 ppb in pesticide manufacturing effluents (Scow et al. 1982). Analysis of 1980 to 1983 effluent sampling data from the STORET database revealed that trichlorophenol (isomers not specified) was found at >10 ppb, the detection limit in 39 of 1,297 samples (Staples et al. 1985). 2,4,6-Trichlorophenol was detected at 0.033 ppb in intake waters and at 0.59 ppb in effluent at a Canadian chemical manufacturing plant on the St. Clair River, the border between Michigan and Ontario (King and Sherbin 1986). It was also detected at 17 ppb in leachate of a nearby landfill for solid wastes, such as chlorinated tars. Activated carbon treatment reduced the concentration in the drain discharging to the St. Clair to 0.17 ppb. The chlorination of water containing trace levels of phenol produced 2- and 4-chlorophenol, with small amounts of 2,4-dichlorophenol and negligible amounts of 2,4,6-trichlorophenol (Joshipura and Keliher 1980). 2,4,6-Trichlorophenol was detected at 2.3 ppb in a one-week composite sampling of municipal wastewater (Gossett et al. 1983).

2,4,6-Trichlorophenol was detected in sewage sludge at 2 of 44 publicly owned treatment works (POTWs) surveyed by the U.S. EPA and in sludges at 67 of 241 POTWs surveyed by various studies in several U.S. cities (Fricke et al. 1985). The EPA survey detected 2,4,6-trichlorophenol at a mean concentration of 2.3 mg/kg (dry weight), while other surveys reported a weighted mean dry weight concentration of 42.3 mg/kg. In a survey of 250 Michigan sewage treatment plant sludges, 2,4,6-trichlorophenol concentrations ranged from 0.19 to 1,300 mg/kg dry weight (mean not reported) (Phillips et al. 1983).

2,4,6-Trichlorophenol in fish was reported at 4.5 mg/kg based on the estimated BCF and water column concentration (Scow et al. 1982). It was not detected in 22 composite samples of fish collected from harbors and tributaries of the Great Lakes (DeVault 1985). Combined values for 2,4,5- and 2,4,6-trichlorophenols in fish livers, collected from the Pacific Ocean 6 km northwest of the discharge zone for the Los Angeles County wastewater treatment plants, were reported at 29 to 629 ppb (wet weight) (Gossett et al. 1983). Wet weight concentrations of 10 ppb for combined trichlorophenols were reported in shrimp (ridgeback prawn, *Sicyonia ingentis*) muscle and 55 ppb in the red pointer crab (*Mursia gaudichaudii*) digestive gland. 2,4,6-Trichlorophenol have been reported in biota of a Canadian stream that received leachate from a chemical waste dump; rock bass (*Ambloplites rupestris*), 7 ppb, and crayfish (*Orconectes propinquus*), 10 ppb (Metcalfe et al. 1984). 2,4,6-Trichlorophenol in fish taken from Finnish lakes near pulp bleaching industries were reported as; in a very polluted lake (sediment concentration, 27.7 ppm, dry weight), 13.6 ppb (average) for pike (*Esox lucius*) and 55.9 ppb (average) for roach (*Leuciscus rutilus*). In the least polluted lake (sediment concentration, 4.68 ppm), 2,4,6-trichlorophenol concentrations were ND (roach) and 0.79 ppb (pike) (Paasivirta et al. 1985).

PENTACHLOROPHENOL

Levels Monitored or Estimated in the Environment

Pentachlorophenol has been detected widely in the environment due to its widespread use by industry, the agricultural sector, and the general public, as a cooling-water algicide and fungicide, herbicide, molluscicide, paint preservative, plywood and fiberboard waterproofing agent, and mud drilling and photographic solution biocide. Pentachlorophenol is now regulated as a restricted-use pesticide and it is no longer available to the general public. Studies previous to the 1990s have detected pentachlorophenol in indoor air, surface waters, ground water, drinking water, soils, rainwater, and a variety of foods, possibly from the past use pattern. However, current contamination of these media by pentachlorophenol is expected to be very limited given the current restricted usage and its limited environmental persistence.

Air

Limited information is available on the levels of pentachlorophenol in ambient air. Estimated atmospheric concentrations of pentachlorophenol, using air models, have been reported (EPA 1980d). A cumulative estimate based on all emission sources was 0.15 to 136 ng/m^3. The lower end of this estimate (0.15 ng/m^3) coincides with the upper end of the range of computed air concentration estimates for Hawaii (0.002 to 0.063 ng/m^3). These estimates were based on pentachlorophenol concentrations in rainwater, where it has been used extensively as an herbicide and wood preservative.

Water

The U.S. EPA STORET database as of March 1979, containing 497 surface water data, revealed that 82% were at the detection limit; 84% of the remaining observations fell between 0.1 and 10 µg/L, with a total range of 0.01 to 100 µg/L (EPA 1979). These data indicate that ambient levels of pentachlorophenol in surface waters are usually <1 µg/L, with much higher levels recorded in more industrialized areas. Pentachlorophenol levels in surface water downstream of source points are: 0.1 to 0.7 µg/L in the Willamette River (Buhler et al. 1973); 9 µg/L in a river below a paper mill (Rudling 1970); 0.1 to 1 µg/L in the Great Lakes (EPA 1980d); <1 µg/L in a river at a sewage discharge site in Sacramento, CA (Wong and Crosby 1978); 0.038 to 10.5 mg/L in a stream running through an industrial district in Pennsylvania (Fountaine et al. 1975); and 0.01 to 0.48 µg/L in streams of Hawaii (Young et al. 1976).

Pentachlorophenol in drinking waters has been reported at 0.04 to 0.28 µg/L in Corvallis, OR, (Buhler et al. 1973); a mean concentration of 0.07 µg/L in 108 samples surveyed by the National Organics Monitoring Survey (NOMS); and <1 to 800 ppb (average, 227 ppb) in seven drinking water wells in Oroville, CA (Wong and Crosby 1978).

Pentachlorophenol was detected in raw effluent from a series of wood-treatment plants at levels ranging from 25 to 150 mg/L (Dust and Thompson 1972) and in influent (1 to 5 ppb) and effluent (1 to 4 ppb) at streams at a sewage plant in Corvallis, OR (Buhler et al. 1973). It was also detected in surface water and groundwater samples collected from a wood-treatment facility in Arkansas (ATSDR 1994).

Data are sparse on actual measurements of pentachlorophenol in soil. Pentachlorophenol concentrations of 3.4 to 654 ppm were reported in soil within 12 inches of treated utility poles (Arsenault 1976).

Other Environmental Media

Levels of pentachlorophenol in food are tested as a part of the U.S. FDA routine monitoring surveys. In 1973 to 1974, 10 out of 360 composite food samples contained pentachlorophenol at 0.01 to 0.03 ppm; 1 in dairy products; 1 in cereals; 1 in vegetables and 7 in sugar (Manske and Johnson 1976). In the following year, 13 out of 240 composites contained pentachlorophenol (0.01 to 0.04 ppm), again mostly in sugars (Jansson and Manske 1977). Random samples testing of Florida foods showed the presence of pentachlorophenol at 1 to 1,000 ppb, mainly in grain products. It was also detected in peanut butter (1.8 to 62 ppb) and chicken (6 to 12 ppb) (Farrington and Munday 1976).

Pentachlorophenol in fish tissue for the years 1976 to 1979 (EPA STORET) ranged from DL to 50 mg/kg. Mean concentrations (mg/kg) by region were as follows: Lake Michigan, 0.002; lower Mississippi, 0.478; Pacific Northwest, 16.38; Alaska, 5.0; western Gulf of Mexico, ND; and south-central-lower Mississippi, ND (EPA 1979).

Pentachlorophenol has been reported in 9 out of 65 samples of children's paints in The Netherlands, ranging from 0.001 to 0.27 mg/L (Van Langeveld 1975).

It should be noted that the data reported above are 13 to 17 years old and more recent data are not available. Use of pentachlorophenol has declined steadily in the past due to regulatory restrictions. It is very likely that the levels of pentachlorophenol detected in other environmental media have decreased since they were reported.

Fish

Some of the commonly detected phenolic compounds in fish tissues include chlorinated phenols, their methylated derivatives, and alkyl phenols. Twelve polyhalogenated compounds were detected in fish from Lake Paijanne, Finland, which has a number of pulp mills located on its shore. The most commonly detected chlorinated phenolic compounds were 2,4,6-trichlorophenol, 4,5,6-trichloroguaiacol, and tetrachloroguaiacol (Table 7.7). Similarly, discharge of pulp mill wastes into Atlantic coastal waters (Canada) resulted in the adulteration of fish tissues with highly chlorinated phenols. The contamination of fish in the Fraser River (Canada) with chlorinated phenols was from wood preservation plants (Table 7.7). Pentachloroanisole (methylated derivative of pentachlorophenol) was detected in fish from the Detroit River (U.S.) and in fish from 15 of 26 rivers samples near the Great Lakes (Kuehl 1981). Lower chlorinated anisoles and pentachlorinated anisole have also been detected in fish from the Arkansas River and in fish exposed to municipal effluents (Veith et al. 1979).

Table 7.7 Average Concentration (µg/kg, Wet Weight) of Chlorinated Phenolic Compounds in Fish Tissue

	Lake Paijanne (Finland)	
Compound	Pike	Roach
2,4,6-Trichlorophenol	15.5	30.3
2,3,4,6-Tetrachlorophenol	15.0	9.0
Pentachlorophenol	6.9	8.8
4,5,6-Trichloroguaiacol	12.4	27.6
Tetrachloroguaiacol	116.2	13.5
Tetrachlorocatechol	12.6	17.9

	Atlantic Coast (Canada) (µg/kg lipid)			
Compound	Tomcod	Smelt	Winter Flounder	Alewife
2,4-Dichlorophenol	2,900	5,050	2,070	ND
2,4,6-Trichlorophenol	1,500	340	750	ND
Pentachlorophenol	3,100	4,800	5,240	820

	Fraser River (British Columbia, Canada)			
Compound	Staghorn Sculpin	Northern Squafish	Largescale Sucker	Dolly Warden
Tetrachlorophenol	30.9	3.5	<3	<3
Pentachlorophenol	59.9	16.8	15.6	34.9

ND = not detected
Sources: Paasivirta et al. 1980; Jones 1981

Tissue concentrations of chlorophenols, chlorocatechols, and chloroguaiacols determined mostly in the 1980s, in fish species such as salmon, sturgeon, and other species are given in Table 7.8. Values for fish species collected in Finland ranged from 1.8 to 123 µg/kg.

Lower chlorinated phenols are more water-soluble and the sodium salts of higher chlorinated phenols, such as those commonly used as wood preservatives, are also water-soluble (Seiler 1991). Chlorinated phenolic compounds tend to partition into sediments in aquatic systems (Allard et al. 1988; Remberger et al. 1986; Moore and Ramamoorthy 1984). Sorption to particulate matter containing organic matter is the major route of removal of chlorophenols from a water column (HWC 1988; Hattemer-Frey and Travis 1989). Volatilizational loss is not a significant fate process for most chlorophenols from aquatic environments. 2-Chlorophenol with a vapor pressure of 2.2. torr at 20°C will have a tendency to volatilize from the water column, but its high aqueous solubility and high solvation, resulting from their high acidic nature, will retard such a loss. Volatilization is inversely related to the degree of chlorination. For example, the volatilizational half-lives of 2,4,6-trichlorophenol and pentachlorophenol, from experimental outdoor ponds, were 48 h and 312 h, respectively (Sugiura et al. 1984).

Chlorinated phenols, due to their sorption potential, bind to soils and sediments (≥95%) where they may undergo biotransformation and bioconcentration in aquatic

Table 7.8 Mean Residue Levels (μg/kg) of Chlorinated Phenols in Fish

Species	Location	Year	DCP	T$_3$CP	T$_4$CP	PCP	References
Cod	St. Croux Estuary N.B., Maine, U.S.	1972				0.82	Zitko et al. 1974
Atlantic salmon	Hatcheries	1973				0.9	Zitko et al. 1974
Winter flounder	Near St. John, N.B., Canada	1977	2,300	1,420		3,000	Bacon and Silk 1978
Smelt	Downstream of pulpmill discharge	1977	5,100	740		2,360	Bacon and Silk 1978
Sturgeon	Downstream of pulpmill discharge	1977	370	28		260	Bacon and Silk 1978
Roach	Finland	1982			32.0	23.9	Paasivirta et al. 1985
Pike	Finland	1982			25.6	13.3	Paasivirta et al. 1985
Baltic salmon (several species)	Finland	1983		1.8–3.0	12.5	2.7	Paasivirta et al. 1985
Pink salmon	Fraser River, B.C., Canada	1986	ND–11.6	ND–28.5	ND–12.5	0.2–5.8	Servizi et al. 1988
Chinook salmon	Fraser River, B.C., Canada	1987			2–27	8–62	Rogers et al. 1988
Eulachon (whole fish)	Lower Fraser River, B.C., Canada	1986, 1988			20.8–21.4	22.1	Rogers et al. 1990

DCP = Dichlorophenol; T$_3$CP = Trichlorophenol; T$_4$CP = Tetrachlorophenol; PCP = Pentachlorophenol, ND = Not Detected.

and terrestrial organisms (HWC 1988). Adsorption of chlorophenols to soils is favored under acidic conditions and in the presence of high organic matter in soil. Since volatilization of chlorophenols from soils is minimal, leaching into groundwater may become an important transport process. Again, this is favored in mineral or basic soils. A lower pH and higher organic content of soil will retard leaching of chlorophenols from soil to groundwater. Biotransformation (aerobic and anaerobic) removes chlorophenols from soils and depends upon soil type, soil moisture content, temperature, mineral nutrients, and organic content of soil. Persistence or resistance to microbial degradation follows the chlorine positions on the phenol ring (Liu and Pacepavicius 1990). In general, chlorophenols with chlorine at the 2 and 4 or 4 ring positions readily biodegrade aerobically. Pentachlorophenol and 2,3,4,5-tetrachlorophenol both aerobically biodegrade more rapidly than most of the di- and trichlorophenols with chlorine atoms at ring positions other than 2,4. Whereas, anaerobic biodegradation favors chlorine positions 2 and 4 only. Non-2,4 chlorine positions are not degraded.

Biotransformations

Microbial Metabolism

Although chlorophenols have antifungal and antimicrobial activities, they are known to be degraded by microorganisms, such as fungi. The following factors have been involved in the metabolism of chlorophenols:

1. An aromatic ring with the halogen atom in the meta position is resistant to degradation.
2. Fungi, including some wood rot fungi, may produce enzymes, tyrosinases, and peroxidases that can degrade pentachlorophenol (PCP). The stability of the benzene ring increases with increasing chlorine atoms in the ring. The primary detoxicifying mechanism is the deactivation of the hydroxyl group. A field study showed that 91% of PCP was removed from low-organic soil by two white-rot fungi (*Phanerochaete chrysosporium* and *P. sordida*) in 6.5 weeks, below optimal conditions of temperature and nutrients (Lamar and Dietrich 1990).
3. The biodegradation rate depends on "the adaptation time." *Ortho-* and *para-* chlorophenols at 1 mg/L, when added to domestic sewage, did not biodegrade up to 30 days at 20°C (Versar 1979), but a similar level of the same chlorophenols was biodegraded at 20°C in polluted waters. It was clear that only specialized microorganisms adapted to chlorophenols will metabolize them. The following hypothetical pathway for the biodegradation of PCP was proposed by Reiner et al. (1978).

Bacteria, in the presence of humic matter in the water column, are reported to have mineralized 3,4-dichlorophenol, 2,4,5-trichlorophenol, and pentachlorophenol (Larsson et al. 1988; Tranvick et al. 1991). Chlorophenols are microbially metabolized in water, sediments, and soils and the rate of degradation, whether aerobic or anaerobic, depends on the type of microbes present, the degree of chlorination, and isomeric positions of chlorine atoms on the phenol ring, soil type, moisture content, temperature, minerals and nutrients, and percent of organic matter (RTP 1994).

Laboratory studies showed that (a) two strains of *Rhodococcus* aerobically degraded chlorinated phenols in pulp bleaching effluents (Haggblom and Salkinoja-Salonen 1991); (b) *Laccase* enzymes from the fungus *Trametes versiclor* (white rot basidiomycetes) partially dechlorinated a variety of chlorophenols, guaiacols, and catechols in chlorinated kraft bleach effluents (Roy-Arcand and Archibald 1991). The disappearance percentage of chlorophenols was: 30.4 to 58.2% (monochlorophenols); 42.9 to 70.7% (dichlorophenols); 72.4% to 100% (trichlorophenols); 40.7 to 67.6% (tetrachlorophenols), and 34 to 64.8% (pentachlorophenol) in 30 minutes. Generally, microbial degradation of chlorinated phenolic compounds occurs faster and greater in sediments than in the water column (Baker and Mayfield 1980). Chlorophenols are biotransformed via reductive dechlorination to simpler chlorophenols.

Metabolism of Pentachlorophenol

Pentachlorophenol (PCP) was shown to be susceptible to accumulation and metabolism by plants. Three percent of the applied ^{14}C-PCP was accumulated by rice plants and half of that amount was extractable in 3 days; remainder was tightly bound to plant material, not extractable by methanol (Hague et al. 1978). About 90% of the extractable portion was the parent compound, PCP; about 9%, unidentified conjugated products; and about 1%, dechlorinated product. Degradation of PCP in soils is characterized by dechlorination and ring cleavage, yielding a variety of products (Kaufmann 1978). Under suitable environmental conditions, biodegradation could reduce the half-life of PCP in soil to 20 days (Hattemer-Frey and Travis 1989). Pentachlorophenol was metabolized more rapidly in flooded or anaerobic soils than aerobic moist soils (Kaufmann 1978). Soya bean and spinach plants were reported to take up PCP from soil and conjugate and metabolize it. Ninety-five percent of PCP had disappeared in 90 days from the soil where these plants were growing. Principal metabolites were methoxy tetrachlorophenol, 2,3,4,6-tetrachlorophenol, pentachloro-anisole, and 2,3,4,6-tetrachloroanisole (Casterline et al. 1985).

Lack of oxidation or conjugation systems for excreting PCP metabolites were reported in studies in the 1960s. It was demonstrated that sulphate conjugation is one of the most common degradation pathway for phenolic compounds in goldfish (Kobayashi and Atitake 1975). Other studies have revealed the presence of both sulphate and glucuronide conjugation systems in fish, as in mammals, for the clearance of chlorophenols. The liver–enzyme activity generally deceases with increasing chlorine content, with activity being the lowest for PCP.

Pentachlorophenol is rapidly metabolized in rats, with rapid elimination from kidneys, liver, and blood. The dechlorination mediated by liver chromasal enzymes is accentuated by pretreatment with inducing agents such as phenobarbital. Hydrolytic dechlorination takes place first, followed by dechlorination to yield trichloro-*p*-hydroquinone (Ahlborg 1977).

Photolysis

With increasing acidity of the compound (decreasing pK_a value), the proportion of the anion compared to the unionized molecule increases. This is significant environmentally since the anion absorbs well beyond 310 nm (sunlight spectrum), leading to photolytic reactions. For many chlorophenols, pK_a values are in the range of 4 to 8 (Table 7.9). The conditions under which they photolyze and their resulting breakdown products are listed in Table 7.9.

Table 7.9 Photolysis of Chlorophenols and their Breakdown Products

Chlorophenol	Light Source	Exposure Conditions	Products
2-MCP	313 nm	Aqueous alkali	2-hydroxy phenol
2-MCP	—	Aqueous solution (1000 mg/L) in presence of H_2O_2	Photodegraded below odor threshold (2 mg/L)
2-MCP	296 nm	Irradiation	Cyclopentadienic acids[1]
3-MCP	—	Irradiation	Resorcinol[1]
4-MCP	—	Irradiation	Hydroquinone and dihydroxy biphenyls[1]
4-MCP	254 nm and >290 nm	Aqueous	Tarry material
	254 and >290 nm	Cyanide in aqueous solution	4-cyano phenol
2,4-DCP	Solar irradiation	10 d, aerated	Totally degraded to a polymer of 2-hydroxy benzene
2,4-DCP	> 290 nm	In the presence of riboflavin (photosensitizer)	Mixture of tetrachlorophenoxy phenols and tetrachloro dihydroxy biphenyls
2,4,6-TCP	—	In the presence of electron acceptor	2,6-dichlorobenzoquinone and 2,6-dichlorohydroquinone
PCP	Sunlight	2% solution of Na salt of PCP	50% degraded in 10 days; major products: chloroanilic acid, tetrachlororesorcinol, 2,5-dichloro-3-hydroxy 6-pentachlorophenoxy-p-benzaquinone, 2,6-dichloro-3-hydroxy-5-(2',4',5',6'-tetrachloro-3'-hydroxyphenoxy)-p-benzaquinone, and 3,5-dichloro-4-(2,3,5,6-tetrachloro-4-hydroxyphenoxy)-6-(2,3,4,5-tetrchloro-6-hydroxyphenoxy)-O-benzoquinone
	Summer sunlight or UV light	Aqueous solution (100 mg/L) 300–450 nm	After 7 days, chlorophenols, tetrachlorodihydroxy benzenes, and nonaromatic fragments such as dicloro maleic acid; after 20 days: hydroxylated trichlorobenzoquinones, trichlorodiols, dichloromaleic acid, chloride ions, and CO_2
	Sunlight	Accidental release into a freshwater lake	2,3,5,6- and 2,3,4,5-tetrachlorophenols

MCP = Monochlorophenol; DCP = Dichlorophenol; TCP = Trichlorophenol; PCP = Pentachlorophenol
[1]Boule et al. 1987; others Versar 1979.

Photolytic breakdown of chlorophenols occurs, to a limited extent, under natural conditions. 2,4,6-Trichlorophenol was degraded photolytically (99%) in an oligotrophic lake and 97% in an eutrophic lake (Yoshida et al. 1987). Dichlorophenols and chloroguaiacols are presumed to be stable in sunlight without photodegradation. Turbidity and color of the water column restricted the entry of the light and, in turn, inhibiting photodegradation (Carey et al. 1984; Environment Canada 1991). As stated earlier, the Pk_a (4.8, 5.0) of pentachlorophenol greatly enhances its photodegradation, particularly at higher pHs; rapid degradation at pH 7.3 compared to slower degradation at pH 3.31 (Wong and Crosby 1981). The products included tetrachlorophenols, trichlorophenols, and other related reactive products (HWC 1988). Exposure to sunlight in the month of July rapidly photodegraded pentachlorophenol in aqueous solution ($t_{1/2}$ = 48 h) and was nondetectable in 10 days (Wong and Crosby 1981).

The pathways for aerobic biodegradation of lower chlorinated phenols is hydroxylation followed by the clevage of the aromatic ring, with elimination of chlorine from the intermediate metabolite (Reineke and Knackmuss 1988; Neilson et al. 1990). In summary, chlorophenols would persist in the environment only if input rates exceed rates of transformation and degradation.

Chlorinated guaiacols and chlorocatechols are also microbially degraded in the natural environment. Rapid degradation of chloroguaiacols by soil microflora was reported when sludge from pulp and paper mills was spread on land (Dargitz et al. 1991). The degradation rate decreased with increase in the chlorination of the phenolic ring. The degradation pathway is similar to that for chlorophenols, namely, hydroxylation of the ring, methylation of phenol groups, oxidative coupling of phenols, oxidative cleavage of the aromatic ring, followed by decarboxylation or hydrolysis. Anaerobic bacterial transformation of chloroguaiacols and catechols to chloroveratroles by O-methylation has been reported (Allard et al. 1985; Remberger et al. 1986). Neilson et al. (1984) reported metabolization of di-, tri-, and tetrachloroguaiacols to the corresponding O-methyl compounds, immediate downstream of pulpmill effluent discharge. For example, 3,4,5-trichloroguaiacol was metabolized to yield 3,4,5-trichloroveratrole, 3,4,5-trichlorocatechol, and 3,4,5-trichlorosyringol.

Under anaerobic conditions, O-demethylation of chloroguaiacols to corresponding catechols was followed by dechlorination. For example, 4,5,6-trichloroguaiacol → 3,4,5-trichlorocatechol → 3,5- and 4,5- dichlorocatechols was observed (Rosemarin et al. 1990). Aerobic demethylation could convert 3,5- and 3,6-dichloroguaiacols to corresponding catechols (Apajalahti et al. 1986). Chlorocatechols, in general, are metabolized via the dechlorination pathway, yielding lesser-chlorinated isomers (Allard et al. 1991; Nielson et al. 1987).

Using fugacity modeling, partitioning of chlorophenols into various environmental compartments has been estimated (Table 7.10).

TOXICITY PROFILE

Acute Toxicity

Toxicity of phenols to aquatic organisms (algae, invertebrates, and fish) generally increases with increasing chlorine content of the benzene ring [Table 7.11 (algae

Table 7.10 Percent Distribution of Chlorinated Phenolic Compounds

Compound	Air	Water	Soil	Sediment	Other Compartments
2-Chlorophenol	89.06	10.71	—	—	0.22
2,4-Dichlorophenol	14.2	76.5	4.8	4.5	0.009
2,4,6-Trichlorophenol	11.5	41.4	24.2	22.7	0.04
2,3,4,6-Tetrachlorophenol	8.3	18.3	38.0	35.4	0.08
Pentachlorophenol	3.1	5.9	47.0	43.8	0.09
Trichloroguaiacol	54.9	25.5	10.1	9.4	0.02

Source: RTP 1994

and invertebrates) and 7.12 (fish)]. The LC_{50} of phenol is about 25 mg/L compared to about 1 mg/L for pentachlorophenol. Similarly, corresponding levels for 3,5-dichlorocatechol and trichlorocatechol and tetrachlorocatechol are 2.9 and 1.1 mg/L, respectively. In general, the *para*-substituted compounds are more toxic than *ortho*- and *meta*-compounds. The LC_{50}s of four *para*-chlorophenols and two *ortho*-chlorophenols are 3.8 to 14 and 8.1 to 58 mg/L, respectively. This could be due to relatively faster transport of *para*-compounds across the gill membrane.

Table 7.11 Acute Toxicity (96 h LC_{50}, mg/L) of Chlorophenols to Aquatic Algae and Invertebrates

	Algae	Invertebrates	
	Freshwater	*Cladoceran*	*Shrimp*
Compound	Chlorophytes*	*Daphnia magna*	*Mysidopsis bahia*
Phenol	10–30	36.4	—
2-MCP	500	4.4	—
4-MCP	4.8	4.4	29.7
2,4-DCP	—	2.6	—
2,4,5-TCP	1.2	2.7	3.8
2,4,6-TCP	5.9	6.0	—
2,3,5,6-TeCP	2.7	0.6	21.9
2,3,4,6-TeCP	0.6	0.3	—
PCP	1.0–2.7	0.7–0.8	0.1–5.1

* *Selanstrum capricornutum* and *Chlorella pyrenoidosa*
MCP = Monochlorophenol; DCP = Dichlorophenol; TCP = Trichlorophenol; TeCP = Tetrachlorophenol; PCP = Pentachlorophenol
Sources: Le Blanc 1980; U.S. EPA 1980b; Buiekema et al. 1979; Adema and Vink 1981; Voss et al. 1980; PAPRICAN 1990; Phipps et al. 1981.

LC_{50}, 96 h values for pentachlorophenol for invertebrates and for several fish species are given in Tables 7.13 and 7.14.

Toxicity studies involving trichlorophenols experienced solubility and volatilizational problems leading to erroneous values for LC_{50} measurements. Actual exposure concentrations were found to be 50 to 60% lower than the design concentrations due to the fate processes and solubility limitations of trichlorophenols. The 96h LC_{50} for tetra- and trichloroguaiacols in rainbow trout were approximately 0.32 and 0.75 mg/L (Leach and Thakore 1977). The 96h LC_{50} values for tetra- and trichloroguaiacols in

Table 7.12 Acute Toxicity (96 h LC$_{50}$, mg/L) of Chlorophenols to Fish Species

Compound	Guppy	Sheepshead Minnow	Salmonids	Bluegill Sunfish	Fathead Minnow	Mixed Species
Phenol	—	—	—	16.4	36	4.4–44.5
2-MCP	—	—	2.9	6.6	12.3	8.1–58.0
4-MCP	49–66	5.4	—	3.8	—	3.8–14.0
2,4-DCP	—	—	2.0	2.0	8.2	2.0–13.7
2,4,5-TCP	6.3–50	1.7	0.5; 0.9	0.4	0.45	0.4–0.9
2,4,6-TCP	3.1–39.8	—	—	0.3	9.0	0.3–9.0
2,3,5,6-TeCP	—	—	<0.5;0.30	0.2	—	—
2,3,4,6-TeCP	—	—	—	0.1	—	0.1–0.5
PCP	—	—	—	0.1	0.2	0.06–1.7

Sources: LeBlanc 1980; Buccafusco et al. 1981; U.S. EPA 1980; Buiekema et al. 1979; Adema and Vink 1981; Voss et al. 1980; PAPRICAN 1990; Phipps et al. 1981; Heitmuller 1981.

Table 7.13 Acute Toxicity (96 h LC$_{50}$, mg/L) of Pentachlorophenol to Invertebrates

Molluscs	Daphnia magna	Crangon crangon	Artemis salina	Snail[a]	Isopod[b]
0.11–0.18	0.68–0.8	0.17–16.0	0.17–16.0	0.22 (22°C)	2.30 (8.6°C)
				0.73 (8.6°C)	4.32 (4.2°C)
				0.138 (3.2°C)	>7.77 (3.2°C)

[a] *Physa gyrina*
[b] *A. racovitzai*

Source: Adema and Vink 1981

the bleak (*Alburnus alburnus*) were 3.9 and 5.2 mg/L, respectively (Renberg et al. 1980). Chloroguaiacols can be transformed by bacteria in the environment to chloroveratroles, which could be more toxic than the parent guaiacol. Mortality threshold concentration for 3,4,5,6-tetrachloroveratrole in the zebra fish was reported to be fivefold less than that for 4,5,6-trichloroguaiacol (Neilson et al. 1984).

Plants

Pentachlorophenol causes inhibition of chlorosis in the macrophyte, *Lemna minor* (0.8 to 1.0 mg/L) at greater concentrations than reported for similar effects in algae. However, no adverse effects to *L. minor* by pentachlorophenol after exposure to 1.11 mg/L for 21 days was reported (Eisler 1989).

Embryo lethality

Invertebrates

A lowest observed effect level (LOEL) and no-observed-effect-level (NOEL) of 1.48 mg/L and 0.74 mg/L, respectively were reported for the embryo lethality of 2,4-dichlorophenol to *Daphnia magna*. The endpoints considered for embryo

Table 7.14 Acute Toxicity (LC_{50}, 96 h, mg/L) of Pentachlorophenol for Various Fish Species

Species	Value	References
Common carp	0.0095	Choudhury et al. 1986
Pinfish	0.031–0.053	EPA 1980c
		Mayer 1987
Bluegill sunfish	0.032–0.215	EPA 1980c
		Johnson and Finlay 1980
		Hedtke and Arthur 1985
		Mayer and Ellersieck 1986
Rainbow trout	0.034–0.121	EPA 1980c
		Johnson and Finlay 1980
		Dominguez and Chapman 1984
		McKim et al. 1987
Flounder	0.050–0.130	Choudhury et al. 1986
Channel catfish	0.054–0.068	Choudhury et al. 1986
		Johnson and Finlay 1980
		Mayer and Ellersieck 1986
Coho salmon	0.055	EPA 1980c
White crappie	0.056–0.075	Cote 1972
Sockeye salmon	0.063–0.068	EPA 1980c
		Webb and Brett 1973
Chinook salmon	0.068–0.078	EPA 1980c
		Johnson and Finlay 1980
		Iwama and Greer 1979
White sucker	0.085	Hedtke et al. 1986
Fathead minnow	0.120–0.350	Hedtke and Arthur 1985
		Cote 1972
		Hedtke et al. 1986
		Crossland and Wolff 1985
Brook trout	0.128	EPA 1980c
Largemouth bass	0.136–0.287	Johansen et al. 1987
		Mathers et al. 1985
		Johansen et al. 1985
Sheepshead minnow	0.223–0.442	Borthwick and Schimmel 1978
Atlantic salmon	0.5	Cote 1972

Sources: WHO 1989; Eisler 1989.

lethality included survival, reproduction, mean total young/adult, and the brood size per adult (Gerisch and Milazzo 1988).

Fish

The larval stage is considered to be the most sensitive life stage in fish to environmental stress. Table 7.15 lists the concentrations of chlorophenols causing lethal effects in embryo-larval tests. It should be noted that the lethality data lends evidence that the larva, rather than the embryo or fry, is the most sensitive early life stage test. The NOEL and LOEL for trichlorophenol to fathead minnows during the embryo stage were 2.1 and >2.1 mg/L, respectively. Smith et al. (1991) reported for the American flagfish (*Jordanella floridae*) a threshold concentration (LOEL) for egg hatchability, 10-day larval survival of >1.76 mg/L, and fry survival after 28-day exposure of 1.34 mg/L. For 2,3,5,6-tetrachlorophenol, threshold concentrations were reported as 0.775 mg/L for egg hatchability; 0.260 mg/L for 10-day larval survival; and LOEL threshold as 0.245 mg/L for 28-day fry survival (Smith et al. 1991).

Table 7.15 NOEL and LOEL Concentrations (mg/L) for Fathead Minnow and American Flagfish Exposed to Chlorophenols

Compound	Egg Hatchability NOEL	Egg Hatchability LOEL	Larval Survival NOEL	Larval Survival LOEL	Fry Survival LOEL	References
MCP	4	>4	4	8.1		LeBlanc 1984
DCP	1.24	>1.24	0.29	0.46		Holcombe et al. 1982
TCP	2.1	>2.1	0.97	2.1		LeBlanc 1984
		>1.76[a]		>1.76[a]	1.34[a]	Smith et al. 1991
TeCP	—	0.775		0.26	0.245	Smith et al. 1991
PCP		0.135		0.075	0.135	Smith et al. 1991
	0.128[a]	0.233[a]	0.073[a]	0.128[a]		Holcombe et al. 1982

Embryo-larval test = 28 days; [a]American Flagfish data

NOEL = No Observed Effect Level: LOEL = Lowest Observed Effect Level; MCP = Monochlorophenol, DCP = Dichlorophenol, TCP = Trichlorophenol, TeCP = Tetrachlorophenol, and PCP = Pentachlorophenol

Chronic Effects

The growth effects on aquatic organisms and the corresponding concentrations of chlorophenols are given in Tables 7.16 to 7.19.

Table 7.16 Growth Reduction Effects of Chlorophenols on Aquatic Organisms

Compound	Marine Phytoplankton (Three Species)	Concentration (mg/L)
3-MCP	20% reduction	2.5
3-MCP	80% reduction	5.0
	Selanastrum Capricornutum	
4-MCP	EC_{50} — 50% reduction	4.1
	Amphipod (*Pontoporeia affinis*) and Isopod (*Mesidotea entomon*)	
4-MCP	100% irreversible reduction of swimming activity	0.1

MCP = Monochlorophenol
Source: Krijgsheld and van der Gen 1986

Table 7.17 Growth Reduction Effects (96 h EC$_{50}$) of Chlorophenols in Algae

	Algae	Concentration (mg/L)	References
2,4,6-TCP	*Scenedesmus subspicatus*	5.6	Geyer et al. 1985
PCP	*Scenedesmus subspicatus*	0.09	Geyer et al. 1985
PCP	*Duanaliella sp.*	0.08–14.0	
PCP	*Chlorella sp.*		
PCP	*Phaeodactylum tricornutum*		Adema and Vink 1981

TCP = Trichlorophenol; PCP = Pentachlorophenol.

Table 7.18 Growth Reduction Effects of 2,4-dichlorophenol in *Daphnia magna*

Test Type	Toxic Endpoint	Concentration (mg/L)
21-day static/renewal	NOEL	1.48
21-day static/renewal	LOEL	2.96
21-day static/renewal	MATC	0.74–1.48

NOEL = No Observed Effect Level: LOEL = Lowest Observed Effect; MATC = Maximum Acceptable Toxic Concentration.

Source: Gerisch and Milazzo 1988

Table 7.19 Growth Reductions in Fish Larvae on Exposure to Chlorophenols

Compound	Species	Exposure	Effect	Concentration (mg/L)	References
2,4-DCP	Fathead minnow	28 day	NOEL	0.77	Holcombe et al. 1982
	Fathead minnow	28 day	LOEL	1.24	Holcombe et al. 1982
TCPs	Fathead minnow	7 day	NOEL	0.361	Norberg-King 1988
	Fathead minnow	7 day	LOEL	0.683	Norberg-King 1988
TCPs	Fathead minnow	30 day (F)	NOEL	0.97	LeBlanc 1984
	Fathead minnow	30 day (F)	LOEL	2.1	LeBlanc 1984
2,4,6-TCP	American flagfish	28 day	EC$_{50}$	0.75	Smith et al. 1991
TeCP	American flagfish	28 day	NOEL	1.035	Smith et al. 1991
PCP	Fathead minnow	7 day (S) renewal	NOEL LOEL	0.179 0.358	Pickering 1988 Pickering 1988
PCP	Fathead minnow	28 day 28 day	NOEL LOEL	0.045 0.073	Holcombe et al. 1982 Holcombe et al. 1982
Commercial PCP	Fathead minnow	90–day partial life cycle study	EC$_{50}$	0.013	Cleveland et al. 1982
Purified PCP	Fathead minnow	90–day partial life cycle study	EC$_{50}$	≥0.085	Cleveland et al. 1982
Dowicide EC-7 (91% PCP)	Fathead minnow	90–day partial life cycle study	EC$_{50}$	0.139	Cleveland et al. 1982

DCP = Dichlorophenol, TCP = Trichlorophenol, TeCP = Tetrachlorophenol and PCP = Pentachlorophenol; NOEL = No Observed Effect Level: LOEL = Lowest Observed Effect;

Mammalian Toxicity

Acute toxicity of chlorinated phenolic compounds on mammalian species are given in Table 7.20.

Table 7.20 Acute Toxicity of Chlorinated Phenolic Compounds to Mammals

Compound	Species	Route of Exposure	Toxicity Endpoint	Concentration mg/kg.bw.	References
PCP, Technical	Rat	Oral	LD_{50}	80–170	Harrison 1959 Fielder 1982
PCP	Human	Accidental poisoning	LD_{100} (min. dose)	29.0	Ahlborg and Thunberg 1980
2,4-DCP	Mammal	Oral	LD_{50}	580–4,000	EPA 1982
Nonchlorinated Phenol	Mammal	Oral	LD_{50}	180–600	RTECS 1978
TCG	Rat, male	Oral	LD_{50}	3,000	Chu et al. 1979
TeCG	Rat, male	Oral	LD_{50}	1,690	Chu et al. 1979
TeCC	Mice, male	Oral	LD_{50}	318.0	Renner et al. 1986
	Mice, F	Oral	LD_{50}	331	Renner et al. 1986
TeCC	Mice, male	IP	LD_{50}	161.0	Renner et al. 1986
TeCC	Mice, F	IP	LD_{50}	167.0	Renner et al. 1986
PCP, Na salt	Rat	Inhal.	LD_{50}	11.7	Fielder 1982 Hoben et al. 1976
PCP, Na salt	Rat	Oral	LD_{50}	29.0	Ahlborg and Thunberg 1980

DCP = Diclorophenol; PCP = Pentachlorophenol; TCG = Trichloroguaiacol; TeCG = Tetrachloroguaiacol; TeCC = Tetrachlorocatechol; F = Female; Inhal = inhalation; IP = Intraperitoneal.

Review of the literature shows that chlorinated phenols affect aquatic organisms through two mechanisms. The primary one is "uncoupling," or intolerance, with oxidative phosphorylation within cells, the main process in biological systems for the transfer and storage of energy (Eisler 1989). The uncoupling of phosphorylation results in a range of related effects such as increased body temperature, respiration, heart rate, and metabolic rate similar to vigorous exercise (Hodson and Blunt 1981). The potency for uncoupling oxidative phosphorylation decreases with the degree of chlorination in chlorophenols. Although chlorinated guaiacols, catechols and vanillins appear to follow the same disruptive mechanism for toxicity, the toxic potency of these other groups seem to be at a lower level (McKague 1981). The second mechanism for toxicity of chlorophenol is through their narcotic action on biological systems, reducing cellular activity, and may be related to interaction with lipo-protein cell membrane systems. For lesser chlorinated phenols, the toxicity is primarily through general cellular narcosis.

REGULATIONS

Pentachlorophenol

Agency	Description	Value	Reference
WHO	Drinking water guideline	10 µg/L	WHO 1984
IARC	Possibly carcinogenic to humans		IARC 1991, cited in ATSDR, 1994

Regulations
OSHA	PEL-TWA (skin)	0.5 mg/m³	OSHA 1974

Guidelines
Air
NIOSH	Immediately dangerous to life or health (IDLH), level	150 mg/m³	NIOSH 1978
ACGIH	TLV-TWA	0.5 mg/m³	ACGIH 1986

Water
U.S. EPA/ODW	Maximum contaminant level goal (MCLG) (proposed)	0.22 mg/L	EPA 1985b
	Health Advisories		EPA 1987e
	One day	1.0 mg/L	
	Ten day	0.3 mg/L	
	Longer-term (child)	0.3 mg/L	
	Longer-term (adult)	1.05 mg/L	
	Lifetime	0.22 mg/L	
U.S. NAS	Safe drinking water level	0.021 mg/L	NAS 1977
	Suggested no adverse effect level (SNARL)		NAS 1986
	Child	0.006 mg/L	
	Adult (technical PCP)	0.007 mg/L	
	Adult (commercial PCP)	0.009 mg/L	
U.S. EPA/OWRS	Ambient water quality criteria		EPA 1980c
	Ingestion of water and organisms	1.01 mg/L	
U.S. EPA	Carcinogenic classification (under review)		
	RfD (oral)	0.03 mg/kg/day	EPA 1988

State Agencies
	Water Quality Guidelines	0.006–0.220 mg/L	FSTRAC 1988, CELDS 1988
	Acceptable Ambient Air concentrations/Standards and Guidelines		
	8 h average	0.005–0.012 mg/m³	NATICH 1987
	24 h average	0.034–8.0 µg/m³	
	Annual average	1.67–25.64 µg/m³	

Limits in Soil (permissible cleanup levels), mg/kg

Location	Residential	Industrial	Reference
Massachusetts	7	43	Sittig 1994
Michigan	11	120	
New Jersey	1,700 (surface)	10,000	
	100 (subsurface)	—	
Oregon	5	50	
Texas	534	47.7	

2,4,6-Trichlorophenol

WHO	Drinking water guideline	0.010 mg/L	WHO 1984

Regulations
Water

U.S. EPA/OSW	Proposed regulatory level based on toxicity	0.30 µg/L	EPA 1986f
IARC	Possibly carcinogenic		IARC 1987
U.S. EPA	RfD (oral)	0.03 mg/kg/day	EPA 1988

Guidelines
Air

U.S. EPA	Cancer potency factor;q* (inhalation)	0.02 mg/kg/day	IRIS 1988

Water

U.S. EPA/OWRS	Ambient Water Quality Criteria		EPA 1980a
	Water and fish consumption	1.2 µg/L	
	Fish consumption only	3.6 µg/L	
	Water (organoleptic effects)	2.0 µg/L	
U.S. NAS	SNARL		NAS 1982
	24 h	17.5 mg/L	
	7 day	2.5 mg/L	

Water/Food

U.S.EPA	Cancer potency factor;q* (oral)	0.02 mg/kg/day	IRIS 1988
U.S.EPA	Probable human carcinogen		IRIS 1988

State Regulations and Guidelines
Air

	Acceptable Ambient Air Concentration		
	Massachusetts	0.16 µg/m^3	NATICH 1988

Water

	Drinking water guideline		FSTRAC 1988
	Kansas	0.017 mg/L	
	Maine	0.700 mg/L	
	Minnesota	0.0176 mg/L	
	New Jersey, MCL (10^{-6} life time cancer risk)	0.001 mg/L	NJDEP 1989

Limits in Soil (permissible cleanup levels), mg/kg

Location	Residential	Industrial	Reference
Massachusetts	39	220	Sittig 1994
Michigan	120	1,400	
New Jersey	62 (surface)	260	
	52 (subsurface)	—	
Oregon	60	500	

REFERENCES

Adema, D.M.M. and Vink, G.J. 1981. A comparative study of the toxicity of 1,1,2-trichloroethane, dieldrin, pentachlorophenol and 3,4-dichloroaniline for marine and freshwater organisms. *Chemosphere.* 10:533–554.

Agriculture Canada 1987. Discussion document, pentachlorophenol, wood preservative, Pesticides Directorate, Agriculture Canada, Ottawa, Canada.

Ahlborg, U.G. 1977. Metabolism of chlorophenols: studies on dechlorination in mammals. Swedish Environmental Protection Board, PM895, Stockholm.

Ahlborg, U.G. and Thunberg, T.M. 1980. Chlorinated phenols; Occurrence, toxicity, metabolism and environmental impact. *CRC Crit. Rev. Toxicol.* 7:1–35.

Ahling, B. and Lindskog, A. 1982. Emission of chlorinated organic substances from combustion. In: Chlorinated Dioxins and Related Compounds, Impact on the Environment. O. Hutzinger, R.W. Frei, E. Merian, and Pocchiari, Eds. Pergamon Press, New York.

Alberta Environment 1989. Water quality in the Wapiti-Smoky River system downstream of a pulp mill, 1983.

Allard, A.S., Hynning, P.A., Lindgren, C., Remberger, M., and Neilson, A.H. 1991. Dechlorination of chlorocatechols by stable enrichment cultures of anaerobic bacteria. *Appl. Environ. Microbiol.* 57:77–84.

Allard, A.S., Remberger, M., and Neilson, A.H. 1985. Bacterial-*O*-methylation of chloroguaiacols: effect of substrate concentration, cell density, and growth conditions. *Appl. Environ. Microbiol.* 49:279–288.

Allard, A.S., Remberger, M., Viktor, T., and Neilson, A.H. 1988. Environmental fate of chloroguaiacols and chlorocatechols, *Environ. Sci. Technol.* 21:131–142.

Apajalahti, J.H.A., Karponoja, P., and Salkinoja-Salonen, M.S. 1986. *Rhodococcus chlorophenolicus* sp. A chlorophenol-mineralizing actinomycete. *Int. J. Syst. Bacteriol.* 36:246–251.

Arsenault, R.D. 1976. Pentachlorophenol and Contained Chlorinated Dibenzodioxins in the Environment. American Wood-Preservers Assoc., Alexandria, VA. 122–147.

ATSDR 1990. Agency for Toxic Substances and Disease Registry. Toxicological profile for 2,4,6-Trichlorophenol, TP-90-28 December 1990, Public Health Service, U.S. Department of Health and Human Services, Atlanta, GA.

ATSDR 1994. Agency for Toxic Substances and Disease Registry. Toxicological profile for pentachlorophenol, TP-93/13, May 1994, Public Health Service, U.S. Department of Health and Human Services, Atlanta, GA.

Axegard, P. 1986. Substituting chlorine dioxide for elemental chlorine makes the bleach plant effluent less toxic. *Tappi J.* 69: 54–59.

Bacon, G.B. and Silk, P.J. 1978. Bioaccumulation of toxic compounds in pulp mill effluents by aquatic organisms in receiving waters. EPS, Canada, CPAR Report 675–1.

Baker, M.D. and Mayfield, C.I. 1980. Microbial and non-biological decomposition of chlorophenols and phenol in soil. *Water Air Soil Pollut.* 13:411–424.

Blackman, G.E, Parke, G.E., and Garton, F. 1955. The physiological action of substituted phenols. I Relationship between chemical structure and physiological activity. *Arch. Biochem. Biophys.* 54:45–54.

Borthwick, P.W. and Schimmel, S.C. 1978. Toxicity of pentachlorophenol and related compounds to early life stages of selected estuarine animals. In: *Pentachlorophenol: Chemistry, Pharmacology and Environmental Toxicology.* Rao, K.R. Ed., p.141. Plenum Press, New York.

Boule, P., Guyon, C., and Lemaire, J. 1982. Photochemistry and environment. IV. Photochemical behaviour of monochlorophenols in dilute solution. *Chemosphere.* 11:1179–1188.

Boule, P., Guyon, C., Tissot, A., and Lamaire, J. 1987. Specific phototransformation of xenobiotic compounds; chlorobenzenes and halophenols. In: *Photochemistry of Environmental Aquatic Systems,* Zika, R.G. and Cooper, W.J., Eds., pp. 10–26, American Chemical Society, Washington, D.C.

Boyle, T.P., Robinson-Wilson, E.F., Petty, J.D., and Weber, W. 1980. Degradation of pentachlorophenol in simulated lentic environment. *Bull. Environ. Contam. Toxicol.* 24:177–184.

Buccafusco, R.J., Ellis, S.J., and LeBlanc, G.A. 1981. Acute toxicity of priority pollutants to bluegill sunfish (*Lepomis machrochirus*). *Bull. Environ. Contam. Toxicol.* 26:446–452.

Buikema, Jr., A.L., McGuinniss, M.J., and Cairns, J. 1979. Phenolics in aquatic ecosystem: a selected review of recent literature. *Marine Environmental Res.* 2:87–181.

Buhler, D.R., Rasmussen, M.E., and Nakane, H.S. 1973. Occurrence of hexachlorophenol and pentachlorophenol in sewage and water. *Environ. Sci. Technol.* 24:929–934.

Burtshell, R.H., Rosen, A.A., Middelton, F.M., and Ettinger, M.E. 1959. Chlorine derivatives of phenols causing taste and odour. *J. Amer. Water Works. Assoc.* 51:205–214.

Carey, J.H., Fox, M.E., Brownlee, B.G., Metcalfe, J.L., and Platford, R.E. 1984. Disappearance kinetics of 2,4- and 3,4-dichlorophenol in a fluvial system. *Can. J. Physiol. Pharmacol.* 62:972–975.

Casterline, J.L., Barnett, N.M., and Ku, Y. 1985. Uptake, translocation and transformation of pentachlorophenol in soybean and spinach plants. *Environ. Res.* 37:101–118.

CCREM. 1987. Canadian Council of Resources and Environment Ministers. Canadian Water Quality Guidelines, March 1987, 6.104–6.109, Ottawa, Ontario, Canada.

Chaudhry, G.R. and Chapalamadugu, S. 1991. Biodegradation of halogenated organic compounds. *Microbial Rev.* 55:59–79.

Choudhury, H., Coleman, J., DeRosa, C.T., and Stara, J.F. 1986. Pentachlorophenol: Health and environmental effects profile. *Toxicol. Ind. Health.* 2:483–571.

Chu, I., Ritter, L., Marino, I., Yagminas, A.P., and Villeneueve, D.C.1979. Toxicity studies on chlorinated guaiacols in rat. *Bull. Environ. Contam. Toxicol.* 22:293–296.

Cirelli, D.P. 1978. Patterns of pentachlorophenol usage in America — an overview. In: Pentachlorophenol, Chemistry, Pharmacology, and Environmental Toxicology. Rao, K.R., Ed., Plenum Press, New York. 13–18.

CIS. 1988. *Directory of World Chemical Producers.* 1989/90 ed. Chemical Information Services, TTD, 568, Oceanside, NY.

Cleveland, L., Buckler, D.R., Mayer, F.L., and Branson, D.R. 1982. Toxicity of three preparations of chlorophenol to fathead minnows: A comparative study. *Environ. Toxicol. Chem.* 1:205–221.

CMR. 1987. Chemical Marketing Reporter. Biocide ranks thin as costs multiply. Schnell Publishing Co., New York.

Cote, R.P. 1972. A literature review of the toxicity of pentachlorophenol and pentachlorophenates. *Environ. Can. EPA Manus. Rep.* 72, 14.

Crosby, D.G. 1981. Environmental chemistry of pentachlorophenol. *Pure Appl. Chem.* 53:1051–1080.

Crossland, N.O. and Wolff, C.J.M. 1985. Fate and biological effects of pentachlorophenol in outdoor ponds. *Environ. Toxicol. Chem.* 4:73–86.

Dargitz, P., Brenzy, R., Joyce, T.W., and Overcash, M.R. 1991. Biotransformation in soil of chlorophenols related to bleach plant effluents. In: TAPPI 1991 Environmental Conference Book 1, pp. 339–345.

DeVault, D.S. 1985. Contaminants in fish from great lakes harbour and tributary mouths. *Arch. Environ. Contam. Toxicol.* 14:587–594.

Doedens, J.D. 1967. Chlorophenols. In : Kirk-Othmer Encyclopedia of Chemical Technology. Vol. 5, 2nd ed., Parolla, E.A., Schetty, G.O., Dankberg, F.L., Kerstein, J.J., and Strauss, L.L., Eds., John Wiley & Sons, Toronto, Ontario, Canada, pp. 325–338.

Dominguez, S.E., and Chapman, G.A. 1984. Effect of pentachlorophenol on the growth and mortality of embryonic and juvenile steelhead trout. *Arch. Environ. Contam. Toxicol.* 13:739–743.

Dust, J.V. and Thompson, W.S. 1972. Pollution control in wood preserving industry: Part 3. Chemical and physical methods of treating wastewater. *Forest Prod. J.* 22:25–30.

Dougherty, R.C. and Piotrowska, K. 1976. Screening by negative chemical ionization mass spectrometry for environmental contamination with toxic residues: Application to human urine. *Proc. Natl. Acad. Sci.* 73:1777–1781.

Duxbury, C.L. and Thompson, J.E. 1987. Pentachlorophenol alters the molecular organization of membranes in mammalian cells. *Arch. Environ. Contam. Toxicol.* 16:367–373.

Eisler, R. 1989. Pentachlorophenol hazards to fish, wildlife and invertebrates: A synoptic review. U.S. Fish and Wildlife Service, Patixent Wildlife Research Center, Laurel, MD.

Engvild, K.C. 1986. Chlorine-containing natural compounds in higher plants. *Phytochemistry.* 25:781–791.

Environment Canada 1991. Effluents from pulp mills using bleaching: Priority List Assessment Report No. 2. Canadian Environmental Protection Act.

EPS. 1981. Environment Canada, Environmental Protection Service. Chlorophenols and their impurities in the Canadian environment, Report # EPS 3-EG-81-2, March 1981.

EPA. 1980a. U.S. Environmental Protection Agency. Ambient water quality criteria for chlorinated phenols. Office of Water Regulations and Standards, Criteria and Standards Division, U.S. EPA, Washington, D.C. EPA-440/5-80-032.

EPA. 1980b. U.S. Environmental Protection Agency. Ambient water quality criteria for 2,4-Dichlorophenol. Office of water regulations and standards, Criteria and Standards Division, U.S. EPA, Washington, DC, EPA-440/5-80-03.

EPA. 1982. U.S. Environmental Protection Agency. Exposure and risk assessment for chlorinated phenols (2-chlorophenol, 2,4-dichlorophenol and 2,4,6-Trichlorophenol), revised, EPA-440/4-85-007.

EPA. 1980c. U.S. Environmental Protection Agency. An Exposure and risk assessment for pentachlorophenol. EPA-440/4-81-021, October 1980.

EPA. 1979. STORET data base, U.S. Environmental Protection Agency, Washington, D.C.

EPA. 1980d. U.S. Environmental Protection Agency. Exposure and risk assessment for pentachlorophenol. Washington, D.C., NTIS PB85-211944. EPA 440/4-81-021.

EPA. 1984. Wood preservatives pesticides: creosote, pentachlorophenol, inorganic arsenicals. Position Document 4. U.S. Environmental Protection Agency, Office of Pesticides and Toxic Substances, Washington, D.C.

Falk, M.R. and Lawrence, M.J. 1973. Acute toxicity of petrochemical drilling fluid components and wastes to fish. *Environ. Canada Fish and Marine Serv., Central Region. Tech. Report.* 73–1.

Farquaharson, M.E., Gage, J.C., and Northcover, J. 1958. The biological action of chlorophenols. *Br. J. Pharmacol.* 13,20–24.

Farrington, D.S. and Munday, J.W. 1976. Determination of trace amounts of chlorophenols by gas-liquid chromatography. *Analyst.* 101:639–643.

Fielder, R.J. 1982. Pentachlorophenol. Health and Safety Commission Advisory Committee on Toxic Substances Review. U.K.

Firestone, D., Ress, J., Brown, N.L. et al. 1972. Determination of polychlorinated dibenzo-p-dioxins and related compounds in commercial chlorophenols. *J. Assoc. Off. Anal. Chem.* 55:85–92.

Fishbein, L. 1973. Mutagens and potential mutagens in the biosphere. I. DDT and its metabolites, polychlorinated biphenyls, polycyclic aromatic hydrocarbons, haloethers. *Total Environ.* 4:305.

Fountaine, J.E., Joshipura, P.B., and Keliher, P.N. 1975. Some observations regarding pentachlorophenol levels in Haverford Township, Pennsylvania. *Water Res.* 10:185–188.

Freiter, E.R. 1979. Chlorophenols. In: Kirk-Othmer Encyclopedia of Chemical Technology. Grayson, M. and Eckroth, D., Eds., Vol. 5: 3rd Ed. New York, NY. John Wiley & Sons, 864–872.
Fricke, C., Clarkson, C., and Lomnitz, E. 1985. Comparing priority pollutants in municipal sludges. *Biocycle.* 26:37–47.
Gerisch, F.M. and Milazzo, D.P. 1988. Chronic toxicity of aniline and 2,4-dichlorophenol to *Daphnia magna Straus. Bull. Enviorn. Contam. Toxicol.* 40:1–7.
Geyer, H., Scheunert, I., and Korte, F. 1985. The effects of organic environmental chemicals on the growth of the alga *Scenedesmus subspicatus*; A contribution to environmental biology. *Chemosphere.* 14:1355–1370.
Gossett, R.W., Brown, D.A., and Young, D.R. 1983. Predicting the bioaccumulation of organic compounds in marine organisms using octanol/water partition coefficients. *Mar. Pollut. Bull.* 14:387–392.
Haggblom, M.M., Apajalahti, J.H.A., and Salkinoja-Salonen, M.S. 1988. Hydroxylation and dechlorination of chlorinated guaiacols and syringols by *Rhodococcus chlorophenolicus*. *Appl. Environ. Microbiol.* 54:683–687.
Haggblom, M. and Salkinoja-Salonen, M. 1991. Biodegradability of chlorinated organic compounds in pulp bleaching effluents. *Water. Sci. Technol.* 24(3/4),161–170.
Hague, A., Schenunert, I., and Korte, F. 1978. Isolation and identification of a metabolite pentachlorophenol $-^{14}C$ in rice plants. *Chemosphere.* 7:65.
Hansch, C.C. and Leo, A. 1979. Substituent Constants for Correlation Analysis in Chemistry and Biology, John Wiley & Sons Inc. New York.
Harrison, D.L. 1959. The toxicity of wood preservatives to stock. Part I. Pentachlorophenol. *NZ. Vet. J.* 7, 89–98.
Hattemer-Frey, H.A., and Travis, C.C. 1989. Pentachlorophenol: Environmental partitioning and human exposure. *Arch. Environ. Contam. Toxicol.* 18:482–489.
Health and Welfare Canada (HWC) 1980. Phenols. In: Guidelines for Canadian Drinking Water Quality 1978. Supporting Documentation. Supply & Services Canada, Ottawa, pp. 471–488.
Health and Welfare Canada (HWC) 1988. Chlorophenols and their impurities: A health hazard evaluation, HWC, Ottawa, Canada, H-46-2/88-110.
Hedtke, S.F. and Arthur, J.W. 1985. Evaluation of a site-specific water quality criterion for pentachlorophenol using outdoor experimental streams. In: *Aquatic Toxicology and Hazard Assessment, Seventh Symposium,* Calswell, R.D., Purdy, R., and Bahner, R.C., Eds., pp. 551–564, ASTM., Philadelphia.
Hedtke, S.F., West, C.W., Allen, K.N., Norberg-King, T.J., and Mount, D.I. 1986. Toxicity of pentachlorophenol to aquatic organisms under naturally varying and controlled environmental conditions. *Environ. Toxicol. Chem.* 5:531–542.
Heitmuller, P.T. 1981. Acute toxicity of 54 industrial chemicals to sheepshead minnows (*Cyprinodon Variegatus*). *Bull. Environ. Contam. Toxicol.* 27:596–604.
Herrick, E.C., Goldfarb, A.S., and Fong, C.V. 1979. Hazards associated with organic chemical manufacturing: Chlorophenols by chlorination of phenol. MITRE Technical Report No. MTR-78W00364-05. The MITRE Corp., McLean, VA.
Hess, P., Asshauer, J., and Holander, H. 1982. The possible formation of tetrachlorodibenzo-p-dioxins in the production of chloroanil. *Ecotox. Environ. Safety.* 6:336–346.
Higgenbotham, G.R., Huang, A., Firestone, D., et al. 1968. Chemical and toxicological evaluation of isolated and synthetic chloro derivatives of dibenzo-p-dioxin. *Nature.* 220:702–703.
Hoben, H.J., Ching, S.A., and Casarett, L.J. 1976. A study of inhalation of pentachlorophenol by rats. *Bull. Environ. Contam. Toxicol.* 15:463–465.

Hodson, P.V. and Blunt, B. 1981. Temperature-induced changes in pentachlorophenol chronic toxicity to early life stages of rainbow trout. *Aquat. Toxicol.* 1:113–127.

Holcombe, G.W., Phipps, G.L., and Fiandt, J.T. 1982. Effects of phenol, 2,4-dimethyl phenol, 2,4-dichlorophenol and pentachlorophenol on embryo, larval, and early-juvenile fathead minnows (*Pimephales promelas*). *Arch. Environ. Contam. Toxicol.* 11:73–78.

IARC. 1979. International Agency for Research on Cancer. Monographs on the Evaluation of Carcinogenic Risks of Chemicals to Humans; Some halogenated hydrocarbons, Vol.20, IARC, Lyon, France.

IARC. 1986a. International Agency for Research on Cancer. Monographs on the Evaluation of Carcinogenic Risk of Chemicals to Humans. 2,3,5- and 2,4,6-trichlorophenol. 20:349–367.

IARC. 1986b. International Agency for Research on Cancer. Monographs on the Evaluation of Carcinogenic Risk of Chemicals to Humans. Occupational exposures to chlorophenols. 41:319–356.

IARC. 1979. International Agency for Research on Cancer. Monographs on the Evaluation of Carcinogenic Risk of Chemicals to Humans. World Health Organization. Lyon, France. Vol. 20, 303–325.

IARC. 1987. IARC Monographs for the Evaluation of the Carcinogenic Risk of Chemicals to Humans. Lyon, France: World Health Organization, International Agency for Research on Cancer. Overall Evaluations of Carbinogenicity: An Updating of IARC Monographs. Vols. 1 to 42.

Iwama, G.K. and Greer, G.L. 1979. Toxicity of sodium pentachloro phenate to juevenile chinook salmon under conditions of high loading density and continuous-flow exposure. *Bull. Environ. Contam. Toxicol.* 23:711–716.

Jansson, R.D. and Manske, D.D. 1977. Pesticide and other chemical residues in total diet samples (XI). *Pestic. Monit. J.* 11:116–131.

Johansen, P.H., Matthers, R.A., and Brown, J.A. 1987. Effect of exposure to several pentachlorophenol concentrations on growth of young-of-year largemouth bass, *Micropterus salmoides*, with comparison to other indicators of toxicity. *Bull. Environ. Contam. Toxicol.* 39(3), 379–384.

Johansen, P.H., Matthers, R.A., Brown, J.A., and Colgan, P.W. 1985. Mortality of early life-stages of largemouth bass, *Micropterus salomoides* due to pentachlorophenol exposure. *Bull. Environ. Contam. Toxicol.* 34:377–384.

Johnson, W.W. and Finlay, M.T. 1980. *Handbook of Acute Toxicity of Chemicals to Fish and Aquatic Invertebrates*. Department of the Interior, U.S. Fish and Wildlife Services, Washington, D.C.

Jolley, R.L., Jones, G., Pitt W.W., and Thompson, J.E. 1978. Chlorination of organics in the cooling waters and process effluents. In: *Water Chlorination: Environmental Impact and Health Effects*. Jolley R.L., Ed.,Vol. I, Ann Arbor Science Publishers, Ann Arbor, MI. pp. 105–138.

Jolley, R.L., Jones, G., and Pitts, W.W. 1978. Chlorination of organics in cooling waters and process effluents. In: *Water Chlorination: Environmental Impact and Health Effects*. Jolley R.L., Ed., Ann Arbor Science Publishers, Ann Arbor, MI. pp. 105–138.

Jones, P.A. 1984. Chlorophenols and their impurities in the Canadian environment, 1983 supplement, Environmental Protection Service, Environment Canada, EPS 3-EP-84-3.

Jones, P.A. 1981. Chlorophenols and their impurities in the Canadian environment, Environmental Protection Service, Environment Canada, EPS 3-EP-81-2.

Joshipura, P.B. and Keliher, P.N. 1980. Chlorinated phenolic compounds: formation, detection, and toxicity in drinking water. *Environ. Sci. Res.* 16:453–460.

JRB Associates. 1980. Level I materials balance — chlorophenols — Final report. Contract #68-01-5793. U.S. Environmental Protection Agency, Washington, D.C.

Kaufmann, D.D. 1978. Degradation of pentachlorophenol in soil, and by soil microorganisms. In: *Pentachlorophenol: Chemistry, Pharmacology and Environmental Toxicology,* Rao, K.R., Ed., pp. 27–39. Plenum Press, New York.

King, L. and Sherbin, G. 1986. Point sources of toxic organics to the Upper St. Clair River. *Water Pollut. Res. J. Can.* 21:433–446.

Kirby, M.K. 1986. Effect of waste discharges on the water quality of Nipigon Bay, Lake Superior, 1983. Water Resources Branch, Ontario Ministry of the Environment, Ontario, Canada, October 1986.

Kobayashi, K. and Akitake, H. 1975. Studies on the metabolism of chlorophenols in fish. I. Absorption and excretion of PCP by goldfish. *Bull. Japan. Soc. Sci. Fisheries.* 41:87–92.

Kovacs, T.G., Voss, R.H., and Wong, A. 1984. Chlorinated phenolics of bleached kraft mill origin. An olfactory evaluation. *Water Res.* 18:911–916.

Krasner, S.W., McGuire, M.J., Jacangelo, J.G., Patania, N.L., Reagan, K.M., and Aieta, E.M. 1989. The occurrence of disinfection byproducts in U.S. drinking water. *J. Am. Water Works.* 81:41–53.

Krijgsheld, K.R. and van der Gen, A. 1986. Assessment of the impact of the emission of certain organochlorine compounds in the environment. Part I. Monochlorophenols and 2,4-dichlorophenol. *Chemosphere.* 25:825–860.

Kuehl, D.W. 1981. Unusual polyhalogenated chemical residues identified in fish tissue from the environment. *Chemosphere.* 10:231–242.

Lamar, R.T. and Dietrich, D.M. 1990. *In situ* depletion of pentachlorophenol from contaminated soil by *Phanerochaete spp. Appl. Environ. Microbiol.* 56:3093–3100.

Land, B. 1974. The toxicity of drilling fluid components to aquatic biological systems (A literature review). *Dept. Environ. Fish Marine Serv. Res. Div. Tech. Report* 487.

Larsson, P., Okla, L., and Tranvik, L. 1988. Microbial degradation of xenobiotic, aromatic pollutants in humic water. *Appl. Environ. Microbiol.* 54:1864–1867.

Leach, J.M. and Thakore, A.N. 1977. Compounds toxic to fish in pulp mill waste streams. *Prog. Water. Tech.* 9:787–798.

LeBlanc, G.A. 1980. Acute toxicity of priority pollutants to water flea (*Daphnia magna*). *Bull Environ. Contam. Toxicol.* 24:684–691.

LeBlanc, G.A. 1984. Interspecies relationships in acute toxicity of chemicals to aquatic organisms. *Environ. Toxicol. Chem.* 3:47–60.

Lee, H.B., Hong-You, R.L., and Fowli, P.J. 1989. Chemical derivitization analysis of phenols. Part VI. Determination of chlorinated phenols in pulp and paper effluents. *J. Assoc. Off. Anal. Chem.* 72:979–984.

Leo, A., Hansch, C., and Elkins,D. 1971. Partition coefficients and their uses. *Chem. Rev.* 7:525–616.

Leuenberger, C., Ligocki, M.P., and Pankow, J.F. 1985. Trace organic compounds in rain. 4. Identities, concentrations, and scavenging mechanisms for phenols in urban air and rain. *Environ. Sci. Technol.* 19:1053–1058.

Leuenberger, C., Ligocki, M.P., and Pankow, J.F. 1985. Trace organic compounds in rain: identities,concentrations and scavenging mechanisms for phenols in urban air and rain. *Environ. Sci. Technol.* 19:1053–1058.

Lindstrom, K., and Mohamed, M. 1988. Selective removal of chlorinated organics from kraft mill total effluents in aerated lagoons. *Nordic Pulp Paper Res. J.* 1:26–30.

Liu, D. and Pacepavicius, G. 1990. A systematic study of the aerobic and anaerobic biodegradation of 18 chlorophenols and three cresols. *Toxic Assess.* 5:367–387.

Mackay, D. 1982. Correlation of bioconcentration factors. *Environ. Sci. Technol.* 16:274–278.

Mannsville Chemical Products Synopsis. 1987. Pentachlorophenol. Ashbury Park, NJ.

Manske, D.D. and Johnson, R.D. 1976. Pesticide and metallic residues in total diet samples (X). *Pestic. Monit. J.* 8:110–124.

Mathers, R.A., Brown, J.A., and Johansen, P.H. 1985. The growth and feeding behaviour responses of largemouth bass (*Micropterus salmoides*) exposed to PCP. *Aquat. Toxicol.* 6:157–164.

Mayer, F.L. 1987. Acute Toxicity Handbook of Chemicals to Estuarine Organisms. U.S. EPA. Report 600/9-87-017.

Mayer, F.L. and Ellersieck, M.R. 1986. Manual of Acute Toxicity: Interpretation and Data Base for 410 Chemicals and 66 Species of Freshwater animals. U.S. Fish and Wildlife Services, Resource Publication 160.

McGuire, M.J., Krasner, S.W., Reagan, K.M., Aieta, E.M., Jacangelo, J.G., Patania, N.L., and Gramith, K.M. 1989. Disinfection by-products in United States drinking waters, Final Report, Vol. I United States Environmental Protection Agency Association of Metropolitan Water Agencies.

McKague, A.B. 1981. Phenolic constituents in pulp mill streams. *J. Chromatogr.* 208:287–293.

McKim, J.M., Schmieder, P.K., Carlson, R.W., Hint, E.P., and Niemi, G.J. 1987. Use of respiratory-cardiovascular response of rainbow trout (*Salmo gairdneri*) in identifying acute toxicity syndromes in fish: Part I. Pentachlorophenol, 2,4-dinitrophenol, tricaine methane sulfonate and 1-octanol. *Environ. Toxicol. Chem.* 6:295–312.

Metcalfe, J.L., Fox, M.E., and Carey, J.H. 1984. Aquatic leeches (*Hirudinea*) as bioindicators of organic chemical contaminants in freshwater ecosystems. *Chemosphere.* 13:143–150.

Milnes, M.H. 1971. Formation of 2,3,7,8-tetrachlorodibenzo-p-dioxin by thermal decomposition. *Nature.* 232:395.

MOE (Ministry of the Environment, Ontario) 1986. Ambient level information on 32 chemicals for hazard assessment, MOE, Ontario, Canada.

MOE (Ministry of the Environment, Ontario) 1988. Ontario water treatment plants annual reports, 1987. Drinking Water Surveillance Program, MOE, Ontario, Canada.

Moore, J.W. and Ramamoorthy, S. 1984. *Organic Chemicals in Natural Waters, Applied Monitoring and Impact Assessment,* Springer-Verlag, New York. 289 p.

Muelder, W.W. and Shadoff, L.A. 1973. The preparation of uniformly labelled c-2,7-dichlorodibenzo-p-dioxin and c-2,3,7,8-tetrachlorodibenzo-p-dioxin. Chlorodioxins-origins and fate. *Am. Chem. Soc. Adv. Chem. Ser.* 120.

National Council for Air and Stream Improvement (NCASI) 1981. Experience with the analysis of EPA's organic priority pollutants and compounds chracteristic of pulp mill effluent. NCASI, Tech. Bull. 343, New York.

National Research Council of Canada (NRCC) 1982. Chlorinated Phenols: Criteria for Environmental Quality, NRCC, Publication No. 18578.

NCI. 1979. National Cancer Institute. Carcinogenesis Testing Program. Bioassay of 2,4,6-trichlorophenol for possible carcinogenicity. Bethesda, MD. DHEW/PUB/NIH-79-1711; NCI-CG-TR-155.

Neeley, W.B., Branson, D.R., and Blau, G.E. 1974. Partition coefficient to measure bioconcentration potential of organic chemicals in fish. *Environ. Sci. Technol.* 8:1113–1115.

Neilson, A.H., Allard, A.S., Hynning, P.-A., Remberger, M., Viktor, T., 1990. The environmental fate of chlorophenolic constituents of bleachery effluents. *TAPPI.* 239–247.

Neilson, A.H., Allard, A.S., Lindgren, C., Remberger, M. 1987. Transformations of chloroguaiacols, chloroveratroles and chlorocatechols by stable consortia of anaerobic bacteria. *Appl. Environ. Microbiol.* 53:2511–2519.

Neilson, A.H., Allard, A.S., Reiland, S., Remberger, M., Viktor, T., and Landner, L. 1984. Tri- and tetra-chloroveratrole metabolites produced by bacterial-*O*-methylation of tri- and tetra-chloroguaiacols: An assessment of their bioconcentration potential and their effects on fish reproduction. *Can. J. Fish. Aquat. Sci.* 41:1502–1512.

NJDEP 1989. New Jersey Department of Environmental Protection, Division of Science and Research, Support document, Trenton, NJ.

Norberg-King, T.J. 1988. An interpolation estimate for chronic toxicity: The ICP Approach. National effluent toxicity assessment center, Environmental Research Laboratory, U.S. EPA, Duluth, MN.

NTP. 1984. National Toxicology Program. Fourth annual report on carcinogens, U.S. Departartment of Health and Human Services, Washington, D.C.

Ontario Ministry of the Environment (OME) 1984. Chlorinated phenols in the environment. Scientific Criteria Document for Standard Development. No. 2-84. Water Resources Branch, Toronto, Ontario, Canada.

Olie, K., Vermeulen, P.L., and Hutzinger, O. 1977. Chlorodibenzo-p-dioxins and chlorodibenzofurans are trace components of fly ash and flue gas of some municipal incinerators in The Netherlands. *Chemosphere.* 6:455–459.

Paasivirta, J., Tenhola, H., Palm, H., and Lammi, R. 1992. Free and bound chlorophenols in kraft pulp bleaching effluents. *Chemosphere.* 24:1253–1258.

Paasivirta, J., Heinola, K., Humppi, T. et al. 1985. Polychlorinated phenols, guaiacols, and catechols in the environment. *Chemosphere.* 14:469–491.

Paasivirta, J., Sarkka, J., Leskijarvi, T., and Roos, A. 1980. Transportation and enrichment of chlorinated phenolic compounds in different aquatic food chains. *Chemosphere.* 9:441–456.

Pearce, P.J. and Simpkins, R.J.J. 1968. Acid strengths of some substituted picric acids. *Can. J. Chem.* 46:241–248.

Phillips, J.H., Zabik, M., and Leavitt, R. 1983. Analytical techniques for the quantitation of substituted phenols in municipal sludges. *Int. J. Environ. Anal. Chem.* 16:81–93.

Phipps, G.L., Holcombe, G.W., and Fiandt, J.T. 1981. Acute toxicity of phenol and substituted phenols to the fathead minnow. *Bull. Environ. Contam. Toxicol.* 26:585–593.

Pickering, Q.H. 1988. Evaluation and comparison of two short-term fathead minnow tests for estimating chronic toxicity. *Water Res.* 22:883–893.

Piet, G.J. and DeGrunt, F. 1975. Organic chloro compounds in surface and drinking water of The Netherlands. European Colloqium. Eur. 5196. Luxembourg 1974. Commission of the European Communities, Luxembourg, 81–92.

Pignatello, J.J., Martinson, M.M., Steiert, J.G., Carlson, R.E., and Crawford, R.L. 1983. Biodegradation and photolysis of pentachlorophenol in artificial freshwater streams. *Appl. Environ. Microbiol.* 46:1024–1031.

Prager, B., Jacobson, P., and Schmidt, P., Eds., 1923. *Beilsteins Handbuch der Organischen Chemie.* Vol. 6; Syst. No. 522, 4th ed. Springer Verlag, Berlin, 190.

Pulp and Paper Research Institute of Canada (PAPRICAN) 1990. Aquatic toxicity equivalency factors for chlorinated phenolic compounds present in pulp mill effluent. PAPRICAN, Pointe Claire, PQ, Canada.

Rappe, C., Gara, S., and Buser, H.R. 1978a. Identification of polychlorinated dibenzofurans (PCDFs) in commercial chlorophenol formulations. *Chemosphere.* 7:981–991.

Rappe, C., Marklund, S., and Buser, H.R. 1978b. Formation of polychlorinated dibenzo-p-dioxins (PCDDs) and dibenzofurans (PCDFs) by burning or heating chlorophenates. *Chemosphere.* 7:269–281.

Registry of Toxic Chemical Substances (RTECS) 1978. Washington, D.C.

Reineke, W. and Knackmuss, H.-J. 1988. Microbial degradation of haloaromatics. *Annu. Rev. Microbiol.* 42, 263.

Reiner, E.A., Chu, J., and Kirsch, E.J. 1978. Microbial metabolism of pentachlorophenol. In: Rao, K.R., Ed., *Pentachlorophenol.* Environmental Science Research Series, Vol. 12. Plenum Press, New York, 67–81.

Remberger, M., Allard, A.S., and Neilson, A.H. 1986. Biotransformation of chloroguaiacols, chlorocatechols and chloroveratroles in sediments. *Appl. Environ. Microbiol.* 51:552–558.

Renberg, L., Svanberg, O., Bengtsson, B.E., and Sudstrom, G. 1980. Chlorinated guaiacols and catechols bioaccumulation potential in bleaks (*Alburnus alburnus*, Pisces) and reproductive and toxic effects on the harpacticoid, *Nitocra spinipes* (crustacea). *Chemosphere.* 9:143–150.

Renner, G., Hopfer, C., and Gokel, J.M. 1986. Acute studies of pentachlorophenol, pentachloroanisole, tetrachlorohydroquinone, tetrachlorocatechol, tetrachlororesorcinol, tetrachlorodimethoxybenzenes and tetrachlorobenzenethiol diacetates administered to mice. *Toxicol. Environ. Chem.* 11:37–50.

Rogers, I.H., Birthwell, I.K., and Kruzynski, G.M. 1990. The Pacific eulachon (*Thaleichthys pacificus*) as a pollution indicator organism inn the Fraser River Estuary, Vancouver, British Columbia, Canada. *Sci. Total. Environ.* 97/98, 713–727.

Rogers, I.H., Servizi, J.A., and Levings, C.D. 1988. Bioconcentration of chlorophenols by juvenile chinook salmon (*Onchorhynchus tshawytscha*) overwintering in the Upper Fraser River: Field and laboratory tests. *Water Pollut. Res. J. Can.* 23:100–113.

Rosemarin, A., Notini, M., Soderstrom, M., Jensen, S., and Landner, L. 1990. Fate and effects of pulp mill chlorophenolic 4,5,6-trichloroguaiacol in a model brackish water ecosystem. *Sci. Total Environ.* 92:69–89.

Roy-Arcand, L. and Archibald, F.S. 1991. Direct dechlorination of chlorophenolic compounds by laccases from *Trametes* (Coriolus) *versicolor. Enzymol. Microbiol. Technol.* 13:194–293.

RTP. 1994. Regulatory Toxicology and Pharmacology. Interpretive Review of the Potential Adverse Effects of Chlorinated Organic Chemicals on Human Health and the Environment. Report of an Expert Panel, 20: Part 2. Academic Press, San Diego.

Rudling, L. 1970. Determination of pentachlorophenol in organic tissues and water. *Water Res.* 4:533–537.

Saarikoski, J. and Viuksela, M. 1981. Influence of pH on the toxicity of substituted phenols to fish. *Arch. Environ. Contam. Toxicol.* 10:747–753.

Salkinoja-Salonen, M.S., Valo, R., Apajalahti, J., Hakulinen, R., Silakoski, L., and Jaakkkola, T. 1984. Biodegradation of chlorophenolic compounds in wastes from wood-processing industry. In: *Current Perspectives in Microbial Ecology,* American Society Microbiology, Klug, M.J. and Reddy C.A., Eds., Washington, D.C. pp. 668–676.

Salkinoja-Salonen, M., Saxelin, M.L., Pere, J. et al. 1981. Analysis of toxicity and biodegradability of organochlorine compounds released into the environment in bleaching effluents of kraft pulping, In: *Advances in the Identification and Analysis of Organic Pollutants in Water,* Keith, L.H., Ed., Ann Arbor Publishing, Ann Arbor, MI. pp. 1131–1164.

Sax, N.I. 1975. *Dangerous Properties of Industrial Materials.* 4th ed., Van Nostrand Reinhold. New York.

Scow, K., Goyer, M., Perwak, J. et al. 1982. *Exposure and Risk Assessment for Chlorinated Phenols (2-chlorophenol, 2,4-dichlorophenol, 2,4,6-trichlorophenol).* Arthur D. Little, Cambridge, MA. EPA 440/4-85-007; NTIS PB85-211951.

Seiler, J.P. 1991. Pentachlorophenol, *Mutat. Res.* 257:27–47.

Servizi, J.A., Gordon, R.W., and Carey, J.H., 1988. Bioconcentration of chlorophenols by early life stages of Fraser River pink and Chinook salmon (*Onchorhynchus gorbuscha,* and *O. Tshawytscha*). *Water Pollut. Res. J. Can.* 23:88–99.

Sithole, B.B. and Williams, D.T. 1986. Halogenated phenols in water at forty Canadian potable water treatment facilities. *J. Assoc. Off. Anal. Chem.* 69:80–810.

Sittig, M. 1980. *Priority Toxic Pollutants, Health Impacts and Allowable Limits.* Noyes Data Corp. NJ. 370 p.

Siuda, J.F. and DeBernardis, J.F. 1973. Naturally occurring halogenated organic compounds. *Lloydia.* 36(2), 107–143.

Smith, A., Bharath, A., Mallard, C., Orr, D., Smith, K., Sutton, J.A., McCarty, L.S., and Ozburn, G.W. 1991. The acute and chronic toxicity of ten chlorinated organic compounds to the American flagfish (*Jordanella floridae*). *Arch. Environ. Contam. Toxicol.* 20:94–102.

Staples, C.A., Werner, A.F., and Hoogheem, T.J. 1985. Assessment of priority pollutant concentrations in the United States using STORET database. *Environ. Toxicol. Chem.* 4:131–142.

Sugiura, K., Aoki, M., Kaneko, S., Daisaku, I., Komatsu, Y., Shibuya, H., Suzuki, H., and Goto, M. 1984. Fate of 2,4,6-trichlorophenol, pentachlorophenol, p-chlorobiphenyl and hexachlorobenzene in an outdoor experimental pond comparison between observations and predictions based on laboratory data. *Arch. Environ. Contam. Toxicol.* 13, 745–748.

The Chemical Daily Co. 1984. *JCW Chemicals Guide 1984–85.* Tokyo: 312.

Tranvik, L., Larsson, P., Okla, L., and Regnell, O. 1991. *In situ* mineralization of chlorinated phenols by pelagic bacteria in lakes of differing humic content. *Environ. Toxicol. Chem.* 10:195–200.

U.S. Tariff Commission. 1951. Synthetic organic chemicals, U.S. production and sales 1950. U.S. Government Printing Office, Report No. 173, second series, 130. Washington, D.C.

USITC. 1975. Synthetic organic chemicals, U.S. production and sale, 1974. U.S. International Trade Commission, Washington, D.C., USITC Publication No. 776. 187.

USITC. 1981. Imports of benzenoid chemicals and products — 1980. U.S. International Trade Commission, Washington, D.C., USITC Publication No. 1163.

Van Langeveld, HEAM. 1975. Determination of pentachlorophenol in toy paints. *J. Assoc. Off. Anal. Chem.* 58, 19–22.

Veith, G.D., Defoe, D.L., and Bergstedt, B.V. 1979. Measuring and estimating the bioconcentration factor of chemicals in fish. *J. Fish Res. Bd. Can.* 36:1040–1048.

Versar. 1979. Water-related environmental fate of 129 priority pollutants. Vol. II. U.S. Environmental Protection Agency, Publication No. EPA-440/4-79-029b. Washington, D.C.

Voss, R.H., Wearing, J.T., Mortimer, R.D., Kovacs, T., and Wong, A. 1980. Chlorinated organics in kraft bleachery effluents. *Pap. Puu.* 12:809–814.

Voss, R.H. and Yunker, M.G. 1983. A study of chlorinated phenolics discharged into kraft mill receiving waters. Report prepared for the council of forest industries of British Columbia, Canada. p.131.

Waggott, A. and Whestland, A.B. 1978. Contribution of different sources to contamination of surface waters with specific persistent organic pollutants. In: *Aquatic Pollutants: Transformation and Biological Effects.* Hutzinger, O. van Lelyveld, L.H., and Zoetman, B.C.J., Eds., *Proceedings of the Second International Symposium on Aquatic Pollutants,* Amsterdam, The Netherlands, September 26–28, 1977. pp. 141–168.

Wannstedt, C., Rotella, D., and Siuda, J.F. 1990. Chloroperoxidase mediated halogenation of phenols. *Bull. Environ. Contam. Toxicol.* 44:282–287.

Weast, R.C., Ed. 1974. *CRC Handbook of Chemistry and Physics, 55th ed.,* CRC Press, Boca Raton, FL.

Webb, P.W. and Brett, J.R. 1973. Effects of sublethal concentrations of sodium pentachlorophenate on growth rate, food conversion efficiency and swimming performance in underyearling sockeye salmon (*Onchorhynchus nerka*). *J. Fish. Res. Bd. Can.* 30:499–507.

WHO (World Health Organization) 1989. Chlorophenols other than pentachlorophenol, WHO, Environmental Health Criteria 93, Geneva.

WHO (World Health Organization) 1987. Pentachlorophenol, WHO, Environmental Health Criteria 71. p. 236, Geneva.

Wong, A.S. and Crosby, D.G. 1978. Photolysis of pentachlorophenol In: Pentachlorophenol: Chemistry, Pharmacology and Environmental Toxicology. Rao, K.R., Ed., Plenum Press, New York, pp.15–19.

Wong, A.S. and Crosby, D.G. 1981. Photodecomposition of pentachlorophenol in water. *J. Agr. Food Chem.* 29, 130–135.

Yoshida, K., Shigeoka, T., and Yamamuchi, F. 1987. Evaluation of aquatic environmental fate of 2,4,6-trichlorophenol with a mathematical model. *Chemosphere.* 15:2531–2544.

Young, H.F., Lau, L., and Konno, S.K. 1976. Water quality monitoring; Kaneohe Bay and selected watersheds July to December 1975 VL. Technical Report No. 98.

Zitko, V., Hutzinger, O., and Choi, P.M.K. 1994. Determination of pentachlorophenol and chorobiphenylols in biological samples. *Bull. Environ. Contam. Toxicol.* 12:649–53.

CHAPTER 8

Chlorinated Dioxins and Furans

Polychlorinated dibenzo-*p*-dioxins (PCDDs) and polychlorinated dibenzofurans (PCDFs) are almost ubiquitous in the environment because of the numerous pathways of their formation in industrial processes, including emissions of incinerator systems, in various wastewaters, and bleach-kraft pulpmill effluents. There are 75 homologues and isomers of PCDDs, ranging from two monochloro dioxins to the fully chlorinated octachloro dioxins (Table 8.1). They vary from each other in their physical properties and toxicities to the ecosystem and human health.

PCDDs exhibit the following characteristics:

1. A relatively stable aromatic nucleus
2. Many isomers with parallel toxic properties
3. An increase in chlorine content in the two benzene rings which increases their environmental stability, lipophilicity, thermal stability, and resistance to acids, bases, oxidants, and reductants
4. Property 3 accounts for the widespread presence of PCDDs in the environment

PHYSICAL AND CHEMICAL PROPERTIES

PCDDs have been detected in trichlorophenol (TCP), tetrachlorophenol (TeCP), and pentachlorophenol (PCP), and in chlorophenol derivatives such as 2,4-dichloro phenoxyacetic acid (2,4-D) and 2,4,5-trichloro phenoxyacetic acid (2,4,5-T). Since combustion of organic materials and seepage from chemical dumps apparently also generate detectable residues in the environment, PCDDs are now known to be ubiquitous contaminants in both aquatic and terrestrial ecosystems.

Table 8.2 lists the chemical and physical properties of the most toxic PCDD, namely, 2,3,7,8-tetrachlorodibenzo-*p*-dioxin (2,3,7,8-T4CDD). 2,3,7,8-T4CDD is stable in the presence of heat, acids, and alkalies but begins to decompose at 500°C. The decomposition is virtually complete within 21 s at 800°C. It is susceptible to photodegradation under UV light, particularly in the presence of a hydrogen-donating solvent.

Table 8.3 summarizes the important properties of the homologues of PCDDs from mono- to octachloro dibenzo-*p*-dioxins. Both vapor pressure and solubility

Table 8.1 A List of PCDD Isomers (Total of 75 Isomers)

Monochloro-(2)		Pentachloro-(14)	
1-		1,2,3,4,6-	1,2,3,8,9-
2-		1,2,3,4,7-	1,2,4,6,7-
		1,2,3,6,7-	1,2,4,6,8-
		1,2,3,6,8-	1,2,4,6,9-
Dichloro-(10)		1,2,3,6,9-	1,2,4,7,8-
		1,2,3,7,8-	1,2,4,7,9-
1,2-	1,8-	1,2,3,7,9-	1,2,4,8,9-
1,3-	1,9-		
1,4-	2,3-	**Hexachloro-(10)**	
1,6-	2,7-		
1,7-	2,8-	1,2,3,4,6,7-	
		1,2,3,4,6,8-	
Trichloro-(14)		1,2,3,4,6,9-	
		1,2,3,4,7,8-	
1,2,3-	1,3,7-	1,2,3,6,7,8-	
1,2,4-	1,3,8-	1,2,3,6,7,9-	
1,2,6-	1,3,9	1,2,3,6,8,9-	
1,2,7-	1,4,6-	1,2,3,7,8,9-	
1,2,8-	1,4,7-	1,2,4,6,7,9-	
1,2,9-	1,7,8-	1,2,4,6,8,9-	
1,3,6-	2,3,7-		
Tetrachloro-(22)		**Heptachloro-(2)**	
1,2,3,4-	1,2,6,9-	1,2,3,4,6,7,8-	
1,2,3,6-	1,2,7,8-	1,2,3,4,6,7,9-	
1,2,3,7-	1,2,7,9-		
1,2,3,8-	1,2,8,9-		
1,2,3,9-	1,3,6,8-	**Octachloro-(1)**	
1,2,4,6-	1,3,6,9-		
1,2,4,7-	1,3,7,8-	1,2,3,4,6,7,8,9-	
1,2,4,8-	1,3,7,9-		
1,2,4,9-	1,4,6,9-		
1,2,6,7-	1,4,7,8-		
1,2,6,8-	2,3,7,8-		

The numbers in parentheses indicate the number of isomers for that homologue of PCDD.

Source: NRCC 1981a

decrease with increase in the chlorine content of PCDDs, whereas, the log octanol/water partition coefficient (log K_{ow}) increases. The decrease in vapor pressure is almost 500,00 fold from mono-CDD to hepta-CDD congener. Similarly, the variation in solubility among the eight congeners of PCDDs is about 11 million-fold (from 295 µg/L to 7.4×10^{-4} µg/L, Table 8.3).

Dibenzofuran is an organic compound that has two benzene rings fused to a central furan ring. Chlorinated dibenzofurans is a class of compounds in which one to eight chlorine atoms are attached to the benzene ring positions of a dibenzofuran structure. There are eight homologues, monochlorinated to octachlorinated CDFs. There are 135 possible CDF isomers, including 4 mono-CDFs, 16 di-CDFs, 28 tri-CDFs, 38 tetra-CDFs, 28 penta-CDFs, 16 hexa-CDFs, 4 hepta-CDFs, and one

Table 8.2 Physical and Chemical Properties of 2378-T4CDD

Chemical name:	2,3,7,8-tetrachlorodibenzo[b,e](1,4)-dioxin
	2,3,7,8-tetrachlorodibenzo-p-dioxin;dioxin
	2,3,7,8-TCDD;
	2,3,7,8-tetrachlorodibenzo-1,4-dioxin
Trade name:	None. The compound is not produced commercially.
Chemical formula:	$C_{12}H_4Cl_4O_2$
Chemical structure:	

Identification Nos.

CAS registry no.:	1746-01-6 (SANSS 1987)
NIOSH/RTECS no.:	HP3500000 (SANSS 1987)
EPA hazardous waste no.:	Not assigned (HSDB 1987)
OHM/TADS no.:	8300192 (SANSS 1987)
DOT/UN/NA/IMCO shipping no.:	Not assigned (HSDB 1987)
Hazardous substances data bank (HSDB) no.:	4151
NCI no.:	NCI-C03714

Property	Value	References
Molecular weight	321.97	
Physical state	Solid at room temperature	
Melting point	305°C	Schroy et al. 1985, 1986
Boiling point	412.2°C (estimated)	Schroy et al. 1985
Solubility water (ng/L)	7.91 (20–22°C);	Adams and Blaine 1986
	19.3 (22°C)	Marple et al. 1986a
Organic solvents (mg/L)	o-Dichlorobenzene, 1400; chlorobenzene, 720; benzene, 570; chloroform, 370; methanol, 10; acetone, 110	Schroy et al.1985
Partition coefficients	Log K_{ow}, 6.15–7.28	EPA 1985
	Log K_{oc}, 6.0–7.39	Schroy et al. 1985
		Jackson et al.1986
		Marple et al. 1986b
Vapor pressure (mm Hg)	1.52×10^{-9} (25°C)	Schroy et al. 1985
	1.4×10^{-9} (estimated, 25°C)	
Henry's law constant atm–m³/mol. 25°C	8.1×10^{-5} (estimated from water solubility and vapor pressure)	Palansky et al. 1986
	6.4×10^{-4} (estimated)	Podoll et al. 1986
Conversion factors vapor	1 ppb = 13.384 µg/m³ (20°C)	

Source: ATSDR 1989.

octa-CDF. Each of these compounds is called a congener and there are 135 CDF congeners, due to molecular asymmetry, compared to 75 CDD congeners. PCDFs are relatively stable under acid and alkali attack, but they start to decompose at 700°C. In general, the melting points increase and vapor pressures and water solubilities of CDFs decrease with an increase in chlorination. These hydrophobic compounds are generally colorless solids and are soluble in nonpolar organic

Table 8.3 Selected Physical and Chemical Properties of PCDDs

Homologue	Mol. Wt.	Vapor Pressure Pa × 10⁻³	Water solubility (µg/L)	Log K_{ow}
M1CDD	218.5	73–75	295–417	4.75–5.00
D2CDD	253.0	2.47–9.24	3.75–16.7	5.60–5.75
T3CDD	287.5	1.07	8.41	6.35
T4CDD	322.0	0.00284–0.275	0.0193–0.55	6.60–7.10
P5CDD	356.4	0.00423	0.118	7.40
H6CDD	391.0	0.00145	0.00442	7.80
H7CDD	425.2	0.000177	0.0024	8.0
O8CDD	460.0	0.000953	0.00004	8.20

Pa × 7.50062 × 10⁻³ = mm of Hg; M1CDD = Monochlorodibenzo-*p*-dioxin; D2CDD = Dichlorodibenzo-*p*-dioxin; T3CDD = Trichlorodibenzo-*p*-dioxin; T4CDD = tetrachlorodibenzo-*p*-dioxin; P5CDD = pentachlorodibenzo-*p*-dioxin; H6CDD = Hexachlorodibenzo-*p*-dioxin; H7CDD = Heptachlorodibenzo-*p*-dioxin and O8CDD = Octachloro dibenzo-*p*-dioxin.
Source: Mackay et al. 1992.

solvents (ATSDR 1994). Table 8.4 lists the chemical identity, physical, and chemical properties of 2,3,7,8-tetrachlorodibenzofuran (2,3,7,8-T4CDF).

Table 8.4 Chemical Identity, Physical, and Chemical Properties of 2,3,7,8-T4CDF

Chemical name:	2,3,7,8-Tetrachlorodibenzofuran
	2,3,7,8-Tetrachlorodiphenylene oxide
Trade name:	None. The compound is not produced commercially
Chemical formula:	$C_{12}H_4Cl_4O$
Chemical structure:	
Identification numbers:	
CAS registry:	51207-31-9
NIOSH/RTECS:	HP5295200
HSDB:	4306
NCI:	C56611

Source: ATSDR 1994

Compound	CAS No.	Formula	Mol. Wt. g/mol.	Melting Point °C	Boiling Point °C
2,8-D2CDF	5409-83-6	$C_{12}H_6OCl_2$	237.1	184	375
2,3,7,8-T4CDF	51207-31-9	$C_{12}H_4OCl_4$	306.0	227	438.3
2,3,4,7,8-T5CDF	57117-31-4	$C_{12}H_3OCl_5$	340.42	196	464.7
1,2,3,4,7,8-H6CDF	70648-26-9	$C_{12}H_2OCl_6$	374.87	225	487.7
1,2,3,6,7,8-H6CDF	57117-44-9	$C_{12}H_2OCl_6$	374.87	232	487.7
1,2,3,4,6,7,8-H7CDF	67562-39-4	$C_{12}HOCl_7$	409.31	236	507.2
1,2,3,4,7,8,9-H7CDF	55673-89-7	$C_{12}HOCl_7$	409.31	221	507.2
O8CDF	39001-02-0	$C_{12}OCl_8$	443.8	258	537

Source: Mackay et al. 1992

Compound	Vapor Pressure* Pa	Water Solubility (μg/L)	Log K_{ow}	Henry's Law Const Pa.m³/mol
2,8-D2CDF	1.46×10^{-2}	14.5	5.44	6.377
2,3,7,8-T4CDF	1.99×10^{-4}	0.419	6.1	1.461
2,3,4,7,8-P5CDF	1.72×10^{-5}	0.236	6.5	0.505
1,2,3,4,7,8-H6CDF	3.08×10^{-6}	0.00825	7.0	1.454
1,2,3,6,7,8-H6CDF	3.61×10^{-6}	0.0177	—	0.741
1,2,3,4,6,7,8-H7CDF	5.74×10^{-7}	0.00135	7.4	1.425
1,2,3,4,7,8,9-H7CDF	5.39×10^{-7}	—	—	—
O8CDF	1.01×10^{-7}	0.00116	8.0	0.191

D2CDF = Dichlorodibenzofuran; T4CDF = Tetrachlorodibenzofuran; P5CDF = Pentachlorodibenzofuran; H6CDF = Hexachlorodibenzofuran; H7CDF = Heptachlorodibenzofuran; O8CDF = Octachlorodibenzofuran.

— = Data not available; * = Liquid
Source: Mackay et al. 1992.

The following values have been reported for the vapor pressure and Henry's law constant for selected PCDFs (Eitzer and Hites 1988; 1989).

Compound	Vapor Pressure mm Hg	Henry's Law Const. atm.m³/mol	Log K_{ow}	Log K_{oc}
1,3,7,8-T4CDF	—	1.48×10^{-5}	—	—
2,3,7,8-T4CDF	9.21×10^{-7}	1.48×10^{-5}	5.82	5.61 (estimated)
1,2,3,4,8-P5CDF	—	2.63×10^{-5}	6.79	—
1,2,3,7,8-P5CDF	2.73×10^{-7}	2.63×10^{-5}	6.79	—
2,3,4,7,8-P5CDF	1.63×10^{-7}	2.63×10^{-5}	6.92	—
1,2,3,4,7,8-H6CDF	6.07×10^{-8}	2.78×10^{-5}	—	—
1,2,3,6,7,8-H6CDF	6.07×10^{-8}	2.78×10^{-5}	—	—
1,2,3,7,8,9-H6CDF	3.74×10^{-8}	2.78×10^{-5}	—	—
1,2,3,4,6,7,8-H7CDF	1.68×10^{-8}	4.1×10^{-6}	7.92	—
O8CDF	—	1.7×10^{-6}	8.20 (7.97)	8.57 (estimated)

Source: ATSDR, TP-93/04, 1994
— = Data not available.

As in the case of PCDDs, a number of physical and chemical properties show a trend relative to the chlorine content of the furan congeners. With the increase in chlorine content, both vapor pressure and solubility decrease, while the log K_{ow} increases. The vapor pressures range almost 150,000-fold, from a value of 1.4×10^{-2} Pa for 2,8-dichlorodibenzofuran, to 1.0×10^{-7} Pa for octachlorodibenzofuran (Table 8.4). The reported aqueous solubilities vary over approximately 12,500-fold from the most soluble 2,8-dichlorodibenzofuran (D2CDF), at 14.5 μg/L, to 1.16×10^{-3} μg/L for octachlorodibenzofuran (O8CDF). The log K_{ow} values range over ~2.55 log units (about 320-fold), from a value of 5.44 for 2,8-D2CDF, to 8.0 for O8CDF (Table 8.4). As a result of hydrophobic tendencies increasing with the increase in chlorine content of furans, including D2CDFs (Table 8.4), all CDFs in the environment would likely be associated with organic particulate matter or dissolved high molecular organic compound.

PRODUCTION AND USE PATTERN

PCDDs and PCDFs are not commercially produced and have no direct use. They are inadvertently formed during the production of 2,4,5-trichlorophenol from 1,2,4,5-tetrachlorobenzene. PCDDs and PCDFs are not imported into North America and Europe. Since 2,4,5-TCP is used in the production of numerous herbicides and preservatives, PCDDs have been reported to be a contaminant of 2,4,5-T, 2,4,5-esters, 2,4-D, Clophen, and Silvex.

PCDFs are produced as undesirable products during the manufacture of PCBs, polychlorinated phenols, and herbicides, such as Agent Orange. They are also formed during the pyrolysis of PCBs, polychlorinated phenols, polychlorinated diphenyl ethers (PCDEs), polychlorinated benzenes, and phenoxy herbicides. Municipal and industrial incinerators also produce PCDFs. PCDFs can also be formed by photolysis of PCBs, PCDEs, and polychlorinated benzenes (van den Berg et al. 1985). Other than in research, there are no known uses of chlorinated dioxins and furans.

SOURCES IN THE ENVIRONMENT

Natural Sources

Chlorinated dioxins and furans were detected in soil and suspended particulate matter in the area of combustion processes. Combustion of pre-existing polychlorinated phenols could not account for all of the PCDDs and PCDFs detected in particulate matter, soil, and dust (Bumb et al. 1980). Therefore, it was concluded that natural combustion processes, such as forest fires and slash-burning, must be considered when accounting for PCDDs and PCDFs in the environment (Bumb et al. 1980).

Chlorinated dioxins were detected near fireplaces, charcoal burners, and cigarettes (Bumb et al. 1980). Woodburning stoves were estimated to produce 10 to 50 g Nordic TEQ annually (Vikelsoe et al. 1991-Danish Surveys). Residential combustion of untreated wood was reported to produce chimney soot containing PCDDs ranging from 2 to 2,500 ppt (Nestrick and Lamparski 1982). Octa-CDDs were detected in ash from heavily PCP-treated wooden boxes burned in the open air, at 27,000 ppt (Chiu et al. 1983). Open air burning produced the highest levels of PCDDs in a variety of samples studied (Clement et al. 1985). Levels of PCDFs around 31,000 ppt were detected when untreated wood was burned in open fire stoves (Clement et al. 1985). In another analysis of particulates in samples of air, ash, and soil, collected from sites of controlled forest fires (Tashiro et al. 1990), pre-burn samples (high-volume air samples) showed the presence of only two congeners of PCDDs at 0.03 pg/m^3. After the test burn, air samples showed the presence of low levels of T4CDD, O8CDD, H7CDF, and O8CDF. However, during the test burn, much higher levels of PCDDs and PCDFs were detected in the air samples; tetra-CDDs, ND-12 pg/m^3; penta-CDDs, ND-80 pg/m^3; hexa- and hepta-CDDs, ND-100 pg/m^3; and O8CDD, 8-200 pg/m^3. Three out of four soil samples, after the test burn, were shown to contain 46, 100, and 270 ppt of O8CDD. The 100 ppt sample also contained 200 and 110 ppt of P5CDD and H7CDD, respectively. No firm conclusions could be drawn as to whether forest fires are significant natural sources of PCDDs and PCDFs in soils.

From 1930 to 1970, sediment cores from Siskiwit Lake on Isle Royale from Lake Superior have shown an increasing trend (DL = 0.4 pg/g) to 800 pg/g. Octa- and hepta-CDFs were the only congeners detected at significant levels in sediment cores (Hites 1990). It was concluded from this study that anthropogenic combustion sources were likely to be the contributing sources, since the forest fires should not have accounted for the trend observed during this time period.

Anthropogenic Sources

The main sources of 2,3,7,8-T4CDD in the environment are from the production and use of certain herbicides and chlorophenols, and the improper disposal of chemical wastes generated during the manufacture of 2,4,5-trichlorophenol and 2,4,5-trichlorophenoxy aceticacid (2,4,5-T), related herbicides, hexachlorophene, and chlorinated benzenes. The bleaching process in the pulp and paper industry releases dioxins and furans into the environment.

Releases from Chlorophenols and Chlorinated Pesticides

The phenoxy herbicide 2,4,5-T, produced prior to 1960, was reported to contain up to 100 µg/g of 2,3,7,8-T4CDD. That level dropped significantly to <0.02 µg/g, before the ban of 2,4,5-T. Agent Orange, a mixture of butyl esters of 2,4,5,T and 2,4-D, produced prior to 1970, contained 0.02 to 54 µg/g 2,3,7,8-T4CDD.

PCDDs and PCDFs are formed as by-products in the manufacture of chlorophenols as a result of the chemical process which requires temperatures higher than 95°C. This is conducive for the formation of PCDDs and PCDFs (Environment Canada 1981). The concentrations of H6CDD, H7CDD, and O8CDD congeners detected in commercial grade tetra- and penta-chlorophenols to using the GC-ECD method, were ND to 38 ppm (average, 12.7 ppm); ND to 39 ppm (average, 12.5 ppm), and ND to 15 ppm (average, 3.9 ppm), respectively (Firestone et al. 1972). Lower chlorinated dioxins and furans were detected in some chlorophenols. Many chlorophenols did not show the presence of 2,3,7,8-T4CDD, possibly due to lack of precursors which are essential in the formation of 2,3,7,8-T4CDD (Environment Canada 1981). Presence of hexa-, hepta-, and octa-CDDs in tri-, tetra-, and pentachlorophenols were confirmed in another study (Woolson et al. 1972). Six of the PCP samples showed the presence of 100 to 1,000 ppm of H6CDD; the technical grade PCP (86% pure) contained >500 times more of H6CDD, H7CDD, and O8CDD than the 95% pure analytical grade PCP (Villanueve et al. 1973). However, around 1974, an improved process by the manufacturer produced chlorophenols considerably lower in PCDD and PCDF content. Average concentrations of hexa-, hepta-, and octa-CDDs were 1, 6.5 and 15 ppm, respectively, and hexa-, hepta-, and octa-CDFs were <1 10 1.8 ppm (Watson and Kobel 1974). A sample of PCP manufactured in the U.S. was analyzed to contain a total PCDD content of 1,000 ppm and PCDFs content of 280 ppm; the levels for tetra- to octa-CDFs were reported to be 0.9, 4, 32, 120, and 130 ppm (Rappe et al. 1978). Commercial grade chlorophenols produced in Canada contained tetra-, hexa-, hepta-, and octa-CDDs at an average concentration of 0.008, 10, 140 and 370 ppm with no P5CDD reported (Agriculture Canada 1984). Effluents from wood-preservation plants were also sources of PCDDS and PCDFs in the range of 21 to 424 and 3 to 99 ng/L, respectively. The annual release of PCDDs and PCDFs from these six wood-preservation plants was estimated to be 1040 g and 580 g, respectively.

The production of chlorophenols has been discontinued in Canada; it is still being imported, for greatly reduced use in the pesticide and wood-preservation industries (Sheffield 1985). The level of PCDFs in commercial chlorophenols varies depending on the level of isomeric congeners; this is due to different degrees of chlorination and methods of synthesis. For example, 2,4,6-trichlorophenol and pentachlorophenol from different countries were shown to contain varying levels of PCDDs and PCDFs (Table 8.5). The predominant PCDFs were (1) 1,2,4,6,8-penta-; (2) 1,2,3,4,6,8-hexa-; (3) 1,2,4,6,7,8-hexa-; (4) 1,2,4,6,8,9-hexa-; (5) 1,2,3,4,6,7,8-hepta-, and (6) 1,2,3,4,6,8,9-hepta-CDFs (Rappe and Buser 1981).

Table 8.5 Concentrations of PCDDs and PCDFs Detected in Commercial Chlorophenols (µg/g)

Congener	PCDFs Tetra	Penta	Hexa	Hepta	Octa	ΣPCDFs	ΣPCDDs
2,4,6-TCP (Sweden)	1.5	17.5	36	4.8	—	60	<3
2,4,6-TCP (U.S.)	1.4	2.3	0.7	<0.02	—	4.6	0.3
PCP (U.S.)	—	—	30	80	80	190	2,625
PCP (U.S.)	≤0.4	40	90	400	260	790	1,900
PCP (Germany)	—	—	0.03	0.8	1.3	2.1	6.8

PCDDs = Polychlorinated dibenzo-p-dioxins; PCDFs = Polychlorinated dibenzofurans; TCP = Trichlorophenol; PCP = Pentachlorophenol.
Source: Rappe and Buser 1981.

Commercial PCP and NaPCP, extensively used for wood preservation, contained trace amounts of PCDFs (Hagenmaier and Brunner 1987). PCDFs tend to volatilize from wood surfaces and contaminate indoor air.

The concentrations of PCDFs, measured in indoor ambient air of a kindergarten building in West Germany, using PCP-treated wood, were as follows:

Congener	Concentration (pg/m³)
TetraCDFs (not including 2,3,7,8-T4CDF)	0.27
1,2,3,7,8-PentaCDF	0.1
PentaCDFs (other than 2,3,7,8-congener)	3.51
1,2,3,4,7,8-HexaCDF	0.37
1,2,3,6,7,8-HexaCDF	0.60
1,2,3,7,8,9-HexaCDF	0.16
HexaCDFs (other than 2,3,7,8-congener)	12.3
1,2,3,4,6,7,8-HeptaCDF	10.7
1,2,3,4,7,8,9-HeptaCDF	0.38
HeptaCDFs (other than 2,3,7,8-congener)	12.2
Octa-CDF	6.0

Source: Mukerjee et al. 1989

Use of certain commercial products can be a source of PCDFs in ambient air. Releases of PCDDs and PCDFs from waste are mainly related to source points such as: Love Canal, a dumping ground for municipal and chemical wastes from 1940 to 1953; a landfill from a trichlorophenol plant in Verona, MO, and Gulf Port, MS, where Agent Orange was stored (Weerasinghe and Gross 1985).

PCDFs are present as contaminants in PCBs and are released into the environment during manufacture, use, and disposal. Table 8.6 lists the levels of PCDFs detected in some commercial aroclors, clophens, and phenoclors. Phenoxy herbicides generally contain higher concentrations of PCDDs than PCDFs. One sample of Agent Orange (a 50:50 mixture of n-butyl esters of 2,4-D and 2,4,5-T) contained one tri-, four tetra- (non-2,3,7,8- isomers), and penta-CDFs, a total of 0.7 µg/g (EPA 1986a).

Table 8.6 Concentrations of PCDFs in Commercial PCBs (µg/g)

Compound	Tri	Tetra	Tetra (non-2378)	Penta	Hexa	ΣPCDFs
Aroclor 1248, 1969[a]	—	0.5	0.3	1.2	—	2.0
Aroclor 1254, 1969[a]	—	0.1	1.4	0.2	—	1.7
Aroclor 1254, 1970[a]	—	0.2	0.9	0.4	—	1.5
Aroclor 1254[b]	0.10	0.25	0.81	0.70	—	1.9
Aroclor 1254 (lot KK 602)[b]	—	0.05	0.02	0.10	—	0.2
Aroclor 1260, 1969[a]	—	0.1	0.5	0.4	—	1.0
Aroclor 1260 (lot AK 3)[a]	—	0.2	0.3	0.3	—	0.8
Aroclor 1260[b]	0.06	0.30	1.10	1.0	1.35	3.8
Aroclor 1016, 1972[a]	—	<0.001	<0.001	<0.001	—	—
Clophen A 60[a]	—	1.4	2.2	5.0	—	8.4
Clophen T 64[b]	0.10	0.30	2.45	1.73	0.82	5.4
Phenoclor DP-6[a]	—	0.7	2.9	10.0	—	13.6

[a] Bowes et al. 1975a
[b] Rappe and Buser 1981

Pulp and Paper Mill Sources

A study in Sweden estimated that bleached pulp kraft mills released a total of 2,3,7,8-T4CDD TEQs (Toxic Equivalency Factors) (Eadon et al. 1986) of 5 to 15g annually. The Canadian Pulp and Paper Mill Association (CPPA) conducted a National Pulpmill Dioxin Survey to calculate the release from 45 bleached kraft mills through their final effluents (CPPA 1990). The 2,3,7,8-T4CDD TEQ (using International Toxic Equivalency Factors, ITEFs) isomer output alone was estimated to range from 185.78 g/year to 235.12 g/year. The lower values were arrived at on the assumption that the output is zero when the concentration in the final effluent was less than the detection limit. The higher value, on the other hand, assumed that the output was equal to the detection limit, when the concentration in the final effluent was less than the detection limit. The total 2,3,7,8-T4CDD TEQ, including all chlorinated dioxin and furan isomers and congeners, was estimated to be between 290.76 (nondetectables assumed to be equal to zero) to 510.90 (nondetectables assumed equal to detection limit) per year. The detection limit varied from 14 to 410 parts per quadrillion (ppq = 10^{-15} g) and, therefore, the upper estimate (assuming the nondetectables to be equal to detection limit) is highly conservative (CPPA 1990). In this study, 11 out of the 45 bleached kraft mills had no detectable concentrations of any PCDDs/PCDFs, whereas, four mills reported no detectable 2,3,7,8-T4CDD/2,3,7,8-T4CDF isomers. It shows that the mean output of PCDDs/PCDFs would only be slightly above the detection limit. Large outputs of PCDDs and PCDFs from a few mills in this study considerably increased the average. The future annual output of 2,3,7,8-T4CDD TEQ from all pulp mills in Canada would be less than 18.1 g. (10 pg/L × 110.16 × 10^6 liters/day × 365 days = 18.09 g/year).

Sludge from seven pulp and paper mills from a different study showed a 2,3,7,8-T4CDD concentration range of ND(1 pg/g) to >400 pg/g (Kuehl et al. 1987).The chlorine bleaching process in the pulp and paper industry has been shown to form PCDDs and PCDFs. Several mills have changed to a nonchlorine bleaching process and/or a chemo-thermo-mechanical pulping process.

Thermal Processes

Combustion Processes

PCDDS and PCDFs are formed during a variety of combustion processes (EPA 1987a,b; Stieglitz and Vogg 1987; Ballschmitter et al. 1988) Incinerators fitted with best available technologies (BATs) to incinerate hazardous chemicals and municipal wastes release only low concentrations of PCDDs and PCDFs (Tiernan et al. 1983; Morita et al. 1987; Acharya et al. 1991). Incinerators not employing BATs, with respect to operating conditions and emission treatment trains, were shown to release significantly higher concentrations of PCDDs and PCDFs in stack and fly ash samples, in the range of 1 to 1,712.0 ppb (Chiu et al. 1983). Power plant fly ash contained 1.0 to 32.0 ppb of PCDDs/PCDFs. Fly ash, a fine particulate matter in emissions, trap almost all of the PCDDs/PCDFs (Olie et al. 1983). About 35,000 tons of fly ash are estimated to be produced for each million tons of waste incinerated. Fly ash collected by emission abatement techniques end up in landfills and only a small amount is leached, <0.4 ppb for each congener group of tetra- to octa-PCDDs/PCDFs (Karasek and Dickson 1987).

The combustion processes can be divided into large and small systems. Large systems include municipal waste incineration (Bonafanti et al. 1990; Brna and Kilgore 1990; des Rosiers 1987; Tiernan et al. 1985; Tong and Karasek 1986), industrial and hazardous waste incineration (des Rosiers 1987; Muto et al. 1991), and power plants and fossil fuels (des Rosiers 1987; Hutzinger and Fiedler 1989). Small combustion systems include home heating and fireplaces (Clement et al. 1985; Safe 1990a), household waste incineration (Harrad et al. 1991), automobile exhaust (Ballschmiter et al. 1986a,b; Marklund et al. 1987), and medical waste incineration (des Rosiers 1987; Glasser et al. 1991; Lindner et al. 1990). Incineration of industrial and hazardous wastes that produce PCDFs include wastes containing PCBs, polychlorinated diphenyl ethers, 2,4,5-trichlorophenol esters (Choudhury and Hutzinger 1982; Sedman and Esparza, 1991) and chlorinated benzenes (Choudhury and Hutzinger 1982; Oberg and Bergstrom 1987), waste oil (Taucher et al. 1992), biosludge from paper and pulp mills (des Rosiers 1987; Mantykoski et al. 1989; Someshwar et al. 1990), polyvinyl chloride (Christmann et al. 1989a), and municipal sewage sludge (Clement et al. 1987a; des Rosiers 1987). The typical concentrations of total tetraCDFs, pentaCDFs, hexaCDFs, heptaCDFs, and octaCDF in fly ash from municipal waste incineration are 79.5, 120.3, 116.3, 108.2, and 42.9 ppb, respectively (Safe 1990a). The corresponding CDF concentrations in soot from home heating oil are 28.9, 16.6, 6.2, 1.8, and 0.3 ppb and in soot from coal/wood burning for home heating are 50.8, 30.0, 11.7, 3.2, and 0.5 ppb. The concentrations of 2,3,7,8-tetraCDF congener in municipal fly ash, soot from heating oil, and soot from coal/wood burning are 2.5, 1.1, and 1.9 ppb, respectively. The combined bottom and fly ash from five state-of-the-art mass-burn municipal waste incinerators with a variety of BATs were analyzed for CDFs. The concentrations of PCDFs (ng/kg or ppt) in ash samples were determined to be 2,3,7,8-tetraCDF, 176-626; 1,2,3,7,8-pentaCDF, 52-194; 2,3,4,7,8-pentaCDF, 43-171; 1,2,3,4,7,8-hexaCDF, 74-654; 1,2,3,6,7,8-hexaCDF, 131-660; 1,2,3,7,8,9-hexaCDF, 36-479; 2,3,4,6,7,8-hexaCDF, 5-124; 1,2,3,4,6,7,8-heptaCDF, 139-1,842; and 1,2,3,4,7,8,9-heptaCDF, 8-119 (ATSDR TP-93/04, 1994)

The mechanisms postulated for the formation of PCDFs in combustion processes are:

1. PCDFs are already present in trace amounts within the fuel and are not destroyed during combustion
2. PCDFs are formed during combustion from precursors (e.g., PCBs, PCPs) present in the fuel
3. *de novo* synthesis from nonchlorinated organic substance and chlorine-containing molecules (Hutzinger and Fiedler 1989).

Details about the mechanisms of PCDF formation in combustion processes are available in the literature (Hutzinger and Fiedler 1989; Jay and Stieglitz 1991; Stieglitz et al. 1989). Studies on the control technologies for the reduction of PCDF emissions from municipal waste incinerators are also available in the literature (Brna and Kilgore 1990; Jordan 1987; Takeshita and Akimoto 1989). A significant reduction of PCDF concentrations in the flue gas, from municipal and industrial waste incinerators and fossil fuel-fired power stations, has been reported by the addition of a mixture of anhydrous calcium hydrate and coke to the flue gas or by treating the flue gas with titanium dioxide catalyst in the presence of ammonia (Hagenmaier et al. 1991).

Accidental Fires or Malfunction of PCB-filled Transformers and Capacitors

Some of the major fires/malfunctions involving PCB transformers and capacitors in the U.S. are: a transformer fire inside an office building in New York, in 1981; a transformer fire inside an office building in Massachusetts, in 1982; a transformer fire adjacent to a high-rise building in California, in 1983; a transformer fire inside an office building in Illinois, in 1983; and a capacitor fire inside an office building in Ohio, in 1984 (des Rosiers and Lee 1986; Hryhorczuk et al. 1986; Stephens 1986; Tiernan et al. 1985). PCDFs were detected in air, soot, or wipe samples from all of these fires. However, it was determined that PCDF levels do not appear to increase in PCB fluids in electrical equipment from normal usage (des Rosiers and Lee 1986). The concentrations of total tetra-, penta-, hexa-, hepta-, and octa-CDFs in air samples from different locations of a building, following a transformer fire in San Francisco, ranged from not detected to 53.9, not detected to 11.0, not detected to 1.3, not detected to 3.7, and not detected to 165.0 pg/m^3 (Stephens 1986). A maximum concentration of 2,3,7,8-tetraCDF inside the building air was 18.5 pg/m^3 (Stephens 1986). The concentration range of 2,3,7,8-tetraCDF in soot samples from other transformer/capacitor fires in the U.S. was 3 to 1,000 µg/g (des Rosiers and Lee 1986). Other international fires/accidents involving PCBs, that lead to the formation of PCDFs and the mechanism of PCDF formation from PCBs, are reported in the literature (Erickson 1989; Hutzinger et al. 1985a,b).

Emissions from Automobile

It has been estimated that the contribution from automobile exhaust to the total burden of PCDDs and PCDFs in the environment at least matches that of emissions from municipal incinerators. Also reported was the presence of small amounts of PCDDs/PCDFs in used motor oil (Ballschmitter et al. 1986a,b). Automobiles using leaded gasoline have been reported to exhaust 100 pg 2,3,7,8-T4CDD TEQs per kilometer (km) (Bingham et al. 1989). Estimated values of PCDDs and PCDFs at 30 to 520 pg/km 2,3,7,8-T4CDD TEQs have also been reported (Marklund et al. 1987a,b).

Metal Refining Processes

The sources in this category include the process for metal refining foundries, metal recovery facilities, steel and copper smelting plants, and magnesium smelting operations. Chlorine sources are believed to be PVC and cutting oil used in metal refining (Rappe 1991); significant quantities of PCDDs/PCDFs (ppb-ppm range) have been reported in the emissions/ash, higher than found in emissions from municipal incineration and automobile exhausts (Oheme et al. 1989a; Rappe et al. 1987). It was estimated that the scrubber plant from a magnesium facility in Southern Norway released 300 to 500 g/y of 2,3,7,8-T4CDD TEQs into the sea (Oheme et al. 1989b). Chlorinated furans were 10 times higher than PCDDs. A secondary copper smelter was shown to release about 800 g/ky 2,3,7,8-T4CDD TEQs (National Dioxin Survey, EPA 1987a).

Other Industrial Processes

Chlorinated dioxins and furans have been found in effluents from petroleum refineries, likely from the catalyst regeneration process. Nonchlorinated aromatic hydrocarbons present in the coker catalyst may be the precursors along with the chlorine from nonrefined oils (Beard et al. 1991). Catalytic high temperature chlorination processes, rather than combustion sources, appear to be the dominant sources of PCDDs and PCDFs in the Dutch aquatic environment, based on their congener profiles in industrial effluent, waste streams, and sediments. Sediment data indicate that production of 1,2,-dichloroethane and allylchloride/epichlorohydrin are a major source of PCDDs and PCDFs in the The Netherlands (Evers 1991). The congener profile of graphite sludge from a Swedish chloralkali plant was similar to that of bleached pulp effluent with an isomer pattern consisting of primarily 2,3,7,8-T4CDF and 1,2,7,8-T4CDF. Very few isomers for the P5CDF and H6CDF isomers were also present. This "chloralkali pattern" could be seen in the soil and surface water runoff from an industrialized area along the Gota River (Kjeller et al. 1991).

Distillation residues from dry cleaning facilities were found to contain concentrations of PCDDs/PCDFs in the ppb range (Towara et al. 1991). Isomer distribution patterns and concentrations of 2,3,7,8-T4CDD/T4CDF in fat and liver tissues of ringed seals and polar bears, from various regions of the Canadian Arctic, indicate long-range atmospheric transport of chlorinated dioxins and furans derived from European and Asiatic sources (Norstrom et al. 1990). In addition, the authors concluded that riverain input of 2,3,7,8-T4CDD from bleached kraft pulp mills was not the source of these chemicals in mammals in the Canadian Southern Arctic region. The quantities of PCDDs and PCDFs deposited were 25 to 50 times greater than estimated emissions from Sweden and from other human activities around the Baltic Sea. Additional sources and the role of long-range transport need to be identified. Similarly, air samples collected as part of the Swedish dioxin survey, showing that concentrations and patterns were dependent on wind direction, emphasizes the need to study long-range atmospheric transport (Rappe 1991).

Analysis of sediment cores from Green Lake, NY, a merimitic lake (i.e., no annual turnover) receiving only atmospheric input, showed detectable levels of PCDDs and PCDFs in samples dating from 1860 to 1865. The concentrations in sediment cores steadily increased to a maximum between the mid-1960s to mid-1970s and appeared to decrease during the 1980s. This data trend is consistent with that reported for Baltic Sea sediments, indicating that the deposition maxima of PCDDs/PCDFs occurred during the 1970s (Smith et al. 1991).

Sewage Treatment Plants

The concentrations of chlorinated dioxins and furans from sewage treatment plants were reported to be 20 to 120 ppt (dry weight, analytical detection limit 1 to 10 ppt) (Darskus and Schlesing 1989). Isomer profiles for sewage sludge indicate it may be a primary source (Rappe, 1991; Oberg et al. 1991).

Analysis of incoming sewage water from a Swedish treatment facility, handling mainly residential wastewaters, found no detectable concentrations of tetra- through hexachlorinated dioxins and furans (detection limits ranged from 0.26 to 20 ppq). The sludge from this sewage treatment plant contained significant concentrations of dioxins and furans. 2,3,7,8-T4CDD and 2,3,7,8-T4CDF were present at 0.72 and 4.2 ppt, respectively, and H7CDD, H7CDF, and the O8CDD/CDF isomers were present at concentrations of 1,500, 160, and 9,100 ppt, respectively (Rappe et al.

1989). Assessment of a second municipal sewage treatment plant in Sweden, handling a greater amount of industrial effluents, indicated similar concentrations and congener distributions of chlorinated dioxins and furans, as observed in the plant that received mainly municipal sewage (Rappe et al. 1989) This suggests that industrial processes were not major contributors.

Chlorinated Drinking Water

Chlorinated dioxins and furans do not appear to be produced in significant quantities during the chlorination of drinking water in water treatment plants. From 4,347 results, obtained for 399 water samples from various locations, in Ontario, only a small percentage had detectable concentrations of O8CDD. No samples were found to contain detectable quantities of 2,3,7,8-T4CDD. The detection limits were in the low ppq range (Jobb et al. 1989).

ENVIRONMENTAL RESIDUES AND EXPOSURE ROUTES

Exposure pathways to 2,3,7,8-T4CDD have changed since late 1970s. Herbicide preparations containing 2,4,5-T had 2,3,7,8-T4CDD as a contaminant and were the primary route of human exposure during their manufacture, use, and disposal. The use of 2,4,5-T was completely banned in 1979 and, thus, the occupational and general population exposure to 2,3,7,8-T4CDD, due to manufacture and use of these herbicides was eliminated. Currently, the important exposure routes are contaminated soil, dump sites, and municipal incinerators. The exposure pathways have changed from manufacturing sources to consumption of foods contaminated with 2,3,7,8-T4CDD (ATSDR 1989).

Air

Ambient air data collection suffers from sampling and analytical difficulties associated with very low levels of 2,3,7,8-T4CDD (Table 8.7).

Table 8.7 Levels of T4CDD Reported in Air

Location	Medium	Concentration	Congener
Bloomington, IN	Ambient air	18–92 fg/m^3	ΣT4CDD
Municipal incinerators	Flue gas	38 fg/m^3	2,3,7,8-T4CDD
Incinerators, Europe	Flue gas	0.05–1.3 ng/m^3	2,3,7,8-T4CDD
Incinerators, U.S.	Flue gas	Max. 3.5 ng/m^3	2,3,7,8-T4CDD
(Urban site, traffic tunnel, downwind site of a municipal incinerator and a site in the vicinity of a dump and metal refinery	Ambient air	0.02–0.08 pg/m^3	2,3,7,8-T4CDD
Sweden, 4 locations	Urban air	0.001–0.009 pg/m^3	2,3,7,8-T4CDD
Office building, Binghampton, NY after a PCB fire	Indoor air	0.23–0.47 pg/m^3	2,3,7,8-T4CDD
After accidental locomotive fire, Sweden	Ambient air	50 pg/m^3	2,3,7,8-T4CDD

fg = femtogram (10^{-18} g)

Sources: ATSDR 1989; Barnes 1983; Eitzer and Hites 1986; Marklund et al. 1986; Ballschmitter 1986a,b; Rappe and Kjeller 1987; EPA 1985; Rappe et al. 1985; Smith et al. 1986.

The PCDFs levels in ambient air in North America show geographical variability based on the emission sources (Table 8.8). The general trend is industrial/auto tunnel>urban>suburban>rural (Eitzer and Hites 1989a). In a given area, seasonal and daily variations are observed (Nakano et al. 1990). PCDFs in air are higher on rainy days with high humidity and on less windy days. Levels are also higher during the winter than in summer, due to the increasing contribution from heating sources (Hunt et al. 1990). In urban/suburban areas, total tetra to octa-CDF congeners are 0.13 to 7.34, 0.09 to 5.10, <0.09 to 12.55, 0.08 to 12.71, and 0.13 to 3.78 pg/m^3, respectively. In rural areas, these PCDFs are nondetectable. Generally, the tetra- and penta-CDFs are higher in the vapor phase, while hepta- and octa-CDFs predominate in the particulate phase. The ratio of vapor/particulate phase increases during the summer when compared to the winter (Hunt et al. 1990). The congener profile follows that of their sources.

Non-2,3,7,8 PCDFs predominate in the air near a municipal solid waste (MSW) incinerator. Among the 2,3,7,8-substituted isomers in the air, the 1,2,3,4,6,7,8-hepta-CDF congener dominated, followed by 2,3,7,8-T4CDF (Tiernan et al. 1989). Considerably higher concentrations of PCDFs are found in the indoor air (ATSDR 1994).

Water

Only a small portion of the PCDDs and PCDFs discharged into surface waters would remain in the dissolved phase due to their very low aqueous solubility (Choudhury and Webster 1989). Removal of PCDDs/DFs from the dissolved aqueous phase includes partitioning onto particulates and binding to dissolved organic matter (Marcheterre et al. 1985) and sediments (Corbet et al. 1988). Other important removal processes are bioptake, photolysis, hydrolysis, and oxidation (Mill 1985) and volatilization (Corbet et al. 1988). It has been estimated that concentrations of PCDDs/DFs in surface waters would range from 30 to 70% of their aqueous solubility (Walters et al. 1989). PCDDs/DFs would likely accumulate in aquatic biota, particularly animals near the top of the food chains.

PCDDs/PCDFs are not reported to be produced during chlorination of drinking water in water treatment plants. In a survey of drinking water (399 water samples producing 434 results) from several locations in Ontario, CA, only a small percentage contained detectable octaCDFs. No samples were found to contain detectable (D.L., low ppq range) 2,3,7,8-T4CDD. Drinking water supplies from 20 communities in New York, showed that total tetraCDFs, at 2.6 ppq (pg/L), and octaCDF, at 0.8 ppq, were the only two congener groups detected in 1 of 20 water supplies (Meyer et al. 1989). The concentration of 2,3,7,8-T4CDF was 1.2 ppq.

2,3,7,8-T4CDD has been detected in aqueous industrial effluents, sediments, and leachates from hazardous waste sites. The concentrations of T4CDDs, including 2,3,7,8-TCDD, in effluents from a trichlorophenol manufacturing facility in the U.S. ranged from nondetected to 100 pg/g (detection limit, 10 to 30 pg/g). The leachate samples from a waste disposal site in Arkansas had a mean 2,3,7,8-TCDD level of 14 ng/L. The sump pump water from residences and leachates from the Love Canal area in New York contained 2,3,7,8-TCDD, ranging from nondetected to 1,560 ng/L. The concentrations of 2,3,7,8-TCDD in sediments from storm sewers, residential

Table 8.8 PCDFs in Ambient Outdoor Air in North America

Location	Sampling Year	PCDF	Concentration (pg/m³)	References
Lake trout, Wisconsin, outdoor	1987	Total tetraCDFs	0.083	Edgerton et al. 1989
		Total pentaCDFs	0.067	
		Total hexaCDFs	0.031	
		Total heptaCDFs	0.012	
		OctaCDF	0.006	
Akron, OH, outdoor	1987	2,3,7,8-tetraCDF	0.200	Edgerton et al. 1989
		Total tetraCDFs	1.23	
		1,2,3,7,8-pentaCDF	0.029	
		2,3,4,7,8-pentaCDF	0.036	
		Total pentaCDFs	0.590	
		1,2,3,4,7,8-hexaCDF	0.083	
		1,2,3,6,7,8-hexaCDF	0.065	
		1,2,3,7,8,9-hexaCDF	0.032	
		total hexaCDFs	0.620	
		1,2,3,4,6,7,8-heptaCDF	0.237	
		Total heptaCDFs	0.383	
		OctaCDF	0.180	
Columbus, OH, outdoor	1987	2,3,7,8-tetraCDF	0.405	Edgerton et al. 1989
		Total tetraCDFs	2.85	
		1,2,3,7,8-pentaCDF	0.045	
		Total pentaCDFs	0.995	
		1,2,3,4,7,8-hexaCDF	0.165	
		1,2,3,6,7,8-hexaCDF	0.141	
		1,2,3,7,8,9,-hexaCDF	0.079	
		Total hexaCDFs	0.785	
		1,2,3,4,6,7,8-heptaCDF	0.335	
		Total heptaCDFs	0.450	
		OctaCDF	<0.260	
Waldo, OH, outdoor	1987	2,3,7,8-tetraCDF	0.130	Edgerton et al. 1989
		Total tetraCDF	0.890	
		1,2,3,7,8-pentaCDF	0.021	
		Total pentaCDF	0.500	
		1,2,3,4,7,8-hexaCDF	0.098	
		1,2,3,6,7,8-hexaCDF	0.014	
		1,2,3,7,8,9-hexaCDF	0.097	
		Total hexaCDFs	0.510	
		1,2,3,4,6,7,8-heptaCDF	0.220	
		1,2,3,4,7,8,9-heptaCDF	0.019	
		Total heptaCDFs	0.290	
		OctaCDF	0.077	
Los Angeles, CA, outdoor	1987	2,3,7,8-tetraCDF	0.021	Maisel and Hunt 1990
		Other tetraCDFs	0.30	
		1,2,3,7,8-pentaCDF	0.077	
		2,3,4,7,8-pentaCDF	0.077	
		Other PentaCDFs	0.41	
		1,2,3,4,7,8-hexaCDF	0.151	
		1,2,3,6,7,8-hexaCDF	0.25	
		2,3,4,6,7,8-hexaCDF	<0.069	
		1,2,3,7,8,9-hexaCDF	<0.083	
		Other hexaCDFs	<0.080	
		1,2,3,4,6,7,8-heptaCDF	<0.190	

Table 8.8 PCDFs in Ambient Outdoor Air in North America (Continued)

Location	Sampling Year	PCDF	Concentration (pg/m³)	References
		1,2,3,4,7,8,9-heptaCDF	<0.018	
		Other heptaCDFs	0.26	
		OctaCDF	0.056	
Niagara Falls, NY, outdoor	1987–88	Total tetraCDF	1.53	Smith et al. 1990b
		2,3,7,8-CDF/unknown isomer	<0.11	
		Total pentaCDFs	0.98	
		Total hexaCDFs	1.45	
		Total heptaCDFs	1.37	
		OctaCDF	0.51	
U.S. and Canada, outdoor ambient air	—	Total tetraCDF	1.09	Wadell et al. 1991
		Total pentaCDFs	0.63	
		Total hexaCDFs	0.72	
		Total heptaCDFs	1.14	
		OctaCDF	0.62	
Bloomington, IN	1986	2,3,7,8/2,3,4,8/2,3,4,6-tetraCDFs	0.048	Eitzer and Hites 1990
		Total tetraCDFs	0.263	
		1,2,3,7,8-/1,2,3,4,8-pentaCDFs	0.017	
		2,3,4,7,8-/1,2,3,6,9-pentaCDFs	0.017	
		Total pentaCDFs	0.20	
		1,2,3,4,7,8/1,2,3,4,6,7-hexaCDFs	0.023	
		1,2,3,6,7,8-/1,2,3,4,7,9-hexaCDFs	0.016	
		2,3,4,6,7,8-hexaCDF	0.015	
		1,2,3,7,8,9-hexaCDF	0.0007	
		Total hexaCDFs	0.113	
		1,2,3,4,6,7,8-heptaCDF	0.039	
		1,2,3,4,7,8,9-heptaCDF	0.005	
		Total heptaCDFs	0.071	
		OctaCDF	0.028	
Southern California, outdoor	1987–89	2,3,7,8-tetraCDF	<0.007–0.482	Hunt et al. 1990
		1,2,3,7,8-pentaCDF	<0.010–1.9	
		2,3,4,7,8-pentaCDF	<0.009–0.110	
		1,2,3,4,7,8-hexaCDF	<0.001–0.27	
		1,2,3,6,7,8-hexaCDF	<0.001–0.800	
		2,3,4,6,7,8,-hexaCDF	<0.001–0.280	
		1,2,3,4,6,7,8-heptaCDF	<0.002–1.58	
		1,2,3,4,7,8,9-heptaCDF	<0.002–0.092	
Windsor, Canada, outdoor	1987–88	Total tetraCDFs	0.21	Bobet et al. 1990
		Total pentaCDFs	0.09	
		Total hexaCDFs	0.10	
		Total heptaCDFs	0.08	
		OctaCDF	0.13	
Walpole Island, Canada, outdoor	1987–88	Total tetraCDFs	<0.05	Bobet et al. 1990
		Total pentaCDFs	<0.07	
		Total hexaCDFs	<0.10	
		Total heptaCDFs	<0.07	
		OctaCDF	<0.14	

Table 8.8 PCDFs in Ambient Outdoor Air in North America (Continued)

Location	Sampling Year	PCDF	Concentration (pg/m³)	References
Bridgeport, CT, outdoor	1987–88	2,3,7,8-tetraCDF	0.078	Maisel and Hunt, 1990
		Total tetraCDFs	0.856	
		1,2,3,7,8-pentaCDF	0.031	
		2,3,4,7,8-pentaCDF	0.047	
		Total pentaCDFs	0.547	
		1,2,3,4,7,8-hexaCDF	0.106	
		1,2,3,6,7,8-hexaCDF	0.039	
		2,3,4,6,7,8-hexaCDF	0.087	
		1,2,3,7,8,9-hexaCDF	0.007	
		Total hexaCDFs	0.580	
		1,2,3,4,6,7,8-heptaCDF	0.212	
		1,2,3,7,8,9-heptaCDF	0.033	
		Total heptaCDFs	0.369	
		OctaCDF	0.211	
Boston, MA, office building, indoor	—	2,3,7,8-tetraCDF	(0.37)[a]–1.4	Kominsky and Kwoka 1989
		Total tetraCDFs	(0.64)[a]–6.2	
		Total pentaCDFs	(0.12)[a]–1.9	
		Total hexaCDFs	(0.39)–1.5[a]	
		OctaCDF	(0.54)–1.8[a]	
Albany, NY, outdoor	1987–88	2,3,7,8-tetraCDF/ unknown isomer	0.89	Smith et al. 1990b
		Total tetraCDFs	3.86	
		Total pentaCDFs	2.00	
		Total hexaCDFs	0.28	
		Total heptaCDFs	<0.34	
		Octa CDF	<0.50	
Binghampton, NY, outdoor	1988	2,3,7,8-tetraCDF/ unknown isomer	0.18	Smith et al. 1990b
		Total tetraCDFs	0.94	
		Total pentaCDFs	0.25	
		Total hexaCDFs	<0.09	
		Total heptaCDFs	<0.14	
		Octa CDF	<0.30	
Dayton, OH, outdoor suburban/roadside	1988	Total tetraCDFs	0.13	Tiernan et al. 1989
		Total pentaCDFs	0.24	
		Total hexaCDFs	0.14	
		Total heptaCDFs	0.11	
		Octa CDF	<0.07	
Dayton, OH, outdoor, municipal solid waste incinerator	1988	2,3,7,8-tetraCDF	0.11	Tiernan et al. 1989
		Total tetraCDFs	1.23	
		1,2,3,7,8-pentaCDF	0.46	
		2,3,4,7,8-pentaCDF	0.53	
		Total pentaCDFs	5.10	
		1,2,3,4,7,8-hexaCDF/ unknown isomer	1.18	
		1,2,3,6,7,8-hexaCDF	2.27	
		1,2,3,7,8,9-hexaCDF	<0.06	
		2,3,4,6,7,8-hexaCDF	<0.41	
		Total hexaCDFs	12.55	
		1,2,3,4,6,7,8-heptaCDF	8.22	
		1,2,3,4,7,8,9-heptaCDF	0.56	
		Total heptaCDFs	12.71	

Table 8.8 PCDFs in Ambient Outdoor Air in North America (Continued)

Location	Sampling Year	PCDF	Concentration (pg/m³)	References
		OctaCDF	3.78	
Dayton, OH, outdoor rural area	1988	Total tetraCDFs	<0.02	Tiernan et al. 1989
		Total pentaCDFs	<0.02	
		Total hexaCDFs	<0.05	
		Total heptaCDFs	<0.07	
		Octa CDF	<0.17	
Utica, NY, outdoor	1988	2,3,7,8-tetraCDF/ unknown isomer	1.15	Smith et al. 1990b
		Total tetraCDFs	7.34	
		Total pentaCDFs	3.16	
		Total hexaCDFs	<0.36	
		Total heptaCDFs	<0.24	
		Octa CDF	<0.61	
Toronto Island, Canada, outdoor	1988–89	Total tetraCDFs	0.404	Steer et al. 1990b
		Total pentaCDFs	0.118	
		Total hexaCDFs	0.204	
		Total heptaCDFs	0.240	
		Octa CDF	0.142	
Dorset, Canada, outdoor	1988–89	Total tetraCDFs	0.164	Steer et al. 1990b
		Total pentaCDFs	0.200	
		Total hexaCDFs	0.074	
		Total heptaCDFs	0.52	
		Octa CDF	0.194	
Windsor, Canada, outdoor	1988–89	Total tetraCDFs	0.733	Steer et al. 1990
		Total pentaCDFs	0.383	
		Total hexaCDFs	0.333	
		Total heptaCDFs	0.550	
		Octa CDF	0.182	

[a] Detection Limit.
Source: ATSDR,TP-93/04, May 1994

sump water, and surface water around the same site were nondetected (detection limit, 10 to 100 pg/g) to 9,570 ng/g (EPA 1985, Lamparski et al. 1986, Tiernan et al. 1985). The concentration of 2,3,7,8-TCDD in the sludge of seven pulp and paper mill wastewaters in the U.S. had concentration levels ranging from nondetectable (<1.0 pg/g) to 414 pg/g (Kuehl et al. 1987; ATSDR 1989).

Table 8.9 lists the 1988 concentrations (pg/g) of PCDDs and PCDFs in surface water sediments immediately below pulp and paper mills in Canada.

Soils and Sediments

Ambient concentrations of 2,3,7,8-T4CDD in most soils are below the detection limits of current analytical methods. In urban soils, the level of 2,3,7,8-T4CDD is in the range of <0.0002 to 0.009 ng/g. In a national dioxin study, the EPA sampled soils from 138 rural and 221 urban sites not associated with sources of 2,3,7,8-T4CDD. Only 17 of the rural and urban soils had detectable levels of 2,3,7,8-T4CDD at a concentration range of 0.2 to 11.2 pg/g (TMN 1987). 2,3,7,8-T4CDD has been

Table 8.9 1988 Concentrations (pg/g) of PCDDs and PCDFs in Sediments Near Canadian Pulp Mills

Congener	1[†]	2*	3*	4*	5*	6*	7*
2378-T4CDD	<3–27	a = <15 b = <15	a = <15 b = <15	a = <15 b = <15	a = <15 b = <15	a = <15 b = <15	a = <15 b = <15
Total T4CDD	<3–782	a = <15 b = <15	a = <15 b = <15	a = <15 b = <15	a = <15 b = <15	a = <15 b = <15	a = <15 b = <15
Total P5CDD	<5–1,110	a = <20 a = <20–28	a = <20 b = <20	a = <20 b = <20	a = <20 b = <20	a = <20 b = <20	a = <20 b = <20
Total H6CDD	<6–16,432	a = <30 b = <30	a = <30 b = <30	a = <30 b = <30	a = <30 b = <30	a = <30 b = <30	a = <30 b = <30
Total H7CDD	<10–3,431	a = <50 b = <50	a = <50 b = <50	a = <50 b = <50	a = <50 b = <50	a = <50 b = <50	a = <50 b = <50
O8CDD	<12–8,022	a = <75 b = <75-572	a = <75 b = <75	a = <75 b = <75	a = <75 b = <75	a = <75 b = <75	a = <75 b = <75
2378-T4CDF	5–3,179	a = <10 b = 10–2,077	a = <10 b = <10–274	a = <10–134 b = <36.6–238	a = <10 b = 68.5–3,168	a = <10 b = 100–642	a = <10 b = <10–2,217
Total T4CDF	5–4,867	a = <10 b = <10–3,227	a = <10 b = <10–406	a = <10–192 b = 53.8–359	a = <10 b = 96.7–4,521	a = <10 b = 125–642	a = <10 b = <10–3,655
Total P5CDF	<3–313	a = <15 b = <15–51	a = <15 b = <15	a = <15 b = <15	a = <15 b = <15	a = <15 b = <15	a = <15 b = <15
Total H6CDF	<6–364	a = <25 b = <25–34	a = <25 b = <25	a = <25 b = <25	a = <25 b = <25	a = <25 b = <25	a = <25 b = <25
Total H7CDF	<6–731	a = <40 b = <40	a = <40 b = <40	a = <40 b = <40	a = <40 b = <40	a = <40 b = <40	a = <40 b = <40
O8CDF	40–378	a = <75 b = <75	a = <75 b = <75	a = <75 b = <75	a = <75 b = <75	a = <75 b = <75	a = <75 b = <75

a = Upstream; b = Downstream

1. Near Canadian pulp mills; 2. MacKenzie, B.C. pulpmills; 3. Prince George, B.C. pulpmills; 4. Quesnel, B.C. pulpmills; 5. Weyerhauser Canada, B.C. pulpmills; 6. Power Consolidated (China) Pulp, Inc. (Celgar) B.C. pulpmill; 7. Crestbrook Forest Industries Ltd. B.C. pulpmill.

Sources: [†]Trudel 1991; *Maah et al. 1989.

detected in samples that originated from certain industrial and, waste disposal sites, and sites involved in the accidental spillage of chemicals containing 2,3,7,8-T4CDD (Table 8.10). It is clear that accidental or improper disposal of still bottom residue from the manufacture of 2,4,5-trichlorophenol (2,4,5-TCP) may produce one of the highest levels of 2,3,7,8-T4CDD in soils.

The maximum 2,3,7,8-T4CDF and 2,3,7,8-substituted T4CDF concentration of 0.3 ppt (ng/kg) and 11.0 ppt, respectively, were determined in sediments from the uncontaminated Elk River in Minnesota (Reed et al. 1990). The maximum concentrations of total penta-, hexa-, hepta-, and octa-CDFs in the sediment samples from the same river were 25, 12, 30, and 23 ppt, respectively. The analyte was not detected in all samples. The concentrations of 2,3,7,8-T4CDF in sediment from the lower Hudson River (New York), Cuyahoga River (Ohio), Menominee River (Wisconsin), Fox River (Wisconsin), Raisin River (Michigan), and Saginaw River (Wisconsin) ranged from 5 to 97 ppt (O'Keefe et al. 1984; Smith et al.1990a). The concentration of 2,3,7,8-T4CDF in sediment from the uncontaminated Lake Pepin in Wisconsin was 1 ppt, while in sediment from Lake Michigan in Green Bay, WI was 24 ppt (Smith et al. 1990a). T4CDF was detected at ≤1,400 ppt in sediment from New Bedford Harbor, MA near a Superfund site (Norwood et al. 1989). The levels of T4CDF in estuarine sediment ranged from 15 ppt for uncontaminated sediment in Long Island Sound (New York) to 4,500 ppt in sediment from an estuary adjacent to a facility which used to produce 2,4,5-T in Newark, NJ (Bopp et al. 1991; Norwood et al. 1989). A survey of harbor sediments near a wood-treatment facility at Thunder Bay (Ontario, Canada), showed that the tetra- and penta-CDFs concentrations were lower than the detection limit, whereas, the levels of higher chlorinated congeners were higher (maximum levels for H6CDF and O8CDF were 6.5 and 400 ppb, respectively, (McKee et al. 1990).

PCDFs were detected in uncontaminated soils in the vicinity of Elk River, MN. The levels (in ppt) were as follows, with the detection limit in parentheses: 2,3,7,8-T4CDF,ND (0.8); total tetraCDFs, ND(0.8) to 1.2; total hexaCDFs, 6.7 to 150; 1,2,3,4,6,7,8-heptaCDF, 26 to 72; total heptaCDFs, 30 to 260; and octaCDF, ND(3) to 270 (Reed et al. 1990). Soils adjacent to a refuse incineration facility in Hamilton, Ontario, Canada, were reported to contain PCDFs (in ppt) as follows: total T4CDFs, ND(0.3) to 71; total pentaCDFs, ND(1.3) to 6.0; total hexaCDFs, ND(1.3); total heptaCDFs, ND(1.3) to 180; and octaCDF,ND(0.8) to 811 (McLauglin et al. 1989). These levels were not considered elevated compared to urban background levels. Similar levels of PCDFs, in soil adjacent to a municipal incinerator in England, were also reported as not being different from background levels (Mundy et al. 1989). In Germany, much higher levels of PCDFs were reported in soil from a PCP-containing waste landfill; the concentrations (in ppt) were: 1,2,3,7,8-/1,2,3,4,8-pentaCDF, 17,000; 2,3,4,7,8-pentaCDF, 7,000; 1,2,3,4,7,8-/1,2,3,4,7,9-hexaCDF, 152,000; 1,2,3,6,7,8-hexaCDF, 48,000; 1,2,3,7,8,9-hexaCDF, 3,000; and 2,3,4,6,7,8-hexaCDF, 24,000 (Hagenmaier and Berchtold 1986).

Table 8.10 Levels of 2,3,7,8-TCDD in Soil from Different Contaminated Areas

Site	Sample History	TCDD Concentration[a,b] (ng/g)	References
Love Canal, NY	Soils outside the dump site	ND (0.001–0.020)	EPA 1985
Jacksonville, AZ	Waste disposal site	ND–2.9	EPA 1985
Midland, MI	Inside DOW facility	0.01–52	Nestrick et al. 1986
St. Louis, MO	Urban sample of no obvious source of contamination	0.12	EPA 1985b
Shenandoah Stables, MO	Contaminated by waste oil	101–33,000	Tiernan et al. 1985; Kimbrough et al. 1984
Timberline Stables, MO	Contaminated by waste oil	30–42	Tiernan et al. 1985; Kimbrough et al. 1984
Bliss Farm, MO	Contaminated by waste oil	382[c]	Tiernan et al. 1985; Kimbrough et al. 1984
Bubbling Springs Ranch, MO	Contaminated by waste oil	76–95	Tiernan et al. 1985; Kimbrough et al. 1984
Minker Resident, MO	Contaminated by waste oil	50[c]	Tiernan et al. 1985; Kimbrough et al. 1984
Times Beach, MO	Contaminated by waste oil	4.4–317	Tiernan et al. 1985; Kimbrough et al. 1984
Urban areas, U.S.	Urban samples of no obvious source of contamination	<0.0002–0.009	Nestrick et al. 1986
New Jersey	Spillage of 2,4,5-TCP still bottom	26,000[c]	Jackson et al. 1986
New Jersey	Scrap yard where used reactor vessels were collected	1,100[c]	Jackson et al. 1986
Lansing, MI	Urban sample	ND(0.0007)–0.003	Nestrick et al. 1986
Gaylord, MS	Urban sample	ND(0.0002)	Nestrick et al. 1986
Detroit, MI	Urban sample	0.0021–0.0036	Nestrick et al. 1986
Chicago, IL	Urban sample	0.0042–0.0094	Nestrick et al. 1986
Akron, OH	Urban sample	0.0063	Nestrick et al. 1986
Nashville, TN	Urban sample	0.0008	Nestrick et al. 1986
Pittsburgh, PA	Urban sample	0.0026	Nestrick et al. 1986
Philadelphia, PA	Urban sample	0.0009	Nestrick et al. 1986
Brooklyn, NY	Urban sample	0.0026	Nestrick et al. 1986
Arlington, VA	Urban sample	ND (0.0003)	Nestrick et al. 1986

[a] ND = Not detected
[b] Values in brackets are detection limits
[c] Only one sample analyzed

Fish

Table 8.10 lists the concentrations of PCDDs and PCDFs in fish downstream of pulp mills. Tetra- and penta-chloro isomers ranged from <detection limit to 527 pg/g, while higher chlorinated congeners were generally lower than the detection limits.

The concentrations of PCDFs in meat, fish, and dairy products, from a supermarket in New York, were 0.14 to 7.0, 0.07 to 1.14, and 0.3 to 5.0 ppt (wet weight), respectively; the concentrations of 2,3,7,8-T4CDF were 0.01 to 0.1, 0.02 to 0.73, and 0.02 to 0.15 ppt (wet weight), respectively (Schecter et al. 1993). The 2,3,7,8-T4CDF is predominant in fish, followed by 2,3,4,7,8-P5CDF. Among the Great Lakes, Lake Erie and Lake Superior are cleaner than the other three lakes in PCDF contamination (Table 8.12). The mean levels of 2,3,7,8-substituted PCDFs (total) in gutted whole fish from the St. Maurice River, Quebec, Canada, collected from the immediate downstream of a kraft mill, was 260 pg/g (ppt); the level decreased with distance. At 95 km downstream, 2,3,7,8-PCDFs (total) was 112 ppt (Hodson et al. 1992).

TOXICITY PROFILE

Invertebrates

2,3,7,8-T4CDD is not acutely toxic to aquatic invertebrates. LC0 for *Daphnia magna* for 1 to 30 days of exposure was 1.33 µg/L (Isenee 1978) and for 9-day exposure to 1 to 21-day-old *Daphnia magna* was 0.2 to 1,030 ng/L (Adams et al. 1986). For snails, LC0 for 48 h exposure was 0.2 µg/L (Miller et al. 1973).

Fish

LC_{50} (28d) for fathead minnows was reported to be 1.7 ng/L (Adams et al. 1986). Delayed mortality was reported for fathead minnows, pike, and rainbow trout after termination of exposure to 2,3,7,8-T4CDD (Helder 1980, 1981). Similar post-exposure mortality has been reported for juvenile coho salmon (Miller et al. 1979). Exposure via intraperitoneal route has yielded LD_{50} values of 3 and 20 for yellow perch and rainbow trout, respectively, over an 80-day post-exposure observation period. A NOEL value of 1 µg/kg has been reported for rainbow trout (Spitsbergen et al. 1986, 1987, 1988a). Similar exposure and 80-day post-exposure observation has yielded LD_{50} values of 3, 3, 5, 10, 11, and 16 µg/kg for yellow perch, carp, bullhead, rainbow trout, largemouth bass, and bluegill, respectively (Kleeman et al. 1988). Lethality data on fish has been mostly reported on 2,3,7,8-T4CDD. When compared to mammalian species, LD_{50} values follow: guinea pig [LD_{50}, 0.6 to 2.1 µg/kg >fish (LD_{50} see above)] = rats (LD_{50}, 22.5 to 45 µg/kg) <hamsters (LD_{50}, 5000 µg/kg) (Schwetz et al. 1973; McConnell et al. 1978; Olson et al. 1980).

No-observed-adverse effect level (NOAEL) of 34 pg/g has been reported for Lake Superior lake trout at the end of the sac-fry (juvenile) stage of development (Cook et al. 1991). The lowest concentration for lethality in sac fry was 55 pg/g.

Table 8.11 Levels of PCDDs and PCDFs (pg/g = ppt) in Fish Near Effluent Discharges from Pulp Mills

Species	Location	Compound	Concentration[a]	References
Longnose sucker (whole fish)	Downstream of Hinton, AB, P & P mill	2378-T4CDD	3–17 (11.8)	RTP 1994
White sucker (whole fish)	Grand Prairie, Wapiti River, AB	2378-T4CDD	13–17 (15)	RTP 1994
Fish	Fraser River, BC, d/s of pulp and timber industries	2378-T4CDD	5.4–19.5 (12.2)	Mah et al. 1989
Fish	Fraser River, BC, d/s P & P mills	2378-T4CDD	2.9–137 (43.3)	Mah et al. 1989
Fish	Thompson River, BC, d/s P & P mills and STP	2378-T4CDD	12.5–60.9 (44.4)	Mah et al. 1989
Fish	Williston Lake, BC, d/s forest industries, d/s	2378-T4CDD	ND–4.6 (2.8)	Mah et al. 1989
	P & P mill	2378-T4CDD	ND–2.5 (2.3)	Mah et al. 1989
Fish	Fraser River, BC, d/s P & P mill	2378-T4CDD	ND–4.4 (2.1)	Mah et al. 1989
Perch	Gulf of Bothnia, Sweden, d/s P & P mill	2378-T4CDD	2.6–19	Rappe et al. 1989
Arctic char	Lake Vettern, Sweden, d/s P & P mills	2378-T4CDD	6.5–25	Rappe et al. 1989
White sucker	North Saskatchewan River, AB, d/s P & P mill	2378-T4CDD	8	RTP 1994
White sucker	North Saskatchewan River, AB, d/s P & P mill	2378-T4CDD	0.7–3.0 (2.3)	RTP 1994
Shrimp	Jackfish Bay, Lake Superior, d/s bleached kraft mill	2378-T4CDD	9	Sherman et al. 1990
Fish	Fraser River, BC, d/s P & P mill	Total T4CDD	ND–4.4 (2.1)	Mah et al. 1989
Fish	Fraser River, BC, d/s P & P mill	Total T4CDD	5.4–19.5 (12.2)	Mah et al. 1989
Fish	Fraser River, BC, d/s two P & P mills	Total T4CDD	2.9–137 (43.3)	Mah et al. 1989
Fish	Thompson River, BC, d/s P & P mill	Total T4CDD	12.5–60.9 (44.4)	Mah et al. 1989
Fish	Williston Lake, BC, d/s forest industries	Total T4CDD	ND–4.6 (2.8)	Mah et al. 1989
Fish	Williston Lake, BC, d/s pulp industries	Total T4CDD	ND–2.5 (2.3)	Mah et al. 1989
Fish	Columbia River, BC, d/s pulp co.	Total T4CDD	ND–10.5 (5.7)	Mah et al. 1989
White sucker	Athabasca River, AB, d/s P&P mill	2378–T4CDF	16–71 (52)	Whittle 1989
Fish	Fraser River, BC, d/s P & P mill	2378–T4CDF	40.8–67.4 (50.4)	Mah et al. 1989
Fish	Fraser River, BC, d/s pulp & timber industry	2378–T4CDF	85–290 180)	Mah et al. 1989
Fish	Fraser River, BC, 5 km d/s two P & P mills	2378–T4CDF	15.8–1,185 (393)	Mah et al. 1989

CHLORINATED DIOXINS AND FURANS

Table 8.11 Levels of PCDDs and PCDFs (pg/g = ppt) in Fish Near Effluent Discharges from Pulp Mills (Continued)

Species	Location	Compound	Concentration[a]	References
Fish	Columbia River, BC, d/s P & P mill	2378–T4CDF	25.8–908 (527)	Mah et al. 1989
Fish	Thompson River, BC, d/s P & P mill and STP	2378–T4CDF	16–704 (369)	Mah et al. 1989
White sucker (whole fish)	Grand Prairie, Wapiti River, AB, d/s P & P mill	2378–T4CDF	17–240 (152)	RTP 1994
White sucker (whole fish)	Grand Prairie, Wapiti River, AB, d/s P & P mill	2378–T4CDF	280–290 (280)	RTP 1994
White sucker (whole fish)	North Saskatchewan River, AB, d/s P & P mill	2378–T4CDF	17.5	RTP 1994
White sucker (whole fish)	North Saskatchewan River, AB, d/s P & P mill	2378–T4CDF	<0.4–13 (7.2)	RTP 1994
Fish	Williston Lake, BC, d/s pulp mill	2378–T4CDF	11.1–485 (33.3)	Mah et al. 1989
Fish	Williston Lake, BC, d/s forest industry	2378–T4CDF	6.2–80.1 (45)	Mah et al. 1989
Fish	Kootenay River, BC, d/s pulp mill	2378–T4CDF	25.6–38.6 (30.4)	Mah et al. 1989
Perch	Gulf of Bothnia, d/s pulp mill	2378–T4CDF	2.1–8.7	Rappe et al. 1989
Arctic char	Lake Vettern, Sweden, d/s two P & P mills	2378–T4CDF	20–75	Rappe et al. 1989
Fish	Kootenay River, BC, d/s pulp mill	Total T4CDF	25.6–38.6 (30.4)	Mah et al. 1989
Fish	Fraser River, BC, d/s P & P mill	Total T4CDF	40.8–67.4 (50.4)	Mah et al. 1989
Fish	Fraser River, BC, d/s pulp and timber industries	Total T4CDF	85–290 (180)	Mah et al. 1989
Fish	Fraser River, BC, 5 km d/s two pulp mills	Total T4CDF	15.8–1,190 (390)	Mah et al. 1989
Fish	Columbia River, BC, d/s pulp mill	Total T4CDF	25.8–908 (527)	Mah et al. 1989
Fish	Thompson River, BC, d/s pulp mill	Total T4CDF	16–704 (369)	Mah et al. 1989
Fish	Williston Lake, BC, d/s pulp mill	Total T4CDF	11.1–48.5 (33.3)	Mah et al. 1989
Fish	Williston Lake, BC, d/s forest industries	Total T4CDF	6.2–80.1 (45)	Mah et al. 1989
White sucker	Jackfish Lake, Lake Superior, d/s BKME	T4CDF	44	Sherman et al. 1990
Shrimp	Jackfish Lake, Lake Superior, d/s BKME	T4CDF	48	Sherman et al. 1990
Clam	Jackfish Lake, Lake Superior, d/s BKME	T4CDF	34	Sherman et al. 1990

Table 8.11 Levels of PCDDs and PCDFs (pg/g = ppt) in Fish Near Effluent Discharges from Pulp Mills (Continued)

Species	Location	Compound	Concentration[a]	References
White suckers and clams	Jackfish Lake, Lake Superior, d/s BKME	Total P5CDD	ND	Sherman et al. 1990
Shrimp	Jackfish Lake, Lake Superior, d/s BKME	Total P5CDD	<7	Sherman et al. 1990
Perch	Gulf of Bothnia d/s pulp mill	1,2,3,7,8-P5CDD	0.6–1.0	Rappe et al. 1989
Arctic char	Lake Vettern, Sweden d/s two pulp mills	1,2,3,7,8-P5CDD	3.9–16	Rappe et al. 1989
Fish	Kootenay River, BC, d/s pulp mill	Total P5CDD	ND	Mah et al. 1989
Fish	Fraser River, BC, d/s P & P mill	Total P5CDD	ND	Mah et al. 1989
Fish	Fraser River, BC, d/s pulp and timber industries	Total P5CDD	ND	Mah et al. 1989
Fish	Fraser River, BC, 5 km d/s two pulp mills	Total P5CDD	ND–1.28	Mah et al. 1989
Fish	Columbia River, BC, d/s pulp mill	Total P5CDD	ND	Mah et al. 1989
Fish	Thompson River, BC, d/s pulp mill & STP	Total P5CDD	ND–4.2 (3.4)	Mah et al. 1989
Fish	Williston Lake, BC, d/s pulp mill	Total P5CDD	ND–6.7 (3.7)	Mah et al. 1989
Fish	Williston Lake, BC, d/s forest industries	Total P5CDD	ND	Mah et al. 1989
Shrimp	Jackfish Bay, Lake Superior, d/s BKME	Total P5CDF	16	Sherman et al. 1990
White sucker clam	Jackfish Bay, Lake Superior, d/s BKME	Total P5CDF	<3	Sherman et al. 1990
Perch	Gulf of Bothnia, Sweden d/s pulp mill	1,2,3,7,8-P5CDF	0.3–0.4	Rappe et al. 1989
Arctic char	Lake Vettern, Sweden, d/s two pulp mills	1,2,3,7,8-P5CDF	2–8	Rappe et al. 1989
Perch	Gulf of Bothnia, Sweden d/s pulp mill	2,3,4,7,8-P5CDF	0.9–1.6	Rappe et al. 1989
Arctic char	Lake Vettern, Sweden, d/s two pulp mills	2,3,4,7,8-P5CDF	7.2–39	Rappe et al. 1989
Fish	Kootenay River, BC, d/s pulp mill	Total P5CDF	ND	Mah et al. 1989
Fish	Fraser River, BC, d/s P & P mill	Total P5CDF	ND	Mah et al. 1989
Fish	Fraser River, BC, d/s pulp and timber industries	Total P5CDF	ND–4.0 (3.33)	Mah et al. 1989
Fish	Fraser River, BC, 5 km d/s two pulp mills	Total P5CDF	ND–25.1 (7.8)	Mah et al. 1989
Fish	Columbia River, BC, d/s pulp mill	Total P5CDF	ND–13.4 (4.5)	Mah et al. 1989
Fish	Thompson River, BC, d/s pulp mill & STP	Total P5CDF	ND–12.7 (4.2)	Mah et al. 1989
Fish	Williston Lake, BC, d/s pulp mill	Total P5CDF	ND	Mah et al. 1989
Fish	Williston Lake, BC, d/s forest industries	Total P5CDF	ND	Mah et al. 1989

Table 8.11 Levels of PCDDs and PCDFs (pg/g = ppt) in Fish Near Effluent Discharges from Pulp Mills (Continued)

Species	Location	Compound	Concentration[a]	References
Fish	Fraser River, BC, all locations as above	Total H6CDD	ND	Mah et al. 1989
Fish	Fraser River, BC, 4-5 km d/s of two pulp mills	Total H6CDD	ND–40.8 (12.5)	Mah et al. 1989
Fish	All locations as above in Fraser River, Columbia River, Kootenay River, Thompson River, Williston Lake, BC	Total H6CDF	ND	Mah et al. 1989
Fish	All locations as above in Fraser River, Columbia River, Kootenay River, Thompson River, Williston Lake, BC	Total H7CDD	ND	Mah et al. 1989
Fish	All locations as above in Fraser River, Columbia River, Kootenay River, Thompson River, Williston lake, BC	Total H7CDF	ND	Mah et al. 1989
Fish	All locations as above in Fraser River, Columbia River, Kootenay River, Thompson River, Williston lake, BC	Total O8CDD	ND	Mah et al. 1989
Fish	All locations as above in Fraser River, Columbia River, Kootenay River, Thompson River, Williston lake, BC	Total O8CDF	ND	Mah et al. 1989

[a]Concentration range, with the average concentration in brackets; d/s = downstream; P&P = pulp and paper; STP = sewage treatment plant; BKME = Bleached kraft mill effluent; AB = Alberta, Canada; BC = British Columbia, Canada.

Chronic effects on exposure to PCDDs include hemorrhaging prior to death, fin necrosis, and morphological changes, such as edema and abdominal swelling; following are examples:

Species	Compound	Route	Conconcentration
Rainbow trout	2,3,7,8-T4CDD	IP	5–10 and 25 µg/kg
Guppies	2,3,7,8-T4CDD	Water	1–10 µg/L
Pike, Rainbow trout, Yellow perch	2,3,7,8-T4CDD	Water	1–10 ng/L
Mosquito fish	2,3,7,8-T4CDD	Water with soil containing	2.4 ng/L
Catfish, fingerling	2,3,7,8-T4CDD	0.1 ppm	4.2 ng/L (lethal)

Sources: Kleeman et al. 1988; Miller et al. 1973; Helder 1981; Spitsbergen et al. 1988a; Yockim et al. 1978

All the above showed the chronic effects prior to lethality.

Table 8.12 Levels of PCDFs (ppt, wet weight) in Aquatic Species Including Fish

Species	Location	PCDF	Concentration	References
Lake trout (*Salvelinus namaycush*) Walleye trout (*S. vitreum vitreum*) Composite	Lake St. Clair	2,3,7,8-tetraCDF	24.8	Zacharewski et al. 1989
		1,2,3,7,8-pentaCDF	3.7	
		2,3,4,7,8-pentaCDF	5.4	
		1,2,3,4,7,8-hexaCDF	0.5	
		1,2,3,6,7,8-hexaCDF	0.5	
		1,2,3,7,8,9-hexaCDF	<0.05	
		2,3,4,6,7,8-hexaCDF	0.9	
		1,2,3,4,6,7,8-heptaCDF	0.5	
		1,2,3,4,7,8,9-heptaCDF	<0.2	
		OctaCDF	0.8	
Lake trout (*Salvelinus namaycush*) Walleye trout (*S. vitreum vitreum*) Composite	Lake Michigan	2,3,7,8-tetraCDF	34.8	Zacharewski et al. 1989
		1,2,3,7,8-pentaCDF	4.9	
		2,3,4,7,8-pentaCDF	10.2	
		1,2,3,4,7,8-hexaCDF	1.4	
		1,2,3,6,7,8-hexaCDF	1.1	
		1,2,3,7,8,9-hexaCDF	<0.05	
		2,3,4,6,7,8-hexaCDF	1.3	
		1,2,3,4,6,7,8-heptaCDF	0.9	
		1,2,3,4,7,8,9-heptaCDF	<0.2	
		OctaCDF	<2.0	
Lake trout (*Salvelinus namaycush*) Walleye trout (*S. vitreum vitreum*) Composite	Lake Huron	2,3,7,8-tetraCDF	22.8	Zacharewski et al. 1989
		1,2,3,7,8-pentaCDF	6.2	
		2,3,4,7,8-pentaCDF	12.8	
		1,2,3,4,7,8-hexaCDF	1.6	
		1,2,3,6,7,8-hexaCDF	1.2	
		1,2,3,7,8,9-hexaCDF	<0.07	
		2,3,4,6,7,8-hexaCDF	1.4	
		1,2,3,4,6,7,8-heptaCDF	0.5	
		1,2,3,4,7,8,9-heptaCDF	<0.1	
		OctaCDF	<0.3	
Lake trout (*Salvelinus namaycush*) Walleye trout (*S. vitreum vitreum*) Composite	Lake Ontario	2,3,7,8-tetraCDF	20.6	Zacharewski et al. 1989
		1,2,3,7,8-pentaCDF	4.7	
		2,3,4,7,8-pentaCDF	20.2	
		1,2,3,4,7,8-hexaCDF	12.7	
		1,2,3,6,7,8-hexaCDF	1.9	
		1,2,3,7,8,9-hexaCDF	<0.1	
		2,3,4,6,7,8-hexaCDF	1.2	
		1,2,3,4,6,7,8-hexaCDF	0.9	
		1,2,3,4,7,8,9-hexaCDF	<0.1	
		OctaCDF	<0.9	
Lake trout (*Salvelinus namaycush*) Walleye trout (*S. vitreum vitreum*) Composite	Lake Erie	2,3,7,8-tetraCDF	11.3	Zacharewski et al. 1989
		1,2,3,7,8-pentaCDF	1.4	
		2,3,4,7,8-pentaCDF	2.7	
		1,2,3,4,7,8-hexaCDF	0.2	
		1,2,3,6,7,8-hexaCDF	0.3	
		1,2,3,7,8,9-hexaCDF	<0.1	
		2,3,4,6,7,8-hexaCDF	0.5	
		1,2,3,4,6,7,8-heptaCDF	0.6	
		1,2,3,4,7,8,9-heptaCDF	<0.2	
		OctaCDF	<1.1	

CHLORINATED DIOXINS AND FURANS

Table 8.12 Levels of PCDFs (ppt, wet weight) in Aquatic Species Including Fish (continued)

Species	Location	PCDF	Concentration	References
Lake trout (*Salvelinus namaycush*) Walleye trout (*S. vitreum vitreum*) Composite	Lake Superior	2,3,7,8-tetraCDF 1,2,3,7,8-pentaCDF 2,3,4,7,8-pentaCDF 1,2,3,4,7,8-hexaCDF 1,2,3,6,7,8-hexaCDF 1,2,3,7,8,9-hexaCDF 2,3,4,6,7,8-hexaCDF 1,2,3,4,6,7,8-heptaCDF 1,2,3,4,7,8,9-heptaCDF OctaCDF	15.7 1.7 2.8 0.5 0.3 <0.06 0.4 0.4 <0.2 <0.8	Zacharewski et al. 1989
Lobster (*Homarus Americanus*) digestive gland	Mipamichi Bay and Limestone Point, NB, Sydney Harbour and Port Morien, N.S.	Total tetraCDFs Total pentaCDFs Total hexaCDFs Total heptaCDFs OctaCDF	189.8 52.2 37.9 <9.1 (2–10)[a]	Clement et al. 1987
Carp (*Cyprinus carpio*)	Lake Ontario	Total pentaCDFs	1,015	Stalling et al. 1987
Coho salmon (*Oncorhynchus kisutsch*)		Total tetraCDFs	327	Stalling et al. 1987
Lobster (*Homarus Americanus*) heptopancreas	Newark Bay New York Bight	2,3,7,8-tetraCDF Total tetraCDFs 1,2,3,7,8-/1,2,3,4,8-pentaCDF 2,3,4,7,8-pentaCDF Total pentaCDFs 1,2,3,4,7,8-/1,2,3,4,7,9-hexaCDF 1,2,3,6,7,8-hexaCDF 1,2,3,7,8,9-hexaCDF 2,3,4,6,7,8-hexaCDF Total hexaCDFs 1,2,3,4,6,7,8-heptaCDF 1,2,3,4,7,8,9-heptaCDF OctaCDF	365.7 1,568.6 79.5 179.2 1,008.4 10.7 <6.0 <3.0 7.0 172.1 <3.8 <3.8 <29.2	Rappe et al. 1991
Striped bass (*Morone saxatilis*) meat	Newark Bay New York Bight	2,3,7,8-tetraCDF Total tetraCDFs 1,2,3,7,8-/1,2,3,4,8-pentaCDF 2,3,4,7,8-pentaCDF Total pentaCDFs 1,2,3,4,7,8-/1,2,3,4,7,9-hexaCDF 1,2,3,6,7,8-hexaCDF 1,2,3,7,8,9-hexaCDF 2,3,4,6,7,8-hexaCDF Total hexaCDFs 1,2,3,4,6,7,8-heptaCDF 1,2,3,4,7,8,9-heptaCDF OctaCDF	68.7 92.5 7.1 30.3 58.5 1.1 0.4 <0.1 <2.6 3.2 1.6 <0.4 <3.0	Rappe et al. 1991

Table 8.11 Levels of PCDFs (ppt, wet weight) in Aquatic Species Including Fish (continued)

Species	Location	PCDF	Concentration	References
Blue Crab (*Callinectus sapidus*) meat	Newark Bay New York Bight	2,3,7,8-tetraCDF	13.3	Rappe et al. 1991
		Total tetraCDFs	148.7	
		1,2,3,7,8-/1,2,3,4,8-pentaCDF	5.5	
		2,3,4,7,8-pentaCDF	7.3	
		Total pentaCDFs	91.9	
		1,2,3,4,7,8-/1,2,3,4,7,9-hexaCDF	2.6	
		1,2,3,6,7,8-hexaCDF	0.6	
		1,2,3,7,8,9-hexaCDF	<0.2	
		2,3,4,6,7,8-hexaCDF	<2.3	
		Total hexaCDFs	9.4	
		1,2,3,4,6,7,8-heptaCDF	3.2	
		1,2,3,4,7,8,9-heptaCDF	<0.9	
		Total heptaCDFs	3.2	
		OctaCDF	<7.1	
Blue Crab (*Callinectus sapidus*) heptopancreas	Newark Bay New York Bight	2,3,7,8-tetraCDF	628.3	Rappe et al. 1991
		Total tetraCDFs	7,049.3	
		1,2,3,7,8-/1,2,3,4,8-pentaCDF	185.7	
		2,3,4,7,8-pentaCDF	391.4	
		Total pentaCDFs	4,219.1	
		1,2,3,4,7,8-/1,2,3,4,7,9-hexaCDF	261.0	
		1,2,3,6,7,8-hexaCDF	43.3	
		1,2,3,7,8,9-hexaCDF	<5.0	
		2,3,4,6,7,8-hexaCDF	9.8	
		Total hexaCDFs	803.3	
		1,2,3,4,6,7,8-heptaCDF	184.6	
		1,2,3,4,7,8,9-heptaCDF	7.1	
		OctaCDF	<51	
Lobster (*Homarus Americanus*) meat	Newark Bay New York Bight	2,3,7,8-tetraCDF	<0.3	Rappe et al. 1991
		Total tetraCDFs	27.1	
		1,2,3,7,8-/1,2,3,4,8-pentaCDF	2.4	
		2,3,4,7,8-pentaCDF	1.8	
		Total pentaCDFs	33.6	
		1,2,3,4,7,8-/1,2,3,4,7,9-hexaCDF	0.4	
		1,2,3,6,7,8-hexaCDF	<0.2	
		1,2,3,7,8,9-hexaCDF	<0.2	
		2,3,4,6,7,8-hexaCDF	<2.0	
		Total hexaCDFs	7.8	
		1,2,3,4,6,7,8-hepta CDF	<0.9	
		1,2,3,4,7,8,9-heptaCDF	<0.9	
		OctaCDF	<7.7	

[a] detection limit

Growth reductions have been reported in a number of different species after short-term exposure to 2,3,7,8-T4CDD. Clinical symptoms resembled the "wasting syndrome," reported for mammalian species (Peterson et al. 1984). Significant changes in feeding pattern and growth in a variety of fish species at different life stages were reported after a short duration exposure to 0.1 to 5.6 ng/L of 2,3,7,8-T4CDD (Helder 1981). Total body weight loss occurred at a administered dose of 1 to 10 µg/kg of 2,3,7,8-T4CDD (Kleeman et al. 1988; Spitsbergen et al. 1988b). Similar effects were not reported on other PCDDs in the literature. Teratogenic and reproductive effects of 2,3,7,8-T4CDD have been reported in fish and fish eggs (Helder, 1981; Wannemacher et al. 1992).

Exposure to 2,3,7,8-T4CDD and TCDFs has been shown to increase the activity of the enzyme EROD (7-ethoxy resorufin-o-deethylase) and increase in cytochrome 450 activity (Van der Weiden et al. 1989; Muir et al. 1990). The ED_{50} (median effective dose) for induction of AHH by 2,3,7,8-T4CDD (by intraperitoneal route) was 0.002 µmol/kg, and by PCBs was 0.37 µmol/kg (Janz and Metcalf, 1991).

LD_{50} values of 2,3,7,8-T4CDD vary several fold due to the different susceptibilities of several species. Table 8.13 lists the marked difference in the acute oral toxicity of 2,3,7,8-T4CDD.

Table 8.13 Oral LD_{50} Values of 2,3,7,8-T4CDD

Mammalian Species	LD_{50} (µg/kg)
Guinea pig	1
Rat	22–45
Monkey	70
Rabbit	155
Mouse	144
Dog	300
Hamster	5000

Source: Hon-Wing Leung and Paustenbach, 1989.

Health Effects of 2,3,7,8-T4CDD

In humans, 2,3,7,8-T4CDD causes chloracne, a severe skin lesion, most commonly on the head and upper body. Chloracne lasts for years after cessation of exposure.

- There is suggestive evidence that 2,3,7,8-T4CDD causes liver damage in humans, as shown by an increase in levels of certain enzymes in the blood. Animal studies have shown severe liver damage in some species.
- Animal exposure to 2,3,7,8-T4CDD results in severe loss of body weight prior to death. There is suggestive evidence that 2,3,7,8-T4CDD causes loss of appetite, weight loss, and digestive disorders.
- Animal studies demonstrated that 2,3,7,8-T4CDD produced toxicity to immune system. It was not demonstrated in humans, but may result in greater susceptibility.

- Some animal species, on exposure to 2,3,7,8-T4CDD, demonstrated adverse reproductive effects, including abnormal abortions. The monkey is a very sensitive mammalian species to this toxic property of 2,3,7,8-T4CDD. This effect has not been demonstrated in humans.
- Some animal species, on exposure to 2,3,7,8-T4CDD during pregnancy, resulted in malformations in the offspring. Low levels of 2,3,7,8-T4CDD have been detected in human milk, but the effects to infants and children are not known.
- There is no adequate evidence for carcinogenic hazard of 2,3,7,8-T4CDD to humans. Certain herbicides mixtures, containing 2,3,7,8-T4CDD as an impurity, provide limited evidence of causing cancer in exposed humans. Based on the positive evidence in animal studies, 2,3,7,8-T4CDD is probably carcinogenic to humans.

Health Effects of PCDFs

Accidental poisoning occurred in Japan and Taiwan in 1960s and 1970s, where many people ate food cooked with rice oil contaminated with PCBs containing PCDFs. The dose was higher than normally found in average human diet.

- Skin and eye irritations, severe ache, darkened skin colors, swollen eyelids with discharge, but not in all people exposed.
- Vomiting and diarrhea, anemia, frequent lung infections, numbness, and other effects to the nervous systems.
- Mild changes in the liver.
- Animals fed PCDFs showed severe body weight loss and severe damage to the stomach, liver, kidneys, and immune systems.
- Birth defects and testicular damage in animals.
- No studies conclusively proved that PCDFs caused cancer in accidentally poisoned humans. There are no cancer studies on animals that were fed or inhaled PCDFs. One study showed negative results on skin cancer on animals when they were treated with skin applications.
- Application of PCDFs, after applying a carcinogen, produced skin cancer; this proved the role of promoting effect of PCDFs.
- Not classified as a carcinogen.

Relative Toxicity of Chlorinated Dioxins and Furans

2,3,7,8-T4CDD is the most toxic isomer and has the greatest amount of toxicity information available. The toxicity substantially varies among the different PCDD and PCDF isomers and congeners. In order to assess the mixture of PCDDs and PCDFs in a scientifically sound basis, the Toxic Equivalency Factors (TEFs) have been developed. The basis for this TEF scheme is to relate the toxic potency of the different PCDD and PCDF isomers and congeners to 2,3,7,8-T4CDD. In other words, the TEF values will convert the concentrations of all isomers and congeners of PCDD and PCDF to the toxicity equivalent of 2,3,7,8-T4CDD.

TEF, as proposed by Safe (1990), is based on structure–activity relationship, laboratory enzyme induction studies, and acute and chronic *in vitro* studies. The TEF scheme developed by Walker and Peterson (1991) was based on the toxicity

Table 8.14 Toxic Equivalency Factors for PCDDs and PCDFs

Isomer	ITEF[a]	Safe[b]	Walker and Peterson[c]
2,3,7,8-T$_4$CDD	1.0	1.0	1.0
1,2,3,7,8-P$_5$CDD	0.5	0.5	0.7
1,2,3,4,7,8-H$_6$CDD	0.1	0.1	0.3
1,2,3,7,8,9-H$_6$CDD	0.1	0.1	—
1,2,3,6,7,8-H$_6$CDD	0.1	0.1	—
1,2,3,4,6,7,8-H$_7$CDD	0.01	0.01	—
O$_8$CDD	0.001	0.001	—
2,3,7,8-T$_4$CDF	0.1	0.1	0.03
2,3,4,7,8-P$_5$CDF	0.5	0.5	0.4
1,2,3,7,8-P$_5$CDF	0.01	0.05	0.3
1,2,3,4,7,8-H$_6$CDF	0.1	0.1	0.3
1,2,3,7,8,9-H$_6$CDF	0.1	0.1	—
1,2,3,6,7,8-H$_6$CDF	0.1	0.1	—
2,3,4,6,7,8-H$_6$CDF	0.1	0.1	—
1,2,3,4,6,7,8-H$_7$CDF	0.01	0.01	—
1,2,3,4,7,8,9-H$_7$CDF	0.01	0.01	—
O$_8$CDF	0.001	0.001	—

[a] ITEF = International TEF Scheme
[b] Safe 1990
[c] Walker and Peterson, 1991. Values are rounded off.

difference between mammalian species and aquatic species, such as rainbow trout, in early life stages. Caution should be exercised in applying the generalized TEF values to all species and circumstances. These TEF values are based on the weight-of-evidence evaluation of currently available toxicity information for a wide range of species. As and when further information becomes available, it is likely these TEF values will be revised.

REGULATIONS

Several studies examined the 2,3,7,8-T4CDD toxicity in two-year rodent chronic bioassays. The major studies include a rat study conducted by Kociba et al. (1978) and rat and mouse studies conducted by the U.S. National Toxicology Program (NTP 1980). In the study by Kociba et al., male and female Sprague-Dawley rats were fed 2,3,7,8-T4CDD in the diet at doses of 0.001, 0.01, and 0.1 µg/kg/day for two years. At 0.001 µg/kg/day, no adverse effects were observed; at 0.01 µg/kg/day, significant increases in hepatocellular nodules were observed in the female rat; and at 0.1 µg/kg/day, toxicity and decreased body weight were observed. Also, significant increases of carcinoma of the hard palate, nasal turbinates, and tongue in male rats and significant increases in hepatocellular carcinomas, neoplastic nodules, and carcinoma of the hard palate, nasal turbinates, and lungs in the female rats were observed. Early mortality was also observed in all groups, but was statistically significant only in the female rats exposed to 0.1 µg/kg/day of T4CDD (Kociba et al. 1978, 1979).

The doses for the U.S. NTP study were 0.01, 0.05, and 0.5 µg 2,3,7,8-T4CDD/kg/week for Osborne-Mendel male and female rats. This study observed significant increases in carcinoma of the thyroid in male rats, at all doses, and in subcutaneous tissue fibroma, at the high dose. In the female rats, significant increases were observed only at high doses in hepatocellular carcinoma and neoplastic nodules, subcutaneous tissue fibrosarcoma, and adrenal cortical adenoma (NTP 1980).

In the NTP (1980) mouse study, male and female B6C3F1 mice were orally dosed with 2,3,7,8-T4CDD at 0.01, 0.05, and 0.5 µg/kg/week for male mice and at 0.04, 0.2, and 2.0 µg/kg/week for female mice. At the high doses, significant increases in heptacellular carcinoma in male mice, and significant increases in heptacellular carcinoma, fibrosarcoma, histiocytic lymphoma, thyroid adenoma, and cortical adema in female mice were observed.

The NTP (1980) study also conducted a skin-painting study in Swiss-Webster mice in which males received 0.001 µg 2,3,7,8-T4CDD 3 days/week/104 weeks. Significant increase of fibrosarcoma in the integumentary systems of female mice was observed.

U.S. agencies, such as the EPA and Centers for Disease Control (CDC), assumed that 2,3,7,8-T4CDD behaves as a tumor initiator in animals and used extrapolation models to estimate human risk and ADI (Allowable Daily Intake). The Ontario Ministry of the Environment (OME), Canada, the State Institute of National Health (SINH) of the The Netherlands, and the Federal Environmental Agency (FEA) of the Federal Republic of Germany concluded that 2,3,7,8-T4CDD does not have initiator activity and applied safety factor approach to the NOAEL (No-Observable Adverse Effect Level). The 1600-fold variation for acceptable intakes estimated by the Americans, Canadians, and Europeans indicates a large scientific uncertainty about 2,3,7,8-T4CDD toxicity. Regulators are attempting to use human epidemiological data to eliminate some limitations inherent to animal data, particularly since techniques to measure T4CDD levels in adipose tissue and serum are now available.

Comparison of Allowable Daily Intakes (ADIs) for T4CDD Set by Governmental Agencies

Agency	Risk Analysis Approach	End Point	ADI (pg/kg/day)
U.S. EPA[a]	Linearized multistage	Cancer	0.0064
U.S. CDC[b]	Linearized multistage	Cancer	0.028-1.428
U.S. FDA[c]	Safety Factor (77)	Cancer	13.0
OME[d]	Safety Factor (100)	Cancer + reproductive	10.0
Federal Republic of Germany, FEA[e]	Safety Factor (100–1,000)	Cancer + reproductive	1.0–10.0
SINH, The Netherlands	Safety Factor (250)	Cancer + reproductive	4.0

[a]EPA 1985; [b]Kimbrough et al. 1984; [c]Cordle 1981; [d]Ontario Ministry of the Environment 1985; [e]Federal Environment Agency, Germany 1984; [f]State Institute of National Health (SINH), The Netherlands, Van der Heijden et al. 1982.

Regulatory Limits for 2,3,7,8-T4CDD

Ambient air µg/m³

State	0.5 h	8 h	24 h	Annual
Arizona	0.042 (1 hr)	0.01		0.00003
Kansas				0.000,000,0303
Maine			0.000,0035	0.000,000,25
Michigan				0.000,000,023
New York				0.000,00,03
Washington				0.000,000,03

Domestic drinking water µg/L
Arizona = 0.000,000,2
Florida = 0.01
Kansas = 0.000,0022
Maine = 0.000,0022
Massachusetts = 0.000,000,03
New Jersey = 0.01
Virginia = 0.000,000,22

Soil Permissible Cleanup Level mg/kg

Location	Residential	Agricultural	Industrial
Arizona	0.000,009	—	—
Canada	0.000,1	0.000,01	—
Massachusetts	0.000,004	—	0.000,02
Michigan	0.000,017	—	0.000,02
New York	0.000,0045	—	—
Washington	0.000,00667	—	—

REFERENCES

Acharya, P., Decicco, S.G., and Novak, R.G. 1991. Factors that can influence and control the emissions of dioxins and furans from a hazardous waste incinerators. *J. Air Water Manage. Assoc.* 41:1605–1615.

Adams, W.J., DeGraeve, G.M., Satourin, T.D., Conney, J.D., and Mosher, G.M. 1986. Toxicity and bioconcentration of 2,3,7,8-TCDD to fathead minnow (*Pimelphales promelas*). *Chemosphere*. 15:1503–1511.

Agriculture Canada 1984. Laboratory Division. Unpublished information.

ATSDR 1989. Agency for Toxic Substances and Disease Registry, U.S. Public Health Service, U.S. Department of Health and Human Services, Atlanta, GA, Toxicological Profile for 2,3,7,8-Tetrachloro-dibenzo-p-dioxin, June 1989.

ATSDR 1994. Agency for Toxic Substances and Disease Registry, U.S. Public Health Service, U.S. Department of Health and Human Services, Atlanta, GA, Toxicological Profile for Chlorodibenzofurans, TP-93/04, May 1994.

Ballschmitter, K., Buchert, H., Niemczyk, R., Munder, A., and Swerev, M. 1986a. Automobile exhaust versus municipal-waste incineration as sources of the polychloro-dibenzodioxins (PCDD) and -furans (PCDF) found in the environment. *Chemosphere*. 17:915–995.

Ballschmitter, K., Buchert, H., and Niemcyzk, R. 1986b. Automobile exhaust versus municipal waste incineration as sources of the polychloro-dibenzodioxins (PCDDs) and -furans (PCDFs) found in the environment. *Chemosphere*. 15:901–915.

Ballschmitter, K., Braunmiller, I., Niemczyk, R., and Swerev, M. 1988. Reaction pathways for the formation of polychloro-dibenzodioxins (PCDD) and dibenzofuran (PCDF) in combustion processes. II. Chlorobenzenes and chlorophenols as precursors in the formation of polychloro-dibenzodioxins and polychloro-dibenzofurans in flame chemistry. *Chemosphere.* 17:995–1005.

Barnes, D.G. 1983. Chlorinated Dioxins Work Group, EPA, Washington, D.C. 12:643–655.

Beard, A., Naikwadi, K.P., and Karasek, F.W. 1991. Mechanisms of formation of PCDDs and PCDFs in the petroleum refining industry. Eleventh International Symposium on Chlorinated Dioxins and Related Compounds. Research Triangle Park, NC.

Bingham, A.G., Edmunds, C.J., Graham, B.W.L., and Jones, M.T. 1989. Determination of PCDDs and PCDFs in car exhaust. *Chemosphere.* 19:669-673.

Bobet, E., Berard, M.F., and Dann, T. 1990. The measurement of PCDD and PCDF in ambient air southwestern Ontario. *Chemosphere.* 20:1439–1445.

Bonafanti, L., Cioni, M., and Rossi, C. 1990. Evaluation of PCDD/PCDF emission from the combined combustion of RDF with coal. *Chemosphere.* 20:1891–1897.

Bopp, R.F., Gross, M.L., and Tong, H. 1991. A major incident of dioxin contamination: sediments of New Jersey estuaries. *Environ. Sci. Technol.* 25:951–956.

Bowes, C.W., Mulvihill, M.J., and Simoneit, B.R. 1975a. Identification of chlorinated dibenzofurans in American polychlorinated biphenyls. *Nature.* 256:305–307.

Brna, T.G. and Kilgore, J.D. 1990. Control of PCDD-PCDF emission from municipal waste combustion systems. *Chemosphere.* 20:1875–1882.

Bumb, R.R., Crummett, W.B., Cutie, S.S., Giedhill, J.R., Hummell, R.H., Kagel, R.O., Lamparski, L.L., Luoma, E.V., Miller, D.L., Nestrick, T.J., Shadoff, L.A., Stehl, R.H., and Woods, J.S. 1980. Trace chemistries of fire: a source of chlorinated dioxins. *Science.* 210:385–390.

Canadian Pulp and Paper Association (CPPA) 1990. 1989 National Dioxin Survey of Pulp Mill Effluents. CPPA. Unpublished data.

Chiu, C., Thomas, R.S., Lockwood, J., Li, K., Halaman, R., and Lao, R.C. 1983. Polychlorinated hydrocarbons from power plants, wood burning and municipal incinerators. *Chemosphere.* 12:607–616.

Choudhury, G.C. and Hutzinger, O. 1982. Mechanistic aspects of the thermal formation of halogenated organic compounds including PCDDs. 2. Thermochemical generation and destruction of dibenzofurans and dibenzo-p-dioxins. *Environ. Toxicol. Chem.* 5:67–93.

Choudhury, G.G., Sundstrom, G., and Ruzo, L.O. 1977. Photochemistry of chlorinated diphenyl ethers. *J. Agric. Food Chem.* 25:1371–1376.

Choudhury, G.C. and Webster, G.R.B. 1989. Environmental photocehemistry of PCDDs. Quantum yields of the direct phototransformation of 1,2,3,7-tetra-, 1,3,6,8-tetra-, 1,2,3,4,6,7,8-hepta- and 1,2,3,4,6,7,8,9- octachlorodibenzo-p-dioxin in aqueous acetonitrile and their sunlight half-lives. *J. Agr. Food. Chem.* 37:254–261.

Christmann, W., Kasiske, D., and Kloppel, K.D. 1989. Combustion of polyvinyl chloride as important source for the formation of PCDD-PCDF. *Chemosphere.* 19:387–392.

Clement, R.E., Tosine, H.M., and Osborne, J. 1987a. Emission of chlorinated organics from a municipal sewage sludge burning incinerator. *Chemosphere.* 16:1895–1900.

Clement, R.E., Tosine, H.M., and Taguchi, V. 1987b. Investigation of American Lobster, *Homarus americanus,* for the presence of chlorinated dibenzo-p-dioxins and dibenzofurans. *Bull. Environ. Contam. Toxicol.* 39:1069–1075.

Clement, R.E., Tosine, H.M., and Ali, B. 1985. Levles of polychlorinated dibenzo-p-dioxin and dibenzofuran in woodburning stoves and fireplaces. *Chemosphere.* 14:815–819.

Cook, P.M., Kuehl, D.W., Walker, M.K., and Peterson, R.E. 1991. Bioaccumulation and toxicity of TCDD and related compounds in aquatic ecosystems. In: Biological Basis for Risk Assessment of Dioxins and Related Compounds, Gallo, M., Scheuplein, R.J., and Van Der Heijden, K.A., Eds., pp. 143-167, Cold Spring Harbor Laboratory Press, Cold Spring Harbor, NY.

Corbet, R.L., Webster, G.R.B., and Muir, D.C.G. 1988. Fate of 1,3,6,8-tetrachlorodibenzo-p-dioxin in an outdoor aquatic system. *Environ. Toxicol. Chem.* 7:167–180.

Cordle, F. 1981. The use of epidemiology in the regulation of dioxins in the food supply. *Regul. Toxicol. Pharmacol.* 1:379–387.

Darskus, R. and Schelessing, H. 1989. *Levels of Polychlorinated Dioxins and Furans in Sewage Sludge.* Dioxin '89. 9th International Symposium on Chlorinated Dioxins and Related Compounds, Toronto, Ontario, Canada.

des Rosiers, P.E. and Lee, A. 1986. PCB fires: correlation of chlorobenzene isomer and PCB homolog contents of PCB fluids with PCDD and PCDF contents of soot. *Chemosphere.* 15:1313–1323.

des Rosiers, P.E. 1987. National Dioxin Study. ACS Symposium Series no. 338:34–53.

Eadon, G., Kaminsky, L., Silkworth, J., Aldous, K., Hilker, D., O'Keefe, D., Smith, R., Gierthy, J., Hawley, J., Kim, N., and DeCaprio. 1986. Calculation of 2,3,7,8-TCDD equivalent concentrations of complex environmental contaminant mixtures. *Environ. Health Perspect.* 70:221–227.

Edgerton, S.A., Czuczwa, J.M., and Rench, J.D. 1989. Ambient air concentrations of polychlorinated dibenzo-p-dioxins and dibenzofurans in Ohio: sources and health risk assessment. *Chemosphere.* 18:1713–1730.

Eitzer, B.B. and Hites, R.A. 1986. Concentrations of dioxins and dibenzofurans in the atmosphere, *Int. J. Environ. Anal. Chem.* 27:215–230.

Eitzer, B.D. and Hites, R.A. 1988. Vapour pressures of chlorinated dixoins and furans. *Environ. Sci. Technol.* 22:1362–1364.

Eitzer, B.D. and Hites, R.A. 1989a. Atmospheric transport and deposition of polychlorinated dibenzo-p-dioxins and dibenzofurans. *Environ. Sci. Technol.* 23:1396–1401.

Eitzer, B.D. and Hites, R.A. 1989b. Polychlorinated dibenzo-p-dioxins and dibenzofurans in the ambient atmosphere of Bloomington, Indiana. *Environ. Sci. Technol.* 23:1389–1395.

Environment Canada 1981. Chlorophenols and their impurities in the Canadian environment. Environmental Impact Control Directorate, March 1981, Economic and technical review report EPS 3-EC-81-2.

EPA. 1985. U.S. Environmental Protection Agency. Health Assessment document for polychlorinated dibenzo-p-dioxins. Washington, D.C., Office of Health and Environmental Assessment. EPA Report No. 600/8-84-014F.

EPA. 1987a. U.S. Environmental Protection Agency. National Dioxin Study Tier 4-combustion sources; Final Report-site 10 secondary copper recovery Cupola Furnace MET-A. U.S. EPA, Research Triangle Park, NC. NTIS PB88-101183.

EPA. 1987b. U.S. Environmental Protection Agency. National Dioxin Study Tiers 3, 4, 5, and 7. U.S. EPA 440/4-87-003. Office of Water Regulations and Standards, Washington D.C.

EPA. 1986a. U.S. Environmental Protection Agency. Health assessment document for polychlorinated dibenzofurans, Cincinnati, OH, Environmental Criteria and Assessment Office. NTIS PB86-221256.

Erickson, M.D. 1989. PCDFs and related compounds produced from PCB fires — a review. *Chemosphere.* 19:161–165.

Evers, E.H.G. 1991. The formation of polychlorinated dibenzo-p-dioxins and dibenzofurans during industrial chlorination processes. Eleventh International Symposium on Chlorinated Dioxins and Related Compounds, Research Triangle Park, NC.

Federal Environmental Agency. 1984. Report on dioxins. Erich Schmidt Verlag, Berlin.

Firestone, D., Ress, J., Brown, N.L., Barron, R.P., and Damico, J.N. 1972. Determination of polychlorinated dibenzo-p-dioxins and related compounds in commercial chlorophenols. *J. Assoc. Off. Anal. Chem.* 55(1):85–92.

Glasser, H., Chang, D.P.Y., and Hickman, D.C. 1991. An analysis of biomedical waste incineration. *J. Air Waste Manage. Assoc.* 41:1180–1188.

Hagenmaier, H., Horch, K., and Fahlenkamp, H. 1991. Destruction of PCDD and PCDF in refuse incineration plants by primary and secondary measures. *Chemosphere.* 23:1429–1438.

Hagenmaier, H. and Berchtold, A. 1986. Analysis of waste from production of sodium pentachlorophenate for polychlorinated dibenzodioxins (PCDD) and dibenzofurans (PCDF). *Chemosphere.* 15:1991–1994.

Hagenmaier, H. and Brunner, H. 1987. Isomer-specific analysis of pentachlorophenol and sodium pentachlorophate for 2,3,7,8-substituted PCDD and PCDF at sub-ppb levels. *Chemosphere.* 16:1759–1764.

Hagenmaier, H., Brunner, H., and Haag, R. 1987. Copper-catalyzed dechlorination-hydrogenation of polychlorinated dibenzo-p-dioxins, polychlorinated dibenzofurans and other chlorinated aromatic compounds. *Environ. Sci. Technol.* 21:1085–1088.

Harrad, S.J., Fernandes, A.R., and Creaser, C.S. 1991. Domestic coal combustion as a source of PCDDs and PCDFs in the British environment. *Chemosphere.* 23:255–261.

Helder, T. 1980. Effects of 2, 3, 7, 8-tetrachloro dibenzo-p-dioxin (TCDD) on early life stages of the pike (*Esox lucitus L.*). *Sci. Total Environ.* 14:255–264.

Helder, T. 1981. Effects of 2,3,7,8-tetrachloro dibenzo-p-dioxin (TCDD) on early life stages of rainbow trout (*Salmo gairdneri*). *Toxicology.* 19:101–112.

Hites, R.A. 1990. Environmental behaviour of chlorinated dioxins and furans. *Ace. Chem. Res.* 23:194–201.

Hodson, P.V., McWhirter, M., and Ralph, K. 1992. Effects of bleached kraft mill effluent in fish in the St. Maurice River, Quebec. *Environ. Toxicol. Chem.* 11:1535–1651.

Hon-Wing Leung and Paustenbach, D.J. 1989. In: The Risk Assessment of Environment and Human Health Hazards: A Textbook of Case Studies. Paustenbach, D.J., Ed., John Wiley & Sons, New York, pp. 689-710.

Hryhorczuk, D.O., Orris, P., and Kominsky, J.R. 1986. PCB, PCDDF and PCDD exposure following a transformer fire; Chicago. *Chemosphere.* 15:1297–1303.

Hunt, G.T., Maisel, B.E., and Hoyt, M. 1990. Ambient concentrations of PCDDs/PCDFs (polychlorinated dibenzodioxins/dibenzofurans) in the South Coast air basin. *NTIS* PB90-169970.

Hunt, G.T. and Maisel, B.E. 1990. Atmospheric PCDDS/PCDFs in wintertime in a northeastern U.S. urban coastal environment. *Chemosphere.* 20:1455–1462.

Hutzinger, O., Blumich, M.J., and Berg, M.D.V. 1985a. Source and fate of PCDDs and PCDFs: an overview. *Chemosphere.* 14:581–600.

Hutzinger, O., Choudhury, G.G., and Chittim, B.G. 1985b. Formation of polychlorinated dibenzofurans and dioxins during combustion, electrical equipment fires and PCB incineration. *Environ. Health Perspec.* 60, 3-9.

Hutzinger, O. and Fiedler, H. 1989. Sources and emissions of PCDD and PCDF. *Chemosphere.* 18:23–32.

Isenee, A.R. and Jones, G.E. 1971. Absorption of translocation of root and foliage applied 2,4-dichlorophenol, 2,7-dichlorodibenzo-p-dioxin, and 2,3,7,8-tetrachlorodibenzo-p-dioxin. *J. Agr. Food Chem.* 19:1210–1214.

Jackson, D.R., Roulier, M.H., Grotten, H.M., Rust, S.W., and Warner, J.S. 1986. Solubility of 2,3,7,8-TCDD in contaminated soils. In: Chlorinated Dioxins and Dibenzofurans in Perspective, Rappe, C., Choudhury, G., and Keith, L.H., Eds., Lewis Publishers, Boca Raton, FL. pp. 185-200.

Janz, D.M. and Metcalfe, C.D. 1991. Nonadditive interactions of mixtures of 2,3,7,8-TCDD and 3,3′,4,4′-tetrachlorobiphenyl on aryl hydrocarbon hydroxylase induction in rainbow trout (*Onchorhynchus mykiss*). *Chemosphere.* 23(4), 467-472.

Jay, K. and Stieglitz, L. 1991. On the mechanism of formation of polychlorinated aromatic compounds with copper (II) chloride. *Chemosphere.* 22:87–96.

Jobb, B., Hunsinger, R., Clement, R., and Thompson, T. 1989. A survey of drinking water supplies in the Province of Ontario for dioxins and furans, Dioxin '89, Ninth International Symposium on Chlorinated Dioxins and Related Compounds, Toronto, Ontario, Canada, 1989.

Jordan, R.J. 1987. The feasibility of wet scrubbing for treating waste-to-energy flue gas. *J. Air Pollut. Control Assoc.* 37:422–430.

Karasek, F.W. and Dickson, L.C. 1987. Model studies of polychlorinated dibenzo-p-dioxin formation during municipal refuse incineration. *Science.* 237:754–756.

Keuhl, D.W., Butterworth, D.C., DeVita, W.M., and Sauer, C.P. 1987. Environmental contamination by polychlorinated dibenzo-p-dioxins and dibenzofurans associated with pulp and paper mill discharge. *Biomed. Environ. Mass Spectrom.* 14:443–447.

Kimbrough, R.D., Falk, H., Stehr, P., and Fries, G. 1984. Health implication of 2,3,7,8-Tetrachlorodibenzo-p-dioxin contamination of residential soil. *J. Toxicol. Environ. Health.* 14:47–93.

Kjeller, L.-O., Kulp, S.-E., De Wit, C., Lexen, K., Hasselsten, I., Rappe, C., Jonsson, P., and Jansson, B. 1991. Sediment, soil and water contamination by polychlorinated dibenzodioxins and dibenzofurans (PCDD/DFs) from sludge from graphite electrodes used in chlorine production. Eleventh International Symposium on Chlorinated Dioxins and Related Compounds, Research Triangle Park, NC, p. 108.

Kleeman, J.M., Olson, J.R., and Peterson, R.E. 1988. Species differences in 2,3,7,8-tetrachlorodibenzo-p-dioxin toxicity and biotransformation in fish. *Fundam. Appl. Toxicol.* 10(2):206–213.

Kociba, R.J., Keyes, D.G., Beyer, J.E., Carreon, R.M., and Gehring, P.J. 1979. Long-term toxicologic studies of 2,3,7,8-tetrachlorodibenzo-p-dioxin (TCDD) in laboratory animals. *Ann. NY Acad. Sci.* 320:397–404.

Kociba, R.J., Keyes, D.G., Beyer, J.E. et al. 1978. Results of a two-year chronic toxicity and oncogenicity study of 2,3,7,8-TCDD in rats. *Toxicol. Appl. Pharmacol.* 46:279–303.

Kominsky, J.R. and Kwoka, C.D. 1989. Background concentrations of polychlorinated dibenzofurans (PCDFs) and polychlorinated dibenzo-p-dioxins (PCDDs) in office buildings in Boston, Massachusetts. *Chemosphere.* 18:599–608.

Kuehl, D.W., Cook, P.M., and Batterman, A.R. 1987. Bioavailability of polychlorinated dibenzo-p-dioxins from contaminated Wisconsin River sediment to carp. *Chemosphere.* 16:667–680.

Lamparski, L.L., Nestrick, T.J., Frawley, N.N., Hummel, R.A., and Kocher, C.W. 1986. Perspectives of a large scale environmental study for chlorinated dioxins: water analysis. *Chemosphere.* 15:1445–1452.

Lindner, G.A., Jenkins, A.C., and McCormack, J. 1990. Dioxins and furans in emissions from medical waste incinerators. *Chemosphere.* 20:1793–1800.

Mackay, D., Shiu, W.Y., and Ma, K.C. 1992. *Illustrated Handbook of Physical and Chemical Properties and Environmental Fate for Organic Chemicals, Monoaromatic Hydrocarbons, Chlorobenzenes and PCBs*. Vols. 1 and 2, Lewis Publishers, Boca Raton, FL.

Mah, F.T.S., MacDonald, D.D., Sheehan, S.W., Tuominen, T.M., and Valiela, D. 1989. Dioxins and Furans in sediment and fish from the vicinity of ten inland mills in British Columbia. Environment Canada, Conservation and Protection, Inland Waters, Pacific and Yukon region, Vancouver, B.C., Canada.

Maisel, B.E. and Hunt, G.T. 1990. Background concentrations of PCDDs-PCDFs in ambient air — a comparison of toxic equivalency factor TEF models. *Chemosphere*. 20:771–778.

Mantykoski, K., Paasivirta, J., and Mannila, E. 1989. Combustion products of biosludge from pulp mill. 1989. *Chemosphere*. 19:413–416.

Marcheterre, L., Webster, G.R.B., Muir, D.C.G., and Grift, N.P. 1985. Fate of ^{14}C-Octachlorodibenzo-p-dioxin in an artificial outdoor pond. *Chemosphere*. 14:835–838.

Marklund, S., Rappe, C., Tysklind, M., and Egeback, K.E. 1987. Identification of polychlorinated dibenzofurans and dioxins in exhausts from cars run on leaded gasoline, *Chemosphere*. 16:29–36.

Marklund, S., Kjeller, L.O., Hansson, M., Tysklind, M., Rappe, C., Ryan, C., Collazo, H., and Dougherty, R. 1986. Determination of PCDDs and PCDFs in incineration samples and pyrolytic products. In: Chlorinated Dioxins and Dibenzofurans in Perspective. Rappe, C., Choudhury, G., and Keith, L.H., Eds. Lewis Publishers, Boca Raton, FL. pp. 79-92.

McConnell, E.E., Moore, J.A., Haseman, J.K., and Harris M.W. 1978. The comparative toxicity of chlorinated dibenzo-p-dioxins in mice and guinea pigs. *Toxicol. Appl. Pharmacol.* 44:335–356.

McKee, P., Burt, A., and McCurvin, D. 1990. Levels of dioxins, furans and other organic contaminants in harbour sediments near a wood preserving plant using pentachlorophenol and creosote. *Chemosphere*. 20:1679–1685.

McLauglin, D.L., Pearson, R.G., and Clement, R.E. 1989. Concentrations of chlorinated dibenzo-p-dioxins(CDD) and dibenzofurans(CDF) in soil from the vicinity of a large refuse incinerator in Hamilton, Ontario, Canada. *Chemosphere*. 18:851–854.

Meyer, C., O'Keefe, P., and Hilker, D. 1989. A survey of twenty community water systems in New York State, USA, for PCDDs and PCDFs. *Chemosphere*. 19:21–26.

Mill, T. 1985. Prediction of environmental fate of tetrachlorodibenzodioxin. In: Dioxins in the Environment. Kamrin, M.A. and Rogers, P., Eds., Hemisphere Publishing, Washington D.C. p. 173.

Miller, R.A., Norris, L.A., and Hawkes, C.L. 1973. Toxicity of 2,3,7,8-tetrachlorodibenzo-p-dioxin (TCDD) in aquatic organisms. *Environ. Health Perspect.* 5:177–186.

Miller, R.A., Norris, L.A., and Loper, B.R. 1979. The response of coho salmon and guppies to 2,3,7,8-tetrachlorodibenzo-p-dioxin (TCDD) in water. *Amer. Fish. Soc. Trans.* 108(4):401–407.

Morita, M., Yashurara, A., and Ito, H. 1987. Isomer specific determination of polychlorinated dibenzo-p-dioxins and dibenzofurans in incinerator related samples in Japan. *Chemosphere*. 16:1959–1964.

Muir, D.C.G., Yarechewski, A.L., Metner, D.A., Lockhart, W.L., Webster, G.R.B., and Friessen, K.J. 1990. Dietary accumulation and sustained hepatic mixed function oxidase enzyme induction by 2,3,4,7,8-pentachlorodibenzofuran in rainbow trout. *Environ. Toxicol. Chem.* 9:1463–1472.

Mukerjee, D., Papke, O., and Kamarus, W. 1989. Indoor air contamination with polychlorinated dibenzo-p-dioxins and dibenzofurans. *Toxicol. Ind. Health*. 5:731–745.

Mundy, K.J., Brown, R.S., and Pettit, K. 1989. Environmental assessment at and around a chemical waste treatment facility, I. *Chemosphere.* 19:381–386.

Muto, H., Saito, K., and Shinada, M. 1991. Concentrations of polychlorinated dibenzo-p-dioxins and dibenzofuran from chemical manufacturers and waste disposal facilities. *Environ. Res.* 54:170–182.

Muto, H. and Takizawa, Y. 1991. The liquid-phase photolyses of tetra- and octa-CDDs and their CDFs in hexane solution. *Chemistry Letters.* 2:273–276.

Nakano, T., Tsuji, M., and Okuno, T. 1990. Distribution of PCDDs and PCBs in the atmosphere. *Atmos. Environ.* 24A:1361–1368.

Narang, A.S., Swami, K., and Narang, R.S. 1991. Pyrolysis and combustion of liquids and solids containing pentachlorophenol. *Chemosphere.* 22:1029–1043.

National Research Council of Canada (NRCC) 1981. Polychlorinated dibenzo-p-dioxins: Criteria for their effects on man and his environment. Associate Committee on Scientific Criteria for Environmental Quality, NRCC Publication No. 18574, Ottawa, Canada, 251 p.

National Toxicology Program (NTP). 1980. Carcinogenesis Bioassay of 2,3,7,8-TCDD in Osborne-Mendel Rats and B6C3F1 Mice (Gavage Study). DHHS Publication (NIH) 82-1765.

Nestrick, T.J. and Lamparski, L.L. 1982. Isomer-specific determination of chlorinated dioxins for assessment of formation and potential environmental emission from wood combustion. *Anal. Chem.* 54:2292–2299.

Norstrom, R.J., Simon, M., and Mair, D.C.G. 1990. Polychlorinated dibenzo-p-dioxins and dibenzofurans in marine mammals in the Canadian North. *Environ. Pollut.* 66:1–19.

Norwood, C.B., Hackett, M., and Pruell, R.J. 1989. Polychlorinated dibenzo-p-dioxins and dibenzofurans in selected estuarine sediments. *Chemosphere.* 18:553–560.

Nestrick, T.J., Lamparski, L.L., Frawley, N.N., Hummel, R.A., and Kocher, C.W. 1986. Perspectives of a large scale environmental study for chlorinated dioxins: overview of soil data. *Chemosphere.* 15:1453–1460.

O'Keefe, P., Hilker, D., and Meyer, C. 1984. Tetrachlorodibenzo-p-dioxins and tetrachloro dibenzofurans in Atlantic Coast striped bass and in selected Hudson River fish, waterfowl and sediments. *Chemosphere.* 13:849–860.

Oberg, T. and Bergstrom, J.G.T. 1987. Emission and chlorination pattern of PCDD/PCDF predicted from indicator parameters. *Chemosphere.* 16:1221–1230.

Oberg, L.G., Glas, B., and Rappe, C. 1991. Biochemical formation of PCDDs and PCDFs, p.19. Eleventh International Symposium on Chlorinated Dioxins and Related Compounds, Research Triangle Park, NC, 1991.

Oheme, M., Mano, S., and Bjerke, B. 1989a. Formation of PCDFs and PCDDs by production processes for magnesium and refined nickel. *Chemosphere.* 18:1379–1389.

Oheme, M., Mano, S., Brevik, E.M., and Knutzen, J. 1989b. Determination of polychlorinated dibenzofuran (PCDF) and dibenzo-p-dioxin (PCDD) levels and isomer patterns in fish, crustacea, mussel and sediment samples from a fjord region polluted by Mg-production. *Fresenius Zeitschrift fur Chemie.* 355:987–997.

Olie, K., Berg, M., and Hutzinger, O. 1983. Formation and fate of PCDD and PCDF from combustion processes. *Chemosphere.* 12:627–636.

Olson, J.R., Holscher, M.A., and Neal, R.A. 1980. Toxicity of 2,3,7,8-tetrachloro dibenzo-p-dioxin in the golden syrian hamster. *Toxicol. Appl. Pharmacol.* 5:67–78.

Ontario Ministry of Environment. 1985. Scientific Criteria Document for Standard Development No. 4-84: Polychlorinated Dibenzo-p-dioxins (PCDDs) and Polychlorinated Dibenzofurans (PCDFs), September.

Palansky, J., Kapila, S., Manahan, S.E., Vanderos, A.F., Malhotra, R.K., and Clevenger, T.E. 1986. Studies on vapor phase transport and role of dispersing medium on mobility of 2,3,7,8-TCDD in soil. *Chemosphere.* 15:1389–1396.

Peterson, R.E., Seefeld, M.D., Christian, B.J., Potter, C.L., Kelling, C.K., and Keesey, R.E. 1984. The wasting syndrome in 2,3,7,8-tetrachlorodibenzo-p-dioxin. Toxicity: Basic features and their interpretation. In: *Biological Mechanisms of Dioxin Action,* Poland, A. and Kimbrough, R.D., Eds. Cold Spring Harbor Laboratory Press, Cold Spring Harbor, NY, Banbury Report 18, pp. 291–308.

Podoll, R.T., Jaber, H.M., and Mill, T. 1986. Tetrachlorodiberodioxin: Rates of volatilization and photolysis in the environment. *J. Environ. Sci. Technol.* 20:490–492.

Rappe, C., Gara, A., and Buser, H.R. 1978. Identification of polychlorinated dibenzofurans (PCDFs) in commercial chlorophenol formulations. *Chemosphere.* 7:981–991.

Rappe, C. and Buser, H.R. 1981. Occupational exposure to polychlorinated dioxins and dibenzofurans. American Chemical Society Symposium Series No. 149, 319–342.

Rappe, C. and Kjeller, L.-O. 1987. PCDDs and PCDFs in environmental samples, air, particulates, sediments and soil. *Chemosphere.* 16:1775–1780.

Rappe, C. 1991. Introduction: Levels, Profiles and Patterns, pp.17, Eleventh International Symposium on Chlorinated Dioxins and Related Compounds, Research Triangle Park, NC.

Rappe, C., Marklund, S., Kjeller, L.O., and Tyskland, M. 1986. PCDDs and PCDFs in emissions from various incinerators. *Chemosphere.* 15:1213–1217.

Rappe, C., Kjeller, L.O., and Andersson, R. 1989. Analyses of PCDDs and PCDFs in sludge and water samples. *Chemosphere.* 19:13–20.

Rappe, C., Kjeller, L.O., Bruckmann, P., and Hackhe, K.H. 1988. *Chemosphere.* 17:3–20.

Rappe, C., Nygren, M., and Marklund, S. 1985. Assessment of human exposure to polychlorinated dibenzofurans and dioxins. *Environ. Health Perspect.* 60:303–304.

Rappe, C., Andersson, R., Bergqvist, P.A., Brohede, C., Hansson, M., Kjeller, L.O., Lindstrom, G., Marklund, S., Nygren, M., Swanson, S.E., Tysklind, M., and Wiberg, K. 1987. Overview on environmental fate of chlorinated dioxins and dibenzofurans. Sources, levels and isomeric pattern in various matrices. *Chemosphere.* 16:1603–1618.

Rappe, C., Bergqvist, P.A., and Kjeller, L.O. 1989. Levels, trends and patterns of PCDDs and PCDFs in Scandinavian environmental samples. *Chemosphere.* 18:651–658.

Rappe, C., Bergqvist, P.A., and Kjeller, L.O. 1991. Levels and patterns of PCDD and PCDF contamination in fish, crabs, and lobsters from Newark Bay and the New York Bight. *Chemosphere.* 22:239–266.

Reed, L.W., Hunt, G.T., and Maisel, B.E. 1990. Baseline assessment of PCDDs/PCDFs in the vicinity of the Elk River, Minnesota generating station. *Chemosphere.* 21:159–172.

Safe, S. 1990. Development of Toxic Equivalency Factors (TEFs) for halogenated aromatic hydrocarbons. *Dioxin '90,* Vol. 1, pp. 329–331, Eco-informa Press, Bayreuth, Germany.

Safe, S. 1990a. Polychlorinated biphenyls (PCBs), dibenzo-p-dioxins (PCDDs), dibenzofurans (PCDFs), and related compounds: Environmental and mechanistic considerations which support the development of Toxic Equivalency Factors (TEFs). *Crit. Rev. Toxicol.* 21:51–88.

Safe, S., Bunce, N.J., and Chittim, B. 1977. Photodecomposition of halogenated aromatic compounds. Keith, L.H., Ed. 1977. Identification and analysis of organic pollutants in water. Ann Arbor Science. Ann Arbor, MI. 35–37

Schecter, A., Startin, J., and Wright, C. 1993. Dioxin levels in food from the United States with estimated daily intake. In: Fiedler, H., Frank, H., and Hutzinger, O., Eds., *Organohalogen Compounds,* Vol. 13, Federal Environmental Agency, Vienna, 93–96.

Schroy, J.M., Hileman, F.D., and Cheng, S.S. 1985. Physical chemical properties of 2,3,7,8-TCDD. *Chemosphere.* 14:877–880.

Schroy, J.M., Hileman, F.D., and Cheng, S.S. 1986. Physical/chemical properties of 2,3,7,8-tetrachlorodibenzo-p-dioxin. In: Bahner, R.C. and Hansen, D.J., Eds. *Aquatic Toxicology and Hazard Assessment: Eighth Symposium.* ASTM-P. 891, Philadelphia, PA. 409–421.

Schwetz, B.A., Norris, J.M., Sparschu, G.L., Rowe, V.K., Gehring, P.J., Emerson, J.L., and Gerbig, C.G. 1973. Toxicology of chlorinated dibenzo-p-dioxins. *Environ. Health Perspect.* 5:87–89.

Sedman, R.M. and Esparza, J.R. 1991. Evaluation of the publc health risks associated with semivolatile metal and dioxin emissions from hazardous waste incinerators. *Environ. Health Perspect.* 94:181–188.

Sheffield, A. 1985. Polychlorinated dibenzo-p-dioxins (PCDDs) and Polychlorinated dibenzofurans (PCDFs): Sources and Releases. Environmental Protection Service, Environment Canada, Ottawa, Canada, EPS 5/HA/2.

Sherman, R.K., Clement, R.E., and Tashiro, C. 1990. The distribution of polychlorinated dibenzo-p-dioxins and dibenzofurans in Jackfish Bay, Lake Superior, in relation to a kraft pulp mill effluent. *Chemosphere.* 20:1641–1648.

Sittig, M. 1980. *Priority Toxic Pollutants, Health Impacts and Allowable Limits.* Noyes Data Corporation, Park Ridge, NJ, 370 p.

Smith, R.M., O'Keefe, P., Aldous, K., Briggs, R., Hilker, D., and Connor S. 1991. Measurement of CDFs and CDDs in air samples and lake sediments at several locations in upstate New York. Eleventh International Symposium on Chlorinated Dioxins and Related Compounds, Research Triangle Park, NC. p. 33.

Smith, L.M., Scwartz, T.R., and Feltz, K. 1990a. Determination and occurrence of AHH-active polychlorinated biphenyls, 2,3,7,8-tetrachlorodibenzo-p-dioxin and 2,3,7,8-tetrachlorodibenzo furan in Lake Michigan sediment and biota. The question of their relative toxicological significance. *Chemosphere.* 21:1063–1085.

Smith, R.M., O'Keefe, P.W., and Aldous, K.M. 1990b. Chlorinated dibenzofurans and dioxins in atmospheric samples from cities in New York. *Environ. Sci. Technol.* 24:1502–1506.

Smith, R.M., O'Keefe, P.W., Hilker, D.R., and Aldous, K.M. 1986. Determination of picogram per cubic meter concentrations of tetrachlorinated and pentachlorinated dibenzofurans and dibenzo-p-dioxin in indoor air by high-resolution gas chromatography high-resolution mass spectrometry. *Anal. Chem.* 58(12), 2414–2420.

Someshwar, A.V., Jain, A.K., and Whittmore, R.C. 1990. The effects of sludge burning on the PCDD/PCDF content of ashes from pulp and paper mill hog fuel boilers. *Chemosphere.* 20:1715–1722.

Spitsbergen, J.M., Schat, K.A., Kleemann, J.M., and Peterson, R.E. 1986. Interactions of 2,3,7,8-tetrachlorodibenzo-p-dioxin (TCDD) with immune responses of rainbow trout. *Vet. Immunol. Immunopathol.* 12:263–280.

Spitsbergen, J.M., Kleeman, J.M., and Peterson, R.E. 1987. Morphologic lesions and acute toxicity in rainbow trout (*salmo gairdneri*) treated with 2,3,7,8-tetrachlorodibenzo-p-dioxin. *J. Toxicol. Environ. Health.* cited in Spitsbergen et al. 1988b.

Spitsbergen, J.M., Kleeman, J.M., and Peterson, R.E. 1988a. 2,3,7,8-tetrachlorodibenzo-p-dioxin toxicity in yellow perch (*Perca flavescens*). *J. Toxicol. Environ. Health.* 23:359–383.

Spitsbergen, J.M., Kleeman, J.M., and Peterson, R.E. 1988b. Morphological lesions and acute toxicity in rainbow trout (*Salmo gairdnerii*) treated with 2,3,7,8-tetrachlorodibenzo-p-dioxin. *J. Toxicol. Environ. Health.* 23:333–358.

Stalling, D.L., Norstrom, R.J., and Smith, L.M. 1985. Patterns of PCDD and PCB contamination in Great Lakes fish and birds and their characterization by principal component analysis. *Chemosphere.* 14:627–643.

Steer, P., Tashiro, C., and Clement, R. 1990. Ambient air sampling of polychlorinated dibenzo-p-dioxins and dibenzofurans in Ontario: preliminary results. *Chemosphere.* 20:1431–1437.

Stephens, R.D. 1986. Transformer fire. *Chemosphere.* 15:1281–1289.

Stieglitz, L., Zwick, G., and Beck, J. 1989. Carbonaceous particles in fly ash as a source for the de novo synthesis of organochloro compounds. *Chemosphere.* 19:283–290.

Stieglitz, L. and Vogg, H. 1987. On formation conditions of PCDDs/PCDFs in fly ash from municipal waste incinerators. *Chemosphere.* 16:1917–1922.

Takeshita, R. and Akimoto, Y. 1989. Control of PCDD and PCDF formation in fluidized bed incinerators. *Chemosphere.* 19:345–352.

Tashiro, C., Clement, R.E., Stocks, B.J., Radke, L., Cofer, W.R., and Ward, P. 1990. Preliminary report: Dioxins and furans in prescribed burns. *Chemosphere.* 20:1533.

Taucher, G.A., Hannah, D.J., and Green, N.J.L. 1992. PCDD, PCDF and PCB emissions under variable operating conditions from a waste oil furnace. *Chemosphere.* 25:1429–1433.

Tiernan, T.O., Taylor, M.L., and Garrett, J.H. 1985. Sources and fate of polychlorinated dibenzodioxins and dibenzofurans and related compounds in human environments. *Environ. Health. Perspect.* 59:145–158.

Tiernan, T.O., Wagel, D.J., and Garrett, J.H. 1989. Laboratory and field tests to demonstrate the efficacy of KPEG reagent for detoxification of hazardous wastes containing polychlorinated dibenzo-p-dioxins (PCDD) and dibenzofurans (PCDF) and soils contaminated with such chemical wastes. *Chemosphere.* 18:835–841.

Tiernan, T.O., Taylor, M.L., Garrett, J.H., Vanness, G.F., Solch, J.G., Deis, D.A., and Wagel, D.J. 1983. Chlorodibenzodioxins, chlorodibenzofurans and related compounds in the effluents from combustion processes. *Chemosphere.* 12:595–606.

TMN (Toxic Materials News). 1987. Survey finds little dioxin at control sites. Industry finds trace amounts in paper products and highest levels at pesticide plants. Sept. 30, p. 301.

Tong, Y. and Karasek, F.W. 1986. Comparison of PCDD and PCDF in flyash collected from municipal incinerators of different countries. *Chemosphere.* 15:1219–1224.

Towara, J., Hiller, B., and Hutzinger, O. 1991. PCDD/DF in distillation residues from dry cleaners. Eleventh International Symposium on Chlorinated Dioxins and Related Compounds, Research Triangle Park, NC. p. 296.

Trudel, L. 1991. Dioxins and furans in bottom sediments near the 47 Canadian pulp and paper mills using chlorine bleaching. Environment Canada, Water Quality Branch, Inland Waters Directorate, Ottawa, Canada.

Van den Berg, M., Olie, K., and Hutzinger, O. 1985. Polychlorinated dibenzofurans (PCDFs): environmental occurrence and physical, chemical and biological properties. *Environ. Toxicol. Chem.* 9:171–217.

Van der Heijden, C.A., Knapp, A.G.A.C., Kramers, P.G.N., and van Logten, M.J. 1982. Evaluation of the carcinogenicity and mutagenicity of 2,3,7,8-tetrachlorodibenzo-1,4-dioxin (TCDD): Classification and No-Effect Level, Report DOC/LCM 300/292. State Institute of National Health, Bilthoven.

Van der Weiden, M.E.J., Craane, L.H.J., Evers, E.H.G., Kooke, R.M.M., Olie, K., Seinen, W., and Van Den Berg, M. 1989. Bioavailability of PCDDS and PCDFs from bottom sediments and some associated biological effects in the carp (*Cyprinus carpio*). *Chemosphere.* 19:1009–1016.

Vikelsoc, J., Cederberg, T., Madsen, H., Grove, A., and Hansen, K. 1991. *Emission of Dioxins from Danish Household Stoves.* 11th International Symposium on Chlorinated Dioxins and Related Compounds. Research Triangle Park, MC. 297.

Villanueve, E.C., Jennings, R.W., Bursa, V.W., and Kimborough, R.D. 1973. A comparison of analytical methods for chlorodibenzo-p-dioxins in pentachlorophenol. *J. Agr. Food. Chem.* 23(6), 1089–1091.

Waddell, D., Chittim, B., and Clement, R. 1991. Database of PCDD/PCDF levels in ambient air and in samples related to the pulp and paper industry. *Chemosphere.* 20:1463–1466.

Walker, M. and Peterson, R.E. 1991. Potencies of polychlorinated dibenzo-p-dioxin, dibenzofuran and biphenyl congeners, relative to 2,3,7,8-tetrachlorodibenzo-p-dioxin, for producing early life stage mortality in rainbow trout (*Onchorhynchus mykiss*). *Aquat. Toxicol.* 21:219–238.

Walters, R.W., Ostazeski, S.A., and Guiseppi-Elie, A. 1989. Sorption of 2,3,7,8-tetrachlorodibenzo-p-dioxin from water by surface soils. *Environ. Sci. Technol.* 23:480–484.

Wannenmacher, R., Rebstock, A., Kulzer, E., Schrenk, D., and Bock, R.W. 1992. Effects of 2,3,7,8-tetrachlorodibenzo-p-dioxin on reproduction and oogenesis in zebrafish (*Brachydanio rerio*). *Chemosphere.* 24:1361–1368.

Watson, W.D. and Kobel, E.H. 1974. Stabilized distillation of pentachlorophenol. Official Gazette of the (U.S.) Patent Office. (U.S.) Patent 3816268.

Weerasinghe, N.C.A. and Gross, M.L. 1985. Origins of polychlorodibenzo-p-dioxin (PCDD) and polychlorodibenzofurans (PCDF) in the environment. In: *Dioxins in the Environment*, Kamrin, M.A. and Rodgers, P., Eds., Hemisphere Publishing, Washington D.C. pp. 133–151.

Woolson, E.A., Thomas, R.F., and Ensor, P.D. 1972. Survey of polychlorinated dibenzo-p-dioxin content in selected pesticides. *Adv. Chem. Ser.* 120:112–120.

Yockim, R.S., Isensee, A.R., and Jones, G.E. 1978. Distribution and toxicity of TCDD and 2,4,5-T in an aquatic model ecosystem. *Chemosphere.* 7(3), 215–220.

Zacharewski, T., Safe, L., and Safe, S. 1989. Comparative analysis of polychlorinated dibenzo-p-dioxin and dibenzofuran congeners in Great Lakes fish extracts by gas chromatography-mass spectrometry and *in vitro* enzyme induction activities. *Environ. Sci. Technol.* 23:730–735.

CHAPTER 9

Prioritization for Regulatory and Monitoring Assessment

HAZARD IDENTIFICATION

Hazard is defined as a set of situations with potential to cause adverse effects on the ecological systems and/or human health. Hazard evaluation is a scientific process which evaluates the probabilities of hazard using the best available scientific data about the dispersion of a chemical in the environment and the associated adverse effects on the ecosystem and human health. Once the hazards of a chemical are determined and their major exposure routes established, the prevention and control of exposures to the chemical and minimization or elimination of the adverse effects is a judgmental process, involving socio-economic and political considerations. Each society will likely have a particular legal, socio-economic, and administrative framework for making regulatory decisions on the management of chemicals in the environment, through measures that are mandatory or advisory.

The basic component in any chemical hazard evaluation is the assessment of its total environmental exposure levels. The environmental exposure level is determined by the physico-chemical properties of the chemical, volume, and its use pattern, (closed or open). For a chemical to be hazardous, it has to be toxic at the environmental exposure level and for the exposure duration. In other words, the dose makes the chemical a hazard. In contrast, there are chemicals whose exposure at any level of concentration will produce genetic toxicity, such as carcinogenicity, mutagenicity, teratogenicity, etc. Hazard identification requires both qualitative and quantitative information on genetic and nongenetic toxicity of chemicals, acting either alone or in mixtures. The information on these adverse effects is derived from toxicological, clinical, and epidemiological data. The physical and chemical properties, combined with the use pattern, may provide information of possible exposure levels and the routes of exposure.

The adverse effects of a chemical makes it intrinsically hazardous, but its actual presence in the environment makes it hazardous. A very toxic chemical at a low volume and closed use has no significant exposure potential. Thus, it is not hazardous in the environment. A chemical with low adverse effects, but with a widespread use pattern, has a high exposure level with multiple pathways. Such chemicals could be much more hazardous.

The different aspects of the hazard identification process are outlined below.

Basic Aspects

1. Assessment of intrinsic properties which determine the intrinsic hazards of a chemical.
2. Evaluation of further data requirements which are necessary for a thorough evaluation of intrinsic properties.
3. Risk assessment, which includes the assessment of the intrinsic risk in the real world, involving factors such as volume, plurality of exposures, environmental dispersion, persistence, bioaccumulation, and the size of the population at risk.
4. Prioritizing chemicals of concern.

Specific Aspects

Apart from physical danger from the chemical of concern due to its flammability or explosivity, properties such as aqueous solubility, lipid solubility, boiling point, relative density with respect to water or air, etc., correlate to ecotoxicological effects. Properties such as vapor pressure and lipid solubility relate to inhalation toxicity, dermal absorption, and dermal toxicity. It is important that physical and chemical properties and toxicological and ecotoxicological effects should be considered in hazard identification and assessment. Fate processes, such as environmental persistence, resistance to degradation, and biomagnification in the food ladder, not only account for their environmental presence and associated potential hazards, but also are essential in the assessment of exposure routes. Table 9.1 lists the tests that need to be conducted in hazard identification for new chemicals under the Toxic Substances Control Act (TSCA).

Following is a brief discussion of some of the physico-chemical properties and their value in hazard identification.

Aqueous Solubility

Water solubility of a chemical determines its residence time in water which, in turn, allows the chemical to undergo long-range transport, as well as effect biotic and abiotic transformations in the aquatic environment. Chemicals with low aqueous solubility will likely undergo either rapid volatilization or sorption to suspended matter in the water column, sediment, or soil, or bioaccumulate in aquatic organisms. Knowledge of a chemical's water solubility and its volatilization rate from the water column (Henry's law constant) will assist in determining the environmental medium (or media) for monitoring the chemical's persistence or degradation and its adverse effects. For organic chemicals, which are sparingly soluble in water, inaccurate water solubility values lead to false–positive toxicity data, in terms of the actual exposure concentrations.

Table 9.1 Testing Protocol for Ecological Hazard Identification of New Chemicals Under the TSCA

I. Chemical fate (transport, persistence)
   A. Transport
      1. Adsorption isotherm (soil)
      2. Partition coefficient (water-octanol)
      3. Water solubility
      4. Vapor pressure
   B. Other physico-chemical properties
      1. Boiling/melting/sublimation points
      2. Density
      3. Dissociation constant
      4. Flammability/explodability
      5. Particle size
      6. pH
      7. Chemical incompatibility
      8. Vapor-phase UV spectrum for halocarbons
      9. UV and visible absorption spectra in aqueous solution
   C. Persistence
      1. Biodegradation
         a. Shake flask procedure following carbon loss
         b. Respirometric method following oxygen (BOD) and/or carbon dioxide
         c. Activated sludge test (simulation of treatment plant)
         d. Methane and CO_2 productions in anaerobic digestion
      2. Chemical Degradation
         a. Oxidation (free-radical)
         b. Hydrolysis (25°C; pH, 5.0 and 9.0)
         c. Photochemical transformation in water

II. Ecological Effects
   A. Microbial effects
      1. Cellulose decomposition
      2. Ammonification of urea
      3. Sulphate reduction
   B. Plant effects
      1. Algae inhibition (fresh and sea water, growth, nitrogen fixation)
      2. Duck weed inhibition (increase in fronds or dry weight)
      3. Seed germination and early growth
   C. Animal effects
      1. Aquatic invertebrates (daphnia) acute toxicity (first instar)
      2. Fish acute toxicity test (LC_{50}, 96 h)
      3. Quail dietary LC_{50} test
      4. Terrestrial mammal test
      5. *Daphnia* life cycle test
      6. *Mysidopsis bahia* life cycle test
      7. Fish embryo–juvenile test
      8. Fish bioconcentration test

Source: U.S. EPA 1979.

Hydrolysis and biodegradation are shown to be the most important fate processes for organic chemicals in the aquatic environment and photodegradation and in the vapor or gaseous phase. In the soil, biodegradation, and to some extent, chemical degradation, are important fate processes for organic chemicals. Therefore, it is recommended that for chemicals of concern in the aquatic or soil environment, hydrolysis and biodegradation tests should be conducted first. If the chemical is not degraded by any process, including photodegradation, the chemical is likely to persist in the environment, leading to bioaccumulation and biomagnification in the food chain.

Biodegradation

Bacteria and fungi are known to metabolically degrade complex organic chemicals to simple end-products. This process can take place in soil, sediment, and the water column. The end-products could vary, depending upon the chemical complexity of the parent chemical structure. Aerobic biodegradation could yield end-products, such as carbon dioxide and water, and may also include the normal metabolic products of aerobic microorganisms, nitrates, and sulphates. The anaerobic degradation could yield methane, carbon dioxide, and common end-products of the metabolic processes of the anaerobic microorganisms.

Carbon Dioxide Evolution Test Methods

This test is recommended to evaluate the extent of biodegradation of organic chemicals to carbon dioxide and water in natural systems. The measure of biodegradation is the ratio of the amount of carbon dioxide produced vs. the theoretical amount that could be produced. Positive results (>50% of the theoretically expected CO_2) show that the test chemical will not persist indefinitely in the soil.

Algal Bioassay

Algal assays provide important information about the chemical interference on the simple photosynthetic organisms found everywhere in the environment. They likely play a significant role in the fixation of energy from sunlight. The important role algae play in the ecosystem follow:

1. The oxygen generated in the photosynthetic process is utilized by aquatic organisms and also becomes a global reservoir for atmospheric oxygen.
2. Algae assists in the breaking down organic wastes.
3. Algae are the foundation of most aquatic food chains and are utilized by herbivores as a major food source.

TOXICITY TESTING

Aquatic Toxicity Tests

Toxic effects may include both acute (mortality) and chronic effects such as behavioral physiological, biochemical, and histological changes. High concentration of a given chemical can cause acute toxicity in relatively short exposure time and cause the toxic endpoint (mortality).

Chronic adverse effects onset over a long exposure time and at concentrations relatively lower than the acutely lethal dose concentration. For detailed discussions on various aspects of toxicity test procedures such as design, selection of species, influence of external factors on toxicity testing, the readers are referred to Fundamentals of Aquatic Toxicology by Rand and Petrocelli (1985).

The exposure of organisms to chemicals under investigation can be conducted by using one or more of the following four systems:

1. Static Test

In this test, organisms are exposed to still, but aerated, water held in a test chamber containing the chemical at the desired exposure concentration. The water is not changed during the test.

2. Recirculation Test

This is similar to the static test, but the test solutions and control water are recirculated through a pump. This maintains the water quality and does not alter the test chemical's concentration. This design is not commonly used because of the higher cost of set-up and maintenance and also due to the uncertainty of the chemical's loss at the filter, pump, etc.

3. Renewal Test

This design is similar to the static test but the test solution and control water are replaced periodically (usually 24h). This is done by either transferring the test species to a freshly prepared chamber or changing the test solutions in the original chambers.

4. Flow-Through Test

Here the test solutions and control water flow in and out of the holding chambers. Metering pumps or diluters control the flow of dilution water and stock solution of the test chemical so that proper proportions of each will be mixed. The flow could be intermittent or continuous.

The static and flow-through designs are the most commonly used exposure chambers for toxicity testing of aquatic organisms.

Avian Toxicity Test

Toxicity testing of birds is somewhat similar to that for aquatic species. Testing is mandatory on two avian species, one species, of wild waterfowl, preferably mallard, and one species of upland game bird, preferably bobwhite, native quail, or ring-necked pheasant. Birds used should be 10 to 17 days old at the beginning of the test period and should be examined each 8-day interval. Birds are fed treated diet for the first 5 days, followed by clean diets for 3 days, with observation. The results are used for setting acceptable application levels and for identifying doses for further testing.

Mammalian Toxicity Test

This test is designed to assess the toxic effects on mammalian species exposed to high concentrations of the test chemical. The test will also identify the target organ and the exposure levels for chronic exposure studies. The battery of tests include primary exposure routes such as oral, inhalation, dermal, etc. An acute oral toxicity test may require up to 50 animals, whereas fewer animals could be used for dermal and inhalation exposure tests. Standard protocols are available with the regulatory agencies and should be used if the data are intended to support a regulatory criteria or registration of a chemical for use. The results are expressed as LD_{50}, or in a range of LD_1 (lowest reported lethal dose), LD_{10}, LD_{30}, or LD_{100}.

Subchronic Mammalian Toxicity Test

This test is designed to assess the adverse effects resulting from repeated exposure to a chemical over a portion of the average life span of the animal. These studies yield information on the sublethal effects at relatively low doses (lower than LD_{50} concentrations) of the test chemical. Results from subchronic studies are used to design dose levels for chronic, reproductive, and carcinogenic studies. They are also used to establish the No Adverse Effect Level (NOEL), which is valuable in risk assessment of a chemical. Subchronic test are 30 to 90 days in duration and might vary depending upon the species, route of exposure, etc. Generally, subchronic exposure does not exceed 10% of the animal's life span. Oral and inhalation testing lasts for three months in shorter-lived animals (rodents) and one year for longer-lived animals (dogs, monkeys); dermal studies take one month or less. Wherever feasible the exposure route for subchronic studies should be the same as for humans.

Chronic Mammalian Toxicity Test

This test covers lifetime toxicity, multigeneration reproduction, and carcinogenicity studies. Chronic testing is designed to reveal a myriad of potential toxic effects of a test chemical. It is used to define an etiology of an adverse effect to identify NOEL for the test chemical and estimate the safety factor at the ambient exposure level of the chemical (Stevens and Gallo 1982). Rodents are used in chronic testing; a typical study will involve three treatment and one control groups, with an equal number of animals in each. The chemical is administered 7d/wk for at least 2 years. A three-month-range-finding study with enough doses is conducted to find a level that will suppress body weight gain by 10% (National Cancer Institutes Approach) (Paustenbach 1988). This dose is referred to as the Maximum Tolerated Dose (MTD) and is chosen as the highest dose and 1/4 MTD and 1/8 MTD are usually chosen as the other two test doses.

Mammalian Developmental Toxicity

This test yields information on the developmental toxicities of the test chemical: (1) teratogenic — chemicals which produce gross structural abnormalities; (2) embryotoxic — chemicals, due to several possible toxic actions, which destroy the embryo; and (3) fetotoxic — chemicals which cause toxic or degenerative effects on fetal tissues and organs. The U.S. EPA guidelines have suggested that embryotoxic and fetotoxic describe a wide range of adverse effects; the two terms differentiate the time when the toxicities are apparent (U.S. EPA 1986a).

Carcinogenicity Hazard Identification

Guidelines are available to conduct adequate testing to identify the carcinogenic hazard of a chemical from regulatory agencies, such as the International Agency for Research on Cancer (IARC) and others. The guidelines prepared by the U.S. EPA, after two years of agency-wide effort, included the work of many scientists from across the research community. These guidelines set the principles and procedures for risk assessment and informed decision makers and the public about the process (U.S. EPA 1986b).

Mutagenicity Hazard Identification

Guidelines were developed for mutagenicity identification by the U.S. EPA (1987). For mutagenic hazard identification, it is essential to conduct:

- Mouse specific-locus test for point mutations and the heritable translocation or germ-cell cytogenetic tests for chromosomal aberrations.
- Alternative evaluations where data are not available.

While mammalian germ-cell assays are presently performed on male animals, a chemical cannot be considered nonmutagenic for mammalian germ-cells unless it is shown to be negative for both sexes.

The following factors are to be considered in evaluating chemicals for mutagenic activity:

1. Genetic end-points (gene mutations or chromosomal aberrations) detected by the system.
2. Sensitivity and predictive value of the test systems for several classes of chemicals.
3. Number of test systems used for detecting each genetic end-point.
4. Consistent results from different test systems and different species.
5. Dose-response relationship.
6. Use of recommended protocols.

PREDICTIVE CAPABILITY OF FATE PROCESSES

Volatilization

One of the important fate processes of organic chemicals is volatilizational loss from surfaces, including water columns. This is due to their low aqueous solubilities, moderate to high vapor pressures, and low polarities. Many organic chemicals, in spite of their low vapor pressures, can volatilize rapidly due to their high activity coefficients in solution. The volatilization of organic chemicals from soil surface is complicated by other variables. There is no simple laboratory measurement that will reliably extrapolate the data to the field soil situation.

Volatilization represents the physical transport of organic chemicals from waterbodies to the atmosphere. Theoretical concepts of volatilization of chemicals from water to the atmosphere have been presented by several studies (Mackay and Wolkoff 1973; Liss and Slater 1974; Mackay and Leinonen 1975; Chiou and Freed 1977; Smith et al. 1980).The theoretical approaches and their limitations have been discussed elsewhere. (Lyman et al. 1982). The transport across a two-layer system can also be expressed assuming that concentrations close to either side of the interface are in equilibrium, as expressed by Henry's law constant (H) (see Chapter 2).

The volatilization rate constant can be calculated using the equation:

$$R_v = -\frac{d[C_w]}{dt} = k_v [C_w]$$

where

$$k_v = -\frac{1}{L}\left[\frac{1}{k_\ell} + \frac{RT}{H_c k_g}\right]^{-1}$$

and

R_v = volatilizational rate of a chemical C (mole/L/hr)
C_w = aqueous concentration of C (mole/L = M)
k_v = volatilizational rate constant hr^{-1}
L = depth (cm)
k_1 = mass transfer coefficient in the liquid phase (cm/h)
H_c = Henry's law constant (torr/M)
k_g = mass transfer coefficient in the gas phase(cm/hr)
R = gas constant (litre-atm./mol/degree)
T = absolute temperature (degree Kelvin)

The results of these volatilizational loss estimates suggest that Henry's constant is the determinant factor for the magnitude of k_v. It should also be noted that a chemical with H_c value greater than 40 torr/M (roughly corresponding to 25% liquid phase mass transfer resistance), the volatilizational half-life of the chemical will be less than 10 days. (Smith et al. 1981).

Four regression equations were developed, based on the properties of a series of chloroorganic compounds ranging from *p*-dichlorobenzene to perchloroethylene, for predicting the distribution profile in selected compartments. These compounds possess a wide range of solubilities and vapor pressures (Table 9.2).

The regression equations used were as follows:

1. % chlorinated compound in air = $-0.247(1/H) + 7.9 \log S + 100.6$
2. % chlorinated compound in water = $0.054(1/H) + 1.32$
3. % chlorinated compound in soil = $0.194(1/H) - 7.65 \log S - 1.93$
4. $\log (t_{1/2}) = 0.0027(1/H) - 0.282 \log S + 1.08$

where

$$H = mm \cdot Hg \cdot m^3/mol = \frac{vapor\ pressure \times molecular\ weight}{solubility\ (ppm)}$$

$$S = (Mm/L) = \frac{solubility}{mol.\ wt.}$$

$t_{1/2}$(hours) = half-life of clearance from fish in test ecosystem

Table 9.2 Physical and Chemical Properties of Selected Chlorinated Compounds

Chlorinated Compound	Mol. Wt.	Vapor Pressure mm. Hg	Water Solubility mg/L
p-Dichlorobenzene	147	1.0	79
Trichlorobenzene	180	0.5	30
Hexachlorobenzene	285	10^{-5}	0.035
Trichlorobiphenyl	256	1.5×10^{-3}	0.05
Tetrachlorobiphenyl	291	4.9×10^{-4}	0.05
Pentachlorobiphenyl	325	7.7×10^{-5}	0.01
DDT	350	10^{-7}	0.0012
Perchloroethylene	166	14	150

Source: Neely 1979

Table 9.3 provides the distribution of chlorinated compounds in various compartments of the simulated ecosystem; the values predicted from the regression equations are in parentheses. The close agreement of data validates the approach to estimate partitioning of chemicals based on their fate processes.

Comparison of theoretical loss rates with the measured loss rates for selected pesticide pairs are given in Table 9.4. Theoretical values were derived from the literature values of vapour pressures (P), and the molecular weight (M), and assuming that loss rate was proportional to P$\sqrt{M}$. The agreement was within a factor of 10

Table 9.3 Distribution of Chlorinated Compounds in a Simulated Ecosystem (Predicted Values are in Parentheses)

Chlorinated Compound	Water %	Soil %	Air %	$t_{1/2}$ from Fish (hr)
p-Dichlorobenzene	1.24 (1.31)	1.28 (0.24)	97.5 (8)	15 (14)
Trichlorobenzene	1.33 (1.34)	2.06 (4.09)	96 (94)	17 (20)
Hexachlorobenzene	3.57 (1.98)	39.4 (31)	56 (68)	162 (164)
Trichlorobiphenyl	1.38 (1.33)	15.2 (26)	83 (71)	96 (134)
Tetrachlorobiphenyl	1.5 (1.34)	17 (27)	81 (71)	104 (139)
Pentachlorobiphenyl	1.5 (1.34)	21 (33)	77 (65)	229 (226)
DDT	1.26 (3.17)	67.5 (46.5)	28 (49)	915 (517)
Perchloroethylene	1 (1.32)	1 (~0)	98 (100)	14 (12)

Source: Neely 1979

are reported in earlier studies (Cull and Dobbs 1982). If the high melting chemicals with vapour pressure difference greater than 10 were not included, then the agreement was within a factor of 3.

Table 9.4 Theoretical Loss Versus Measured Loss for Selected Pesticide Pairs

Pesticide Pairs	Theoretical Ratio (TR)	Measured Ratio (MR)	MR/TR
Parathion ethyl/Parathion methyl	0.60	0.64	1.1
Dinoseb/Dimethoate	6.2	7.7	1.2
Dibutyl phthalate/hexachlorobenzene	1.2	1.0	0.83
Hexachlorobenzene[a]/Dieldrin[a]	5.0	3.0	0.60
Trifluralin/Dieldrin	40	21.0	0.53
Atrazine[b]/Dieldrin	0.091	0.037	0.41
Parathion ethyl/Dieldrin	1.7	0.62	0.36
Dinoseb/Parathion methyl	5.8	2.0	0.34
Dibutyl phthalate/Dieldrin	6.0	0.74	0.12
Atrazine/di(2-ehtylhexyl)phthalate[b]	0.31	1.1	3.5
Trifluralin/Dimethoate	15	39	2.6
Atrazine[b]/Monuron[b]	0.62	1.9	3.1
Picloram[b]/Dieldrin	0.018	0.14	7.8

[a] Average value from two experiments
[b] Only a small amount of chemical was volatilized during the experiment and the log (amount remaining) vs. time regression had a large 95% confidence limit; therefore, the loss rate ratio may be suspect.

Source: Reprinted with permission from Dobbs, A.J. et al. 1984; *Chemosphere*, 12(5/6), 687–692, copyright (1984), with kind permission from Elsevier Science Ltd., U.K.

Sorption

Many organic chemicals, particularly the nonpolar and sparingly water-soluble chlorinated compounds, tend to sorb (adsorb and absorb) strongly to sediments, suspended solids, and soils. Sorption determines the fraction of the organic compound available for other fate processes. The conventional sorption equilibrium constant has been modified to take into account the correlation of sorption with the organic content of the sorbent. Sorption coefficients are relatively constant at low aqueous concentrations of the chemical, but tend to decrease with the increase in aqueous concentration of the chemical.

Sorption depends upon the pH and the type of chemical interaction, ionic or nonionic. Sorption of chemicals like 2,4-D, 2,4,5-T, dicamba, chloroamben, and picloram increase with decrease in pH due to the formation of nonionized surfaces and nonionized form of the chemical. Sorption coefficients (K_{oc}) relate well with other accumulation factors, such as n-octanol/water partition coefficients (log K_{ow}) and bioconcentration factors (BCFs) for the biota. Sorption coefficients also provide a measure of leachability of organic compounds, which is valuable in environmental impact assessments. Compounds having K_{oc} values equal to or greater than 1,000 are quite strongly bound to organic matter of soil/sediment and, thus, are considered immobile. Chemicals with K_{oc} values <100 are moderately to highly mobile. Correlation of K_{oc} with other related parameters such as water solubility, K_{ow}, and BCF are discussed in Chapter 2.

Water Solubility

Aqueous solubility of chlorinated organic compounds is one of the most important parameters which influences the fate and transport processes of these chemicals in the environment. Highly water-soluble compounds have relatively longer residence time in the water column and readily undergo biodegradation, and photolytic, and hydrolytic reactions. These chemicals have low sorption capacities for soils and sediments and a low tendency to bioaccumulate. Factors such as temperature, salinity, dissolved organic matter, and pH play a major role in the aqueous solubility of organic compounds.

There are about 18 different regression equations available in the literature to correlate water solubility to octanol/water partition coefficients (Lyman et al. 1982). Most equations cover two-thirds of the chemicals within a factor of 10. Large variations greater than a factor of 100, were associated with nitrogen-containing compounds and almost all were overestimated. Correlation of log K_{ow} with WS (water solubility) using data set from literature for a mixed class of aromatics and chlorinated hydrocarbons has been reported (Banerjee et al. 1980).

Bioconcentration

Bioconcentration represents the accumulation of a chemical in biota; the bioconcentration factor (BCF) is the ratio of the concentration of a chemical in biota to its concentration in the surrounding water column. This process has been discussed in Chapter 2. Some recommended regression equations for predicting BCF values from physical properties, such as octanol-water partition coefficient (k_{ow}), water solubility (WS), and sorption coefficient normalized to organic carbon (k_{oc}), are given in Table 9.5. The comparison of estimated BCFs with field-observed BCFs is given in Table 9.6. Also, comparisons of BCFs estimated from k_{ow}, WS, and, k_{oc}

with laboratory measured BCFs are given in Table 9.7. Variability between measured and estimated BCF values could be due to the variability in biological uptake and inaccuracy in the measurement.

Table 9.5 Some Recommended Regression Equations for Estimating Log BCF[a]

Equation	n	r^2	Chemical Classes	Range of Independent Variable	Species Used
Log BCF = 0.76 log k_{ow} − 0.23[a]	84	0.823	Wide range	7.9 to 8.1 × 10^6	Fathead minnow Bluegill sunfish Rainbow trout Mosquito fish
Log BCF = 2.791 − 0.564 log WS[b]	36	0.49	Wide range	0.001 to 50,000 ppm	Brook trout Rainbow trout Bluegill sunfish Fathead minnow Carp
Log BCF = 1.119 log k_{oc} − 1.579[b]	13	0.757	Wide range	<1 to 1.2 × 10^6	Various

[a] Flow-through studies; n = number of chemicals used in the regression; r = correlation coefficient for the regression equation; [a]Veith et al. 1980; [b]Kenaga and Goring 1980.

Source: Reprinted from Lyman, W.J. et al. 1982, originally published by McGraw-Hill Book Company, copyright (1982) American Chemical Society.

All of the methods described for estimating BCF are based on correlations between a measured or calculated physical or chemical property of an organic compound and BCF. The accuracy and precision of the techniques that are used to measure the correlating property limit the accuracy of the BCF estimates. Therefore, estimated BCF values should be treated only as indicators of the potential for an organic compound to build up in aquatic biota. The estimated BCF of a given chloroorganic compound should be compared with benchmark accumulators such as DDT, PCBs, and also against known non/weak accumulators, such as chloroform, 1,2-dichlorobenzene, or 1,2-dichloroethane as a check.

The higher BCF observed in field conditions (Table 9.6) may be due to mixed exposure to fish from water and food sources. Migration pattern, uptake, and depuration rates, exposure duration, seasonal variations in ambient environment, etc. are likely to cause discrepancies in measurements of BCF.

Table 9.7 provides a comparison between laboratory-measured BCF and values of BCF estimated from correlations with k_{ow}, solubility (S), and k_{oc}. Variability between measured and estimated BCF values could be due to variability in biological uptake and inaccuracy in the measurement.

Photolysis

Photochemical reactions are recognized as important fate processes for organic compounds in the environment. Structural changes in an organic molecule, induced by electromagnetic radiation in the near ultraviolet-visible (240 to 700 nm), are

Table 9.6 Estimated and Field Evaluated BCF Values

Compound and Location	Log k_{ow}	Log WS (ppm)	Estimated BCF	Ambient Water Concentration	Concentration in Fish (Duration and Species)	Observed BCF	References
Aroclor 1016 (Hudson River)	5.88	—	17,000	Mean 0.17 µg/L	2.6 µg/g (mean of 18 fish 3 species, 14 day exposure)	15,000	Skea et al. 1979
DDT (Hamilton Lake, Ontario)	5.75	—	14,000	4.5 ng/L	0.14 µg/g (Alewife)	31,000	Waller and Lee 1979
DDE (Hamilton Lake, Ontario)	5.69	—	12,000	37.4 ng/L	0.23 µg/g (Smelt) 0.46 µg/g (Alewife)	51,000 12,000	Greichus et al. 1973
Dieldrin (Hamilton Lake, Ontario)	—	−1.66	5,300	3.1 ng/L	1.36 µg/g (Smelt) 0.94 µg/g (Sculpin) 0.04 µg/g (Alewife)	36,000 25,000 13,000	Waller and Lee 1979
PCB (Aroclor 1254) (Two lakes in South Dakota)	6.47	—	49,000	<0.5 µg/L	0.11 µg/g	>220	Greichus et al. 1973
Lindane (limestone quarry)	3.89	—	530	25–13 ng/L	~27.3–13.3 ng/g (Trout, 3–7 fish per sample)	~1,090	Hamelink et al. 1977
Trifluralin (Wabash River)	5.33	—	6,600	~1.8 µg/L	10.46 µg/g (237 Sauger, residue in fat)	5,800	Spacie and Hamelink 1979

Source: Reprinted with permission from Lyman, W.J. et al. 1982, originally published by McGraw-Hill Book Company, Copyright (1982), American Chemical Society.

Table 9.7 Comparison of Estimated BCFs with Measured BCFs

Chloroorganic Compound	Parameter Used in Estimation			BCF Estimated from			Measured BCF	Reference
	K_{ow}(1,2)	S(ppm) (3)	K_{oc}(3)	K_{ow}[a]	WS[b]	K_{oc}[++]		
Carbontetrachloride	437	800	—	17	14	—	30	4
p-Dichlorobenzene	2,400	79	—	220	53	—	215	5
1,2,4-Trichlorobenzene	17,000	30	—	970	91	—	2,800	1
Methoxychlor	20,000	0.003	80,000	1,100	16,000	8,100	8,300	1
Pentachlorophenol	126,000	14	900	4,400	140	53	770	1
Hexachlorobenzene	170,000	0.035	3,910	5,600	4,100	280	18,500	1
Heptachlor	275,000	0.030	—	8,000	4,500	—	9,500	1
DDT	562,000	0.0017	23,800	14,000	23,000	27,000	29,400	1
Arochlor 1254	2,950,000	0.01	42,500	49,000	8,300	4,000	100,000	1
Chlordane	1,000,000	0.056	—	21,000	120	—	37,800	1

[a] Equation log BCF = 0.76 log k_{ow} − 0.23; [b] log BCF = 2.791 − 0.564 log WS; [++] log BCF = 1.119 log k_{oc} − 1.579

(1) Veith et al. 1979; (2) Veith et al. 1980; (3) Kenaga and Goring 1980; (4) Baherjee et al. 1980; (5) Neely et al. 1974

Source: Lyman et al. 1982

called photochemical reactions. Absorption of light by the organic molecule is a prerequisite for any photochemical reaction to take place.

The rate of disappearance of an organic compound by direct photolysis is given by

$$-\frac{dc}{dt} = K_p[C] = k_a v \Theta [C]$$

where,
K_p = first order rate constant, Θ = reaction quantum yield, and k_a = rate constant for light absorption by the chemical.

Environmental photolysis is determined by measuring Θ at a single wavelength (λ) in the laboratory. Sunlight intensity (I_λ) data, as a function of time of day, season, and latitude, are available in the literature. The rate constant in sunlight, $k_{p(s)}$ is given by

$$k_{p(s)} = \Theta \Sigma I_\lambda \cdot \varepsilon_\lambda$$

and the half-life in sunlight is given by

$$(t_{1/2})_s = \ln 2 / k_{p(s)}$$

For screening purposes, the upper limit of $k_{p(s)}$ can be calculated by assuming $\Theta = 1$. If the calculated rate constant is small compared to other fate processes, no additional photolytical measurement are required.

EXPOSURE ASSESSMENT

Exposure has been defined as the contact between a chemical and humans, as well as the ecosystem dwellers. Exposure assessment is critical in quantitative risk assessment of chemicals and exposure values are used in making regulatory decisions. Toxicity is defined as the intrinsic property of a chemical; exposure level is an extrinsic property. Toxicity is rather universal in nature and changes only when new information becomes available. Conversely, exposure is very much region-specific and regulatory options and decisions are based on how to reduce and control the exposure to the chemical. Various options are used depending upon the magnitude of health risk and the choice of an alternative. They range from (1) total ban, such as the case of banning chlorinated pesticides and insecticides and switching over to a degradable and non-persistent group of chemicals; (2) restricted use in a closed environment, such as PCBs in capacitors and mercury in thermostats, where it is not easy to find substitutes with similar properties; and to (3) recommendation of advisories such as consumption limit on contaminated fish and other such aquatic organisms.

The U.S. EPA has published general guidelines for carrying out exposure assessments (Table 9.8).

Table 9.8 General Outline for Exposure Assessments

1. Introduction
   - Purpose
   - Scope

2. General estimation on chemicals
   a. Identity
      - Name, synonyms, formula and structure
      - Chemical Abstract Service (CAS) number
      - Grade, contaminants present, and additives
      - Other descriptive or identifying characteristics
   b. Chemical and physical properties

3. Sources
   - Production and distribution
   - Uses ,
   - Disposal patterns
   - Potential environmental releases

4. Environmental pathways and fate processes
   - Transport and transformation
   - Identification of major pathways of exposure
   - Predicted distribution in environment

5. Measures or estimated concentration

6. Exposed populations
   Human population: Size, characteristics, location, and habits
   Non-human populations: Size, characteristics, location, and habits

7. Integrated Exposure Analysis
   - Identification of exposed population and pathways of exposure
   - Human dosimetry and biological measurements
   - Development of exposure scenarios and profiles
   - Evaluation of uncertainty

Source: U.S. EPA, *Fed. Regist.* 51, No.185, 34042–34054, 1986.

The following are the three possible reasons for conducting an exposure assessment (EA):

1. Most EAs are carried out as part of risk assessment to provide the exposure levels for individuals or populations to estimate the risk.
2. Some EAs are used as risk reduction evaluation tools. They can become useful tools for predicting consequences of a variety of regulatory optional actions.
3. An EA is used to determine whether there is a significant exposure as a prerequisite for testing of a chemical based on substantial exposure.

Scoping decides the general outline of the exposure assessment and clear scoping avoids unnecessary expenditure. Questions like "what should be included or excluded from the assessment?" should be addressed collectively, since they are interrelated.

Examples are:

1. Humans vs. non-humans: should both be included in the assessment or only one?
2. Individuals vs. subpopulations vs. general populations.
3. Geographic boundaries: examples are a series of sites, regions, or nationwide when assessing "all plants that make chemical x".
4. Route of exposure: the route of exposure is the means by which the chemical enters the organisms. For humans, the normal routes are ingestion (via food, drinking water), inhalation, and dermal absorption.
5. Media of exposure: the exposure of chemicals can be limited to one medium or could be from multimedia, which include air, water, food, soil, etc. Volatility, lipid solubility, and water solubility are the principal properties that control the multimedia potential of a chemical.
6. Exposure settings: the total exposure can be separated into ambient environment, waste disposal sites, drinking water, occupational environment, consumer products, food items, accidental spills, etc. The boundaries for each exposure setting can be defined and can set the structure and limit for data collection in exposure assessment.
7. Depth of detail: since exposure assessment is used in regulatory decision making, the level of accuracy and detail has to match with the importance of the decision being made.

Influence of Fate Processes on Environmental Pathways

The chemical, when released to an environmental medium, may undergo one or more fate processes, such as physical transformation, chemical transformation, and/or biotransformation.

The following factors should be addressed before exposures are evaluated.

- The predominant fate process of the chemical in each environmental medium and its residence time.
- Concentration changes in each medium with time to focus on the critical medium of exposure.
- The multimedia distribution potential of the chemical.
- The toxicity of the breakdown products and their fate processes.

Based on this information, primary exposure routes can be identified.

RANKING PROTOCOLS

Ranking protocols provide systematic selection of a manageable number of chemicals for assessment using readily available data sources and expert judgement. The fundamental requisite in a ranking system is scoring for each parameter, which reduces the available information to number and scores. The absence of information, common for many chemicals, leads to estimating the score based on scientific expert judgment.

A vector scoring system to prioritize environmental contaminants was developed by the Ontario Ministry of the Environment (OME), Canada (OME 1988). In this system, chemicals are given numerical scores for several parameters (called vector elements) which describe their environmental behavior, exposure potential, and adverse effects on biota in the environment, including humans. If information is not available on a chemical, an asterisk (*) is substituted for the score for that vector element. Various other score modifiers can be applied under specific circumstances. For example, questionable data are identified by a question mark (?); an exclamation mark denotes a worst-case-scenario value. Individual element scores can be combined in specific ways to give a priority ranking for groups of chemicals. The methods of combining element scores can be adjusted to meet the needs of specific users of the scoring system (OME 1990).

The scoring system consists of three phases; each phase requires more specific information on the chemical than the previous phase. Chemicals are ranked in each phase and passed on to the next phase according to their priority ranking. Phase I of the Vector Scoring System simply determines which chemicals, that may be present in a given environment, are actually considered in the scoring system.

Phase II of the Vector Scoring System contains the following nine elements, divided into three major groups:

I. Elements describing exposure ("E" elements):
    P2E1 — Sources
    P2E2 — Releases
    P2E3 — Environmental distribution
    P2E4 — Environmental transport
    P2E5 — Environmental persistence
    P2E6 — Bioaccumulation
II. Elements describing adverse effects ("T" elements):
    P2E7 — Acute lethality
    P2E8 — Other toxicity
III. Element describing aesthetic properties:
    P2E9 — Undesirable aesthetic properties

Scores from zero to three are assigned for each element, based on increasing severity of specific criteria. The criteria for Phase 2 elements rely heavily on chemical and physical properties and on adverse effects indicators that are readily available from summary-type data sources, such as books and review articles (Table 9.9).

Once scores are generated for each element, based on available information, the chemical is placed into a high, medium, or low priority list, or lists indicating a lack of information or chemicals with undesirable aesthetic properties. Those chemicals on the high priority list enter Phase 3 of the scoring system first, followed by those on the medium priority list. Those on the undesirable aesthetic properties list are the automatic candidates for regulatory assessment in Phase 3.

Table 9.9 Scoring Criteria for Phase 2 Vector Exposure Elements

| Element Number | Units | \multicolumn{4}{c}{Scoring Criteria} |
		0	1	2	3
P2E1	kg/year	<5	5–300	300–10,000	>10,000
P2E2	% release	0	0–3	3–30	>30
	narrative	Not used or imported	Used in closed systems with no routine releases	Most converted to another product; or largely restricted to industrial uses; or shipped in large batches	Most released directly into the environment; or used in an open, dispersive manner
P2E3	Measurement basis	Not yet detected	Infrequently detected at specific locations	Frequently detected but only at specific locations	Frequently detected throughout
	Release basis	No known release	Few release sites at a few locations	Few release sites, not concentrated in a few locations	Many release sites, throughout
P2E4	Narrative	<5% of releases partitions into other media; or vapor pressure ≤1kPa; solubility, ≤100 mg/L	≥ one receiving medium containing 5–10% of amount released or vapor pressure ≤1kPa; solubility ≤100 mg/L	≥ one receiving medium containing 10–20% of amount released or vapor pressure >1kPa; solubility >100 mg/L	> two media partitions containing >20% of amount released; or vapor pressure >1kPa; solubility >100mg/L
P2E5	$t_{1/2}$ (day)	< 10	10 to <50	50 to <100	>100
	narrative	Not persistent	Slightly persistent	Moderately persistent	Very persistent
P2E6	BCF	≤20	20–500	500 to 15,000	>15,000
	Log K_{ow}	≤2.0	2.0–4.0	4.0 to 6.0	>6.0

Source: OME 1988

Phase 3 of the scoring system is made up of the following 15 elements in three major categories:

I. Elements Describing Exposure Parameters ("E" elements):
   P3E1 — Environmental concentrations in air
   P3E2 — Environmental concentrations in water
   P3E3 — Environmental concentrations in soil
   P3E4 — Environmental concentrations in sediment
   P3E5 — Environmental concentrations in animals
   P3E6 — Environmental concentrations in plants
   P3E7 — Frequency of dispersion
II. Elements Describing Adverse Effects ("T" elements)
   P3E8 — Acute lethality
   P3E9 — Sublethal effects on nonmammalian species
   P3E10 — Sublethal effects on plants
   P3E11 — Sublethal effects on mammals
   P3E12 — Teratogenicity
   P3E13 — Genotoxicity/mutagenicity
   P3E14 — Carcinogenicity
III. Elements Describing Undesirable Properties:
   P3E15 — Undesirable aesthetic properties

More effort and resources are required to collect information for Phase 3 than Phase 2 (i.e., primary reference sources from a variety of databases provide the main information sources).

Phase 3 element scores range from 0 to 10 for each element. Once the scores are generated, the chemical is placed on one of five lists; high, medium, low, inadequate information, and undesirable aesthetic properties. Chemicals placed on the high priority list receive first assessment for regulatory consideration. A simplified version of the Vector Scoring System (OME 1990) is currently used for screening and categorizing environmental contaminants for several programs due to limited database available for most contaminants.

REFERENCES

Banerjee, S., Yalkowsky, S.H., and Valrani, S.C. 1980. Water solubility and octanol-water partition coefficients of organics. *Environ. Sci. Technol.* 14:1227–1229.

Chiou, C.T. and Freed, V.H. 1977. Chemodynamics Studies on Benchmark Industrial Chemicals. MSF/RA-770286. MTIS PB 274263.

Cull, M.R. and Dobbs, A.J. 1982. Volatilization of chemicals—relative loss rates and the estimation of vapour pressures. *Environ. Pollut.* (Series B). 3:289–298.

Dobbs, A.J., Hart, G.F, and Parsons, A.H. 1984. The determination of vapour pressures from relative volatilization rates. *Chemosphere.* 12(5/6), 687–692.

Greichus, Y.A., Greichus, A., and Emerick, R.J. 1973. Insecticides, polychlorinated biphenyls and mercury in wild cormorants, their eggs, food and environment. *Bull. Environ. Contam. Toxicol.* 9:321–328.

Hamelink, J.L., Waybrant, R.C., and Yant, P.R. 1977. In: *Fate of Pollutants in the Air and Water Environments,* Part 2. Suffet, R., Ed., John Wiley & Sons, New York.

Kenaga, E.E. and Goring, C.A.I. 1980. Relationship between water solubility, soil sorption, octanol-water partitioning and bioconcentration of chemicals in biota. In: Eaton, J.G., Parish, P.R., and Hendricks, A.C., Eds. *Proceedings of the 3rd Symposium on Aquatic Toxicology,* American Society of Testing Materials, Philadelphia, pp.78–115.

Liss, P.S. and Slater, P.G. 1974. Flux of gases across the air-sea interface. *Nature.* 247:181–184.

Lyman, W.J., Reehl, W.F., and Rosenblatt, D.H. 1982. *Handbook of Chemical Property Estimation Methods — Environmental Behaviour of Organic Compounds.* McGraw-Hill, New York.

Mackay, D. and Wolkoff, A.W. 1973. Rate of evaporation of low solubility contaminants from water bodies to atmosphere. *Environ. Sci. Technol.* 7:611–614.

Mackay, D. and Leinonen, P.J. 1975. Rate of evaporation of low solubility contaminants from water bodies to atmosphere. *Environ. Sci. Technol.* 9:1178–1180.

Neely, W.B., Branson, D.R., and Blau, G.E. 1974. Partition coefficients to measure bioconcentration potential of organic chemicals in fish. *Environ. Sci. Technol.* 8:1113–1115.

Neely, W.B. 1979. An integrated approach to assessing the potential impact of organic chemicals in the environment. In: Analyzing the Hazard Evaluation Process, *Proceedings of a Workshop, Waterville, New Hampshire, August 14 -18, 1978,* Dickson, K.L., Maki, A.W., and J. Cairns, J., Eds., American Fisheries Society, Washington, D.C. pp. 74–82.

OME (Ontario Ministry of the Environment) 1990. Scoring system — A scoring system for assessing environmental contaminants, January 1990.

OME (Ontario Ministry of the Environment) 1988. Vector scoring system for the prioritization of environmental contaminants, Final Report-I, OME, Ontario, Canada, 84p.

Paustenbach, D.J., Ed. 1988. *The Risk Assessment of Environmental and Human Health Hazards: A Textbook of Case Studies,* John Wiley & Sons, New York.

Rand, G.M. and Petrocelli, S.R., 1985. *Fundamentals of Aquatic Toxicology,* Hemisphere Publications, Washington, D.C. 666 p.

Skea, J.C., Simonin, H.A., Dean, H.J. et al. 1979. Bioaccumulation of Aroclor 1016 in Hudson River fish. *Bull. Environ. Contam. Toxicol.* 22:332–336.

Smith, J.H., Bomberger, D.C. Jr., and Haynes, D.L. 1981. Volatilization rates of intermediate and low volatility chemicals from water. *Chemosphere.* 10:281–289.

Smith, J.H., Bomberger, D.C., and Haynes, D.S. 1980. Prediction of the volatilization rates of high-volatility chemicals from natural water bodies. *Environ. Sci. Technol.* 14:1332–1337.

Spacie, A. and Hamelink, J.T. 1979. Dynamics of trifluralin accumulation in river fishes. *Environ. Sci. Technol.* 13:817–822.

Stevens, K.R. and Gallo, M.. 1982. In: Hayes, A.W., Ed., *Principles and Methods of Toxicology,* Raven Press, New York.

U.S. EPA (Environmental Protection Agency) 1979. Toxic Substances Control Act (TSCA)-Premanufacture Testing of New Chemical Substances, *Fed. Regist.* 44, (1979), 16240-16292.

U.S. EPA (Environmental Protection Agency) 1986. Guidelines for Health Assessment of Suspect Developmental Toxicants, *Fed. Regist.* CFR 2984, No. 185, 34028-34041.

U.S. EPA 1986. Guidelines for Exposure assessment, *Fed. Regist.* 51, No. 185, 34042-34054.

U.S. EPA (Environmental Protection Agency) 1987. Guidelines for Mutagenicity Risk Assessment, Office of Health and Environmental Assessment, Washington, D.C. EPA-600/8-87/045, pp.2-1 to 2-9.

Veith, G.D., Defoe, D.L., and Bergstedt, B.V. 1979. Measuring and estimating the bioconcentration factor of chemicals in fish. *J. Fish. Res. Bd. Can.* 36:1040–1048.

Veith, G.D., Macek, K.J., Petrocelli, S.R., and Carroll, J. 1980. An evaluation of using partition coefficients and water solubility to estimate bioconcentration factors for organic chemicals in fish. In: *Aquatic Toxicology,* ASTM STP 707. pp.116-129.

Waller, W.T. and Lee, G.T. 1979. Evaluation of observations of hazardous chemicals in Lake Ontario during the International Field Year for the Great Lakes. *Environ. Sci. Technol.* 13:79–85.

CHAPTER 10

The Regulatory Decision-Making Process

RISK ASSESSMENT

Risk is defined as the expected frequency of adverse effects resulting from exposure to chemicals. Risk may be expressed in absolute terms, as risk due to exposure to a specific chemical, or in relative terms, comparing the risk of the exposed population to the unexposed. Risk levels associated with certain commonplace activities, natural occurrences, voluntary activities, and consumption of natural products have been compiled (Wilson and Crouch, 1987; Ames et al. 1987). Risk assessment was originally developed as part of the actuarial techniques of the insurance industry to estimate probabilities of events that result in claims. Then it was extended to the engineering sector to estimate the probabilities of catastrophic failures of engineered systems, such as aircraft and nuclear power plants. Health risk assessment estimates the probabilities of diseases among the population exposed to a range of toxic chemicals and combinations such as cigarette smoke, dietary patterns, and industrial emissions. The ecological risk assessment deals specifically with adverse effects on the ecosystem, which includes plants and animals and ecosystem properties. Human health risk assessment evaluates the adverse effects on humans resulting from exposure to a chemical and predicts the expected frequency of the effect over a life-time period.

Hazard assessment was used before risk assessment. This approach, which was in use from late 1970s to middle 1980s, calculated a margin of safety by comparing the toxicological endpoint of interest (usually an estimate of safe concentration) to an estimated exposure concentration. An expert judgment is made on the adequacy of the margin of safety based on the amount of quality toxicological data. The margin of safety or uncertainty factors applied to safety decisions are basically meant to take into account the variability in specific responses, life stages, short- and long-term biological effects, and test methods.

Both epidemiological and laboratory animal studies collect data from samples drawn from large populations and, therefore, only population averages or statistics can be estimated from these studies. The importance of these two data as key

Definitions and Components of Risk Assessment

	Component	Definition
1.	Hazard identification	The determination of the existence of causal link, or the lack of it, to particular health effects.
2.	Dose-response relationships	The determination of a relationship between the magnitude of exposure and the probability of occurrence of adverse health effects.
3.	Exposure assessment	The determination of the degree of human exposure before and after the introduction of regulatory controls.
4.	Risk characterization	The description of the nature and often the magnitude of risk to human health with associated level of uncertainty.

Source: U.S. NAS 1983

contributors to statistical risk assessment is supported by a recent publication from the U.S. Department of Health and Human Services (U.S. HSS 1986).

Quantifying risk to sensitive/susceptible subpopulations (such as hypersensitive individuals, pregnant women, infants and children, older age groups, etc.) and combining these risks, with estimated risks for the larger population to arrive at a number for the total population risk, is a key challenge for risk assessors.

1. **Hazard identification** — Hazard is defined as a set of circumstances with a potential for causing adverse health effects in humans. For a chemical to be hazardous, it has to be toxic at the environmental exposure level and for the duration of the exposure. In brief, the dose makes the chemical a hazard. There are chemicals whose exposure at all levels of concentrations will produce a particular type of toxicity called genotoxicity. Hazard identification requires both qualitative and quantitative information on genetic and nongenetic toxicity of chemicals, acting either singly or in groups. The adverse effects are the intrinsic properties of a chemical, but its presence in the environment makes it hazardous.
2. **Dose-response evaluation** — This process correlates the dose of a chemical and the incidence of an adverse effect in exposed populations. Factors such as intensity and duration of exposure, age, and sex are taken into consideration. The dose-response curve obtained at the observable range (high dose) is extrapolated to ambient exposure levels; other extrapolations involve interspecies extrapolation of data, i.e., from animals to humans and from one laboratory animal species to wildlife species. These limitations should be described in any dose-response assessment.
3. **Exposure assessment** — This process measures or estimates the intensity, frequency, and duration of animal or human exposure to a chemical present in the ambient environment. In the case of new chemicals, estimates are obtained using benchmark concept and relevant fate process models. Exposure assessment should also identify the possible routes of exposure, size, nature, and classes of exposed populations, and the uncertainties associated with the estimates. Since sources are identified in the process, the assessment should also include control options and available technologies for controlling and reducing the exposure to a chemical.
4. **Risk characterization** — This process estimates the incidence of an adverse health effect under different exposure settings for humans and animals. This is arrived at by linking exposure and dose-response assessments.

STRENGTH AND BOUNDARY OF INFORMATION

Laboratory Data

The variables that could have an effect on the quality of the toxicity data include biotic and abiotic factors and influences by other chemical and methodology differences. The choice of species used in any bioassay (both short- and long-term) is critical in the applicability of the results to the natural environment. The species should be site-specific to where the chemical is to be used. Although field data on various terrestrial and aquatic organisms can be valuable, it is difficult to control all the variables that operate in the natural environment; therefore, laboratory studies in controlled environments might produce more consistent results that are reasonably relevant to the natural environment, provided the critical variables have been taken into account in the studies. Epidemiological studies are very specific, difficult to delineate cause-and-effect relationships, and expensive. In every case, however, it is necessary to ensure that the information obtained is sufficient, relevant, and capable of assisting in proper assessment.

Data on Chemical

It is necessary to collect sufficient information on the chemical, including production level, volume, use pattern, and physical and chemical properties, for assessing its behavior in the environment and its possible adverse effects on the ecosystem and human health.

Table 10.1 outlines some of the basic data requirements that are necessary for the evaluation of a test chemical. The assessment of a chemical requires an evaluation of its environmental behavior and fate after it enters the environment. It should include the resident compartment of the environment and biological systems that would be exposed. The possible degradation, transformation, and bioaccumulation potential of the chemical should also be evaluated. This information, together with the results obtained from a well-developed battery of tests, will assist in setting the necessary boundary of information.

RISK COMMUNICATION

It is well recognized that the manner in which people perceive a risk determines how they respond to it and in what context and, accordingly, results in the public input into the final regulatory decisions. It is hard to have regulations when the public ignores serious risks and recoils in fear from less serious ones. The task of risk communication is not just conveying information, but it is to alert people when they ought to be alerted and reassure them when they ought to be reassured (Sandman, 1986). It is important to clearly explain the environmental risk to the public who are most likely to be affected.

Risk statements about chemicals can be presented to the public through various media. The two most influential and powerful of all methods of communication are

Table 10.1 Basic Data Requirement for the Toxicological Evaluation of Chemicals

Data Requirements	Details
Usage/Release	Annual quantities
	Location
	How used
	Method of transportation
Physico-chemical factors	Structure
	Melting/boiling points
	Density
	Solubility
	Vapor Pressure
	Sorption coefficient (k_{oc})
	Dissociation
	Henry's law constant
	Octanol/water partition coefficient
	Purity and type of contaminants
Bioaccumulation potential	Persistence
	Half-life: water, soil, air
	Bioavailability
	Bioconcentration factor
Transformation potential	Hydrolysis
	Redox reactions
	Biotic transformation/degradations
	Photolytic breakdown
Disposal	Disposal techniques
Testing	Genotoxic potential
	Nongenotoxic potential

print and television. Also, the manner in which the information is presented determines the public's reaction. The public does not realize that present-day equipment is capable of measuring background levels that have been present, but undetected, in the environment for a long time. The public's view of risk is affected by their attitude towards science and technology. A well-informed society might understand the scientific and technical aspects of a particular situation and offer suggestions.

The way in which the public perceives risk has been influenced by their intense interest regarding good health and longevity. Health has been rated, in North America and other parts of the world, as consistently more important, politically, than any clean water, clean air, or the preservation of forests or the earth's crust (Burger 1980). Public awareness about environmental hazards has increased steadily. As a result, they maintain a reasonable understanding about the various chemicals released into the environment and their potential impacts. The public is usually provided with some financial assistance from the regulatory agencies that enables them to acquire independent experts capable of answering and/or explaining, in detail, questions and technical information that may assist the public's understanding and participation in consultative functions of the regulatory process.

Risk Comparison

The two basic types of risk comparisons are: (1) comparison of risk of diverse activities; and (2) the comparison of risk of similar activities (Covello 1989). The comparison of risk of diverse activities involves comparing the chemical risk to that of a diverse set of chemicals or activities. For example, the risk of chemicals and industrial processes are compared to the risks of smoking, driving, flying, and dietary habits, such as drinking diet soft drinks and eating charcoal-broiled steaks (Wilson 1979). The second type of risk comparison involves comparing the risk of a chemical to that of a similar set of chemicals or processes. For example, we compare foods with synthetic chemical additives and pesticide residues to natural foods that may contain natural toxins (Gold et al. 1987). Major limitations of risk comparisons include (1) failure to understand the broad set of qualitative dimensions that determine people's concern about the acceptability of technologies and associated health and environmental risks, and (2) exclusion of the costs and benefits of available alternative technologies. Notwithstanding the limitations, well-constructed and documented risk comparisons effectively communicate risk information (Covello et al. 1988). Risk comparisons can provide a benchmark against which the magnitude of new or unfamiliar risks can be compared, and they can provide information to people about the range and magnitude of risks to which they are exposed (Covello 1989).

Due to an increased environmental awareness, well-organized and informed groups of environmentalists now represent all sectors and professions of society. These groups have enhanced their ability to identify, analyze, articulate, and manage their concerns with exceptional effectiveness at both the technical and political levels. Environmental issues particularly related to health and safety are important social issues and the public is addressing their concern through well-organized, proactive groups. The results of these proactive processes has led to the development of new or updated legislation and regulatory changes in management of toxic chemicals. Both government and industry find themselves having to listen to the public and include them in their decision-making processes.

The goals of risk communication are to develop a well-informed society that is involved, interested, thoughtful, reasonable, solution oriented and collaborative; it should not try to diffuse public concerns or replace actions (Covello and Allen 1988).The seven cardinal rules of risk communication are given below (Covello and Allen 1988):

1. Accept and involve the public as a legitimate partner.
2. Plan carefully and evaluate effort.
3. Listen to the public's specific concerns.
4. Be honest, open, and frank.
5. Coordinate and collaborate with other credible sources.
6. Meet the needs of the media.
7. Speak clearly and with comparison.

RISK MANAGEMENT AND PUBLIC PARTICIPATION

Risk management issues have become matters of great interest and importance to society as a whole. Industrial accidents such as the Exxon Valdez oil spill, the chemical leak in Bhopal, the Chernobyl nuclear disaster, and the PCB fire in Quebec affected the public's confidence regarding industrial controls and safety. These accidents led to the creation of public fear regarding the risk of various industrial and technological developments to both the environment and human health.

Risk assessment is a scientific process, whereas, risk management is a judgmental process based on socio-economic factors. It involves various activities that are necessary to reach risk management decisions. This is followed by risk communication to the affected public and, finally, the determination of the risk level that is acceptable to the affected community.

Participants

Participants should include interest groups and individuals who are likely to be affected by the proposed developmental activity. Limits set on the public involvement may be crucial towards the assessment of risk and the assignment of priorities to this risk; for example, people living near the location of a controversial facility are likely to view risk in a significantly different way from others living further away. Choosing the most effective method of representation is important in order to express their views and concerns. Interest groups are, in some cases, disorganized and the lead role might be taken by an individual who might not be representing the need of the community. A representative cross-section of the community would ensure that vocal special interest individuals or groups do not obscure the community. The public should be represented in terms of age, sex, education, professional status, income, and place and length of residence in the area of concern.

It is essential to include all public or special interest groups as they emerge, since most of those citizens or agencies who desire to be involved seek an organization of like-minded people (Grima 1985). However, those seeking influence may fail to deal with yet-to-emerge vocal participants who have a stake in the outcome of the involvement process (Wengert 1961). For this reason, it might be justified to include representative individuals selected at random from various sectors of the community in question.

Participatory Approaches

Three main approaches to participation have been identified (Grima 1986), according to the manner in which public should be brought into process.

1. Participation through election or appointment of public representatives to different levels of decision-making and administrative bodies (Edmond 1975). It was suggested that the legislation should provide the best institutional form for public participation because it enables the public to participate through its representatives.
2. Quasi-judicial review and arbitration of administrative decisions. This type of participation is reactive and defensive, involving a restricted group.

3. Develop specific mechanisms of participation for decision-making through public consultation and public hearings. This approach tends to constitute a direct, non-discriminatory relation within a process leading to a decision.

No single method or approach for public involvement would satisfy all the requirements for information exchange, consensus building, consultation, and interest involvement. It is prudent to apply various techniques in order to arrive at a workable solution. Some of these include door-to-door campaigning, public meetings, task forces, telecommunications, workshops, hearings, petitions, and "dog-and-pony shows". In all cases, however, techniques should enhance consensus and understanding, encourage dialogues, feedback, display flexibility, and ensure honesty. The principles that should be considered during the design and execution of a public information process are as follows:

- The process must be capable of meeting the public's needs.
- The process has to be open and responsive.
- The risk must be fully defined and explained with directions and honesty.
- All of the public should receive equal treatment.
- The public should be allowed to choose the methods of communication with regulatory agencies.
- The public should be involved early in the process.
- Participation should occur in a climate of trust and cooperation.
- Participation should be integrated with decision-making.
- Public participation should not only meet the public's needs, but it should also appear to meet the public's needs (Ramamoorthy and Baddaloo 1991).

ACCEPTABLE RISK

No single definition clearly explains the term "acceptable risk," but legal definitions have emerged from court decisions on environmental cases concerning setting regulatory limits for chemicals. In the court decisions on the benzene and vinyl chloride cases, the term "acceptable risk" was used to define "safe" by courts. It meant that the societal criteria of the law has been met to assure no significant harm or risk. Acceptable risk, in the court's view, involves a judgmental decision based on three factors: (1) scientific criteria; (2) statutory criteria, and (3) the risks that are acceptable to society. The use of the term "acceptable risk," in direct relationship to the concept of "safe," indicates that it is used as a generic term with broad legal application. For example:

Where a statute directs a balancing of risk and benefits or consideration of feasibility, as in the Toxic Substance Act, the elements that go into a determination based on the consideration of scientific data and the benefits, costs, and technological feasibility relevant under the statutory criteria, the "acceptable" level of risk represents the outcome of that balancing act. (Barnard 1990).

The legal specifications of the term "acceptable" are basically judgmental and involve a case-by-case approach.

Acceptable Level

Following a decision involving vinyl chloride, the U.S. Circuit Court of Appeals proposed that the U.S. EPA must establish a safe level of emissions that will result in acceptable exposure without regard to cost or technical feasibility (NRDC vs. EPA 1987). The judge further indicated that the U.S. EPA could not, under any circumstances, consider cost and technological feasibility at this stage of analysis. It was suggested that the above interpretation by the court was an indirect level of a *de manifests* (i.e., a ceiling above which events are inherently unsafe and should be regulated without regard to cost) to establish an acceptable level (Travis and Frey 1988)

A review of 132 U.S. federal regulatory decisions for suspected carcinogens (Travis et al. 1987), showed that, for small populations, every chemical with an individual lifetime cancer risk above $\sim 10^{-3}$ had been historically regulated. For larger populations, the risk level dropped to 10^{-4}. This population-based *de manifests* level has been considered an appropriate method for establishing risk level because it represents the level of risk that regulatory agencies have deemed acceptable in the past (Travis and Frey 1988). Thus, it is not possible to set regulatory risk levels without some knowledge of past regulatory decisions on analogous risk situations.

Defining acceptable risk and exposure standards requires scientific information, as well as an appreciation of its limits. It also requires a good understanding of the context of the risk and willingness, by the agencies as well as the critics, to deal openly with such difficult value-laden decision processes (Dwyer and Ricci 1989). A recently proposed "relative decision-making" technique (Owen and Jones 1990) compares hazard estimates of individual substances and complex mixtures to one or more well-established reference compounds in a relative potency framework. This concept is called Rapid Screening of Hazard (RASH). This process uses existing toxicity data without the use of theoretical models or prior categorization as carcinogen or noncarcinogen. For specific rules and details for matching toxicological endpoints, readers are referred to Jones et al. (1985). The risk of exposures to hazardous chemicals are compared with the risk of ingesting chlorinated drinking water. These chemicals are suspected carcinogens analyzed by the U.S. EPA Carcinogenic Assessment Group (CAG) and by Owen and Jones (1990). A considerable spread was noticed with risk values varying almost ± 3 to 4 orders of magnitude. The relative potency factors reflect a fairly high degree of stability when considering large data sets. It is inferred, from this analysis, that current methods have unexplainable inconsistencies in the regulation of a group of carcinogens. The approaches and analysis of the RASH-based relative potency approach could offer a viable alternative for a consistent level of regulation of hazardous chemicals or substances. RASH could also be used as a screening tool to prioritize chemicals and might improve consistency, reduce uncertainties, and bolster public confidence in the regulatory decision-making process.

Ideally, the best benchmark for comparative hazard evaluation would be a substance whose real longterm hazard was known with certainty and accepted by the consuming public. True hazards cannot be determined from estimates of safety factors or from extrapolation models. Chlorinated drinking water is one such benchmark standard. All environmental hazards are compared to chlorinated drinking water (Owens and Jones 1990). This hazard zone is conceptually equivalent to the U.S. EPA list of Generally Recognized as Safe (GRAS) food additives (*U.S. Fed. Reg.* 1977). Comparison to the GRAS zone of generally acceptable hazards indicates the relative hazard of a chemical or substance in the environment (Figure 10.1). The GRAS zone of hazards are plotted on the right of the log axis, and hazards from exposures to established U.S. EPA regulatory levels of drinking water contamination are plotted on the left side of the log axis. It can be seen that the GRAS zone of acceptable hazards are 2 to 6 orders of magnitude lower than the chosen cigarette smoke reference standard (a mixture of numerous chemicals reflecting diverse biological mechanisms, represents "worst case" complex mixture). Hazards from vinyl chloride and PCBs are 2 to 3 orders of magnitude below the GRAS zone (chlorinated drinking water), whereas the hazard from Chromium (VI) is nearly 2 orders of magnitude above the GRAS zone. The latter hazard was roughly equivalent to the hazard from an intake of a liter of coffee or tea per day. This analysis leads one to consider that vinyl chloride and PCBs are over regulated and Chromium (VI) is under regulated, relative to other commonly acceptable hazards. A comparative hazards evaluation might improve regulatory consistency.

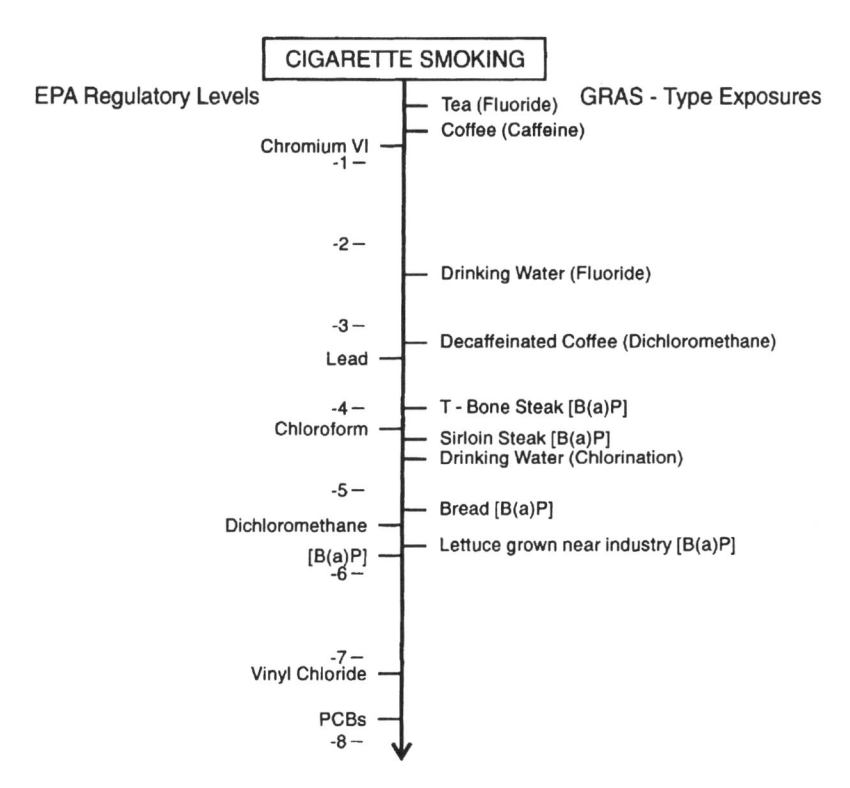

Figure 10.1 Comparison of hazards relative to cigarette smoking (log scale). Environmental hazards and lifestyle exposures are normalized to the reference standard of smoking a pack of cigarettes daily. (From Owen, B.A. and Jones, T.D. 1990. *Reg. Toxicol. Pharmacol.* 11, 132–148. With permission.)

REFERENCES

Barnard, R.C. 1990. Some regulatory definitions of risk: interaction of scientific and legal principles. *Reg. Toxicol. Pharmacol.* 11:201–211.

Burger, E.J. 1980. How citizens think about risks to health. *Risk Analysis.* 8:309–313.

Covello, V.T., Sandman, P., and Slovic, P. 1988. Risk Communication, Risk Statistics and Risk Comparisons, Chemical Manufactureres Association, Washington, D.C.

Covello, V.T. and Allen, F.W. 1988. Seven cardinal rules of risk communication, United States Environmental Protection Agency, Washington, D.C. 1-4.

Covello, V.T. 1989. Communicating right-to-know information on chemicals. *Environ. Sci. Technol.* 23:1444–1449.

Dwyer, J.P. and Ricci, P.F. 1989. Coming to terms with acceptable risks. *Environ. Sci. Technol.* 23:145–146.

Edmond, P. 1975. Participation and the Environment: A strategy for democratizing Canada's environmental protection laws. *Osgood Hall Law Journal* 13:783–837.

Gold, L.S., Slone, T.H., Stern, B.R., Manley, N.B., and Ames, B.N. 1992. Rodent carcinogenisis: Setting priorities. *Science.* 258:261–265.

Grima, A.P. 1986. In: *The Role of Public Participation in the Environmental Impact Process, Environmental Impact Assessment in Canada.* Plewwes, M. and Whitney, J.B.R., Eds., I.E.S., University of Toronto, Ontario, Canada, EE-5.

Grima, A.P. 1985. In: Whitney, J.B.R., and MacLaren, V.W., Eds., *Environmental Impact Assessment: The Canadian Experience,* Institute of Environmental Studies, University of Toronto, Ontario, Canada, 33-51.

Jones, T.D., Walsh, D.J., Watson, A.P., Owen, B.A., Barnthouse, L.W., and Saunders, D.A. 1985. Chemical scoring by a rapid screening of hazard (RASH) method. *Risk Analysis.* 8:99–118.

NRDC vs. EPA 1987. National Resources Defense Council.

Owen, B.A. and Jones,T.D. 1990. Hazard evaluation for complex mixtures: Relative comparisons to improve regulatory consistency. *Reg. Toxicol. Pharmacol.* 11:132–148.

Ramamoorthy, S. and Baddaloo, E. 1991. *Evaluation of Environmental Data for Regulatory and Impact Assessment,* Elsevier Science Publishers, Amsterdam, The Netherlands.

Sandman, P.M. 1986. Explaining environmental risk. United States Environmental Protection Agency (U.S. EPA), Office of Toxic Substances, Washington, D.C. 27 p.

Travis, C.C. and Hattemer-Frey, H.A. 1988. Determining an acceptable level of risk. *Environ. Sci. Technol.* 22:873–876.

Travis, C.C. Richter, S.A., Crouch, E.A.C., Wilson, R., and Klema, E.O. 1987. Cancer risk management — A review of 132 federal regulatory decisions. *Environ. Sci. Technol.* 21:415–420.

U.S. Federal Register 42 (March 1977). 14640–14659.

U.S. National Academy of Sciences (NAS) 1983. Risk assessment in the federal government; managing the process. Committee of the Institutional Means for Risk assessment of Risks to Public Health, Commission on Life Sciences, Washington, D.C.

U.S. Department of Health and Human Services (HSS). 1986. Task Force on Health Risk Assessment, Federal Policy and Practices, Auburn House, Dover, MA.

Wengert, N. 1961. Resource development and public interest — A challenge for research. *Nat. Resources J.* 1:207–233.

Wilson, R. and Crouch, E.A.C, 1987. Risk assessment and comparisons: an introduction. *Science.* 236:267–270.

Wilson, R. 1979. Analyzing the daily risks of life. *Technol. Rev.* 81:40–46.

Glossary of Terms

Acute Exposure — Exposure to a chemical for a duration of 14 days or less.
ACGIH — American Conference of Governmental Industrial Hygienists.
BAT — Best Available Technology.
Bioconcentration Factor (BCF) — The quotient of the concentration of a chemical in aquatic organisms at a specific time, or during a discrete time period of exposure, divided by the concentration in the surrounding water at the same time or during the same period.
BW — Body weight.
CAS — Chemical Abstracts Service.
Cancer Effect Level (CEL) — The lowest dose of a chemical in a study, or group of studies, that produces significant increases in the incidence of cancer or tumours between the exposed population and its appropriate control.
Carcinogen — A chemical capable of inducing cancer.
Ceiling Value — A concentration of a chemical or substance that should not be exceeded, even instantaneously.
Chronic Exposure — Exposure to a chemical for longer than six months for aquatic organisms, 365 days or more for mammalian species, and 104 weeks for rodent carcinogenic bioassay.
DOT/UN/NA/IMCO — Department of Transportation/United Nations/North America/International Maritime Dangerous Goods Code.
Developmental Toxicity — The occurrence of adverse effects on the developing organism that may result from exposure to a chemical prior to conception (either parent), during prenatal development, or postnatally to the time of sexual maturation. Adverse developmental effects may be detected at any point in the life span of the organism.
Embryotoxicity and Fetotoxicity — Any toxic effect on the conceptus as a result of prenatal exposure to a chemical, the distinguishing feature between the two terms is the stage of development during which the insult occurred. The terms, as used here, include malformations and variations, altered growth, and *in utero* death.
EPA — U.S. Environmental Protection Agency.

EPA Health Advisory — An estimate of acceptable drinking water levels for a chemical substance based on health effects information. A health advisory is not a legally enforceable federal standard, but serves as technical guidance to assist federal, state, and local regulatory agencies.

HSDB — Hazardous Substances Data Bank.

IARC — International Agency for Research on Cancer.

Immediately Dangerous to Life or Health (IDLH) — The maximum environmental concentration of a contaminant from which one could escape, within 30 minutes, without any impairing symptoms or irreversible health effects.

Intermediate Exposure — Exposure to a chemical for a duration of 15 to 364 days, as specified in the toxicological profiles.

Immunological Toxicity — The occurrence of adverse effects on the immune system that may result from exposure to environmental contaminants.

In Vitro — Isolated from the living organisms and artificially maintained, as in a test tube.

In Vivo — Occurring within the living system.

ITEFs — International Toxic Equivalency Factors.

Lethal Concentration$_{Lo}$ (LC$_{Lo}$) — The lowest concentration of a chemical in air or water which has been reported to have caused death in humans or other organisms.

Lethal Concentration$_{50}$ (LC$_{50}$) — A calculated concentration of a chemical in air or water to which exposure for a specific length of time is expected to cause death in 50% of defined experimental animal population.

Lethal Dose$_{Lo}$ (LD$_{Lo}$) — The lowest dose of a chemical introduced by a route other than inhalation that is expected to cause death in humans or animals.

Lethal Dose$_{50}$ (LD$_{50}$) — The dose of a chemical introduced by a route other than inhalation to cause death in 50% of the experimental animal population.

Lethal Time$_{50}$ (LT$_{50}$) — A calculated period of time within which a specific concentration of a chemical is expected to cause death in 50% of the experimental animal population.

Lowest-Observed-Adverse-Effect-Level (LOAEL) — The lowest dose of a chemical in a study, or group of studies, that produces statistically or biologically significant increases in frequency or severity of adverse effects between the exposed population and its appropriate control.

MCL — Maximum Contaminant Level.

MCLG — Maximum Contaminant Level Goal.

Malformations — Permanent structural changes that may adversely affect survival, development, or function.

Minimal Risk Level — An estimate of daily exposure to a dose of chemical that is likely to cause **no** appreciable risk of adverse noncancerous effects over a specified duration of exposure.

Mutagen — A substance that causes mutations. A mutation is a change in the genetic material in a body cell. Mutations can lead to birth defects, miscarriages, or cancer.

NCI — National Cancer Institute.

Neurotoxicity — The occurrence of adverse effects on the nervous system following exposure to a chemical.

NIOSH — National Institute for Occupational Safety and Health.

GLOSSARY OF TERMS

No-Observed-Adverse-Effect-Level (NOAEL) — The dose of a chemical at which there are no statistically or biologically significant increases in frequency or severity of adverse effects observed between the exposed population and its appropriate control. Effects may be produced at this dose, but they are not considered to be adverse.
NPDES — National Pollutant Discharge Elimination System.
Octanol-Water Partition Coefficient (K_{ow}) — The equilibrium ratio of the concentrations of a chemical in n-octanol and water, in dilute solution.
ODW — Office of Drinking Water.
OERR — Office of Emergency and Remedial Response.
OGWDW — Office of Groundwater and Drinking Water.
OPP — Office of Pesticides Program.
OSHA — Occupational Safety and Health Administration.
OSW — Office of Solid Waste.
OTS — Office of Toxic Substances.
OWRS — Office of Water Regulations and Standards.
OHMS/TADS — Oil and Hazardous Materials/Technical Assistance Data System.
PCB — Polychlorinated biphenyls.
PCDDs — Polychlorinated dibenzo-*p*-dioxins.
PCDFs — Polychlorinated dibenzofurans.
Permissible Exposure Limit (PEL) — An allowable exposure level in workplace air averaged over an 8-hour shift.
PICs — Products of Incomplete Combustion.
q^* — The upper-bound estimate of the low dose slope of the dose-response curves determined by the multistage procedure. The q^* can be used to calculate an estimate of carcinogenic potency, the incremental excess cancer risk per unit of exposure (usually μg/L for water, mg/kg/day for food, and μg/m^3 for air).
REL — Recommended Exposure Level.
Reference Dose (RfD) — An estimate (with uncertainty spanning perhaps an order of magnitude) of the daily exposure of the human population to a potential hazard that is likely to be without risk of deleterious effects during a lifetime. The RfD is operationally derived from the NOAEL (from animal and human studies) by a consistent application of uncertainty factors that reflect various types of data used to estimate RfDs and an additional modifying factor, which is based on a professional judgement of the entire database on the chemical. The RfDs are not applicable to non threshold effects such as cancer.
Reportable Quantity (RQ) — The quantity of a hazardous substance that is considered reportable under CERCLA. Reportable quantities are (1) 1 pound or greater or (2) for selected substances, an amount established by regulation either under CERCLA or under Section 311 of the U.S. Clean Water Act. Quantities are measured over a 24-hour period.
Reproductive Toxicity — The occurrence of adverse effects on the reproductive systems that may result from exposure to a chemical. The toxicity may be directed to the reproductive organs and/or the related endocrine system. The manifestation of such toxicity may be noted as alterations in sexual behaviour, fertility, pregnancy outcomes, or modifications in other functions that are dependent on the integrity of this system.

RTECS — Registry of Toxic Effects of Chemical Substances.

SNARL — Suggested No Adverse Response Level.

Sorption Coefficient (K_{oc}) — The ratio of the amount of a chemical sorbed per unit weight of organic carbon in the soil or sediment to the concentration of the chemical in solution at equilibrium.

Sorption Ratio (K_d) — The amount of the chemical sorbed by sediment or soil divided by the amount of chemical in the solution phase, which is in equilibrium with the solid phase, at a fixed solid/solution ratio. It is usually expressed as micrograms of chemical sorbed per gram of soil or sediment.

Short-Term Exposure Limit (STEL) — The maximum concentration to which workers can be exposed upto 15 minutes continually. No more than four excursions are allowed per day, and there must be at least 60 minutes interval between exposure periods. The daily TLV-TWA may not be exceeded.

Target Organ Toxicity — This term covers a broad range of adverse effects on target organs or physiological systems (e.g., renal, cardiovascular etc.,) extending from those arising through a single limited exposure to those assumed over a lifetime of exposure to a chemical.

Teratogen — A chemical that causes structural defects that affect the development of an organism.

Threshold Limit Value (TLV) — A concentration of a substance to which most workers can be exposed without adverse effect. The TLV may be expressed as a TWA, as a STEL, or as a CEL (Ceiling Exposure Limit).

Time-Weighted Average (TWA) — An allowable exposure concentration averaged over a normal 8-hour workday or 40-hour workweek.

Toxic Dose (TD_{50}) — A calculated dose of a chemical, introduced by a route other than inhalation, which is expected to cause a specific toxic effect in 50% of a defined experimental animal population.

Uncertainty Factor (UF) — A factor used in operationally deriving the RfD from experimental data. UFs are intended for (1) the variation in sensitivity among the members of the human population, (2) the uncertainty in extrapolating animal data to the case of human, (3) the uncertainty in extrapolating from data obtained in a study that is of less than lifetime exposure, and (4) the uncertainty in using LOAEL data rather than NOAEL data. Usually each of these factors is set equal to 10.

Index

A

Absorption. *See* Sorption
Absorption efficiency of polychlorinated biphenyls, 142
Acceptable level, development of, 350–351
Acceptable risk, development of, 349
ACGIH. *See* Regulations
Acute exposure, 353
Acute toxicity
  chlorinated aliphatic hydrocarbons, 54–60
  chlorinated aromatic hydrocarbons, monocyclic, 110–115
  chlorinated aromatic hydrocarbons, polycyclic, 150–156
  chlorinated phenols, 255–259, 261
  dioxins and furans, 297
  mammalian tests, 326
Adsorption. *See* Sorption
Aerosols, chlorinated aliphatic hydrocarbons from, 33, 40
Agent Orange, 275, 281
Agricultural runoff, chlorinated biocides in, 185–186, 204, 211
Air. *See also* Indoor air
  chlorinated aliphatic hydrocarbons in
    bromoform and chlorodibromomethane, 44–45
    carbon tetrachloride, 44
    chloroethane, 45
    chloroform, 43
    chloromethane, 40–41
    1,2-dichloroethane, 46
    1,1-dichloroethene, 45
    1,2-dichloroethene, 47
    dichloromethane, 42
    1,1-dichloromethane, 45
    hexachloroethane, 52
    1,1,2,2-tetrachloroethane, 50
    tetrachloroethylene, 50, 51
    1,1,1-trichloroethane, 46
    1,1,2-trichloroethane, 47–48
    trichloroethylene, 48
    vinyl chloride, 53
  chlorinated aromatic hydrocarbons, monocyclic in, 96
  chlorinated aromatic hydrocarbons, polycyclic in, 134–135, 150, 155
  chlorinated biocides in
    aldrin and dieldrin, 197
    chlordane, 191–192
    chlorfenvinphos, 211
    chloropyrifos, 208
    DDT, DDD, and DDE, 184–185
    dichlorvos, 209
    endosulfan, 203
    endrin, 202
    heptachlor, 199
    γ-hexachlorocyclohexane, 186, 189
    methoxychlor, 206–207
    mirex, 204
    toxaphene, 200
  chlorinated compounds in, modeling the distribution of, 330
  chlorinated phenols in, 244–245, 246
  dioxins and furans in, 288–289
  regulatory limits for contaminants in. *See* Regulations
Aldrin. *See also* Chlorinated biocides
  environmental residues and exposure routes, 197
  physico-chemical properties, 181
  production and use pattern, 178, 179, 183
  regulations, 216
Algae. *See also* Plants, aquatic
  acute toxicity
    chlorinated aliphatic hydrocarbons, 60–61
    chlorinated aromatic hydrocarbons, monocyclic, 110, 111, 112, 114
  chlorinated aliphatic hydrocarbons in, 40
  chlorinated aromatic hydrocarbons in, 144
  chlorinated biocides in, 187
  growth effects, chlorinated phenols, 260
  polychlorinated biphenyls in, 139, 150
Algal bioassays, 324
Aliphatic hydrocarbons. *See* Chlorinated aliphatic hydrocarbons
Ambient air concentrations. *See* Environmental residues and exposure routes; Regulations
American Conference of Governmental Industrial Hygienists (ACGIH). *See* Regulations
Anthropogenic sources. *See also* Natural sources
  chlorinated aliphatic hydrocarbons, 33, 39–40
  chlorinated aromatic hydrocarbons, monocyclic, 93–94, 106
  chlorinated aromatic hydrocarbons, polycyclic, 132–133
  chlorinated phenols, 240–241
  dioxins and furans, 281–288
Aquatic plants. *See* Plants, aquatic
Aquatic toxicity tests, 324–325
Aroclor products. *See also* Chlorinated aromatic hydrocarbons, polycyclic; Polychlorinated biphenyls
  BCF values, estimated vs. observed, 333, 334
  chemical identification, 128
  composition, 130
  dietary exposure, 146–148
  metabolic degradation of, 143–144
  PCDF contamination in, 283
  physical and chemical properties, 126, 127, 129
  in soil and sediments, 139
  structure and nomenclature, 125
  toxicity. *See* Toxicity profiles, chlorinated aromatic hydrocarbons, polycyclic
  use pattern, 131, 133
Aromatic hydrocarbons, monocyclic, 110. *See also* Chlorinated aromatic hydrocarbons, monocyclic

Atmospheric concentrations. *See* Air; Regulations
Automobile emissions
  chlorinated aliphatic hydrocarbons from, 40
  dioxins and furans from, 286
Avian toxicity test, 325–326

B

BCF. *See* Bioconcentration factor
Benzene
  acceptable risk, development of, 349
  acute toxicity, 111
  environmental residues and exposure routes, 105
Bioaccumulation, 17
  chlorinated biocides, 212
  data, strength and boundary of information, 346
  hexachlorobenzene, 106
  polychlorinated biphenyls, 149
Bioassays
  algal, 324
  strength and boundary of information, 345
Bioavailability, of polychlorinated biphenyls, 155
Bioconcentration
  chlorinated organic compounds, 16–19
  chlorinated phenols, 250, 252
  modeling, 331–334
Bioconcentration factor (BCF), 3
  chlorinated aliphatic hydrocarbons, 28–30
  chlorobenzenes, 106–108
  correlation with normalized organic carbon and octanol-water coefficient, 4, 17–19, 20
  definition, 353
  measurement of, 16
  modeling, estimated vs. observed, 331–332, 333, 334
  polychlorinated biphenyls, 139, 144, 145, 146
  sorption coefficient and, 331
Biomagnification, 322
  chlorobenzenes, 108, 113
  polychlorinated biphenyls, 145–146, 149
Biotransformation. *See also* Metabolic transformation
  aromatic monocyclic compounds, 105
  chlorinated phenols, 250, 252–253
  chloroguaiacols, 257
  hazard identification and, 324
  microbial, 148–149, 252–253
Birds
  acute toxicity, polychlorinated biphenyls, 153
  avian toxicity test, 325–326
  chlorinated aromatic hydrocarbons in, 146, 147
  chlorinated biocides in, 195–196, 206
  chlorobenzenes in, 108
  polychlorinated biphenyls in, 143, 144, 146, 147
Bleached pulp kraft mills. *See* Pulp mills
Breakdown. *See also* Degradation rate; Metabolic transformation; Photolysis
  hydrolysis, 21–22, 323, 324
  ionization, 19–21
  microbial transformation, 148–149, 252–253
Breastfeeding, polychlorinated biphenyls and, 159

Bromodichloromethane
  acute toxicity, 55
  physico-chemical properties, 28
  production and use pattern, 35
Bromoform
  acute toxicity, 55
  environmental residues and exposure routes, 44–45
  regulations, 67
Bromomethane, acute toxicity, 55

C

Capacitors
  chlorinated aromatic hydrocarbons, polycyclic from, 132
  dioxins and furans from, 285–286
Carbon dioxide evolution test methods, 324
Carbon tetrachloride
  acute toxicity, 55
  BCF values, estimated vs. observed, 334
  environmental residues and exposure routes, 44
  half-life in fish tissue, 54
  physico-chemical properties, 28
  production and use pattern, 31, 32, 34
  regulations, 66
  sources in the environment, 33, 39, 40
Carcinogenicity. *See* Acute toxicity; Nonlethal effects
Carcinogenicity hazard identification, 327
Chemical data, strength and boundary of information, 345, 346
Chemical synthesis
  chlorinated aliphatic hydrocarbons from, 33, 40
  chlorinated aromatic hydrocarbons, monocyclic from, 94–95, 106
  dioxins and furans from, 281–282
Children. *See* Populations at high risk
Chlordane. *See also* Chlorinated biocides
  BCF values, estimated vs. observed, 334
  environmental residues and exposure routes, 191–196
  physico-chemical properties, 181
  production and use pattern, 177–178, 179
  regulations, 215–216
Chlordecone
  environmental residues and exposure routes, 205–206
  physico-chemical properties, 182
  production and use pattern, 183–184
Chlorfenvinphos. *See also* Chlorinated biocides
  delayed neurotoxicity of, 213
  environmental residues and exposure routes, 210–211
  physico-chemical properties, 182
  production and use pattern, 184
Chlorinated aliphatic hydrocarbons
  environmental residues and exposure routes, 40–54
  physico-chemical properties, 27–31
  production use and pattern, 31–33
  sources in the environment, 33–40

INDEX

structure, 27
toxicity profile, 54–80
Chlorinated aromatic hydrocarbons, monocyclic
   environmental residues and exposure routes, 96–105
   physico-chemical properties, 94
   production and use pattern, 94–95
   sorption, 15
   sources in the environment, 93–94
   structure, 93
Chlorinated aromatic hydrocarbons, polycyclic. *See also* Polychlorinated biphenyls
   environmental residues and exposure routes, 134–150, 154–161
   physico-chemical properties, 126–127, 129
   production and use pattern, 127–128, 130–132
   regulations, 162–163
   sorption, 15
   sources in the environment, 132–133
   synthesis, 125
   toxicity profile, 150–161
Chlorinated biocides
   dioxins from, 281–282
   environmental residues and exposure routes, 184–211
   physico-chemical properties, 180–181
   production and use pattern, 177–179
   regulations, 213–222
   structure, 177, 212
   toxicity profile, 212–213
Chlorinated organic compounds
   physico-chemical processes, 8–13
   physico-chemical properties, 3–8
   transformation processes, 19–24
   volatilization from soil, 13–19
Chlorinated phenols
   contaminants in, 237, 275, 281–282
   environmental residues and exposure routes, 241–255, 256
   growth effects, 259–260
   metabolic transformation, 101
   physico-chemical properties, 235, 236–237
   production and use pattern, 235, 237–239
   regulations, 262–263
   release of dioxins from, 281–282
   sources in the environment, 239–241
   structure, 235
   toxicity profiles, 255–261
Chlorination of drinking water
   chlorinated aliphatic hydrocarbons from, 33
   dioxins and furans from, 288
Chlorine, percent in Aroclor formulations, 130
Chloroacetechols
   degradation, 255
   environmental residues and exposure routes, 244
   mammalian toxicity, 261
Chlorobenzene. *See also* Chlorinated aromatic hydrocarbons, monocyclic
   acute toxicity, 111
   environmental residues and exposure routes, 97–105

physico-chemical properties, 94
production use and pattern, 94
Chlorobutadienes, sources in the environment, 40
Chlorobutatrienes, sources in the environment, 40
Chlorodibromomethane
   environmental residues and exposure routes, 44–45
   physico-chemical properties, 28
   production and use pattern, 35
   regulations, 67
Chlorodifluoromethane, production and use pattern, 32
Chloroethane
   environmental residues and exposure routes, 45
   physico-chemical properties, 28
   production and use pattern, 32, 35
   regulations, 68
Chlorofluorocarbons, 39, 40
Chloroform
   acute toxicity, 55, 56, 57
   environmental residues and exposure routes, 43–44
   half-life in fish tissue, 54
   hazard comparison, 351
   physico-chemical properties, 28
   production and use pattern, 32, 34
   regulations, 65
   sources in the environment, 33, 39, 40
Chloroguaiacols
   acute toxicity, 256–257, 262
   biotransformation, 257
   degradation, 255
   environmental residues and exposure routes, 244
Chloromethane
   environmental residues and exposure routes, 40–41
   physico-chemical properties, 28
   production and use pattern, 32, 33, 34
   regulations, 63
   sources in the environment, 33
Chlorophenols. *See* Chlorinated phenols
Chloropropadienes, sources in the environment, 40
Chloropropenes, sources in the environment, 40
Chloropyrifos. *See also* Chlorinated biocides
   delayed neurotoxicity of, 213
   environmental residues and exposure routes, 208–209
   physico-chemical properties, 182
   production and use pattern, 184
   regulations, 221
Chloroveratoles, 255, 257
Chromium VI, hazard comparison, 351
Chronic mammalian toxicity test, 326
Cigarette smoke
   chlorinated aliphatic hydrocarbons from, 33, 40
   dioxins and furans from, 280
Clausius-Clapeyron equation, 6
Coal, chlorinated aliphatic hydrocarbons from burning, 33
Combustion. *See also* Wood burning
   cigarette smoke, 33, 40, 280
   coal burning, 33
   incinerators, 132, 133, 241, 284–285

Communication of risk, 345–346, 347
Community groups, 347, 348
Comparison of risk, 347
Cost vs. acceptable level of risk, 350
"Cox Chart," 6

D

2,4-D, 275, 281
Daphnia magna
  acute toxicity
    chlorinated aliphatic hydrocarbons, 60
    chlorinated aromatic hydrocarbons, 144
    chlorinated phenols, 257
    chlorobenzenes, 110
  dioxins and furans in, 297
  growth effects of chlorinated aliphatic hydrocarbons in, 61
Data, strength and boundary of information, 345, 346
DDD. *See* DDT, DDE, and DDD
DDE. *See* DDT, DDE, and DDD
DDT, DDE, and DDD. *See also* Chlorinated biocides
  BCF values, estimated vs. observed, 333, 334
  environmental residues and exposure routes, 184–186, 187–188
  historical background, 177
  malaria cases before and after, 177, 178
  physico-chemical properties, 180, 329
  production and use pattern, 179, 185
  regulations, 213–214
Decision-making, 343–351
Degradation. *See also* Degradation rate; Metabolic transformation; Photolysis
  chlorinated phenols, 253
  chloroguaiacols, 255
  hazard identification and, 323, 324
  hydrolysis, 21–22, 323–324
  ionization, 19–21
  microbial transformation, 148–149, 252–253
  polychlorinated dibenzo-p-dioxins, 275
Degradation rate, 23–24
  chlorinated aromatic hydrocarbons, monocyclic, 104
  polychlorinated biphenyls, 133
Dermal exposure, chlorfenvinphos, 211. *See also* Environmental residues and exposure routes; Regulations
Desorption, 14
Developmental toxicity, 353. *See also* Nonlethal effects
Dibromochloroethane, acute toxicity, 54
1,2-Dibromo-3,chloropropane, physico-chemical properties, 30
1,2-Dichlorobenzene
  acute toxicity, 110, 111, 112, 114
  bioconcentration factors, 107
  environmental residues and exposure routes, 96–99, 101–105
  physico-chemical properties, 94
  production use and pattern, 94, 95

1,3-Dichlorobenzene
  acute toxicity, 111, 112, 113, 115
  bioconcentration factors, 107
  environmental residues and exposure routes, 96, 97, 98, 99, 101, 103, 104
  physico-chemical properties, 94
  production use and pattern, 95
1,4-Dichlorobenzene
  acute toxicity, 110, 111, 112, 113, 114, 115
  bioconcentration factors, 107
  environmental residues and exposure routes, 96–99, 101–104, 109
  physico-chemical properties, 94
  production use and pattern, 94, 95
Dichlorobenzenes
  acute toxicity, 110
  BCF values, estimated vs. observed, 334
  environmental residues and exposure routes, 103
  physico-chemical properties, 329
  production use and pattern, 94
  regulations, 116
1,1-Dichloroethane
  environmental residues and exposure routes, 45–46
  mammalian toxicity, 56, 62
  physico-chemical properties, 29
  production and use pattern, 35
  regulations, 68
1,2-Dichloroethane
  acute toxicity, 55, 57, 58, 59, 60
  environmental residues and exposure routes, 46
  growth effects, 60, 61
  half-life in fish tissue, 54
  mammalian toxicity, 56, 62
  physico-chemical properties, 29
  production and use pattern, 31, 32, 35
  regulations, 69
  reproductive effects, 61, 62
  sources in the environment, 39, 40
Dichloroethanes, acute toxicity, 56
1,1-Dichloroethene
  environmental residues and exposure routes, 45
  physico-chemical properties, 29
  production and use pattern, 32, 36
  regulations, 70
1,2-Dichloroethene
  environmental residues and exposure routes, 47
  production and use pattern, 32, 36
  regulations, 71
1,2-Dichloroethene (cis), physico-chemical properties, 29
1,2-Dichloroethene (trans), physico-chemical properties, 29
1,1-Dichloroethylene
  acute toxicity, 55
  mammalian toxicity, 62
1,2-Dichloroethylene, acute toxicity, 55
Dichloromethane
  acute toxicity, 55, 56
  environmental residues and exposure routes, 42–43
  hazard comparison, 351

INDEX

physico-chemical properties, 28
production and use pattern, 32, 34
regulations, 64
sources in the environment, 33, 39
Dichlorophenols. *See also* Chlorinated phenols
  acute toxicity, 256–257
  embryotoxicity, 257, 259
  environmental residues and exposure routes, 242, 243, 244, 245, 246, 247
  growth effects, 260
  physico-chemical properties, 236
  production and use pattern, 238
2,4-Dichlorophenoxyacetic acid (2,4-D), 275, 281
1,1-Dichloropropane, acute toxicity, 55
1,2-Dichloropropane
  acute toxicity, 55
  physico-chemical properties, 30
1,3-Dichloropropane, acute toxicity, 55
1,3-Dichloropropene, acute toxicity, 55
Dichlorvos. *See also* Chlorinated biocides
  delayed neurotoxicity of, 213
  environmental residues and exposure routes, 209–210
  physico-chemical properties, 182
  production and use pattern, 184
  regulations, 222
Dieldrin. *See also* Chlorinated biocides
  environmental residues and exposure routes, 197–199
  physico-chemical properties, 181
  regulations, 217
Dietary exposure
  breastfeeding, 159
  chlorinated aliphatic hydrocarbons
    chloroform, 43–44
    hexachloroethane, 53
    tetrachloroethylene, 52
    1,1,1-trichloroethane, 46
    trichloroethylene, 49
    vinyl chloride, 54
  chlorinated biocides
    chlordane, 193
    chlordecone, 206
    chlorfenvinphos, 211
    chloropyrifos, 209
    DDT, DDE, and DDD, 187
    dichlorvos, 210
    dieldrin, 198–199
    endosulfan, 204
    endrin, 203
    γ-hexachlorocyclohexane, 190–191
    methoxychlor, 208
    mirex, 206
    toxaphene, 201
  chlorinated phenols, 249
  hexachlorobenzene, 109–110
  polychlorinated biphenyls, 146–148, 154–155, 157
  tobacco products, 33, 40, 204, 280
Dioxins and furans. *See also* Polychlorinated dibenzo-*p*-dioxins; Polychlorinated dibenzo-*p*-furans
  chlorination of drinking water and, 288

environmental residues and exposure routes, 288–297
physico-chemical properties, 277–279
production and use pattern, 280
from pulp mills, 275, 283–284, 287, 298–301
regulations, 307–309
sources in the environment, 280–288
structure, 275–277
teratogenic effects, protection by polychlorinated biphenyls, 156
toxicity profile, 297–307
Disappearance, rate of
  chlorinated phenols, 253
  organic compounds, 22
Disposal testing, 346
Dose-dependent availability, hexachlorobenzene, 113
Dose-reponse evaluation, 344
Drinking water. *See* Water
Dry cleaning
  chlorinated aliphatic hydrocarbons from, 40
  dioxins and furans from, 287
Dry deposition rate, chlorobenzenes, 96

E

Ecological risk assessment, 343
Effluent
  chlorinated aliphatic hydrocarbons from, 40
  chlorinated phenols in, 241, 242, 244, 245, 246, 248–249
  dioxins and furans in, 281–282, 286, 287, 298–301
Electronic industry, chlorinated aliphatic hydrocarbons from, 33
Elimination half-life
  chlorinated aliphatic hydrocarbons in fish, 54
  chlorinated aromatic hydrocarbons in aquatic organisms, 103
  polychlorinated biphenyls, 143, 158–159
Embryos. *See* Embryotoxicity; Populations at high risk
Embryotoxicity
  chlorinated aromatic hydrocarbons, monocyclic, 110, 113
  chlorinated phenols, 257–259
  definition, 353
  mammalian toxicity tests, 327
Endosulfan. *See also* Chlorinated biocides
  environmental residues and exposure routes, 203–204
  physico-chemical properties, 182
  production and use pattern, 178, 183
  regulations, 219
Endrin. *See also* Chlorinated biocides
  environmental residues and exposure routes, 202–203
  physico-chemical properties, 182
  production and use pattern, 178
  regulations, 218
Environmental fate
  chlorinated aliphatic hydrocarbons, 39
  hazard identification and, 323

influence on environmental pathways, 337
modeling, 328–335
Environmentalists, 347, 348
Environmental recycling, polychlorinated biphenyls, 133
Environmental residues and exposure routes
chlorinated aliphatic hydrocarbons, 40–54
chlorinated aromatic hydrocarbons, monocyclic, 96–105, 106–110
chlorinated aromatic hydrocarbons, polycyclic, 134–150, 154–155
key studies, 156–161
chlorinated biocides, 184–211
chlorinated phenols, 241–255, 256
dioxins and furans, 288–297
in exposure assessment, 335–337
EPA. See Regulations
Epidemiology, strength and boundary of information, 345
Explosivity, 322
Exposure
hazard identification and, 321–322
multiple pathway, 322, 337
Exposure assessment, 335–337, 344
Exposure routes. See Environmental residues and exposure routes

F

Fate processes. See Transformation processes
Fetal toxicity, hexachlorobenzene, 113
Fetuses. See Fetal toxicity; Populations at high risk
Field data, strength and boundary of information, 345
Fish. See also Dietary exposure
acute toxicity
chlorinated aliphatic hydrocarbons, 55–56, 59–60
chlorinated aromatic hydrocarbons, monocyclic, 110, 111–113, 114–115
chlorinated phenols, 257
dioxins and furans, 297–305
polychlorinated biphenyls, 150–151, 152–153
bromobenzenes in, 103
chlorinated aromatic hydrocarbons in, 142–146, 147, 150
chlorinated biocides in
chlordane, 194
DDT, DDE, and DDD, 188
dieldrin, 198
endrin, 203
γ-hexachlorocyclohexane, 190
mirex, 206
toxaphene, 201, 202
chlorinated phenols in, 247, 249–252
chlorobenzenes in, 98, 100–101, 102–103, 107, 109
dioxins and furans in, 297–305
embryotoxicity, chlorophenols, 258–259
growth effects
chlorinated aliphatic hydrocarbons, 61
chlorinated phenols, 260

half-life of chlorinated aliphatic hydrocarbons in, 54
hexachlorobenzene in, 101, 104
polychlorinated biphenyls in, 142–143, 147, 150
reproductive effects, chlorinated aliphatic hydrocarbons in, 62
trichloroethylene in, 49
Flammability, 322
Flow-through test, 325
Fly ash. See Incinerators
Food. See Dietary exposure
Food chain. See Bioconcentration factor; Biomagnification
Foreign regulations. See International Workplace Limits
Forest fires. See Wood burning
Fugacity modeling
chlorinated aromatic hydrocarbons, polycyclic, 149–150
chlorinated phenols, 255, 256
Fundamentals of Toxicology, 325
Fungi
chlorinated phenols and, 239, 252, 253
as source of chloromethane, 33
Fungicides, hexachlorobenzene exposure from, 106
Furans. See Dioxins and furans; Polychlorinated dibenzo-p-furans

G

Generally Recognized as Safe (GRAS), 351
Glossary of terms, 353–356
Groundwater. See Water
Growth effects
chlorinated aliphatic hydrocarbons, 60–61
chlorinated aromatic hydrocarbons, monocyclic, 114–115
chlorinated phenols, 259–260
dioxins and furans, 305
mammalian toxicity tests, 326
Guidelines. See Regulations

H

Half-life. See also Elimination half-life; Kinetic half-life; Volatilization half-life
of chemicals under degradation, 24
in crops, chlorinated biocides, 198
in soil, chlorinated phenols, 253
Hazard assessment, 343, 350–351. See also Risk assessment
Hazard identification, 321–324, 344
HCE. See Hexachloroethane
HCH. See Hexachlorocyclohexane
Henry's law constant, 3
chlorinated aliphatic hydrocarbons, 28–30
chlorinated aromatic hydrocarbons, polycyclic, 129
chlorinated biocides, 180–182
furans, 279
volatilization and, 9, 10, 13
volatilization rate constant and, 328

INDEX

Hepatic conditions. *See* Nonlethal effects; Populations at high risk
Heptachlor. *See also* Chlorinated biocides
   BCF values, estimated vs. observed, 334
   environmental residues and exposure routes, 199–202
   physico-chemical properties, 181
   production and use pattern, 177–178, 183
Heptachlor epoxide
   physico-chemical properties, 181
   in water, 199
Heptachlorinated dibenzo-*p*-furans. *See* Polychlorinated dibenzo-*p*-furans
α-Heptachlorocyclohexane, 180
β-Heptachlorocyclohexane, 180
γ-Heptachlorocyclohexane. *See also* Chlorinated biocides
   environmental residues and exposure routes, 186
   physico-chemical properties, 180
Herbicides. *See also* Chlorinated biocides
   hexachlorobenzene exposure from, 106
   release of dioxins from, 281
Hexachlorinated dibenzo-*p*-furans (PCDF). *See* Polychlorinated dibenzo-*p*-furans
Hexachlorobenzene, 106–110
   acute toxicity, 110, 111, 112, 113, 114, 115
   bioconcentration factors, 107, 334
   dietary exposure, 109–110
   environmental residues and exposure routes, 97, 98, 100, 102, 103, 104, 105
   physico-chemical properties, 94, 329
   production use and pattern, 94, 95, 106
   regulations, 116
   sources in the environment, 106
Hexachlorocyclohexane (HCH). *See also* Chlorinated biocides
   BCF values, estimated vs. observed, 333, 334
   physico-chemical properties, 180
   production and use pattern, 177, 179
   regulations, 214–215
Hexachloroethane (HCE)
   acute toxicity, 55, 56, 57, 58, 59, 60
   environmental residues and exposure routes, 52–53
   growth effects, 60, 61
   half-life in fish tissue, 54
   physico-chemical properties, 30
   production and use pattern, 38
   regulations, 78
Hexachlorohexatrienes, sources in the environment, 40
Hexachloromethane, acute toxicity, 56
High risk populations, 158–160, 344
Human body burden, hexachlorobenzene, 109
Human exposure. *See* Dietary exposure; Indoor air; Mammals; Regulations
Human health risk assessment, 332, 343–344, 348
Hydrolysis
   chlorinated organic compounds, 21–22
   hazard identification and, 323–324

Hydrophilicity of polar groups, solubility in water and, 5
Hypersensitive individuals. *See* Populations at high risk

I

IARC. *See* Regulations
Immediately dangerous to life or health (IDLH), 354. *See also* Regulations
Immunocompromised individuals. *See* Populations at high risk
Immunological toxicology, 354
Incinerators
   chlorinated aromatic hydrocarbons, polycyclic from, 132, 133
   chlorinated phenols from, 241
   dioxins and furans from, 284–285
Indoor air
   chlorinated biocides in
      chlordane, 191–192
      chloropyrifos, 208, 209
      dichlorvos, 209
      dieldrin, 197
      methoxychlor, 207
   dioxins and furans in, 282
   polychlorinated biphenyls in, 155, 160
   polychlorinated dibenzofurans in, 282
Industrial processes
   chlorinated aliphatic hydrocarbons from, 33, 39
   chlorinated aromatic hydrocarbons, polycyclic from, 132
   dioxins and furans from, 286–287
   wastewater from. *See* Effluent
Industrial products. *See* Production and use pattern
Ingestion. *See* Dietary exposure; Environmental residues and exposure routes; Regulations
Inhalation, volume of air breathed per day, 135. *See also* Air; Environmental residues and exposure routes; Indoor air; Regulations
Insecticides, 238. *See also* Chlorinated biocides
Insects, chlordane in, 196
Interest groups, 347, 348
International Agency for Reseach on Cancer (IARC). *See* Regulations
International Workplace Limits
   aldrin, 216
   carbon tetrachloride, 66
   chlordane, 215
   chlorinated aromatic hydrocarbons, monocyclic, 115, 116
   chloroform, 65
   chloropyrifos, 221
   dichlorvos, 222
   dieldrin, 217
   endosulfan, 219
   endrin, 218
   hexachlorocyclohexane, 214
   methoxychlor, 220
   tetrachloroethylene, 76
   1,1,1-trichloroethane, 72
   vinyl chloride, 79

Invertebrates
  acute toxicity
    chlorinated phenols, 257
    chlorobenzenes, 110
    dioxins and furans in, 297, 303, 304
    embryotoxicity, chlorinated phenols, 257–258
    reproductive effects, chlorinated aliphatic hydrocarbons, 61
Invertebrates, freshwater
  acute toxicity
    chlorinated aliphatic hydrocarbons, 57
    polychlorinated biphenyls, 150, 151
  chlorinated aromatic hydrocarbons in, 144
  polychlorinated biphenyls in, 139, 144
Invertebrates, marine
  acute toxicity
    chlorinated aliphatic hydrocarbons, 58
    polychlorinated biphenyls, 151
    chlorinated aromatic hydrocarbons in, 139, 141, 144, 145
    chlorinated biocides in, 187
    polychlorinated biphenyls in, 139, 141, 144, 145
Ionization, of chlorinated organic compounds, 19–21

K

Kinetic half-life, of chlorinated organic compounds, 21–22

L

Laboratory data, strength and boundary of information, 345
Landfills
  chlorinated aliphatic hydrocarbons from, 40
  chlorinated aromatic hydrocarbons, polycyclic from, 132, 133
  dioxins and furans from, 283
LC_{50}, See Acute toxicity
LD_{50}, See Acute toxicity
Leo's fragment constant method, 7–8
Lindane. See Hexachlorocyclohexane
Lowest Observed Adverse Effects Level (LOAEL)
  chlorinated aromatic hydrocarbons, polycyclic, 156–157
  definition, 354
Lowest Observed Effect Level (LOEL)
  chlorinated aliphatic hydrocarbons, 56, 61, 62
  chlorinated phenols, 257, 259, 260

M

Malaria, 177, 178
Mammalian toxicity tests, 326–327
Mammals
  acute toxicity
    chlorinated aliphatic hydrocarbons, 56, 61–62
    chlorinated aromatic hydrocarbons, monocyclic, 113
    chlorinated phenols, 261
    dioxins and furans, 305
    polychlorinated biphenyls, 154–156, 157
  chlorinated aromatic hydrocarbons in, 147
  chlorinated biocides in, 187, 190
  exposure through breastfeeding, 159
  polychlorinated biphenyls in, 143, 147
Manufacturing processes
  chlorinated aromatic hydrocarbons, monocyclic from, 94–95, 106
  chlorinated aromatic hydrocarbons, polycyclic from, 132
  dioxins and furans from, 286
Market Basket Surveys. See Dietary exposure
Mass transfer
  phase resistances to, 10
  streams, 11
Media, risk communication and the, 345–346
Metabolic transformation. See also Biotransformation
  aromatic monocyclic compounds, 105
  chlorinated organic compounds, 23–24
  chlorinated phenols, 253
  chlorobenzenes, 97
  chlorophenols, 101
  hazard identification and, 324
  microbial, 148–149, 252–253
Metal refining process, dioxins and furans from, 286
Methoxychlor. See also Chlorinated biocides
  BCF values, estimated vs. observed, 334
  environmental residues and exposure routes, 206–208
  physico-chemical properties, 182
  production and use pattern, 178
  regulations, 220
Methyl chloride, sources in the environment, 39
Methylene chloride, sources in the environment, 39
Methylene dichloride. See Dichloromethane
Methylmercury, 1
Microbial transformation
  chlorinated aromatic hydrocarbons, 148–149
  chlorinated phenols, 252–253
Minamata disease, 1
Minimal risk level, 354
Mink. See Mammals
Mirex. See also Chlorinated biocides
  environmental residues and exposure routes, 204–206
  physico-chemical properties, 182
  production and use pattern, 178, 183–184
  regulations, 220
Modeling
  fate processes, 328–335
  fugacity, 149–150, 255, 256
  photolysis, 332, 335
  sorption, 330–331
  two-film gas transfer, 10
  volatilization, 328–330
  water solubility, 331
Monitoring
  exposure assessment, 335–337
  hazard identification, 321–324
  predictive capability of fate processes, 328–335

INDEX

ranking protocols, 337–340
toxicity testing, 324–327
Monochlorobenzenes
 environmental residues and exposure routes, 97
 production use and pattern, 94, 95
 regulations, 115
Monochlorobiphenyls, 126, 129
Monochlorophenols
 acute toxicity, 256–257
 embryotoxicity, 259
 growth effects, 259
 physico-chemical properties, 236
Monocyclic aromatic hydrocarbons, 61. *See also* Chlorinated aromatic hydrocarbons, monocyclic
Monsanto. *See* Aroclor products
Montreal Protocol, 40, 62
Multiple pathway exposure, 322, 337
Municipal wastewater. *See* Effluent; Sludge
Mutagenicity hazard identification, 327

N

National Institute of Occupational Safety and Health (NIOSH). *See* Regulations
Natural sources. *See also* Anthropogenic sources
 chlorinated aliphatic hydrocarbons, 33, 39, 40
 chlorinated aromatic hydrocarbons, monocyclic, 95
 chlorinated aromatic hydrocarbons, polycyclic, 132
 chlorinated phenols, 239–240
 dioxins and furans, 280–281
Neonates. *See* Populations at high risk
Neurotoxicity, 213, 354
NIOSH. *See* Regulations
NOEC. *See* No Observed Adverse Effect Level
NOEL. *See* No Observed Adverse Effect Level
Nonlethal effects
 chlorinated biocides, 212
 dioxins and furans, 301, 305–306
 mammalian toxicity tests, 326–327
 polychlorinated biphenyls, 156, 157, 159, 160
Nonpoint sources, polychlorinated biphenyls, 133
No Observed Adverse Effect Level (NOAEL;NOEC;NOEL)
 chlorinated aliphatic hydrocarbons, 56, 61, 62
 chlorinated aromatic hydrocarbons, monocyclic, 110
 chlorinated aromatic hydrocarbons, polycyclic, 156–157
 chlorinated phenols, 257, 259, 260
 definition, 355
 dioxins and furans, 297
 tests to identify, 326
Normalized organic carbon, 4
 chlorinated aliphatic hydrocarbons and, 28–30
 correlation with bioconcentration factor, 17

O

Occupational exposure, 158–160. *See also* Indoor air; Regulations

Occupational Safety and Health Administration (OSHA). *See* Regulations
OCDD. *See* Dioxins and furans; Polychlorinated dibenzo-p-dioxins
OCDF. *See* Dioxins and furans; Polychlorinated dibenzo-p-furans
Oceanic sources, chlorinated aliphatic hydrocarbons, 33, 40
Octachlorodibenzo-p-dioxin (OCDD). *See* Dioxins and furans; Polychlorinated dibenzo-p-dioxins
Octachlorodibenzo-p-furan (OCDF). *See* Dioxins and furans; Polychlorinated dibenzo-p-furans
Octanol-water partition coefficient, 4–5
 chlorinated aliphatic hydrocarbons, 28–30
 chlorinated organic compounds, 7–8
 chlorinated phenols, 236–237
 correlation with bioconcentration factor, 17–19, 20
 definition, 355
 sorption coefficient and, 331
 water solubility and, 331
Oil drilling fluids, chlorinated phenols from, 240–241
Oral exposure. *See* Air; Breastfeeding; Dietary exposure; Environmental residues and exposure; Indoor air
Ortho-chlorophenol, 238
OSHA. *See* Regulations
Ozone-depletion, 40

P

PCB. *See* Polychlorinated biphenyls
PCDD. *See* Dioxins and furans; Polychlorinated dibenzo-p-dioxins
PCDF. *See* Dioxins and furans; Polychlorinated dibenzo-p-furans
PEL (permissable exposure level), 355
Pentachlorobenzene
 acute toxicity, 110, 111
 environmental residues and exposure routes, 103
 physico-chemical properties, 94
 production use and pattern, 95
Pentachlorobiphenyl, 329
Pentachloroethane
 acute toxicity, 55, 60
 growth effects, 60
 half-life in fish tissue, 54
Pentachlorophenol. *See also* Chlorinated phenols
 acute toxicity, 256–258
 BCF values, estimated vs. observed, 334
 dioxin and furan contaminants in, 281–282
 embryotoxicity, 259
 environmental residues and exposure routes, 242, 243, 244, 245, 246, 248–255
 growth effects, 260
 physico-chemical properties, 23
 production and use pattern, 237, 238–239
 regulations, 262
Perchloroethylene, 329

Permissable exposure level (PEL), 355. *See also*
  Regulations
Pesticides. *See also* Chlorinated biocides
  BCF values, estimated vs. observed, 333, 334
  chlorophenols as, 241, 248
  dioxins from, 281–282
  hexachlorobenzene exposure from, 106
  sorption, 14–15
  synergistic influence with PCBs, 131
  toxicity, chlorine and, 212–213
Petrochemical drilling fluids, chlorinated phenols
  from, 240–241
Phase resistance to mass transfer, 10
Phellinus pomaceus, 33
Phenols. *See* Chlorinated phenols
Photodegradation. *See* Photolysis
Photolysis
  chlorinated organic compounds, 22–23
  chlorinated phenols, 254–255
  chlorobenzenes, 96, 101, 103
  DDT, DDE, and DDD, 184
  indirect (disappearance rate), 22, 253
  modeling, 332, 335
Physico-chemical processes, 8–13. *See also*
  Bioconcentration; Sorption; Volatilization
Physico-chemical properties. *See also* Octanol-
  water partition coefficient; Vapor pressure;
  Water solubility
  chlorinated aliphatic hydrocarbons, 27–31
  chlorinated aromatic hydrocarbons,
    monocyclic, 94
  chlorinated aromatic hydrocarbons, polycyclic,
    126–127, 129
  chlorinated biocides, 180–181
  chlorinated organic compounds, 3–8
  chlorinated phenols, 235, 236–237
  data, strength and boundary of information, 345,
    346
  dioxins and furans, 277–279
  hazard identification and, 322, 323
Plants, aquatic. *See also* Algae
  acute toxicity
    chlorinated phenols, 257
    chlorobenzenes, 110
  chlorinated aromatic hydrocarbons in, 139
  polychlorinated biphenyls in, 139, 141
Plants, terrestrial, chlorinated phenol uptake by,
  253
Polychlorinated biphenyls (PCB). *See also*
  Chlorinated aromatic hydrocarbons,
  polycyclic
  acute toxicity, 150–156
  in air, 134–135, 150
  in aquatic plants, 139, 141
  BCF values, estimated vs. observed, 333, 334
  biomagnification, 145–146, 149
  in birds, 143, 144, 146, 147
  dietary exposure, 146–148, 157
  environmental residues and exposure routes,
    134–150, 154–161
  in fish, 140, 142–143, 147
  formation, 126
  hazard comparison, 351

in invertebrates, 139, 141, 144, 145
in mammals, 143, 147
nonlethal effects, 156, 157, 159
occupational exposure, 158–160
PCDF as contaminant in, 280, 283
physico-chemical properties, 126–127, 129
precipitation samples, 136, 138
production and use pattern, 127–128,
  130–132
regulations, 158–160, 162–163
in soil and sediments, 136, 138–139, 140–141,
  150
solubility in water, 5
sources in the environment, 132–133
synthesis, 125
toxicity profile, 150–161
transformers and capacitors with, 285–286
in water, 135–136, 137, 150
Polychlorinated dibenzo-*p*-dioxins (PCDD)
  in air, 288
  as contaminant in chlorophenols, 237, 275,
    281–282
  degradation, 275
  in fish, 297–298, 300–301, 305
  isomers, 276
  in soil and sediments, 293–296
  sources in the environment, 280–288
  in water, 289, 293
Polychlorinated dibenzo-*p*-furans (PCDF),
  125–126
  in air, 289, 290–293
  as contaminant in chlorophenols, 237, 275,
    281–282
  as contaminant in polychlorinated biphenyls,
    280
  in fish, 297, 298–299, 300–304
  formation, 126
  in indoor air, 282
  in polychlorinated biphenyls, 280, 283
  in soil and sediments, 294–295
  sources in the environment, 280–288
  toxicity, 154
  in water, 289, 293
Polycyclic aromatic hydrocarbons. *See* Chlorinated
  aromatic hydrocarbons, polycyclic
Populations at high risk
  polychlorinated biphenyls and, 158–160
  risk assessment and, 344
Postnatal toxicity, hexachlorobenzene, 113
Precipitation samples
  chlorinated aromatic hydrocarbons in, 138
  chlorinated biocides in
    chlordane, 192
    chloropyrifos, 208
    DDT, DDE, and DDD, 186
    dieldrin, 197
    endosulfan, 203
    α-hexachlorocyclohexane, 189
    mirex, 204
    toxaphene, 200
  chlorinated phenols in, 242, 243, 244–245
  polychlorinated biphenyls in, 136, 138
Predictive capability of fate process, 328–335

INDEX

Production and use pattern
 chlorinated aliphatic hydrocarbons, 31–33
 chlorinated aromatic hydrocarbons, monocyclic, 94–95
 chlorinated aromatic hydrocarbons, polycyclic, 127–128, 130–132
 chlorinated biocides, 177–179
 dioxins and furans, 280
Public awareness, role of risk communication in, 345–346, 347
Public participation
 risk communication and, 345–346, 347
 risk management and, 348–349
Pulp mills
 chlorophenols from, 241, 242, 244, 245–246
 dioxins and furans from, 275, 281, 283–284, 287, 298–301

Q

QAQPS. *See* Regulations
Quayle's parachor, 5

R

Radiation, solar, 22–23. *See also* Photolysis
Rain. *See* Precipitation samples
Ranking protocols, 337–340
Rapid Screening of Hazard (RASH), 350
Rats. *See* Mammals
Recirculation test, 325
Reference dose (RfD), 355. *See also* Acute toxicity
Regulations, 2
 acceptable level, development of, 350–351
 acceptable risk, development of, 349
 chlorinated aliphatic hydrocarbons, 63–80
  International Workplace Limits, 65, 66, 72, 76, 79
 chlorinated aromatic hydrocarbons, monocyclic, 115–116
  International Workplace Limits, 115, 116
 chlorinated aromatic hydrocarbons, polycyclic, 158–160, 162–163
 chlorinated biocides, 213–222
  International Workplace Limits, 214, 215, 216, 217, 218, 219, 221, 222
 chlorinated phenols, 262–263
 decision-making process, 343–351
 dioxins and furans, 307–309
 exposure assessment and, 335
 prioritization for regulatory assessment, 321–340
Renewal test, 325
Reportable quantity (RQ), 355. *See also* Regulations
Reproductive toxicity
 chlorinated aliphatic hydrocarbons, 61, 62
 chlorinated aromatic hydrocarbons, monocyclic, 110
 definition, 355
 dioxins and furans, 305
 mutagenicity hazard identification, 327
 polychlorinated biphenyls, 150, 156

Reptiles, chlorinated biocides in, 202, 206
Risk
 acceptable, 349, 350–351
 definition, 343
Risk assessment, 322, 343–344, 348
Risk characterization, 344
Risk communication, 345–346, 347
Risk comparison, 347, 350–351
Risk management, 348–349
Risk perception, 346
Runoff
 agricultural, 185–186, 204, 211
 urban, 186, 192, 200, 202

S

Scoring systems to prioritize environmental contaminants, 337–340
Sediments. *See* Soil and sediments
Settling particles. *See* Soil and sediments
Sewage sludge, 247, 285, 287–288
Shellfish, chlorinated aliphatic hydrocarbons in, 61, 62
Short-term exposure limit (STEL), 356. *See also* Regulations
Sick building syndrome. *See* Indoor air
Sludge, 247, 285, 287–288
Smoke. *See also* Wood burning
 cigarette, 33, 40, 204, 280
 coal, 33
Snow. *See* Precipitation samples
Soil and sediments. *See also* Sorption; Volatilization, from soil
 chlorinated aliphatic hydrocarbons in, 41, 42, 43, 45, 47, 52
 chlorinated aromatic hydrocarbons, monocyclic in, 97, 104–105
 chlorinated aromatic hydrocarbons, polycyclic in, 136, 138–139, 140–141, 145, 150
 chlorinated biocides in
  aldrin and dieldrin, 198
  chlordane, 192
  chlordecone, 205
  chlorfenvinphos, 211
  chlorpyrifos, 209
  DDT, DDD, and DDE, 185–186, 187
  dichlorvos, 210
  endosulfan, 204
  endrin, 202–203
  heptachlor, 199
  γ-hexachlorocyclohexane, 188, 190
  methoxychlor, 207
  mirex, 205
  toxaphene, 200–201
 chlorinated compounds in, modeling distribution of, 330
 chlorinated phenols in, 240, 249, 256
 dioxins and furans in, 280–281, 287, 293–296
 regulatory limits for contaminants in. *See* Regulations
 volatilization from, 13–19
Solar radiation, 22–23. *See also* Photolysis
Solubility. *See* Water solubility

Solvents, chlorinated aliphatic hydrocarbons from, 33, 40
Solvent-water partition coefficient, 8
Sorption, 14–16
  chlorinated phenols, 250, 252
  chlorobenzenes, 98, 104–105, 109
  modeling, 330–331
Sorption coefficient, 14, 331
  correlation to organic carbon and water solubility, 4
  definition, 356
  estimation from other related parameters, 15
Sorption ratio, 356
Sources in the environment
  chlorinated aliphatic hydrocarbons, 33–40
  chlorinated aromatic hydrocarbons, monocyclic, 93–94, 95, 106
  chlorinated aromatic hydrocarbons, polycyclic, 132–133
  chlorinated phenols, 239–241
  dioxins and furans, 280–288
  nonpoint, 133
Static test, 325
STEL (short-term exposure limit), 356. *See also* Regulations
Stormwater runoff, chlorinated biocides in, 186, 192, 200, 202
Structure-activity relationships, octanol-water partition coefficient and, 7
Substrate utilization, rate of, 23
Surface water. *See* Water
Suspended sediments. *See* Soil and sediments

T

2,4,5-T, 275, 281
Target organ toxicity, 356
TCDD. *See* Dioxins and furans; Polychlorinated dibenzo-*p*-dioxins
TCDF. *See* Dioxins and furans; Polychlorinated dibenzo-*p*-furans
TCE. *See* Trichloroethylene
TEF. *See* Toxic Equivalency Factors
TEQ. *See* Toxic Equivalency Factors
Terrestrial plants, chlorinated phenol uptake by, 253
Tests, toxicity, 324–327
1,2,4,5-Tetrabromobenzene, environmental residues and exposure routes, 103
1,2,3,4-Tetrachlorobenzene
  environmental residues and exposure routes, 103
  physico-chemical properties, 94
  production use and pattern, 95
1,2,3,5-Tetrachlorobenzene
  acute toxicity, 111
  environmental residues and exposure routes, 103
  physico-chemical properties, 94
1,2,4,5-Tetrachlorobenzene
  acute toxicity, 110, 111
  environmental residues and exposure routes, 103, 105
  physico-chemical properties, 94
  production use and pattern, 95
Tetrachlorobiphenyl, 329
Tetrachlorodibenzo-*p*-dioxins (TCDD). *See* Dioxins and furans; Polychlorinated dibenzo-*p*-dioxins
Tetrachlorodibenzo-*p*-furans (TCDF). *See* Dioxins and furans; Polychlorinated dibenzo-*p*-furans
1,1,1-Tetrachloroethane, production and use pattern, 31, 32
1,1,1,2-Tetrachloroethane, acute toxicity, 55, 60
1,1,2,2-Tetrachloroethane
  acute toxicity, 55, 59, 60
  environmental residues and exposure routes, 50
  growth effects, 60
  half-life in fish tissue, 54
  physico-chemical properties, 30
  production and use pattern, 37
  regulations, 74
  reproductive effects, 61
Tetrachloroethylene
  acute toxicity, 55, 56, 57, 58, 59, 60
  environmental residues and exposure routes, 50–52
  growth effects, 60, 61, 62
  half-life in fish tissue, 54
  mammalian toxicity, 62
  physico-chemical properties, 30
  production and use pattern, 31, 32, 37
  regulations, 76–77
  reproductive effects, 61
  sources in the environment, 39, 40
Tetrachlorophenols. *See also* Chlorinated phenols
  acute toxicity, 256–257
  embryotoxicity, 259
  physico-chemical properties, 236, 237
  production and use pattern, 238
Threshold limit value (TLV), 356. *See also* Regulations
Time-weighted average (TWA), 356. *See also* Regulations
Tissue concentrations. *See* Bioaccumulation; Bioconcentration factor; Uptake
Tobacco products
  contaminants in, 33, 40, 280
  endosulfan in, 204
Toluene, acute toxicity, 111
Total Diet Study. *See* Dietary exposure
Toxaphene. *See also* Chlorinated biocides
  physico-chemical properties, 181
  production and use pattern, 178, 183
Toxic Equivalency Factors (TEF)
  dioxins and furans, 283–284, 286, 306–307
  polychlorinated biphenyls, 160
Toxicity. *See also* Acute toxicity; Embryotoxicity; Growth effects; Nonlethal effects; Reproductive toxicity; Toxicity profiles
  data, strength and boundary of information, 345
  endpoints and hazard evaluations, 350
  fetal, hexachlorobenzene, 113
  neural, 213, 354

INDEX 369

postnatal, hexachlorobenzene, 113
target organ, 356
Toxicity profiles
　chlorinated aliphatic hydrocarbons, 54–62
　chlorinated aromatic hydrocarbons, monocyclic, 110–115
　chlorinated aromatic hydrocarbons, polycyclic, 150–161
　chlorinated biocides, 212–213
　chlorinated phenols, 255–261
　dioxins and furans, 297–307
Toxicity testing, 324–327
Toxic Substances Control Act, testing protocol for hazard identification, 323
Transformation processes. *See also* Biotransformation; Metabolic transformation; Photolysis
　chlorinated organic compounds, 19–24
　data, strength and boundary of information for, 346
　hydrolysis, 21–22, 323, 324
　ionization, 19–21
　microbial transformation, 148–149, 252–253
Transformers
　chlorinated aromatic hydrocarbons, polycyclic from, 132–133
　dioxins and furans from, 285–286
1,2,3-Tribromobenzene, environmental residues and exposure routes, 103
1,3,5-Tribromobenzene, environmental residues and exposure routes, 103
1,2,3-Trichlorobenzene
　acute toxicity, 110, 112, 114
　bioconcentration factors, 107
　environmental residues and exposure routes, 97, 102, 103
　physico-chemical properties, 94
　production use and pattern, 95
1,2,4-Trichlorobenzene
　acute toxicity, 110, 111, 112, 114
　bioconcentration factors, 107, 334
　environmental residues and exposure routes, 97, 98, 100–105
　physico-chemical properties, 94
　production use and pattern, 95
1,3,5-Trichlorobenzene
　acute toxicity, 112
　bioconcentration factors, 107
　environmental residues and exposure routes, 97, 102
　physico-chemical properties, 94
1,4-Trichlorobenzene, environmental residues and exposure routes, 105
Trichlorobenzenes
　environmental residues and exposure routes, 96, 98
　physico-chemical properties, 329
Trichlorobiphenyl, 329
1,1,1-Trichloroethylene, mammalian toxicity, 62
1,1,1-Trichloroethane
　acute toxicity, 55, 59, 60

environmental residues and exposure routes, 46
growth effects, 60
half-life in fish tissue, 54
mammalian toxicity, 56
production and use pattern, 32, 36
regulations, 72–73
reproductive effects, 61
sources in the environment, 40
1,1,2-Trichloroethane
　acute toxicity, 55, 57, 58, 59, 60
　environmental residues and exposure routes, 47–48
　growth effects, 60, 61
　physico-chemical properties, 29
　production and use pattern, 32, 36
　regulations, 73
　reproductive effects, 61
1,1,1-Trichloroethene, physico-chemical properties, 29
1,1,2-Trichloroethylene, half-life in fish tissue, 54
Trichloroethylenes (TCE)
　acute toxicity, 55, 56, 57, 58, 59, 60
　environmental residues and exposure routes, 48–49
　growth effects, 60, 61, 62
　mammalian toxicity, 62
　physico-chemical properties, 30
　production and use pattern, 32, 34, 37
　regulations, 75
　sources in the environment, 40
Trichlorofluoromethane, production and use pattern, 32
Trichloromethane, acute toxicity, 56
Trichlorophenols. *See also* Chlorinated phenols
　acute toxicity, 256–257
　embryotoxicity, 258–259
　environmental residues and exposure routes, 242, 243, 244, 245, 246, 247
　growth effects, 260
　physico-chemical properties, 236
　production and use pattern, 237–238
　regulations, 263
2,4,5-Trichlorophenoxyacetic acid, 275, 281
1,2,3-Trichloropropane, physico-chemical properties, 30
Trophic levels. *See* Bioconcentration factor; Biomagnification
2,4-D, 275, 281
2,4,5-T, 275, 281

U

Uncertainty factor, 356
Uptake, 17
　chlorinated phenols, 253
　chlorobenzenes, 98, 106
　polychlorinated biphenyls, 142
Urban runoff, chlorinated biocides in, 186, 192, 200, 202
U.S. Environmental Protection Agency (EPA). *See* Regulations

V

Vapor pressure
  calculation of, 6–7
  chlorinated aliphatic hydrocarbons, 28–30
  chlorinated aromatic hydrocarbons,
    monocyclic, 94, 96, 101
  chlorinated aromatic hydrocarbons, polycyclic,
    129
  chlorinated biocides, 180–182
  chlorinated organic compounds, 6–7
  dioxins, 278
  furans, 279
  volatilization and, 328
Vector Scoring System, 339–340
Velsicol Company. See Heptachlor
Vinyl chloride
  acceptable level, development of, 350
  acceptable risk, development of, 349, 351
  environmental residues and exposure routes,
    53–54
  physico-chemical properties, 30
  production and use pattern, 31, 32, 38
  regulations, 79–80
  sources in the environment, 39
Volatilization, 13, 330
  chlorinated organic compounds, 13–19
  chlorinated phenols, 250
  chlorobenzenes, 97–98
  γ-hexachlorocyclohexane, 186
  modeling, 328–330
  polychlorinated biphenyls, 133
  from soil, 13–19
  theoretical vs. measured, 13, 330
  from water to air, 8–13. See also Henry's law
    constant
Volatilization half-life
  chlorinated phenols, 250
  chlorobenzenes, 98
  in modeling, 329
  priority pollutants, in lakes and rivers, 12
Volatilization rate constant
  equation, 8–9, 328
  priority pollutants, 12
Volcanic eruptions
  chlorinated aliphatic hydrocarbons from, 40
  chlorinated aromatic hydrocarbons, monocyclic
    from, 94
  chlorinated aromatic hydrocarbons, polycyclic
    from, 132

W

Water. See also Precipitation; Water solubility
  chlorinated aliphatic hydrocarbons in
    bromoform and chlorodibromomethane, 45
    carbon tetrachloride, 44
    chloroethane, 45
    chloroform, 43
    chloromethane, 41
    1,1-dichloroethane, 45, 46
    1,2-dichloroethane, 46
    1,1-dichloroethene, 45
    1,2-dichloroethene, 47
    dichloromethane, 42
    hexachloroethane, 52–53
    1,1,2,2-tetrachloroethane, 50
    tetrachloroethylene, 50–51
    1,1,1-trichloroethane, 46, 48
    trichloroethylene, 48–49
    vinyl chloride, 53
  chlorinated aromatic hydrocarbons, monocyclic
    in, 96–101
  chlorinated aromatic hydrocarbons, polycyclic
    in, 135–136, 137, 150
  chlorinated biocides in
    aldrin and dieldrin, 197
    chlordane, 192
    chlordecone, 205
    chlorfenvinphos, 211
    chloropyrifos, 208–209
    DDT, DDE, and DDD, 185
    dichlorvos, 210
    endosulfan, 204
    endrin, 202
    heptachlor, 199
    γ-hexachlorocyclohexane, 186, 189, 190
    methoxychlor, 207
    mirex, 205
    toxaphene, 200
  chlorinated compounds in, modeling
    distribution of, 330
  chlorinated phenols in, 241–242, 244, 245–246,
    248–249, 256
  chlorination of, 33, 288
  dioxins and furans in, 288
  regulatory limits for contaminants in. See
    Regulations
Water solubility
  chlorinated aliphatic hydrocarbons, 28–30
  chlorinated aromatic hydrocarbons,
    monocyclic, 94, 96
  chlorinated aromatic hydrocarbons, polycyclic,
    126, 127
  chlorinated organic compounds, 3–6
  chlorinated phenols, 235, 236–237, 250
  correlation with bioconcentration factor, 17–19
  correlation with octanol-water coefficient, 4
  dioxins, 278
  estimates of, 4–6
  furans, 279
  hazard identification and, 322–324
  modeling, 331
  polychlorinated biphenyls, 127, 129, 135
Wildlife. See Birds; Fish; Invertebrates; Mammals;
  Reptiles
Wood burning
  chlorinated aliphatic hydrocarbons from, 33
  chlorinated aromatic hydrocarbons, monocyclic
    from, 94
  dioxins and furans from, 280–281, 285
Wood preservatives, 238–239, 240, 281–282

NOV 13 1998